Springer Proceedings in Physics 191

Indexed by Scopus

The series Springer Proceedings in Physics, founded in 1984, is devoted to timely reports of state-of-the-art developments in physics and related sciences. Typically based on material presented at conferences, workshops and similar scientific meetings, volumes published in this series will constitute a comprehensive up to date source of reference on a field or subfield of relevance in contemporary physics. Proposals must include the following:

- Name, place and date of the scientific meeting
- A link to the committees (local organization, international advisors etc.)
- Scientific description of the meeting
- List of invited/plenary speakers
- An estimate of the planned proceedings book parameters (number of pages/articles, requested number of bulk copies, submission deadline).

Please contact:

For Americas and Europe: Dr. Zachary Evenson; zachary.evenson@springer.com
For Asia, Australia and New Zealand: Dr. Loyola DSilva; loyola.dsilva@springer.com

H. Paul Urbach · Deren Li · Dengyun Yu
Editors

Proceedings of the 8th International Symposium of Space Optical Instruments and Applications

ISSOIA 2023, 15–17 November, Beijing, China

Springer

Editors
H. Paul Urbach
Optics Research Group
Delft University of Technology
Delft, The Netherlands

Deren Li
Wuhan University
Wuhan, China

Dengyun Yu
China Aerospace Science and Technology
Beijing, China

ISSN 0930-8989 ISSN 1867-4941 (electronic)
Springer Proceedings in Physics
ISBN 978-981-97-6717-5 ISBN 978-981-97-6718-2 (eBook)
https://doi.org/10.1007/978-981-97-6718-2

This Springer imprint is published by the registered company Springer Nature Singapore Pte Ltd.
The registered company address is: 152 Beach Road, #21-01/04 Gateway East, Singapore 189721, Singapore

Contents

Dynamic Simulation Analysis and Application of Deployable Hood Deployment

Chen Yong[1](✉), Chen Xin[1], Zhou Zhitao[2], Zhang Yifan[1], Ding Ling[1], Wang Jian[1], Wang Bing[1], Zhang Bo[1], and Yang Furao[1]

[1] Beijing Institute of Space Mechanics & Electricity, Beijing, China
475028635@qq.com
[2] Shanghai Satellite Engineering Research Institute, Shanghai, China

Abstract. The effect of deployable hood deployment on the attitude of the whole star is studied, and the deployment strategy is optimised. The theoretical modelling of the unfolding of a single unfolding plate was carried out, and the analytical solution of the rotation angle and time relationship was obtained; the dynamics simulation of the single unfolding plate was carried out in conjunction with the engineering practice of a space remote sensing camera, and a simulation model and parameter settings closer to the actual situation were established. The single unfolding plate unfolding test mechanism was constructed to carry out the unfolding test, and parameters such as driving moment and resisting moment were measured, and parameter corrections were made to the theoretical model and the simulation model. The corrected results show that the difference between the theoretical calculation, simulation and test results is within 5%. The corrected simulation model can be applied to the unfolding dynamics analysis of the deployable hood, which provides a reference basis for the selection of the unfolding mode and the precise attitude control.

Keywords: deployable visor · deployable plate deployment · dynamics simulation · parameter correction · whole star attitude

1 Introduction

In space remote sensing cameras, the deployment mechanism of the deployable hood is a crucial part. Such a mechanism must be in a contracted state during rocket launch and be able to withstand complex mechanical environments such as shock, vibration and overload. In addition, it needs to achieve a highly reliable deployment locking function in the vacuum, radiation and cryogenic environments of space. Historically, there have been dozens of cases in which satellites could not work properly due to mechanical failures of the unfolding mechanism [1–7]. Therefore, in the design of space remote sensing cameras, it is of great significance to strengthen the simulation study of hood dynamics for the design and testing of the deployment mechanism of deployable hoods.

Researchers at home and abroad have carried out in-depth theoretical modelling and simulation analyses of the unfolding mechanism of deployable visors. The dynamics of

H. P. Urbach et al. (Eds.): ISSOIA 2023, SPP 191, pp. 1–13, 2024.
https://doi.org/10.1007/978-981-97-6718-2_1

the orderly unfolding of a solar wing was systematically studied by Rong Jili et al. [8]. Starting from 2000, Amold et al. [9] started to analyse the unfolding dynamics errors of flexible mechanisms. By 2020, the theoretical studies have become quite mature and many large applications are available for engineering purposes [10, 11]. In this paper, the actual engineering situation of a space remote sensing camera is investigated: in terms of theoretical modelling, the literature [12, 13] has established the model through the virtual work method and the energy conservation method, respectively, but there is no analytical solution. Through derivation, it is found that the model proposed in this paper can obtain an analytical solution in the case of a single unfolding plate, and there is no cumulative error, so it can more accurately and quickly analyse the degree of influence of each design variable on the unfolding time; in terms of simulation and analysis, most of the literature adopts the rigid-body dynamics simulation, but its parameter setting and modelling method are quite different from the actual situation. In this paper, the modelling method is closer to the actual situation, the flexible characteristics of the deployable hood are considered in the modelling process, the parameter settings are corrected by the experimental values, and the simulation model is more accurate and has been successfully applied in engineering practice.

2 Theoretical Modelling of Expanded Plate Expansion

In this paper, we consider the motion of the unfolding visor as a single-degree-of-freedom rigid body motion. The main purpose is to develop a theoretical model to determine the relationship between time t and the rotation angle Φ in order to derive the time required to turn the unfolding plate from $\Phi = 0°$ to $\Phi = 180°$ (Fig. 1).

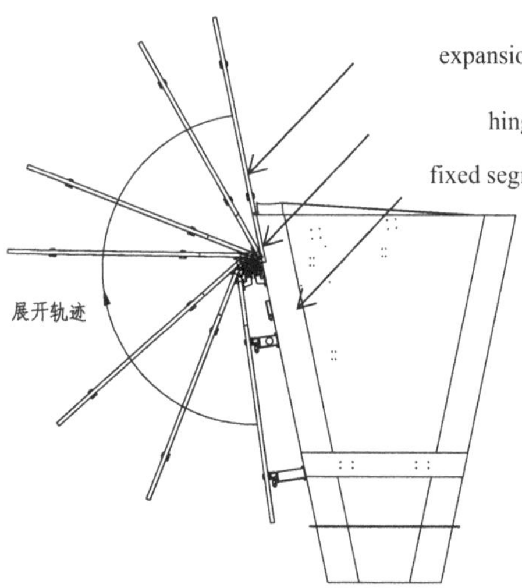

Fig. 1. Schematic diagram of the unfolding trajectory of the hood spreading plate

Figure 2 shows the fixed segment of the hood as a rigid body, where x and y axes are the coordinate system fixed on the fixed segment of the hood. OA represents the

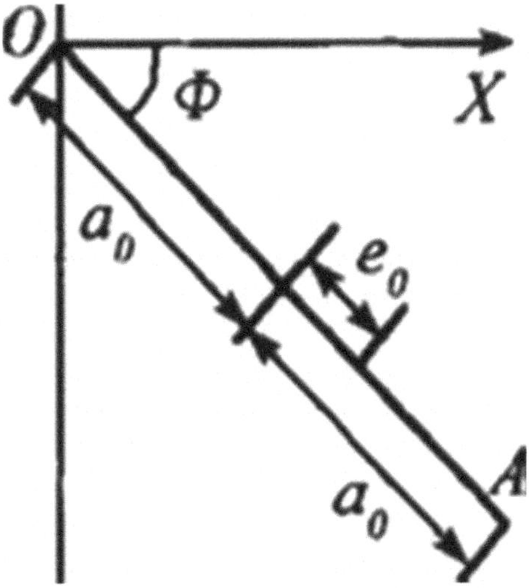

Fig. 2. Theoretical model of single unfolded plate unfolding

unfolded plate of the hood, which has a length of $2a_0$, and a mass of m_0. Point O is the articulation point between the fixed segment of the hood and the unfolded plate, where there is a pre-tensioned coil spring. Due to the hinge connection distance and the pre-embedded block, the centre of mass of the deployable hood does not coincide with the geometric centre, and we set its eccentricity to be e_0.

The coordinates of the centre of mass of the spreading plate are:

$$\begin{cases} x_0 = (a_0 + e_0)cos\Phi \\ y_0 = -(a_0 + e_0)sin\Phi \end{cases} \tag{1}$$

Where: Φ, x_0, y_0 are functions of time t. Here Eq. (1) is first differentiated once and twice with respect to time t:

$$\begin{cases} x\prime_0 = -(a_0 + e_0)sin\Phi \cdot \Phi\prime \\ y\prime_0 = -(a_0 + e_0)cos\Phi \cdot \Phi\prime \end{cases} \tag{2}$$

$$\begin{cases} x''_0 = -(a_0 + e_0)[sin\Phi \cdot \Phi'' + cos\Phi \cdot (\Phi\prime)^2] \\ y''_0 = -(a_0 + e_0)[cos\Phi \cdot \Phi'' + sin\Phi \cdot (\Phi\prime)^2] \end{cases} \tag{3}$$

Let the radius of gyration of the unfolded plate around the centre of mass be ρ_0, then the force of inertia can be deduced from the mechanical equation as:

$$\begin{cases} F_{x0} = m_0(a_0 + e_0)[sin\Phi \cdot \Phi'' + cos\Phi \cdot (\Phi\prime)^2] \\ F_{y0} = m_0(a_0 + e_0)[cos\Phi \cdot \Phi'' + sin\Phi \cdot (\Phi\prime)^2] \end{cases} \tag{4}$$

The moment of inertia is:

$$M_0 = -m_0{\rho_0}^2\Phi'' \tag{5}$$

The driving coil spring at the hinge joint O is a planar volute spring in the shape of an Archimedean helix, which produces a driving torque of:

$$T_{s0} = T_{s0}(d) + k_{s0}\Phi \tag{6}$$

Where: T_{s0} represents the driving torque (total driving torque), $T_{s0}(d)$ represents the value of the preload moment of the driving coil spring in the closed state, and k_{s0} is the spring constant of the coil spring, i.e. the elastic stiffness coefficient.

The articulation point O is also affected by many moments of resistance. In performing the mathematical modelling, we calculate these many types of drag moments uniformly. The main sources of the resistance moments include the friction of the joint bearings in the hinge, the resistance of the cable between the unfolding plate and the fixed section of the hood, and the resistance caused by the different axes of the two hinges due to thermal expansion and contraction. Through empirical values and experimental measurements, we find that there is an approximately linear relationship between the value of the resisting moment and the unfolding angle, and the larger the angle of rotation, the larger the resisting moment. In order to simplify the problem, we can model the resisting moment as a coiled spring unit:

$$T_{r0} = T_{r0}(d) + k_{r0}\Phi \tag{7}$$

Where: T_{r0} is the total resisting moment, T_{r0} (d) is the total resisting moment in the unfolded plate stowed state, and k_{r0} is the elastic stiffness coefficient.

Let an imaginary displacement angle δ_Φ, then the imaginary displacement at the centre of mass of the deployable hood is:

$$\begin{cases} \delta x_0 = -(a_0 + e_0)sin\Phi \cdot \delta_\Phi \\ \delta y_0 = -(a_0 + e_0)cos\Phi \cdot \delta_\Phi \end{cases} \tag{8}$$

According to the principle of virtual work, it can be obtained:

$$F_{x0}\delta x_0 + F_{y0}\delta y_0 + (M_0 + T_{s0} + T_{r0})\delta_\Phi = 0 \tag{9}$$

Bringing Eqs. (4) to (8) into Eq. (9), the collation leads to the following equation:

$$-c_0\Phi'' + k\Phi + \mathrm{T} = 0 \tag{10}$$

Among them:

$$\mathrm{c_0} = \mathrm{m_0}[\rho_0^2 + (\mathrm{a_0} + \mathrm{e})_0^2];\ \mathrm{k} = \mathrm{k_{s0}} + \mathrm{k_{r0}};\ \mathrm{T} = \mathrm{T_{s0}(d)} + \mathrm{T_{r0}(d)}$$

Eq. (10) is a linear second-order differential equation that can be solved analytically:

$$\Phi = c_1 \mathrm{e}^{\sqrt{a \cdot t}} + c_2 \mathrm{e}^{-\sqrt{a \cdot t}} - \mathrm{b/a} \tag{11}$$

where: $a = k/c_0$, $b = T/c_0$, c_1, c_2 are arbitrary constants.

Eq. (11) is the relation curve between angle Φ and time t with initial boundary conditions:

$$\begin{cases} t = 0, \Phi = 0 \\ t = 0, \Phi\prime = 0 \end{cases} \tag{12}$$

The value of the constants c_1, c_2 can be found by bringing the initial boundary condition (12) into equation (11).

By substituting the boundary conditions of the equations, the driving torque, the resisting torque, the geometrical dimensions of the deployable hood, the mass and other parameters, we can find the variation curve of the angle of rotation Φ of the deployable plate with the time of deployment, t. The curve can be calculated as follows.

3 Simulation Modelling

Microgravity impact analysis on mechanisms: There are numerous micro-convex bodies of varying sizes on the contact surfaces of kinematic pairs. Ideally, contact only occurs at the tops of the micro-convex bodies, but in reality, there is a significant deviation in various contact parameters from the ideal case. Therefore, establishing a gap dynamics model considering the actual rough surfaces under microgravity conditions is of great importance for the precise analysis of space mechanisms. Due to the complexity of ground thermal environment and microgravity simulation, reasonable simplifications were made during modeling based on the structural technical state and thermal load characteristics, in order to reduce the time and effort required for model establishment and numerical calculations. The simplifications mainly include: ① the connection of the boom is simplified to the spatial origin of the suspension device; ② all welds and screw connections are simplified to fixed forms; ③ the weight of the mechanism product hanging is simplified to a vertically downward force acting on the corresponding suspension point on the long rail. The calculation precision is improved by adjusting functions such as default values of correlation centers, smoothness, transition, and crossover angle centers. In the simulation modelling process, we combined the actual needs of a space remote sensing camera project to make the modelling method more compatible with the actual situation. Before building the simulation model, we need to build the models of space remote sensing camera, deployable hood and connecting hinge, and set the material properties. Then the model is imported into the simulation platform through the x_t format, and then the boundary conditions are added to complete the modelling. As shown in Fig. 3.

Fig. 3. Modelling of the hinge-mounted hood in the deployed and retracted states

3.1 Deployable Hood Modelling

The deployable hood and its deployment plate are made of aerospace aluminium honeycomb sandwich structure, while the outer skin and edges are made of carbon fibre composites. Since the skin of the unfolded panel is made of anisotropic structure and material and has a large area, we cannot ignore the influence of its flexibility on the unfolding process when modelling the unfolded panel. In the modelling process, we first establish the finite element model of the unfolded plate and calculate the modal state. Then, we attach the geometric dimensions, nodal inertia, modal mass, modal stiffness and other information of the unfolded plate to the model, so that the unfolded plate can be analysed as a composite flexible body. After importing the flexible body, we cannot directly apply moments or kinematic partials to the finite element mesh nodes, as this will lead to simulation errors. Instead, we need to create massless rigid joints during modelling, and then use MPC to connect the mesh and joints that need to be connected.

3.2 Modelling of Resisting Moment and Driving Moment

The resistance and driving moments are modelled as follows: the resistance moments can be experimentally determined to roughly reveal their sources and characteristic laws. In the previous mathematical model, we have simplified the resisting moment into a coil spring unit for modelling. In this modelling process, we use the coil spring moment (Torsion Spring) to model the resisting moment. The torque is calculated using a specific formula where T_{r0} (d) = −0.63 N−m, k_{r0} = −0.23 N−m/ rad.

The unfolding mechanism mainly includes the coil spring driven hinge mechanism, the driving coil spring of this mechanism is pre-tightened when the unfolding plate is closed, and when the unfolding plate is released, its pre-tensioning force is the main driving force; when the unfolding plate is fully unfolded, the coil spring still has a certain residual torque. The preload force and elastic stiffness design of the coil spring are related to the resisting torque. According to the aerospace industry standard, T_{s0} and T_{r0} should satisfy the following relationship:

$$STM = \frac{T_{s0}}{T_{r0}} - 1 \tag{13}$$

STM is the static torque margin, which is generally calculated using the value at the time of unfolding, when the resistance moment is the largest and the driving moment is the smallest. According to Eq. (13) and the resistance moment can design the driving moment of the coil spring, and further determine the geometry, material, and performance of the coil spring. STM is not the larger the better, if the value is too large, the unfolding plate unfolding moment will cause a huge impact on the star, which is easy to cause the satellite to flip, roll sideways, and even structural damage.

In the modelling process of this paper, a rotary vice (Revolute) is used to connect the male and female hinges of the hinge; and the driving torque is simulated by a coil spring moment (Torsion Spring). The torque calculation formula is shown in Eq. (6), where: T_{s0} (d) = 1.2 N−m, k_{s0} = −0.16 N−m/ rad. At this time, STM = 1.11, which meets the design requirements.

3.3 Modelling of the Hinge Mechanism

The hinge mechanism mainly includes a connection portion of the hinge to the star body, the deployable shade, and a hinge locking portion (Fig. 4).

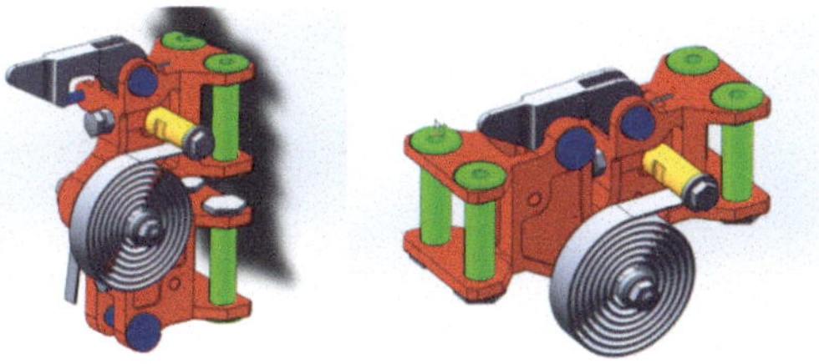

Fig. 4. Schematic diagram of the hinge in the deployed and retracted states

The accuracy of the modelling of the hinge's connection section with the deployable visor and star will directly affect the magnitude of the impact force of the unfolding plate unfolding, the force transfer and the system damping. In this paper, the collision force (Contact Force) is used to simulate the collision of aluminium and aluminium between the hinge and the unfolding plate and the fixed section of the visor. Parameters are set using recommended values: k (stiffness) = 23 kN/m, c(damping) = 1000N-s/m, e = 1.35, Mus = 0.72, Mud = 0.35,d is the amount of deformation of the collision body, which requires a large number of experiments and analyses to be determined, and implicitly nonlinear analysis with collision is carried out in the design analysis, namely Define the material properties, cell properties, boundary conditions (initial velocity), contact body, etc. of the finite element mesh of the collision body, and then perform the nonlinear analysis. The results of the analysis show that the maximum displacement of the collision part during unfolding of the unfolded plate d = 0.142 mm.

In order to simulate the process of unfolding the plate by 180° and then locking it, the literature [14] uses the method of applying a moment at the hinge equivalent to that of the buffer, the magnitude of which is defined by the following function:

$$\mathrm{STEP}(\mathrm{AZ}(\mathrm{marker\_1},\ \mathrm{marker\_2}), 179.8\mathrm{d}, 0, 179.9\mathrm{d}, 1.0)$$

BISTOP(AZ (marker_1, marker_2), WZ (marker_1, marker_2), 179.9d, 180.1d, 1×105, 1.5, 1.5×103, 0.ld) STEP is a step function constructed with cubic polynomials.

BISTOP is a bilateral collision force function with the format:

$$\mathrm{BISTOP}(q, d(q)/d(t), q0, q1, k, e, C, d).$$

Where: q is the displacement variable; d(q)/d(t) is the velocity variable; q_0 and q_1 are the lower and upper queue values of the displacement variable, respectively; k is the stiffness coefficient; e is the restoring force coefficient; c is the damping coefficient; and d is the damping maximum displacement. In the function application of this paper, AZ(marker_1,marker_2) returns the angular difference between two points about Z axis under the specified coordinate system, and WZ(marker_1,marker_2) returns the angular velocity difference. The simulation analysis results are shown in Fig. 5:

The results of this modelling approach are biased towards idealisation, with symmetrical lower and upper queue values for the displacement variables, which does not correspond to the actual situation, and the variables such as damping and maximum displacement are estimated values, which require a large number of experiments to calibrate the parameters in order to satisfy the engineering applications.

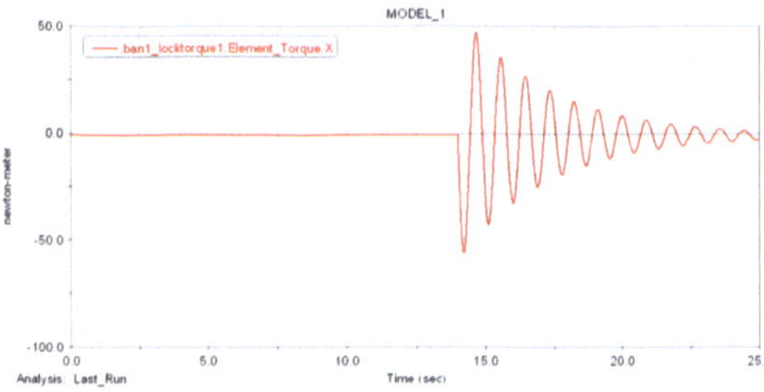

Fig. 5. Curve of impact moment with time at method I connection

This paper also considers a modelling method closer to the actual situation: detailed modelling of the locking mechanism, including simulation of the relative motion of the locking shaft and the male hinge, and the locking shaft entering the locking groove for the collision locking process. This modelling method only requires the parameters of contact force and deformation of the collision body for simulation. The parameters are set as follows: k(stiffness) = 1 × 105 N/m, c(damping) = 1 300 N - s/ m, e = 1.35, Mus = 0.23, Mud = 0.13. The elastic stiffness coefficient of the locking torsion springs and the preload force directly determine whether the locking mechanism can be reliably locked and will not be ejected. In the process of many simulations, we successfully determined the appropriate spring constant and preload force of the locking torsion spring.

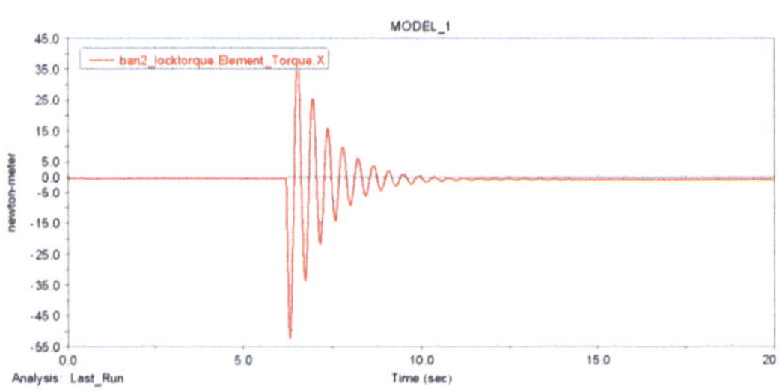

Fig. 6. Variation curve of impact moment with time at the connection of method 2

By comparing the two methods, we find that the former has more ideal curve characteristics, and its peak impact moment is about 53.8Nm; while the latter only needs to master the parameters of the collision force between materials for simulation. In view of the nonlinear stiffness of the hinge due to the hinge gap, we need to take the nonlinear part of the displacement into account when calculating the maximum displacement. The curve obtained in this way changes more drastically, as shown in Fig. 6, the maximum impact moment reaches 43.1 Nm. It is worth noting that the curve trend is closer to the experimental results.

3.4 Pressure Release Mechanisms

Prior to the launch of the satellite, the unfolding plate is pressed against the side of the visor body by means of a compression lever. Once in orbit, the release is achieved by means of a detonator and a cutter. The usual shear compression release mechanism does not provide a separation impulse at the moment of release, so the release can be achieved by editing a simulation script during the modelling process. For cameras with multiple unfolding plates, this approach allows the release time of each unfolding plate to be controlled separately (Fig. 7).

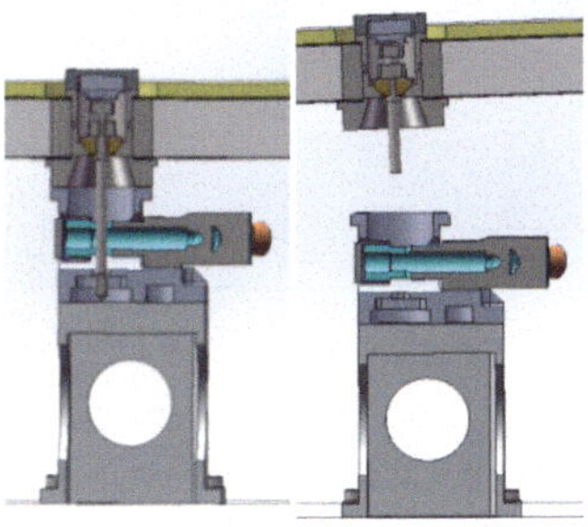

Fig. 7. Schematic diagram of hinges in deployed and retracted states

Using the example of 5 unfolding panels spaced 4 s apart, the script language is as follows:

SIMULATE/DYNAMIC, END = 4.0, STEPS = 100 DEACTIVATE/ MOTION, ID = 1.

SIMULATE/DYNAMIC, END = 8.0, STEPS = 100 DEACTIVATE/MOTION, ID = 2 SIMULATE/DYNAMIC, END = 12.0, STEPS = 100.

This is modelled by adding a Motion to the joint to constrain the unfolding plate and then using a scripting language to disable the Motion at a defined point in time in order to simulate the compression release mechanism.

3.5 Determination of Roll-Out Time

After the unfolding plate is unfolded by 180°, there will be attenuation oscillation under the action of the locking mechanism, and when the unfolding plate reaches 180° for the first time, the peak value of the oscillation is larger, which is not suitable to be used as the basis for judgement of the unfolding. In this paper, the second time the plate reaches 180° is used as the basis for the judgement of the unfolding of the plate. In the modelling process, sensors can be used to return the exact moment of unfolding. The specific approach is to establish two sensors for the determination of AZ (marker_1, marker_2), when the unfolding plate is equal to 180°, continue the simulation. When the simulation is completely finished, the unfolding time can be determined by looking at Sensor_2.

3.6 Examples

In the example section, the designed and predicted resisting moment, unfolding moment, collision force parameters, and geometric parameters are brought into the established model. Where: mass of unfolded plate m0 = 3.92 kg, plate length $2a_0 = 1.5$ m, centre of mass eccentricity $e_0 = 0.05$ m. The results obtained are shown in Fig. 8.

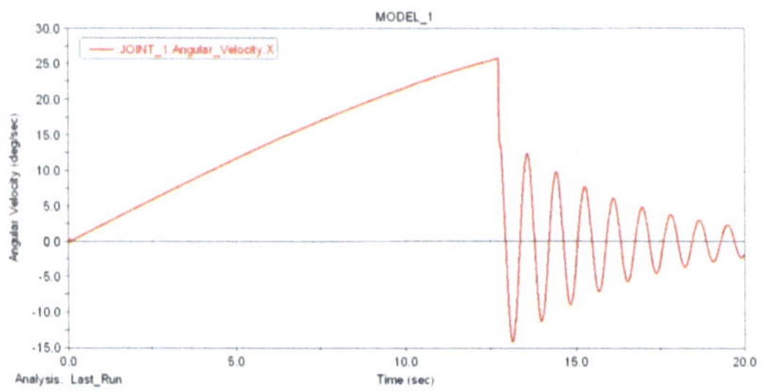

Fig. 8. Theoretical model and simulation results

From Fig. 8, it can be seen that the unfolding angle and unfolding time derived from the theoretical modelling have an approximate exponential relationship. When the turning angle changes from $\Phi = 0°$ to $\Phi = 180°$, it takes a total of 14.02 s. From Fig. 5(b), it can be seen that the unfolding time of the unfolding plate is 4.082 s, which is more in line with the results of the theoretical modelling.

4 Dynamics tests

In this paper, a test prototype of a single unfolded panel was constructed using carbon skinned aluminium honeycomb panels.

The mass of the spreading plate is 3.92 kg (heavier than that of the designed aerospace spreading plate); the dimensions of the spreading plate are 1500 mm (length) × 1000 mm (width) × 25 mm (thickness); the eccentricity of the centre of mass is 60 mm.

Through the testing of this test prototype, we can verify the accuracy of the theoretical calculations and simulation modelling, thus providing more reliable data for the detailed design and overall star attitude analysis.

4.1 Test Prototype

The spreader plate is mounted on the spreader frame by means of a pulley and a flexible lanyard. The pulleys on the spreader frame can be moved with the movement of the lanyard with very little friction. To counteract the effect of gravity, counterweights and chains are used. The transducer should be equal to the dead weight of the test specimen and the lanyard should pass through the centre of gravity of the test specimen to simulate vacuum weightlessness (Fig. 9).

The unfolding plate is mounted by means of coil-spring actuated hinges on a fixed segment, which has a high degree of stiffness and is used to simulate a satellite body.

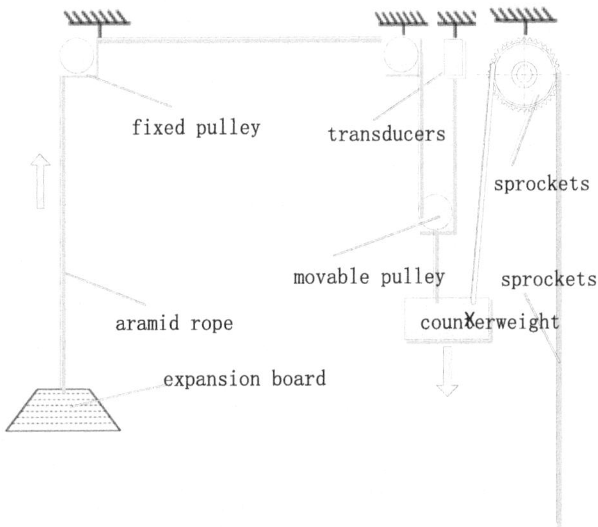

Fig. 9. Unfolding test of single unfolded plate

The mass of the test unfolding plate is 3.92 kg. The parameters of the coiling spring, the unfolding moment and the resisting moment can be measured by means of a certain test method [15–17] and a test apparatus, and the unfolding time is measured by means of a transducer.

4.2 Modelling Amendments

In the test prototype, the data of the hinges, connecting parts and the unfolding plate of the test prototype may have errors with the design values. Therefore, we need to correct the parameters of the theoretical model and the simulation model by the actual measured values, and compare the corrected model with the actual test results to check the accuracy of the model building. In the test prototype, the preload force of the driving coil spring can be adjusted. With other conditions unchanged, the time for the test prototype to unfold will be different by using different preload force, so that we can carry out multiple sets of tests.

During the actual measurement, it was found that the static torque was greater than the estimated value and the spring constant of the driving torsion spring was greater than the design value. Therefore, we need to adjust the preload of the drive torsion spring so that its static torque margin STM > 1 (Table 1).

The adjusted drive torque preload is at least 3.22 N–m. The error between the theoretically derived model as well as the simulation model and the experimental values does not exceed 5%, which indicates that the established model is more accurate and can be applied in engineering design.

After analysis, the main source of error was that the resisting moment did not exactly correspond to the spring unit characteristics. The nonlinearity of the moment of resistance was measured by a special method. The analysis shows that the nonlinearity of the

Table 1. Comparison of Measured Values and Design Values

	design value	measured value
Resistance moment T_{ro} (d)/(N−m)	0.30	0.42
Resistance moment k_{ro} (d)/(N−m-rad)$^{-1}$	−0.05	−0.05
Driving torque T_{ro} (d)/(N−m)	3.0	3.22
Driving torque k_{ro} (d)/(N−m−rad)$^{-1}$	0.21	0.17

resisting moment is due to the fact that the rotation axis of the hinge of the test specimen is not completely perpendicular to the ground, as well as the complex factors such as the air resistance, the resistance of the rope and the truss slide, and so on. And in both the theoretical derivation model and simulation modelling, we regard the drag moment as linearly varying.

5 Conclusion

In this paper, firstly, the display of single unfolding plate is theoretically modelled, and the analytical solution of the relationship between rotation angle Φ and time t is obtained. Secondly, the dynamics simulation of the single unfolding plate is carried out, and the flexible display plate and collision contact hinge are used for modelling, so that the simulation model established is closer to the actual situation. Combined with the engineering practice of a space remote sensing camera, the design of the resisting moment and driving moment satisfies Eq. (13), and the single unfolding plate display test organisation constructed by ourselves was tested. According to the parameters of driving moment and resisting moment measured in the test, the theoretical model and simulation model were corrected. The corrected results show that the error between the theoretical calculation, simulation and test results is within 5%. The corrected simulation model can be applied to the analysis of the unfolding dynamics of the deployable hood, which provides a reference basis for the selection of the unfolding mode and the precise control of the attitude.

References

1. Wise, M.A., Saleh, J.H., Haga, R.A.: Health scorecard of spacecraft platforms:Track record of on-orbit anomalies and failures and preliminary comparative analysis. Acta Astronautiea **68**, 253–268 (2011)
2. Miyoshi, K.: Aerospace mechanisms and tribology technology case study. Tribol. Int. **32**, 673–685 (1999)
3. Takano, T., Michihiro, N., Miyishi, K., et al.: Characteristics verification of a deployable onboard antenna of 10 m maximum diameter. Acta Astronautica **51**(11), 771–778 (2002)
4. Homma, M., Hama, S., Hamamoto, N., et al.: Experimental plan of ETS-8 in orbit: mobile communication and navigation. Acta Astronautica **53**, 477–484 (2003)
5. Meguro, A., Tsujihata, A.: Technology status of the 13 m aperture deployment antenna reflectors for engineering test satellite. Acta Astronautica **47**, 147–152 (2000)

6. Meguro, A., Shintate, K., Usui, M., et al.: In-orbit deployment characteristics of large deployable antenna reflector onboard Engineering Test Satellite VIII. Acta Astronautica **65**, 1306–1316 (2009)
7. Meguro, A., Harada, S., Watanabe, M.: Key technologies for high-accuracy large mesh antenna reflectors. Aeta Astronautica **53**, 899–908 (2003)
8. Rong, J., Xiang, Y., Xin, P., et al.: Vibration control and experimental study of a double-link robotic arm considering joint flexibility. J. Astron. **38**(10), 1024–1031 (2017)
9. Arnold, M., Brüls, O., Cardona, A.: Error analysis of generalisedalpha group time integration methods for constrained mechanical systems. Numerische Mathematik **129**(1),149–179 (2015)
10. Garcia de Jalon, J.: Twenty-five years of natural coordinates. Multibody Syst. Dyn. **18**, 15–33 (2007)
11. Shabana, A.A.: Flexible multibody dynamics: review of past and recent developments. Multibody Sys. Dyn. **1**(2), 159–223 (1997)
12. Zhang, H., Liu, H., Hu, Z., et al.: In-orbit load analysis of solar cell wing with flexible characteristics. J. Astron. **40**(2), 139–147 (2019)
13. Murinan, Tan, S., Wu, Z., et al.: Stability analysis of attitude-vibration coupling of ultra-large flexible space structures. J. Astron. **39**(7), 709–714 (2018)
14. Hu, H., Tian, Q., Zhang, W., et al.: Nonlinear dynamics and control of large-scale mesh-frame deployable space structures. Adv. Mech. **43**(4), 390–414 (2013)
15. Yang, H., Zhang, Z., Jiang, B.: Modelling, control and fault tolerance of multi-body mechanical systems. J. Nanjing Univ. Aeron. Astron. **49**(5), 612–621 (2017)
16. Garcia-Vallejo, D., Escalona, J.L., Mayo, J., et al.: Describing rigidflexible multibody system using absolute coordinates. Nonlinear Dyn. **34**, 75–94 (2003)
17. Gerstmayr, J., Shabana, A.A.: Analysis of thin beams and cables using the absolute nodal co-ordinate formulation. Nonlinear Dyn. **45**(1–2), 109–130 (2006)
18. Luo, H.T., Chen, Z.C., Leng, Y.Q., et al.: Rigid-flexible coupling dynamics simulation of 3 -RPS parallel robot based on ADAMS and ANSYS. Appl. Mech. Mater. **290**, 91–96 (2013)
19. Linkins, P.W.: Finite element appendage equations for hybrid coordinate dynamics analysis. J. Solids Struct. **8**, 709–731 (1972)
20. Wang, Y., Shui, X.: Analysis of ADAMS-based flexible body and its application in solar array deployment. Mech. Des. Manufac. **9**, 146–148 (2006)
21. Ma, X., Wang, T., Wang, B., et al.: Dynamic analysis of flexible attachment deployment for large and complex spacecraft. China Space Sci. Technol. **20**(4), 1–7 (2000)
22. Huang, Y., Fan, K., Jiang, G.: Influence of low-temperature torsional resistance moment of RF cable on satellite deployment locking mechanism. Spacecraft Environ. Eng. **30**(3), 245–249 (2013)
23. Guo, H., Liu, R., Dang, Z.: Analysis of the influence of flexible cables on the performance of articulated extension arms with space rods. J. Harbin Inst. Technol. **42**(1), 55–59 (2010)

Optical System Design of Dual-Band and Small Remote Sensing Camera

Xiaobo He, Zhile Wang, Xinmin Guo(✉), and Lixin Bo

School of Astronautics, Harbin Institute of Technology, Harbin 150080, China
guoxm@hit.edu.cn

Abstract. Traditional visible light aerospace remote sensing cameras are easily affected by clouds, rain and fog, resulting in a decline in their Earth observation capabilities. To solve this problem, this paper designed a visible light and shortwave infrared dual-band remote sensing camera optical system, which can achieve the observation of the Earth in the visible light (0.4 μm–0.8 μm) and short-wave infrared (0.8 μm–2.5 μm) range. The actual working platform of aerospace remote sensing cameras requires miniaturization and lightweight of the camera, so the structure of the optical system adopts an off-axis three-mirror type. It can be lightweight and miniaturized due to optical path folding, and has the advantages of long focal length and no chromatic aberration. The coaxial three-mirror optical system design method is improved by off-axis diaphragm, and the improved method is used to design the off-axis three-mirror optical system. The focal length of the optical system is 600 mm and the total length is only 214 mm. The design results show that the MTF (modulation transfer function) value of each band at the cut-off frequency of 50 lp/mm is greater than 0.4, and the MTF value of each band at the cut-off frequency of 75 lp/mm is greater than 0.25; The diameter of the spot diagram of each band in the entire field of view is smaller than the Airy disk diameter. The RMS (root mean square) value of the system wave aberration in the visible light band is less than $\lambda/16$ ($\lambda = 0.633$ μm), and the RMS value of the system wave aberration in the short-wave infrared band is less than $\lambda/50$ ($\lambda = 2.000$ μm). The performance of the off-axis three-mirror optical system designed in this paper is close to the diffraction limit, which can meet the application needs of miniaturization and lightweight of remote sensing cameras and improve the observation capabilities of remote sensing cameras.

Keywords: Aerospace remote sensing · Dual-band · Miniaturization · Off-axis three-mirror · Optical system design

1 Introduction

Aerospace remote sensing technology not only plays an important role in defense fields such as military reconnaissance and missile interception, but is also widely used in economic and livelihood fields such as natural disaster early warning, terrain mapping, and climate change prediction [1, 2]. An important factor restricting the development of aerospace remote sensing technology is the performance of remote sensing cameras [3].

H. P. Urbach et al. (Eds.): ISSOIA 2023, SPP 191, pp. 14–21, 2024.
https://doi.org/10.1007/978-981-97-6718-2_2

The development of high-resolution, multi-band remote sensing cameras has become a key issue in promoting the development of remote sensing technology. There are three structural types of remote sensing camera optical systems, refractive, catadioptric and reflective. Since the remote sensing camera works on a satellite platform, its performance must meet the satellite platform's requirements for miniaturization and lightweight [4]. Furthermore, remote sensing cameras have long working distances and high resolution requirements, so the focal length is generally long. If a refractive optical path is used, the system will be too long. In addition, the refractive system mainly uses lenses for imaging, which will increase the weight of the system. Reflective optical systems have become the main optical path type for remote sensing cameras due to their lack of chromatic aberration, collapsible optical paths, and the ability to achieve large apertures [5–7]. Among reflective optical systems, off-axis three-mirror optical systems are widely used in various types of remote sensing cameras because they have no central obstruction, simple structure, and foldable optical paths to improve space utilization [8, 9].

At present, multi-band and high-resolution detection is still a technical challenge for remote sensing cameras, this paper determines the initial structure of the optical system through the design theory of coaxial three-mirror systems, and uses the aperture off-axis method to process the coaxial three-mirror system off-axis, the tertiary mirror uses high-order aspherical surfaces to better compensate for the asymmetric aberration of the off-axis system. Moreover, since the actual working platform has restrictions on the total length of the system, a plane mirror is installed between the secondary mirror and the tertiary mirror to refract the optical path, effectively shortening the back focus of the optical system. Finally, a dual-band off-axis three-mirror remote sensing camera optical system that can achieve visible light and short-wave infrared detection is designed.

2 Design Method

2.1 Structural Analysis of Optical System

The structural form of the optical system of aerospace remote sensing cameras can be divided into two types: refractive and reflective. The working bands of this system are visible light band and short-wave infrared band. Its band range is wide and the field of view is small. Therefore, the main issues in the design process are Correction of chromatic aberration. If a refractive structure is used, multiple lenses or a doublet lens group will be used to compensate for chromatic aberration, which increases the complexity of the system and is not conducive to lightweighting the optical system. In reflective optical systems, the direction of the emitted light only depends on the incident angle, so there is no chromatic aberration problem. In addition, the reflective optical system only uses a single reflective surface to reflect light. Compared with a double-sided lens, it will greatly reduce the weight of the optical system. Therefore, this system adopts a reflective structure. Reflective structures are further divided into coaxial and off-axis types. The design purpose of this system includes realizing low-light remote sensing in low-illumination environments at night. The target energy is weak during low-light remote sensing. The coaxial structure has a central obstruction and will lose light. Energy. In order to improve the utilization rate of light energy, this system finally adopts an off-axis structure.

2.2 Coaxial Three-Mirror Initial Structure Calculation

Firstly, the initial structure of the off-axis three-mirror optical system is calculated through the coaxial three-mirror optical system design theory, and then the off-axis three-mirror optical system is obtained by moving the aperture off-axis. The optical path of the coaxial three-mirror optical system is shown in Fig. 1

$M1, M2, M3$ refer to primary mirror, secondary mirror and tertiary mirror respectively. l_2, l_2', l_3, l_3' are the object distance and image distance of the secondary mirror and the third mirror, h_1, h_2, h_3 are the half-apertures of the primary mirror, secondary mirror, and tertiary mirror respectively, f_1' is the focal length of the primary mirror.

The design assumes that the object is at infinity, i.e. $l_1 = \infty, u_1 = 0$, Reflection system object side and image side refractive index can be regarded as opposite numbers, $n_1 = n_2' = n_3 = 1, n_1' = n_2 = n_3' = -1$. The structural parameters of the coaxial three-mirror include: The radii of curvature of the three mirrors are denoted as R_1, R_2, R_3; The distance between the mirrors is expressed as D_1, D_2; The conic coefficients of the three mirror surfaces are expressed as $k_1 = -e_1^2, k_2 = -e_2^2, k_3 = -e_3^2$.

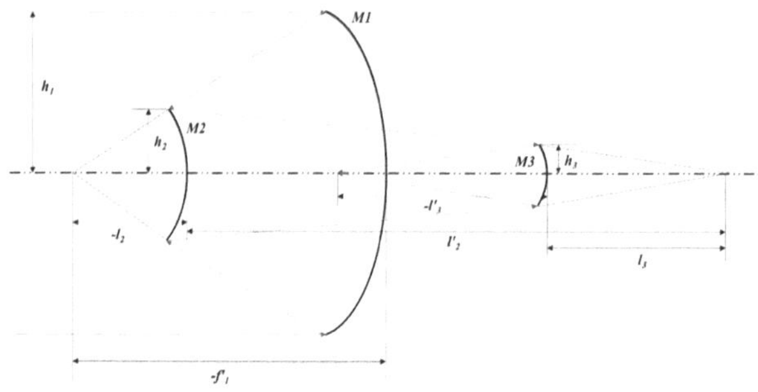

Fig. 1. Optical path of coaxial three-mirror optical system

α_i is the obstruction ratio. β_i indicates the magnification of the mirror. From Fig. 1, the expression of α_i and β_i can be obtained as

$$\begin{array}{l} \alpha_1 = \frac{l_2}{f_1'} \approx \frac{h_2}{h_1} \quad \alpha_2 = \frac{l_3}{l_2'} \approx \frac{h_3}{h_2} \\ \beta_1 = \frac{l_2'}{l_2} = \frac{u_2}{u_2'} \quad \beta_2 = \frac{l_3'}{l_3} = \frac{u_3}{u_3'} \end{array} \tag{1}$$

The relationship between structural parameters and contour parameters can be deduced from Gaussian optical formula:

$$\begin{array}{l} R_1 = \frac{2}{\beta_1 \beta_2} f' \qquad R_2 = \frac{2\alpha_1}{\beta_2(1+\beta_1)} f' \qquad R_3 = \frac{2\alpha_1\alpha_2}{1+\beta_2} f' \\ D_1 = \frac{R_1}{2}(1-\alpha_1) \; D_2 = \frac{R_1}{2}\alpha_1\beta_1(1-\alpha_2) \end{array} \tag{2}$$

From Eq. (2), the radius of curvature of the three mirrors in the coaxial three-mirror and the spacing between the mirrors can be solved for by giving $\alpha_1, \alpha_2, \beta_1, \beta_2, f'$. In order to ensure imaging quality and processability, the surface types of the three mirrors are assumed to be quadratic surfaces. Therefore, the primary monochromatic aberration of the entire system can be expressed as:

$$\begin{cases} S_{\mathrm{I}} = \sum hP + \sum h^4 K \\ S_{\mathrm{II}} = \sum yP - J\sum W + \sum h^3 yK \\ S_{\mathrm{III}} = \sum \frac{y^2}{h}P - 2J\sum \frac{y}{h}W + J^2\sum \phi + \sum h^2y^2K \\ S_{\mathrm{IV}} = \sum \frac{\Pi}{h} \\ S_{\mathrm{V}} = \sum \frac{y^2}{h}P - 3J\sum \frac{y^2}{h^2}W + J^2\sum \frac{y}{h}\left(3\phi + \frac{\Pi}{h}\right) - J^3\sum \frac{1}{h^2}\Delta\frac{1}{n^2} + \sum hy^3K \end{cases} \tag{3}$$

Normalized for the height of light incidence, the focal length of the optical system can be obtained:

$$\begin{array}{lll} P_1 = -\frac{\beta_1^3\beta_2^3}{4} & P_2 = -\frac{\beta_2^3(1+\beta_1)(1-\beta_1)^2}{4} & P_3 = -\frac{(1+\beta_2)(1-\beta_2)^2}{4} \\ W_1 = \frac{\beta_1^2\beta_2^2}{2} & W_2 = \frac{\beta_2^2(1-\beta_1)(1+\beta_1)}{2} & W_3 = \frac{(1-\beta_2)(1+\beta_2)}{2} \\ \Pi_1 = \beta_1\beta_2 & \Pi_2 = -\beta_2(1+\beta_1) & \Pi_3 = 1+\beta_2 \\ \Phi_1 = -\beta_1\beta_2 & \Phi_2 = \frac{\beta_2(1+\beta_1)}{\alpha_1} & \Phi_2 = -\frac{1+\beta_2}{\alpha_1\alpha_2} \\ y_1 = 0 & y_2 = \frac{\alpha_1 - 1}{\beta_1\beta_2} & y_3 = \frac{\alpha_2(\alpha_1-1)+\beta_1(1-\alpha_2)}{\beta_1\beta_2} \end{array} \tag{4}$$

In Eq. (3), K1, K2, and K3 related to the quadratic surface coefficients are:

$$K_1 = \frac{e_1^2}{4}\beta_1^3\beta_2^3 \quad K_2 = -\frac{e_2^2}{4}\frac{(1+\beta_1)^3\beta_2^3}{\alpha_1^3} \quad K_3 = \frac{e_3^2}{4}\frac{(1+\beta_2)^3}{\alpha_1^3\alpha_2^3} \tag{5}$$

Setting S_{I}, S_{II}, S_{III}, S_{IV} and S_{V} to 0, the five equations can be solved for the quadratic surface coefficients of the three mirrors by simplifying them.

2.3 Off-Axis Handling

Because of the optical structure of the coaxial three-mirror optical system, the central obstruction problem will always exist, which will greatly reduce the MTF value of the imaging system. In order to solve the obstruction problem of the coaxial three-mirror, the coaxial three-mirror needs to be processed off-axis, so as to obtain the off-axis three-mirror optical system. The off-axis method used in this paper is diaphragm off-axis, which means that the primary mirror is set up as an aperture diaphragm, and the primary mirror is off-axis or tilted to eliminate obstruction.

3 Optical System Design Results

After optimizing the calculated initial structure of the coaxial three-mirror optical system, the obtained structural parameters of the off-axis three-mirror optical system are shown in Table 1. The conic coefficients of the primary, secondary and tertiary mirror surfaces are -0.898, -3.588 and -0.374 respectively. The surface of the tertiary mirror is an even-order aspherical surface, and its fourth-order coefficient is -1.069×10^{-10}, sixth-order coefficient is 8.688×10^{-14}.

Table 1. Optical system structural parameters.

Optical system comment	Diameter/mm	Radius/mm	Thickness/mm
Primary mirror	197.902	−342.340	−130.000
Secondary mirror	42.306	−132.470	128.000
Plane mirror	19.814	infinity	−104.000
Tertiary mirror	61.104	140.500	169.998

In order to achieve visible light and short-wave infrared dual-band imaging, a common aperture dual optical path is used for imaging, i.e., a beam-splitting element is added after the tertiary mirror to split the optical path into two. Therefore, this system uses a beam splitter as the light splitting element. The beam splitter is coated with SiO_2 as the base material and reflects short-wave infrared light and transmits visible light to achieve dual-band imaging. The layout of optical system is shown in Fig. 2.

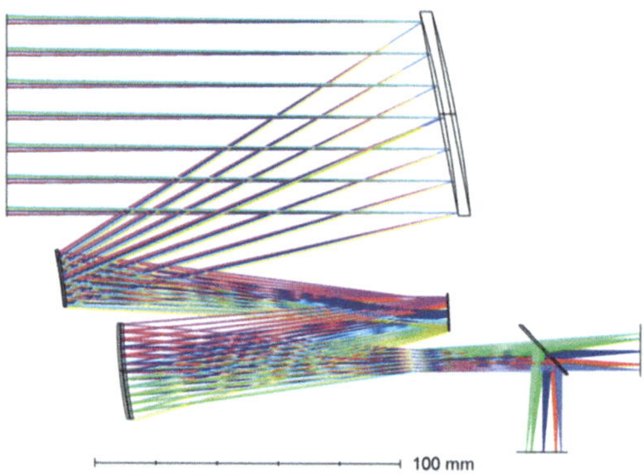

Fig. 2. Layout of off-axis three-mirror optical system

4 Analysis of Image Quality of Optical Systems

In geometric optics, the image formed by an ideal point should be an ideal point under ideal conditions. However, in the actual imaging process, the imaging beam will produce diffraction when passing through the optical system, and the image formed by the point should be a diffraction spot. In practice, an aberration-free system is a diffraction-limited system, that is, the image quality is only affected by the diffraction effect. By comparing the proximity of the imaging diffusion spot of the optical system to the diffraction Airy disk, the aberrations of the optical system can be inferred.

In Fig. 3a, the radius of the Airy disk is 7.594 μm and the maximum RMS value of the spot diagram radius of each field of view is 3.255 μm. It can be seen that the spot diagram of each field of view are located within the Airy disk. In Fig. 3b, the radius of

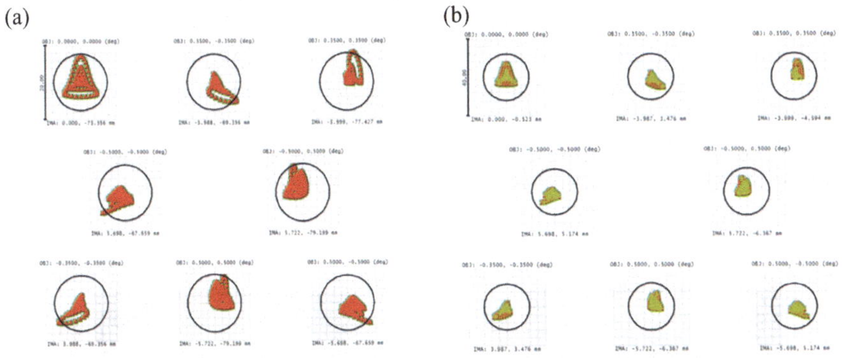

Fig. 3. Spot diagram of optical system. (a) VIS band and (b) SWIR band

the Airy disk is 11.99 μm; the maximum RMS value of the diffuse spot radius in each field of view is 3.620 μm, which is smaller than the Airy disk radius. Therefore, the designed system is close to a diffraction limited system.

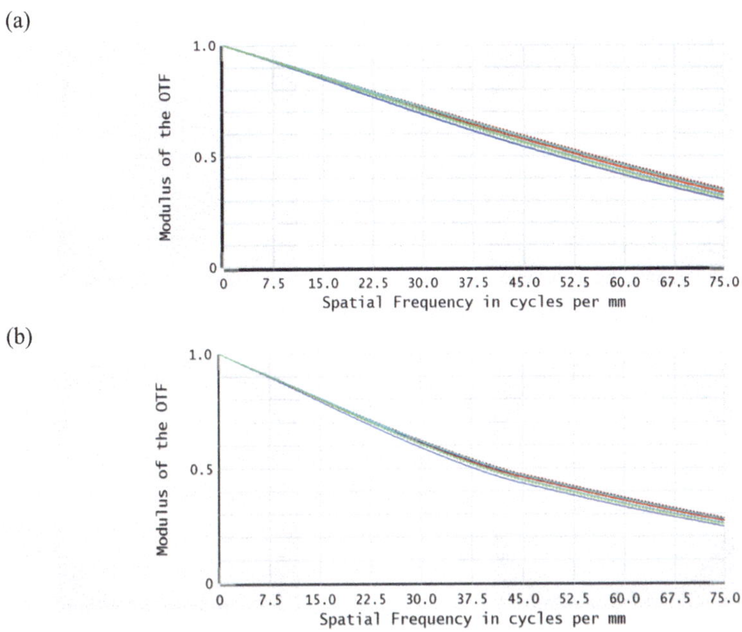

Fig. 4. MTF curve of optical system. (a) VIS band and (b) SWIR band

As shown in Fig. 4, in each field of view within the visible band, the MTF values were greater than 0.5 at 50 lp/mm and greater than 0.3 at 75 lp/mm; in each field of view within the short-wave infrared, the MTF values were greater than 0.4 at 50 lp/mm and greater than 0.25 at 75 lp/mm.

Select three fields of view (0°, 0°), (0.35°, 0.35°) and (0.5°, 0.5°) to analyze the wave aberration of the optical system. The results are shown in Fig. 5. In the visible light band, the maximum PV value of wave aberration is 0.321λ (λ = 0.633 μm), and the RMS value is within 0.062λ; in the short-wave infrared band, the maximum PV value is 0.1016λ (λ = 2.000 μm), and the maximum RMS value is 0.0196 λ. It shows that the aberration of the optical system has been well corrected.

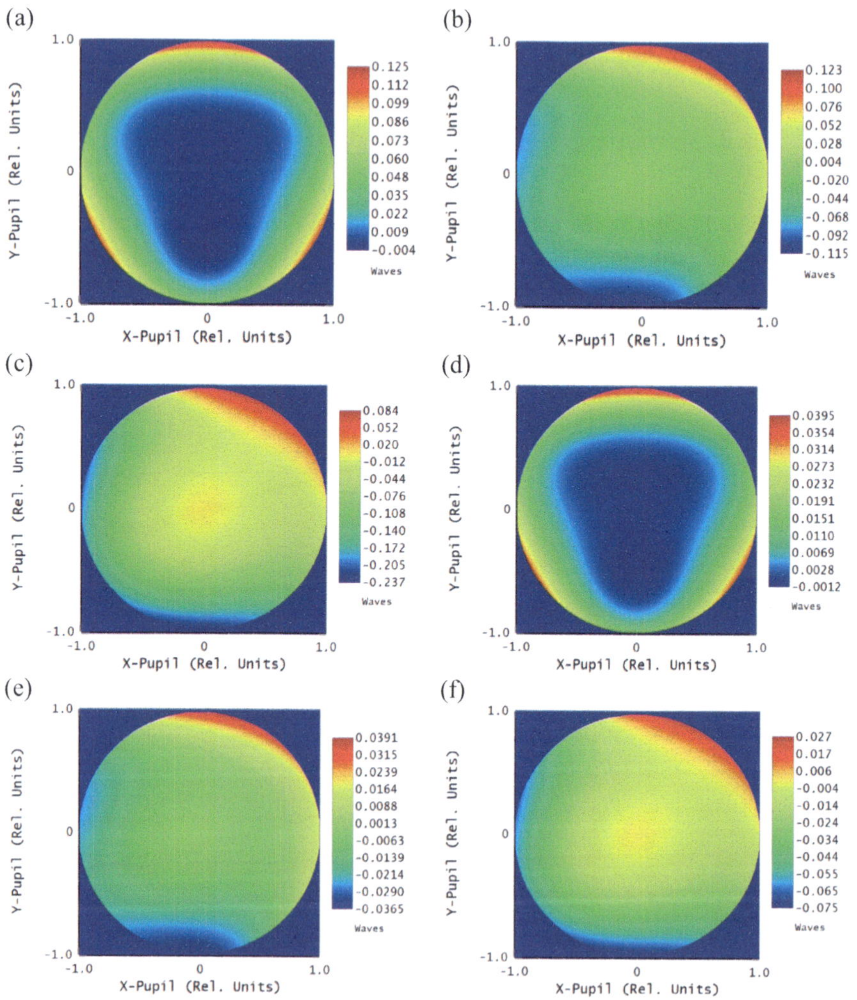

Fig. 5. Wave aberration of optical system. (a) VIS band, FOV (0°, 0°), (b) VIS band, FOV (0.35°, 0.35°) (c) VIS band, FOV (0.5°, 0.5°), (d)SWIR band, FOV (0°, 0°), (e) SWIR band, FOV (0.35°, 0.35°), (f) SWIR band, FOV (0.5°, 0.5°).

5 Conclusion

In this paper, using the method of off-axis aperture of the coaxial three-mirror optical system, an off-axis three-mirror optical system for remote sensing cameras is designed. Among the three mirrors of this system, only the tertiary mirror is a high-order aspherical surface, and both the primary mirror and the secondary mirror are secondary aspherical mirrors, which greatly improves the system's processability and adjustability. The total length of the system is 214 mm (only 1/3 of the focal length) and the imaging quality is close to that of a diffraction-limited system. This system solves the problem of high-resolution detection in the visible light and short-wave infrared bands of remote sensing cameras for earth observation. The design results of this paper provide theoretical foundation and technical support for the research of multi-band remote sensing cameras.

Acknowledgment. This work was supported by the Space Debris and Near-Earth Asteroid Defense Research Project (KJSP2020010304).

References

1. Nakano, T., Tamagawa, Y.: Configuration of an off-axis three-mirror system focused on compactness and brightness. Appl. Opt. **44**(5), 776–783 (2005)
2. Mendenhall, J.A., Lencioni, D.E., Hearn, D.R., et al.: EO-1 advanced land imager in-flight calibration. In: SPIE Conference on Earth Observing Systems III, pp. 416–422 (1998)
3. Chesters, D., Jenstrom, D.: GATES: a small, agile imaging satellite prototype for GOES-R. In: Conference on GOES-8 and Beyond, pp. 30–39 (1996)
4. Korsch, D.: Design and optimization technique for three-mirror telescopes. Appl. Opt. **19**(21), 3640 (1980). https://doi.org/10.1364/AO.19.003640
5. Fuerschbach, K., Rolland, J.P., Thompson, K.P.: Theory of aberration fields for general optical systems with freeform surfaces. Opt. Express **22**(22), 26585–26606 (2014)
6. Zhu, J., Hou, W., Zhang, X., et al.: Design of a low F-number freeform off-axis three-mirror system with rectangular field-of-view. J. Opt. **17**(1) (2015)
7. Cao, C., Liao, S., Liao, Z., et al.: Initial configuration design method for off-axis reflective optical systems using nodal aberration theory and genetic algorithm. Opt. Eng. **58**(10) (2019)
8. Qu, Z., Zhong, X., Zhang, K., et al.: Automatic compact-volume design strategy for unobscured reflective optical systems based on conicoid surfaces. Opt. Commun. **533** (2023)
9. Meng, Q.Y., Wang, H.Y., Liang, W.J., et al.: Design of off-axis three-mirror systems with ultrawide field of view based on an expansion process of surface freeform and field of view. Appl. Opt. **58**(3), 609–615 (2019)

A Combined Optical Telescope and Laser Rangefinder Target Localization Method Based on a Single Device

Shengji Dai and Zezheng Song(✉)

Astronaut Center of China, Beijing, China
zhubaodaddy@163.com

Abstract. With the gradual increase in the number of air and space targets, so does the demand for observation, localization and management of space targets. Compared with radar equipment, optical telescopes are characterized by simple structure and convenient operation, but as a single device, it cannot provide target positioning information.

This paper introduced a joint target localization method based on a single device of optical telescope and laser rangefinder, the main principle of which is to determine the celestial sphere coordinates of space targets by obtaining two rows of orbital roots of space targets and converting them into the first geocentric coordinates in the geocentric coordinate system. Then, based on the positional information of the optical telescope, the second geocentric coordinate of the optical telescope and the second geocentric coordinate of the optical telescope are combined to determine the geocentric coordinates of the space target, the pointing information of the optoelectronic telescope is obtained and corrected by using the geocentric coordinates and the preset pointing error correction model, and the distance of the spatial target relative to the optoelectronic telescope is measured by fusing the laser rangefinder to complete the joint target positioning of the optical telescope and the laser rangefinder.

Keywords: Single device · Optical telescope · Laser rangefinder · Joint location

1 Technical Field

This paper belongs to the field of aerospace measurement and control technology, and specifically involves a combined optical telescope and laser rangefinder target localization method based on a single device.

2 Background Technology

At present, with the increasing variety of space targets, such targets cover a wide range, including artificial satellites, stars, space debris, and re-entry space targets. With the gradual increase in the number of space targets, the demand for observation, localization,

H. P. Urbach et al. (Eds.): ISSOIA 2023, SPP 191, pp. 22–29, 2024.
https://doi.org/10.1007/978-981-97-6718-2_3

and management of space targets is also increasing. At present, there are two main types of observation methods for space targets: optical telescope detection and radar detection. Compared with radar equipment, optical telescopes have the characteristics of simple structure and convenient operation, but they also have the disadvantages of only providing goniometric information and cannot provide target positioning information as a single device. The development of laser rangefinder has the advantages of simple structure, accurate ranging, etc., combined with the telescope can effectively make up for the optical telescope cannot measure the shortcomings of the distance, and radar equipment, as well as realize a single device to complete the target positioning independently.

3 Solutions

To solve the above problems existing in the related technology, this paper provides a combined optical telescope and laser rangefinder target localization method based on a single device. The technical problems to be solved in this paper are realized by the following schemes:

1. Calculate a forecast position of the space target in space according to the space target forecast model;
2. Convert the space target position into a telescope observation coordinate system position;
3. Guide the telescope equipment to observe the space target;
4. Continuously correct the pointing of the laser rangefinder through the off-target amount of the telescope, and obtain the measured angle and range.

3.1 Orbit Prediction Based on SGPA4 Correction Model

Orbital forecasting selects the two-row root TLE and SGPA4 orbital forecasting model for observation and forecasting, and the two-row root is a set of descriptive parameters of the space target orbital model [1]. The main tasks to be accomplished for satellite orbit data generation are:

① read and settle the TLE orbital roots first;
② correct the parameters by using the SGPA4 model;
③ determine the celestial coordinate system of the space target by solving the Kepler equation.

The specific calculation formula is as follows:
First, correct the satellite orbital root:

$$a_1 = a = \left(\frac{k_e}{n_0}\right)^{2/3} \tag{1}$$

$$\delta_1 = \frac{3}{2} \times \frac{k_2}{{a_1}^2} \times \frac{(3cos^2 i - 1)}{(1 - e^2)^{3/2}} \tag{2}$$

$$a_0 = a_1 = \left(1 - \frac{1}{3}\delta_1 - {\delta_1}^2 - \frac{134}{81}{\delta_1}^3\right) \tag{3}$$

$$\delta_0 = \frac{3}{2} \times \frac{k_2}{{a_0}^2} \times \frac{(3cos^2 i - 1)}{(1 - e^2)^{3/2}} \tag{4}$$

$$n_0 = \frac{n_0}{1 + \delta_0} \tag{5}$$

$$a_0 = \frac{a_0}{1 - \delta_0} \tag{6}$$

After performing the atmospheric drag, gravity correction, and adding the long-period term, iterate by substituting $E + \omega$ into the Kepler equation, if:

$$U = IL_T - \Omega \tag{7}$$

Then:

$$(E + \omega)_{i+1} = (E + \omega)_i + \Delta(E + \omega)_i \tag{8}$$

In the formula:

$$\Delta(E + \omega)_{i+1} = \frac{U - a_{yN}\cos(E + \omega)_i + a_{xN}\sin(E + \omega)_i - (E + \omega)_i}{-a_{yN}\sin(E + \omega)_i + a_{xN}\cos(E + \omega)_i + 1} \tag{9}$$

$$(E + \omega)_1 - U \tag{10}$$

Calculate the short-period term:

$$r = a(1 - ecosE) \tag{11}$$

$$\dot{r} = k_e \frac{\sqrt{a}}{r} esinE \tag{12}$$

$$rf = k_e \frac{\sqrt{p_L}}{r} \tag{13}$$

$$cosu = \frac{a}{r}[\cos(E + \omega) - a_{xN} + \frac{a_{yN}(\mathrm{esin}E)}{1 + \sqrt{1 - {e_L}^2}}] \tag{14}$$

$$sinu = \frac{a}{r}[\sin(E + \omega) - a_{yN} + \frac{a_{xN}(\mathrm{esin}E)}{1 + \sqrt{1 - {e_L}^2}}] \tag{15}$$

$$u = tan^{-1}(\frac{sinu}{cosu}) \tag{16}$$

$$\Delta r = \frac{k_2}{2p_L}\left(1 - \theta^2\right)cos2u \tag{17}$$

$$\Delta u = -\frac{k_2}{4{p_L}^2}\left(7\theta^2 - 1\right)sin2u \tag{18}$$

$$\Delta\Omega = -\frac{3k_2\theta}{2{p_L}^2} sin2u \tag{19}$$

$$\Delta i = -\frac{3k_2\theta}{2{p_L}^2} sini_0 cos2u \tag{20}$$

$$\Delta \dot{r} = -\frac{k_2 n}{p_L}\left(1-\theta^2\right) sin2u \tag{21}$$

$$\Delta \dot{rf} = -\frac{k_2 n}{p_L}[\left(1-\theta^2\right) cos2u - \frac{3}{2}\left(1-3\theta^2\right)] \tag{22}$$

Calculate the orbital elements:

$$r_k = r\left[1 - \frac{3}{2}k_2\frac{\sqrt{1-{e_L}^2}}{{p_L}^2}\left(3\theta^2-1\right)\right] + \Delta r \tag{23}$$

$$u_k = u + \Delta u \tag{24}$$

$$\Omega_k = \Omega + \Delta\Omega \tag{25}$$

$$i_k = i_0 + \Delta i \tag{26}$$

$$\dot{r}_k = \dot{r} + \Delta \dot{r} \tag{27}$$

$$\dot{rf}_k = \dot{rf} + \Delta \dot{rf} \tag{28}$$

Calculate the unit direction vector:

$$U = Msinu_k + Ncosu_k \tag{29}$$

$$V = Mcosu_k + Nsinu_k \tag{30}$$

$$M = \begin{cases} M_x = -sin\Omega_k cosi_k \\ M_y = cos\Omega_k cosi_k \\ M_z = sini_k \end{cases} \tag{31}$$

$$N = \begin{cases} N_x = cos\Omega_k \\ N_y = sin\Omega_k \\ N_z = 0 \end{cases} \tag{32}$$

Position and velocity of the space target:

$$r = r_k U \tag{33}$$

$$\dot{r} = \dot{r}_k U + (\dot{rf})_k V \tag{34}$$

This position and velocity coordinate is the celestial reference sphere.

The TLE orbital roots are corrected by the SGP4 model to compute the celestial coordinates of the space target at any instant in time.

3.2 Coordinate Transformation

After calculating the orbit, the celestial coordinates of the space target relative to the center of the earth can be obtained, and to obtain the station coordinates of the satellite relative to the station, the corresponding coordinate transformations need to be carried out.

After the orbit calculation, the celestial sphere coordinates of the space target relative to the center of the earth can be obtained, and to obtain the coordinates of the satellite relative to the station, the corresponding coordinate transformations need to be carried out.

(1) Conversion of celestial coordinates to geocentric coordinates.

The formula for converting the celestial coordinate system to geocentric coordinate system is:

$$\begin{bmatrix} X \\ Y \\ Z \end{bmatrix}_{M(t_0)} = ABCD \begin{bmatrix} X \\ Y \\ Z \end{bmatrix}_D \tag{35}$$

In the formula: $A = RY(-xp) \cdot R(yp)$ is the pole-shift transformation matrix. (xp, yp) is the polar shift coordinates of the instantaneous geodesic; $B = Rz(GAST)$ is the time-angle transformation matrix, and $GAST$ is the instantaneous Greenwich mean sidereal time. C is the nutation transformation matrix, and D is the precession transformation matrix.

(2) Conversion of geocentric coordinates to station coordinates.

For the station equipment, it is necessary to calculate the azimuth and pitch angle of the target relative to the station according to the position of the station and the target position. First of all, the station coordinates are converted from latitude and longitude to geocentric coordinates:

$$\begin{cases} X_c = (N + H)cosBcosL \\ Y_c = (N + H)cosBsinL \\ Z_c = [\big(N(1 - e^2) + H\big)sinB \end{cases} \tag{36}$$

In the formula: (X_c, Y_c, Z_c) is the geocentric coordinates of the observation device, L, B, and H are distributed as the longitude, latitude, and elevation of the observation device N is the radius of the ellipsoid's dyadic circle, e is the first eccentricity of the ellipsoid.

$$N = \frac{a}{\sqrt{1 - e^2 sin^2 B}} \tag{37}$$

$$e^2 = \frac{a^2 - b^2}{a^2} \tag{38}$$

In the formula, a and b are the half-length and half-short axes of the Earth, respectively. The horizon coordinates of the space target relative to the station equipment can be derived:

$$\begin{bmatrix} Y \\ Z \\ X \end{bmatrix} = \left(\begin{bmatrix} 1 & 0 & 0 \\ 0 & cosL & -sinL \\ 0 & sinL & cosL \end{bmatrix} \begin{bmatrix} cosB & sinB & 0 \\ -sinB & cosB & 0 \\ 0 & 0 & 1 \end{bmatrix} \right)^{-1} \begin{bmatrix} Z - Z_c \\ X - X_c \\ Y - Y_c \end{bmatrix} \tag{39}$$

3.3 Positioning of the Optical Telescope Axis System.

Selection of a ground-level photoelectric telescope, the telescope is guided to track the target according to the solved spatial target position information.

According to the calculated space target position information is used to guide the telescope to track the target. The horizontal telescope is a three-axis tracking structure, the three axes are vertical axis, horizontal axis, and visual axis. In the vertical and horizontal axes, according to the axis angle encoder is used to measure the azimuth angle A and pitch angle E pointed by the telescope's view axis. After the telescope tracks the target to acquire the image, the interpretation of the target is offset from the center of the visual axis. It is the off-target quantities Δx and Δy [2], synthesized encoder data, telescope parameters, etc. Using the resulting parameters for axial error model correction. The correction model uses a physical parameter correction model and obtains the azimuth A and pitch E of the target relative to the station.

Axial positioning combines the off-target amount of the target, the telescope pointing position, the exposure center time, and the error correction amount. Obtain the axis system measurement data of the target under test.

Assuming that the coordinates of the center of the CCD image are (x_0, y_0) the center of mass of the target star image is (x_s, y_s), the scale of the image is (S_x, S_y) and the pointing of the telescope system is (A_0, E_0), then the azimuth angle A_s and elevation angle E_s of the target are calculated as:

$$\begin{cases} E_s = E_0 \pm (y_s - y_0) \cdot S_y \\ A_s = A_0 \pm (x_s - x_0) \cdot S_x / cos(E_s) \end{cases} \tag{40}$$

In the formula, the plus and minus signs are selected about the direction of increasing azimuth and elevation angles on the detector image.

The basic steps of axial localization are to first read the encoder data. Get the telescope pointing data (azimuth angle A and elevation angle E); Then the off-target amount $(\Delta x, \Delta y)$ of the space target is calculated by CCD image processing. The ΔA and ΔE between the space target and the telescope are obtained according to the CCD scale; According to the telescope pointing model correction, get the azimuth A and pitch angle E relative to the station. Among them, the key problem is the pointing error correction, which is described in detail below.

The pointing error can be decomposed into azimuth error ΔA and pitch error ΔE [3].

$$\begin{cases} \Delta A_i = \widehat{A} - A = f(A, E) + \varepsilon_i \\ \Delta E_i = \widehat{E} - E = g(A, E) + \eta_i \end{cases} \tag{41}$$

In the formula, $i = 1,2,\ldots, n$, where n is the observation sample $(\widehat{A}, \widehat{E})$ and (A,E) are distributed as the actual and ideal observations of the target in azimuth and pitch. $f(A, E)$ and $g(A, E)$ are the fitting functions of A_i and E_i respectively, which are called the telescope pointing correction model, ε_i and η_i are the observation random errors. The telescope-pointing process is guided by the control system. By obtaining the position information of the goniometric component and the control commands generated by the control system, the pointing model is corrected and pointed to the target position by the operation of the pointing model in the pointing process.

The process of obtaining the pointing error correction model for the telescope is as follows:

(1) Observation of precisely positioned stars in space using an optoelectronic telescope to obtain a series of actual observations and corresponding ideal data uniformly distributed over the whole sky area. Thus, the pointing error data at each position are obtained.

(2) Construct the pointing error correction model $\delta_A = f(A, E)$, $\delta_E = g(A, E)$, the selection of the error model and the stability of the model determine the accuracy of the pointing error model. The factors that have a large impact on the pointing error are the optic axis illumination difference, the azimuth pitch vertical error, the base tilt error, the pitch bearing error, the azimuth bearing error optic axis wobble error. The total pointing error expression is:

$$\Delta A = -\gamma secE + \eta tanE + \psi_X tanEsinA + \psi_Y tanEcosA + \xi_{B1} sin2E + \xi_{B2} cos2E + \xi_{B3} cos2AtanEsinA + \xi_{B4} cos2AtanEcosA \tag{42}$$

$$\Delta E = \psi_X cosA - \psi_Y sinA + \xi_{B3} cosAcos2A - \xi_{B4} sinAcos2A - \delta_T cotE \tag{43}$$

In the formula, γ is the optic axis illumination difference, η is the azimuth pitch vertical error, ψ_X is the base tilt error (east), ψ_Y is the base tilt error (north), $\xi_{B1} sin2E$ is the pitch bearing error (sin), $\xi_{B2} cos2E$ is the pitch bearing error (cos), $\xi_{B3} cos2A$ is the azimuth bearing error (north-south). $\xi_{B4} cos2A$ is the azimuth bearing error (east-west), δ_T is the apparent axis wobble error.

(3) In this paper, the objective is to improve the pointing accuracy of the telescope. Therefore, the sum of squares of the residuals of the model correction is selected as the objective function, and the parameters of the pointing error correction model are solved precisely. That is, to solve the following system of equations.

$$\begin{cases} \sum_{i=1}^{n} [f(A, E) - \Delta A]^2 = min\Sigma \\ \sum_{i=1}^{n} [g(A, E) - \Delta E]^2 = min\Sigma \end{cases} \tag{44}$$

The problem of solving the above objective function is the least squares estimation problem.

(4) The correction coefficients and correction model are introduced into the telescope control system. Then the telescope can point to the target orientation more accurately. To achieve the purpose of improving the pointing accuracy.

3.4 Laser Ranging

Based on the target position calculated from telescope observations, the pointing of the laser ranging system is continuously corrected. The laser emits a laser pulse at a space target and then returns, and the receiving telescope of the ground station receives the reflected echoes. At the same time, the time timer measures the time interval between round trips of the laser pulse (denoted by t, which is multiplied by the speed of light c. The distance from the space target to the station (denoted by R) can be precisely converted. The distance from the space target to the station (denoted by R) can be accurately converted. Then R and t are related as follows:

$$R = \frac{1}{2} \cdot c \cdot t \tag{45}$$

In summary, the observation of space targets is carried out through optical telescopes and laser ranging systems to obtain angular and ranging measurements, and to accomplish the localization of space targets.

References

1. Yu, Y., Guo, L., Li, D.: Applications of SGP4/SDP4 model in satellite orbit forecasting with horizontal photoelectric tracking equipment. Chin. J. Sci. Inst. (34), 1407–1410 (2013)
2. Wu, R.: Research on optoelectronic observation and information processing technologies of LEO space object, pp. 23–27 (2015)
3. Huang, L.: Research in pointing error and shafting technologies for large telescope. University of Chinese Academy of Sciences, pp. 33–46 (2016)

Research and Analysis on Operation Mode of Commercial Optical Remote Sensing Satellite

Lyu Zhen[1(✉)], Zhang Yun Feng[2], and Liang Yin[2]

[1] Beijing Institute of Spacecraft Environment Engineering, Beijing 100094, China
lvzhen_hq@163.com

[2] Beijing Spacecrafts Manufacturing Co., Ltd., Beijing 100094, China

Abstract. Since the implementation of the 14th Five-Year Plan, the special implementation of high-resolution optical satellites and the vigorous rise of commercial satellite remote sensing services have enriched the data sources and image sources of optical remote sensing satellites. The application side's requirements for data services are more timely and accurate, and the associated data applications are becoming more mature, with the supply capacity and service capacity of enabling many social industries. The key to the operation mode of commercial optical remote sensing satellite is to further meet the growth demand of the client through sustainable and rich business service mode, and then improve the supporting role of China's space infrastructure and optical remote sensing satellite for economic and social development.

Keywords: optical remote sensing satellite · commercial operation · mode research

In 2015, The State Council issued the National Medium and Long-term Development Plan for Civil Space Infrastructure (2015–2025), which actively encourages private capital and enterprises to participate in the development and operation of commercial remote sensing satellites. Vigorously support the commercial operation of sub-meter and high-resolution remote sensing satellites and constellations, so as to realize the market-oriented and industrialized development of more commercial remote sensing satellite products and services. The strong support of national policies and the active guidance of industry authorities have brought unprecedented opportunities for the development of commercial remote sensing satellite enterprises. The "Gaojing-1" 03 and 04 satellites independently developed and operated by the China Academy of Space Technology were launched into space, opening a new chapter in the operation of commercial remote sensing optical remote sensing and microwave remote sensing satellites by Chinese aerospace enterprises.

1 Market Status Analysis of Commercial Remote Sensing Satellites

Under the impact of the era of "Internet+", big data, cloud computing and 6G, the demand for commercial remote sensing satellite white market has gradually increased, the commercial remote sensing satellite industry has developed rapidly, and a large

H. P. Urbach et al. (Eds.): ISSOIA 2023, SPP 191, pp. 30–38, 2024.
https://doi.org/10.1007/978-981-97-6718-2_4

number of commercial remote sensing satellite enterprises have emerged to participate in the competition in the remote sensing satellite market. Including Digital Globe Company and Planet Labs company in the United States, ImageSat International company in Europe, UrtheCast company in Canada, and China's Changguan SI Wei dimensional World View and Obit Company.

These enterprises have the right to use and operate commercial remote sensing satellites, to the image or image data receiving, transmission and analysis as the main commercial object, in the form of integrated resources to meet the needs of consumers in all walks of life, and ultimately achieve the value-added benefits of commercial operations. The international market demand for commercial remote sensing satellites is gradually increasing at a compound annual rate of nearly 10%, which stimulates the increase of bulk sales revenue. It is estimated that by 2025, the global commercial remote sensing satellite data market revenue will be close to 5 billion US dollars. In such a strong international market demand and sales revenue, will promote the future development of China's commercial remote sensing satellite operation, out of a commercial space with Chinese characteristics Road.

2 Practical Exploration of Commercial Optical Remote Sensing Satellites in China

According to date released by the China Association of Remote Sensing Application in 2022: Chinese commercial remote sensing information enterprises have reached more than 20,000, with more than 400,000 employees, with an annual output value of 250 billion yuan. China's commercial remote sensing satellite industry is moving forward from the introduction period to the growth period.

2.1 Practice of Commercial Optical Remote Sensing Satellite in China

Since 2015, under the active guidance of the national macro policy, Traditional domestic space enterprises represented by China Academy of Space Technology, commercial space enterprises represented by Beijing 21st Century Space Technology Application Co., LTD., scientific research institutions represented by Changchun Institute of Optics, Fine Mechanics and Physics of the Chinese Academy of Sciences, and other scientific research enterprises and institutions are intensively carrying out the development of commercial satellites, and gradually forming a multi-format and multi-form business The space industry and satellite research and development constitute a force. Among them, Beijing 21st Century Space Technology Application Co., Ltd. developed and launched the Beijing-1 and Beijing-2 commercial remote sensing satellites; Jilin-1 commercial remote sensing satellite was developed and launched by Changguang Satellite Technology Co., LTD.; Beijing Siwei Commercial remote sensing Aerospace Technology Co., Ltd. developed and launched Gaojing-1 commercial remote sensing satellite; Zhuhai Obit Aerospace Technology Co., LTD. developed and launched the Zhuhai No. 1 commercial remote sensing satellite, which has completed the launch mission and been put into operation. These commercial remote sensing satellite companies and their commercial remote sensing satellite constellations jointly promote the development of Chinese commercial remote sensing satellite industry.

The commercial operation mode of Chinese remote sensing satellites is mainly divided into two types. Self-owned satellite operators obtain first-hand remote sensing data by independently developing and launching commercial remote sensing satellites and conducting networking. According to the business planning of the enterprises, they may directly sell remote sensing data or process remote sensing data to provide data services, technical services and software agents with higher profits. The representatives of such enterprises are: Changguang Satellite Technology Co., LTD. Beijing Siwei Commercial remote sensing Aerospace Technology Co., LTD., etc. No own satellite operators mainly through the purchase of data after the deep processing, the formation of unique products to sell, such as aerospace Hongtu PIE-Engine remote sensing cloud platform, GEOVIS Map's digital earth products, hypermap software GIS software and so on.

Form the perspective of satellite remote sensing data application, Chinese satellite remote sensing application service enterprises have regional and industrial characteristics. In terms of regions, Beijing and Wuhan have the largest concentration of remote sensing enterprises, which is related to the existence of remote sensing technology research and development institutions in the two places. In terms of industries, there are the most remote sensing enterprises in land use and remote sensing surveying and mapping, Which is related to the huge market in these two fields.

Form the perspective of users, for a long time, the main body of remote sensing application in China mainly comes from the government, including natural resource management, disaster monitoring, ecological protection, etc. Remote sensing data distribution is also mainly based on relevant scientific research institutions, the data acquisition channels are limited, and the application value of remote sensing data has not been fully tapped.

2.2 Development Trend of Commercial Remote Sensing Satellites

2.2.1 Constellations

A single commercial remote sensing satellite has limited coverage and less data. Only by building a high-density satellite network can it occupy the blue ocean of the commercial space market. To realize the construction of constellation, satellite remote sensing and low altitude remote sensing technology need to be fully distributed, and low-altitude remote sensing technology with low cost, high mobility characteristics of collaborative observation with satellites, make up for the shortcomings of remote sensing satellites in space observation, in order to master more abundant observation data. Traditional single and double stars have long been unable to meet the needs of commercial optical remote sensing satellites, and the future constellation development model will be the development direction of commercial optical remote sensing satellites.

2.2.2 Marketization

With the development and maturity of satellite platforming, payload modularization and satellite integration technologies, remote sensing satellites have been fully applied in basic surveying and mapping, agriculture, forestry and other fields, and have also emerged in the fields of transportation, wildlife protection, hydrological survey, electric power and railway safety. Early remote sensing satellites did not pay attention to mining the commercial value of data, and now major satellite operators take the initiative to provide personalized services to in-depth mining the commercial value behind the data to occupy the market. The application market of commercial remote sensing satellites is guided by users' needs in resolution, timeliness and positioning accuracy, and customized algorithms are carried out for the industry to realize the mining of large-capacity complex data to enhance the value of data. The US Digital Earth company combined machine big data algorithm to extract more than 10 million buildings in Australia to form a 3D stereoscopic image map. It would take three to five years to accomplish this goal using traditional manual methods, which greatly reduces the time for data and information to be updated and provide suggestions for urban planning and construction.

2.2.3 Data Sharing

Earth remote sensing has been developed in the fields of meteorology, ocean, resources, environment and communication in China. For the observation of satellite constellation, each satellite constellation can obtain a large amount of data, and the remote sensing data of a single satellite constellation is difficult to meet the needs of users in different dimensions, which also causes a waste of various comprehensive information processing systems and ground station resources. If an open, converging and collaborative data sharing platform is built, it can receive multi-satellite and multi-source remote sensing data, display the operation of each remote sensing satellite through the platform in real time, and receive satellite data in time to realize information sharing and resource interworking. Breaking the status quo of remote sensing data dispersion, using big data processing, artificial intelligence and other methods to analyze and process the data obtained by the platform, the data collection will be faster, the cost will be greatly reduced, and the dog can realize cross-industry, cross-field and interdisciplinary data sharing, so that the remote sensing data and scientific research technology practice have been deeply applied.

2.3 Commercial Optical Remote Sensing Satellites Are an Important Thrust for Building an Aerospace Power

According to the statistics of the Ministry of Natural Resources, in 2019, the satellites photographed a total of 14,897 orbits and obtained 605,000 effective high-resolution image data with a total amount of 2866TB. Satellite remote sensing big data, derived from land, ocean, atmosphere and human activities, is not only a new strategic resource for the country, but also an important field of competition among countries.

In the more than 50 years since the satellite was launched. The number of inventions that humans have made. More than in the last 2,000 years combined. The development level of the space satellite industry reflects the country's industrial strength. It has become a core industry and a high-end advanced industry that can represent "intelligent manufacturing in China". Satellite manufacturing industry not only creates value, but also is one of the most influential achievements in modern society and national economic development among high-tech industries in the world today.

All countries in the world attach great importance to the development of commercial remote sensing satellites, and constantly improve the research and development technology and application of results of optical remote sensing satellites. This has promoted the industrial development of space remote sensing internationally, and the emphasis on remote sensing satellite technology has indeed made various countries achieve more and more significant social and economic benefits, which also represents that satellite remote sensing is entering a new upsurge of development. The global satellite and application industry has grown rapidly, with an average annual growth rate of more than 10% since the beginning of the 21st century. Developing and improving independent space infrastructure has increasingly become a strategic choice for developed countries and regions to pursue space leadership, seize the commanding heights of economic and technological competition, develop emerging industries, and safeguard their security interests.

In December 2020, the Gaofen-14, developed by the China Academy of Space Technology, The satellite was successfully launched, and its resolution can reach the "decimeter level", reaching the world's advanced level, that is, the smallest pixel color block on the "photo" taken by the "camera" on the satellite can cover the object as long as 1 dm. In order to identify the target of a specific person, it is also necessary to quickly and efficiently locate a specific face image from the massive remote sensing image information. Optical remote sensing technology can help it achieve the effect of "finding a needle in a haystack".

Commercial optical remote sensing satellites play an important role the field of natural resources management and environmental monitoring. The remote sensing data of natural resources generally has the characteristics of medium and low resolution, wide coverage area and short revisit period. Notable achievements in natural resource management have been made in agriculture and forestry.

Agriculture has always been the lifeblood of Chinese national economy. The application of remote sensing technology in agriculture can prevent agriculture distributed in a wide area, and high-resolution earth observation can master the investigation and monitoring of forestry resources in a short time, providing a flexible and fast way for the promotion of digital forestry。The remote sensing data in the field of environmental monitoring has the characteristics of high time resolution, high spectral resolution and long time series. China has encountered environmental problems in the process of rapid urban development. Using remote sensing data to carry out research on urban heat island, atmosphere, haze and other environmental problems will accelerate the pace of ecological civilization construction and play an important role in building a beautiful China, also an important support for the construction of aerospace power.

3 Search on the Operation Mode of China's Commercial Optical Remote Sensing Satellite

3.1 Improve the Policy System of Commercial Operation and Development of Optical Remote Sensing Satellites

Improve the policy system of Chinese commercial optical remote sensing satellite operation, standardize and guide the sound development of domestic commercial remote sensing operators from the policy level. At the same time, under the premise of ensuring national information security and diplomatic legitimate rights and interests, formulate the operation policy of domestic commercial optical remote sensing satellites; Through appropriate adjustment of import and export policies, relax the transaction control of optical remote sensing data; Formulate policies to protect the rights and interests of domestic commercial remote sensing satellite participants, clarify the responsibilities of operators, distributors and value-added service providers, increase the protection of operators' interests, and enhance the international strength of China's commercial optical remote sensing satellite operation in the process of constantly improving the policy system of domestic commercial optical remote sensing satellites.

3.2 Improve the Construction of Industrial Chain

Promote the further expansion of market-oriented operation of domestic commercial remote sensing satellites, so as to improve the industrial chain construction of domestic commercial remote sensing satellites. The vast majority of users in the industrial chain of commercial remote sensing satellites are government departments. In recent years, the government's procurement mode has changed from "purchasing satellites" to "purchasing services", which makes the operators of commercial optical remote sensing satellites consolidate the back-end business foundation and have an independent market development space. However, commercial remote sensing satellite users can be penetrated by cross-industry enterprises, especially with the Internet and big data operation and other related enterprises as potential users, so that the optical remote sensing satellite market from the government to the enterprise and the public field further expanded.

In the industrial chain of commercial optical remote sensing satellites, distributors should be guided to actively derive data analysis, software development and application to achieve the purpose of extending their own service value. Starting from the self-positioning of remote sensing data distributors, data analysis and integration should be carried out from multiple angles and levels while obtaining data, so as to compete reasonably with value-added service providers and promote the smooth operation efficiency of the whole. At the same time, it is also necessary to encourage value-added service providers to expand their business scope, achieve the expansion of the application scope of remote sensing satellite data, enrich their business scope, in order to create more profit space to improve the purpose of industrial chain construction.

3.3 Promote High-Resolution Technology Innovation

In order to achieve the rise of domestic commercial optical remote sensing satellite operating market position, it is necessary to promote the iterative update of high resolution commercial optical remote sensing satellite technology. Most of China's Gaofen special engineering satellites are in high orbit and are used for national defense construction; Commercial remote sensing satellites are mostly located in medium and low orbits, and the launch cost is lower than that of Gaofen special satellites. The technical ideas of opening some Gaofen special engineering satellite technologies to the high-resolution technology of commercial remote sensing satellites can be adopted, and some orbital restrictions can be opened to promote the technological upgrading of high-resolution commercial optical remote sensing satellites, and provide technical support for the operation of domestic commercial optical remote sensing satellites.

3.4 Promote Management Innovation in the Field of Operation

Different from the traditional enterprise's operation of products and services, the operation mode of commercial remote sensing satellites is a complex management structure, which is applied in various fields and can drive the rise of many emerging industries, such as smart cities, smart transportation and other industries. The operation mode of domestic commercial optical remote sensing satellite is very similar to the concept of the Internet of everything in the Internet field. The Internet can be used to further develop the integration and decentralized business of commercial optical remote sensing satellite data, forming an Internet platform of commercial remote sensing satellite data trading market to meet the needs of different users in a timely manner. In addition, according to the user's preference for satellite data browsing and subscription on the Internet platform, the target user demand database can be established by means of big data analysis, and the relevant information of data products and services can be proactively provided to the user, and the demand tracking of potential users can be carried out.

3.5 Improve the Commercialization Degree

Improving the commercialization of China's optical remote sensing satellites is a prerequisite for the development of the satellite operation market. To improve the commercialization degree, one part reduces the pressure on the development and production of commercial remote sensing satellites, and the other part develops the commercial value of operating remote sensing satellites in space, that is, improving the ability to understand customer needs with operators. Improving the commercialization of optical remote sensing satellites can be accomplished in the following ways: opening up the restrictions on the commercialization of optical remote sensing satellites, allowing satellites that do not involve military satellite data to participate in the market competition; To provide technical support for the development and production of commercial optical remote sensing satellites to private enterprises, and reduce the original investment in research and development as much as possible; Promote local participation in the development of commercial satellite projects to support the commercial operation of remote sensing satellites by local enterprises; Regular public service the government's procurement demand for commercial remote sensing satellite products and services makes the commercial operation of domestic optical remote sensing satellites more targeted.

3.6 Attracting International Capital Attention

The operation of commercial optical remote sensing satellites requires a large amount of capital support and operation. Capital is the basis to support the operation mode of commercial optical remote sensing satellite. On the premise of ensuring the security of remote sensing information and permitted by government policies, the financing of commercial optical remote sensing satellite enterprises should be increased, so as to use more levers to effectively mobilize social capital to participate in the industrial development of commercial remote sensing satellites. While receiving the injection of state capital and social capital, It is also necessary to actively encourage commercial optical remote sensing satellite enterprises to attract the attention of international capital, on the one hand to reduce the pressure on the operating cost of domestic commercial remote sensing satellite, on the other hand, it can extract foreign technological advantages from the introduction of capital, promote the innovation of application technology, achieve the purpose of landing domestic commercial remote sensing satellite technological achievements, and improve the overall commercial operation efficiency.

4 Conclusion

Chinese commercial satellite started late, the industrial chain is not mature, compared with the world's advanced countries and international leading aerospace enterprises are still in the growth stage. The development of commercial optical remote sensing satellites should not only be based on China's national conditions, but also aim at the international commercial space market, and find the combination of the two. While strengthening the top-level design, promoting the transformation of business model, tapping potential customers and occupying the blank area of the space market, we should also pay attention to technological innovation, and develop commercial optical remote sensing satellites with high resolution, wide coverage, high density and deep application to occupy the front end of the market. Promote the establishment of commercial standards and off-the-shelf product systems, promote standardized, modular and integrated design concepts, effectively reduce costs, and achieve high stability, light miniaturization and low cost of commercial remote sensing satellites. Cultivate a new commercial remote sensing data application market, guide the data market to expand to the public, explore a new model for the development of China's commercial remote sensing satellites, and help the commercial space industry move towards a new journey of openness, cooperation and diversity.

References

1. Teng, S.H., Li, H., Chen, D.: Development status of commercial remote sensing projects and management thoughts on commercial remote sensing projects for space enterprises. Mod. Econ. Inform. 8 (2020)
2. Yang, J.X., Ying, H.Y.: Application and development trend of commercial remote sensing satellites. Sci. Technol. Inform. **18** (2020)
3. Li, Y.: Research on the operation model of domestic commercial remote sensing satellite based on value network, Master Thesis of Harbin Engineering University (2019)
4. Zhang, S.H., Xu, D.L.: Development, mapping and spatial geographic information of commercial remote sensing satellites of the United States **12** (2016)
5. Campbell, J.B.: Introduction to Remote Sensing, pp. 57–59. The Guilford Press (2023)
6. Xinhua news network. Commercial remote sensing satellites meet the development window period
7. Li, L.: System design and on-orbit performance of super view-1. Space Int. **2018**(9) (2018)

8. Zhao, Z.Y.: Jilin-1 Chinese Largest Commercial remote sensing satellites constellation. Space Explor. **2021**(9) (2021)
9. Chang, G.J.: Technical analysis of high resolution commercial optical remote sensing satellite. Space Craft Eng. **2009**(2) (2009)
10. Qu, H.S., Jin, G., Zhang, Y.: Next View program and progress in optical remote sensing satellites. Chin. J. Opt. Appl. Opt. **02**(6) (2009)

An Adaptive Detection Method of Spatial Circular Feature Based on Arc Segment Under Different Lighting Conditions

Yibing Ding[1(✉)], Zongzheng Zhang[2], Yuxuan Cai[2], Zhenhua Su[1], Haiming Wang[1], Ying Luo[1], Shengping Gong[2], and Hao Wen[3]

[1] DFH Satellite Co., Ltd., Beijing 100094, China
dingyibing@buaa.edu.cn
[2] School of Astronautics, Beihang University, Beijing 100191, China
[3] College of Aerospace Engineering, Nanjing University of Aeronautics and Astronautics, Nanjing 211106, China

Abstract. Accurate recognition of spacecrafts is an important prerequisite for on-orbit service (OOS). For missions such as on-orbit operation, rendezvous and docking, structures with special shape such as docking ring or engine nozzle are usually selected as recognition objects, whose circular feature can help to accurately locate and determine the potential operating points for subsequent tasks. Different lighting conditions in space are easy to interface with visible light imaging, resulting in different brightness, loss of details and blurring of some key features. This paper proposed an adaptive detection method of spatial circular feature based on arc segment to solve the problems of accurate recognition of spacecraft's docking ring under low illumination and local over-exposure conditions. Firstly, we designed an adaptive image enhancement pre-processing method based on multi-scale Retinex with chromaticity preservation (MSRCP). Images obtained by visible camera under different lighting conditions were enhanced by MSRCP and then processed by adaptive threshold method for segmentation, filtering and edge detection. Then an ellipse detection algorithm based on arc segment features was designed to extract the contour of the target's docking ring from the edge curve. Finally, a ground simulation experimental system was built, and we carried out some experiments under different lighting conditions. The experimental results showed that our method can effectively improve the image quality under low illumination and local over-exposure conditions. The average brightness, standard deviation and information entropy of images in different conditions were improved. The circular feature detection algorithm showed excellent detection results on the enhanced images under different lighting conditions. This paper provides a certain reference for the accurate recognition of targets in on-orbit service missions.

Keywords: On-Orbit Servicing · Rendezvous and Docking · Adaptive Threshold · Arc Feature · Ellipse detection

H. P. Urbach et al. (Eds.): ISSOIA 2023, SPP 191, pp. 39–53, 2024.
https://doi.org/10.1007/978-981-97-6718-2_5

1 Introduction

Humans have launched 14,828 spacecrafts into space by March 2023. Simultaneously, there are only 7480 spacecrafts operating normally in orbit, and there are a significant number of malfunctioning or failed spacecrafts, which threaten space safety, waste orbital resources [1]. The key to recycling or reusing a failed spacecraft is the accurate recognition of its typical structural components, which usually include solar panel, communication antenna and bracket, star-arrow docking ring, and engine nozzle [2]. Docking ring and engine nozzle are often designed to be a standard circular shape and are usually presented in an elliptical shape in the camera image. Therefore, the detection of circular feature structures such as docking ring is essentially a study of ellipse detection algorithms.

However, due to the change of lighting conditions during the imaging of the visible camera and the high reflectivity of multi-layer insulation (MLI) materials on the spacecraft's surface, it is easy to cause the image to be over-exposure or too dark, which interferes with the extraction of the contour of the docking ring. High-precision detection of failed satellite components still has many challenges.

A real-time docking ring detection method was presented in [3] that includes arc extraction, classification, and parameter estimation for docking ring detection. In [4], a pose estimation method using grayscale point cloud images to overcome the issues caused by high radiation, intense light, and MLI materials during close-range detection missions. The approach integrated a fast localization module to improve real-time performance. A lightweight ellipse detection and tracking algorithm suitable for robotic applications was presented in [5], which is capable of handling local illumination variations. In [6], an adaptive, robust and effective edge detector for real-time applications was designed to automatically adapt to the environmental changes. The detector's threshold parameter can be automatically adjusted according to the two-dimensional entropy without requiring any priori experiences.

In this paper, an adaptive spatial circular feature detection method under different lighting conditions was proposed to accurate recognition the docking ring of spacecraft. First, we designed an adaptive threshold image enhancement algorithm based on multi-scale Retinex with chromaticity preservation (MSRCP) for low-quality images acquired under dark light and local over-exposure conditions. We have realized the enhancement of image average brightness, standard deviation and information entropy and other indicators under different lighting conditions. Second, based on arc segment feature, we proposed an ellipse detection method, which could handle different images well even the images with partial missing ellipse contour features. The comparative experiment showed that the accuracy and real-time performance of the method have reached a better level. This paper provides a certain reference for the accurate recognition of targets in on-orbit service missions.

2 Adaptive Image Enhancement Pre-processing

Due to the special lighting conditions in space, acquisition equipment, and other factors, the images obtained by visible camera often show distortion, loss of detail, low illumination, or local over-exposure. Many researchers often skip the pre-processing part of the images and the experimental results often relied on the ideal laboratory conditions with manual parameterization. Enhancement of low-quality images can provide higher quality image and more useful information for target recognition. In this section, an image enhancement pre-processing method was designed, which mainly includes the MSRCP-based enhancement processing, adaptive threshold segmentation, and edge detection.

2.1 Gain Processing

In 1971, Land introduced the Retinex theory of image perception to elucidate the phenomenon of perceived color constancy exhibited by objects under varying illumination conditions [7]. The Retinex theory posits that a given image (S) can be conceptualized as the result of processing the reflection image (R) and the illumination image (L) through pixel-wise multiplication, as illustrated in Fig. 1. The reflectance component encapsulates the intrinsic characteristics of the image, which remain consistent across diverse lighting conditions. Conversely, the illumination component embodies external lighting intensities. Image enhancement algorithms grounded in Retinex theory offer several notable advantages, including heightened color fidelity, enhanced contrast, and minimized color distortion. Furthermore, such algorithms are characterized by ease of implementation, rendering them highly accessible.

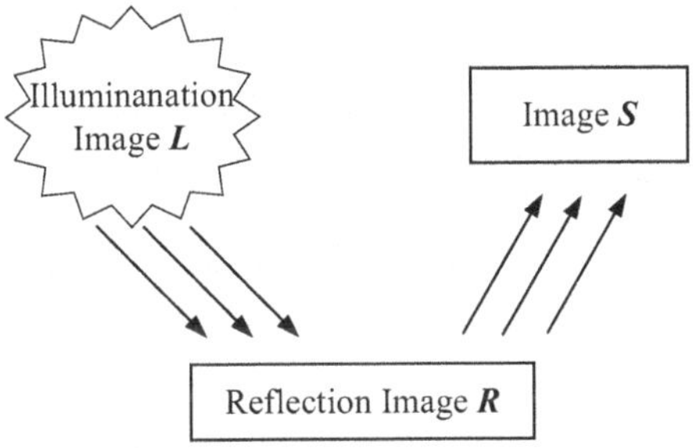

Fig. 1. The schematic diagram of Retinex theory.

The crux of such algorithms lies in employing Gaussian kernels as the surrounding function to estimate the illuminance component. Notable examples include the Single-Scale Retinex (SSR) algorithm [8], as shown below:

$$\log R_i(x, y) = \log S_i(x, y) - \log\left[S_i(x, y) * G(x, y)\right] \tag{1}$$

where $G(x, y)$ is the Gaussian center surround function and * is the convolution calculation. Building upon this foundation, multi-scale Retinex algorithm (MSR) involves performing Gaussian convolution calculations on the image using multiple scale-weighted summation [9], which is shown as follows:

$$\log R_i(x, y) = \sum_{j=1}^{n} W_j\left\{\log S_i(x, y) - \log\left[S_i(x, y) \otimes G(x, y)\right]\right\} \tag{2}$$

in which n represents the number of scale parameters, usually with a value of three and W_j is the weight corresponding to the j_{th} filter function, and satisfies the condition $\sum_{j=1}^{n} W_j = 1$.

MSR convolves each color channel separately, potentially leading to a discrepancy in the *R, G, B* scaling relationship and resultant color distortion. To address this, the multi-scale Retinex with color restoration (MSRCR) algorithm was introduced, incorporating a color restoration factor [10]. However, this approach still poses a risk of color inversion, particularly evident in images with saturated colors. To mitigate this, the multi-scale Retinex with chromaticity preservation (MSRCP) algorithm was proposed [11]. MSRCP applies MSR to the intensity channel and adjusts the output color to maintain chromaticity akin to the original image, thereby preserving color fidelity.

The MSRCP algorithm begins by calculating the mean value of each pixel component in the image and then applies Gaussian low-pass filtering to the original image. Subtracting the low-frequency portion yields the reflection image in the logarithmic domain. MSR results are derived through pixel-level summation of these reflection images, followed by linear processing and automatic color level balancing. Subsequently, data is mapped to each color channel based on the original *R, G, B* scale to generate the integrated output image. The flowchart is depicted in Fig. 2 below:

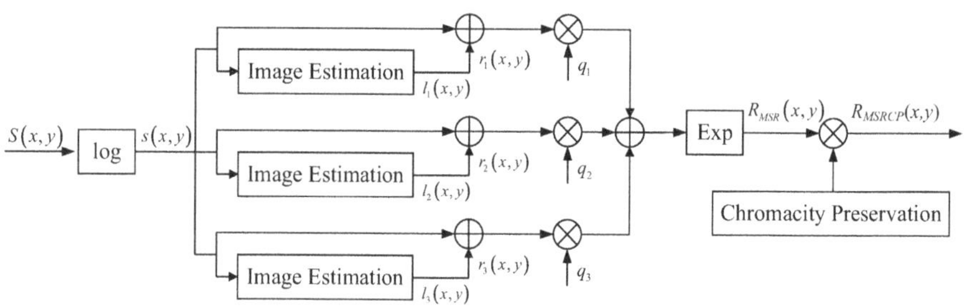

Fig. 2. The main flow chart of MSRCP algorithm.

2.2 Filtering and Threshold Segmentation

The enhanced image often retains various forms of noise. Nonlinear filtering, such as median filtering, effectively removes salt-and-pepper noise and speckle noise without blurring the edge details of the image, making it a preferred method for noise reduction. The median filter is denoted as follows:

$$g(x, y) = med\{f(x-k, y-k)\}, \{k, l \in F\} \tag{3}$$

where f is the pixel value of the current pixel, g is the pixel value reordered by the median, and the pixel point $(x-k, y-k)$ is in the sampling window.

Following filtering, the image comprises both the relevant target object and irrelevant background components. Image segmentation is required to partition the image into distinct regions with unique properties, facilitating the extraction of the relevant portion. Considering constraints such as the computational capacity of onboard systems and the availability of training samples, we employ the OTSU algorithm for segmentation. The image is segmented with a binary method after solving for the optimal threshold through the OTSU algorithm:

$$g(x, y) = \begin{cases} 1, \text{if} f(x, y) > \text{Optimal Threshold} \\ 0, \text{other} \end{cases} \tag{4}$$

OTSU is highly sensitive to noise and target size, yielding optimal segmentation results only for images exhibiting a single peak inter-class variance. It struggles with significantly uneven size ratios between the target and background, such as in cases of uneven illumination, reflection, or complex backgrounds, where the inter-class variance criterion function may exhibit double or multiple peaks, or where there is substantial overlap between the gray levels of the target and background. The previous enhancement processing step significantly enhances image quality and addresses issues encountered by the OTSU algorithm, particularly in cases of over-exposure or excessive darkness.

2.3 Edge Detection

Following the preceding image segmentation process, the identification of the elliptical characteristics of the docking ring necessitates the extraction of edge points. To achieve this, the double threshold non-maximum suppression Canny edge detection algorithm is employed.

Initially, the directional derivatives G_x and G_y are computed using the Sobel operator. The image's gradient magnitude G and direction θ are then determined as follows:

$$\begin{cases} G = \sqrt{G_x^2 + G_y^2} \\ \theta = \tan^{-1} \dfrac{G_y}{G_x} \end{cases} \tag{5}$$

Next, non-maximum suppression is applied, wherein the gradient magnitude of each pixel is compared with that of its neighboring pixels along the gradient direction. Pixels with a greater magnitude are retained, while others are set to zero.

Subsequently, the resulting image undergoes double-threshold hysteresis processing. High and low thresholds are determined based on the gradient difference histogram and intra-class variance minimization. The zero-crossing point of the difference histogram indicates local extrema. The high threshold is established by locating the appropriate zero-crossing position, with half of this value designated as the low threshold.

Pixels exceeding the high threshold are identified as strong edges, while those below the low threshold are suppressed. Pixels with gradient amplitudes falling between the thresholds are considered weak edges. Weak edge pixels are then examined for connections to strong edge pixels. If a connection is found, the weak pixel is retained until all such connections are exhausted. This process yields an image containing numerous edge curves.

3 Ellipse Detection Based on Edge Arc Support Line Segments

The docking ring, a prominent feature on the spacecraft's surface, serves as a crucial identifier for the target. Given that a circle's projection in space typically forms an ellipse on the image plane, detecting ellipses becomes paramount in achieving recognition of docking ring.

3.1 Arc Support Line Segment Screening and Combination

To recognize the docking ring, it is essential to filter out elliptical edge segments from the image, which typically contains numerous edge curves. These edge curves primarily consist of two types of line segments: those originating from regions with consistent level-line angles, resembling straight lines, and those from arc supported regions exhibiting curved distributions [12]. The latter, referred to as arc support line segments, are characterized by concave and convex shapes and indicate the direction of the ellipse center, as illustrated in Fig. 3.

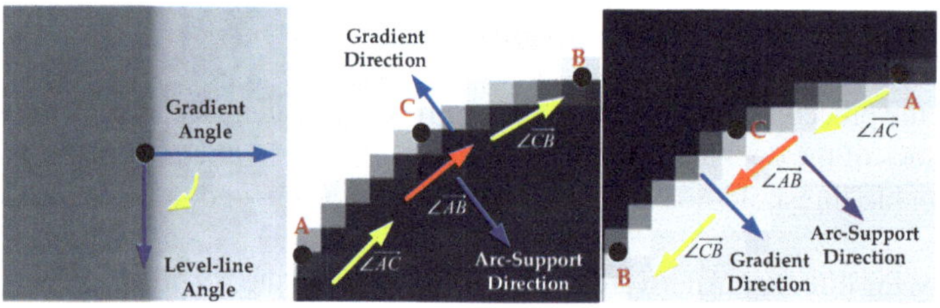

Fig. 3. Schematic diagram of arc support line segment.

The gradient direction within a local area of the image reflects the pattern of light and dark transitions. When the gradient direction aligns with the arc-support direction, the polarity of the arc-support line segment is deemed positive; otherwise, it is negative. Near the circle's edge, the interior is consistently either lighter or darker than the exterior. If the interior is lighter, the corresponding arc support line segment's polarity is positive; otherwise, it is negative. Evidently, arc support line segments belonging to the same circle must exhibit consistent polarity.

As shown in Fig. 4, the ideal elliptical arc should possess three key characteristics: adequate length to avoid significant detection errors or potential undetectability, acute angles between adjacent line segments, and consistent concavity or convexity along the arc segment.

Multiple arc segments are screened out based on the criteria. Subsequently, pairwise comparisons are conducted to determine whether these arc segments belong to the same ellipse. Upon confirmation, an initial set of ellipse arc segments is formed, paving the way for the generation of multiple sets of ellipse arc candidates.

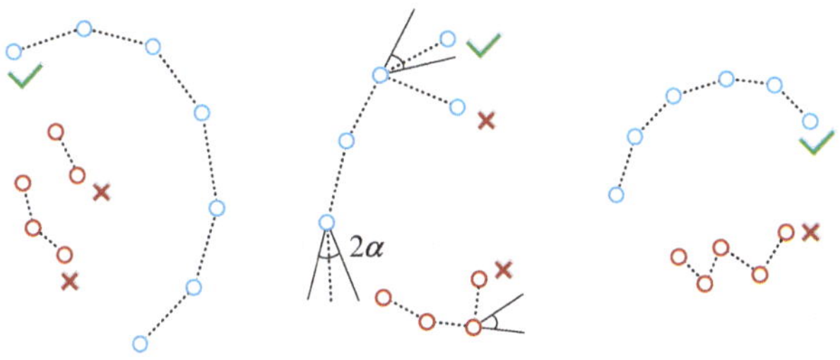

Fig. 4. The three conditions for determining an ideal elliptical arc segment.

3.2 Initial Ellipse Determination by Contour Fitting

Two common methods for fitting ellipse contours are clustering-based approaches (e.g., Hough-based methods) and least-squares fitting. Clustering methods involve time-consuming voting mechanisms, while least-squares fitting minimizes residuals between points and the ellipse. However, quadratic constraints in ellipse fitting problems can lead to iterative inefficiencies. Drawing from Fitzgibbon [13], a non-iterative algorithm solves the positive eigenvector of the characteristic system, improving computational efficiency.

The ellipse is a special form of conic curve, and the conic curve representation of an ellipse in space is as follows:

$$Ax^2 + 2Bxy + Cy^2 + 2Dx + 2Ey + F = 0 \tag{6}$$

at the same time, the equation of the ellipse can be described by the parameter $\left[a, b, x_c, y_c, \theta\right]^T$, where a, b are the semi-major axis and semi-minor axis, respectively, (x_c, y_c) denotes the coordinates of the center of the ellipse, and θ is the angle between the half-length axis and the positive direction of the x-axis of the coordinate system.

Elliptic fitting involves calculating the parameters of the ellipse equation by combining arc segments. The least squares method is employed to derive the conic curve expression of the ellipse, from which the five ellipse parameters are obtained.

$$\begin{cases} x_c = \dfrac{CD - BE}{B^2 - AC} \\ y_c = \dfrac{AE - BD}{B^2 - AC} \\ a = \sqrt{\dfrac{2\left(Ax_c^2 + 2Bx_c y_c + Cy_c^2 - F\right)}{A + C - \sqrt{(A - C)^2 + 4B^2}}} \\ b = \sqrt{\dfrac{2\left(Ax_c^2 + 2Bx_c y_c + Cy_c^2 - F\right)}{A + C + \sqrt{(A - C)^2 + 4B^2}}} \\ \theta = -\dfrac{1}{2}\tan^{-1}\dfrac{-2B}{A - C} \end{cases} \tag{7}$$

Assuming that the coordinates of the pixel points of each line segment in the arc segments are (x_i, y_i), define the matrix $\mathbf{D}$ as follows:

$$\mathbf{D} = \begin{bmatrix} x_1^2 & x_1y_1 & y_1^2 & x_1 & y_1 & 1 \\ x_2^2 & x_2y_2 & y_2^2 & x_2 & y_2 & 1 \\ \vdots & \vdots & \vdots & \vdots & \vdots & \vdots \\ x_n^2 & x_ny_n & y_n^2 & x_n & y_n & 1 \end{bmatrix} \quad (8)$$

Define vector $\mathbf{a} = [A, B, C, D, E, F]^T$ and vector $\mathbf{X} = \left[x^2, xy, y^2, x, y, 1\right]$, and the general equation of an ellipse can be expressed as $\mathbf{aX} = 0$.

To determine the optimal fitting ellipse, the goal is to minimize the value of $\sum_{i=1}^{N} (\mathbf{aX}_i)^2$, the constraint $\Delta = B^2 - 4AC < 0$ needs to be considered. Under the appropriate scaling scale, the above inequality constraints can be transformed into the equality constraint $\Delta = B^2 - 4AC = -1$.

Define the constraint matrix $\mathbf{C}$ as follows:

$$\mathbf{C} = \begin{bmatrix} 0 & 0 & -1 & \cdots & 0 \\ 0 & 2 & 0 & & \\ -1 & 0 & 0 & & \vdots \\ \vdots & & & \ddots & \\ 0 & & \cdots & & 0 \end{bmatrix} \quad (9)$$

The constraints are rewritten in the following:

$$\mathbf{a}^\mathbf{T}\mathbf{Ca} = 1 \quad (10)$$

Constrained ellipse fitting problem can be reduced to minimize $\|\mathbf{Da}\|^2 = \mathbf{a}^\mathbf{T}\mathbf{D}^\mathbf{T}\mathbf{Da}$ under the above constraints. The minimization can be solved by considering the rank-efficient generalized eigenvalue system $\mathbf{S}^{-1}\mathbf{C}$ where $\mathbf{S} = \mathbf{D}^\mathbf{T}\mathbf{D}$ is the scatter matrix. by introducing Lagrange multiplier λ and differentiating, we can obtain a simultaneous system of equations:

$$\begin{cases} \mathbf{D}^\mathbf{T}\mathbf{Da} = \lambda\mathbf{Ca} \\ \mathbf{a}^\mathbf{T}\mathbf{Ca} = 1 \end{cases} \quad (11)$$

The coefficients of the general equation of the desired ellipse can be obtained by solving the positive eigenvalue vector of the Eq. (11), and then the five parameters of the ellipse can be obtained from Eq. (7).

3.3 Affine Transformation Verification of Candidate Ellipse

After contour fitting, multiple candidate ellipses are obtained. However, among these candidates, some may be spurious or redundant, necessitating validation. In this study, we employ affine transformation to assess the eligibility of each ellipse, as illustrated in

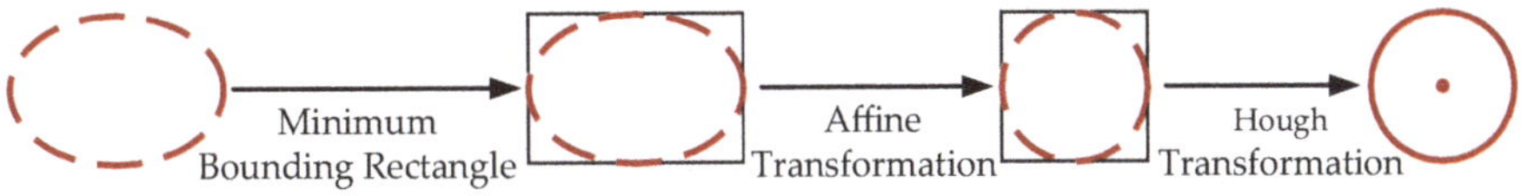

Fig. 5. The schematic diagram of affine transformation.

Fig. 5. Affine transformation is a linear transformation that preserves straightness and parallelism of two-dimensional graphics during the process of mapping from one set of two-dimensional coordinates to another.

The affine transformation expression in the two-dimensional plane is as follows:

$$\begin{bmatrix} x' \\ y' \end{bmatrix} = \begin{bmatrix} a & b \\ c & d \end{bmatrix} \begin{bmatrix} x \\ y \end{bmatrix} + \begin{bmatrix} m \\ n \end{bmatrix} \tag{12}$$

where (x', y') is the transformed coordinate, (x, y) is the coordinate before transformation, and (a, b, c, d, m, n) are the transformation coefficients.

We define the set of candidate ellipse arc segments as S_i^1, and assume the coordinates of the minimum bounding rectangle as (x_i^1, y_i^1), (x_i^2, y_i^2), (x_i^3, y_i^3), (x_i^4, y_i^4). We consider the distance between points (x_i^1, y_i^1) and (x_i^2, y_i^2) as the width of a square. The vertex of the square, $(x_i^{3'}, y_i^{3'})$, $(x_i^{4'}, y_i^{4'})$, can be determined from this width. Then, the coefficients of affine variation (a, b, c, d, m, n) are determined according to Eq. (12).

Then, according to the affine transformation coefficients, the arc segments is affinely transformed to the set, and the set of arc segments is used to fit the circle by Hough transform. The difference between the center of the circle and the center of the ellipse is compared. If the difference is less than the set threshold, the ellipse is valid, otherwise it is eliminated.

Next, the arc segments S_i^1 are transformed to the set S_i^2 using the affine transformation coefficients. Then, the set of arc segments S_i^2 is utilized to fit a circle via the Hough transform. The discrepancy between the centers of the circle and the ellipse is evaluated, and if the difference is below the predefined threshold, the ellipse is deemed valid; otherwise, it is discarded.

4 Experiment and Analysis

The experimental conditions are set as follows: the target model remains static, with the camera optical axis perpendicular to the docking ring plane. Imaging of the target model is conducted under three lighting conditions at varying distances and angles. Normal illumination refers to natural daylight conditions within the laboratory, low illumination denotes scenarios during nighttime with all room lights switched off, and local overexposure describes scenes where an LED flash illuminates the docking ring plane from a certain angle and distance. Distance is defined as the straight-line distance between the camera optical center and the center of the docking ring (with the camera and docking ring center positioned at the same horizontal level), while angle is defined as the angle between the camera optical axis and the normal vector of the docking ring plane.

4.1 Image Enhancement Experiment

Comparative experiments employing different enhancement algorithms were conducted on images captured under three lighting conditions at varying angles and distances. Analysis was performed on several typical scenes. Herein, we present the experimental results for a distance of 1m and an angle of 0°, as depicted in Fig. 6.

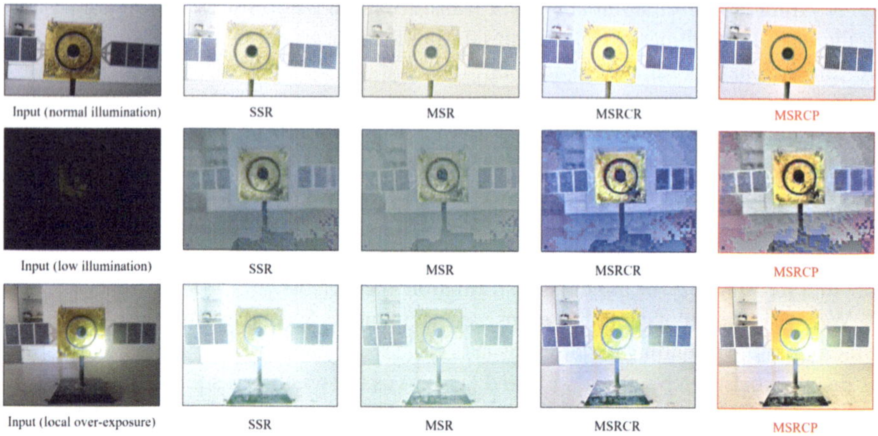

Fig. 6. The schematic diagram of affine transformation.

From the Fig. 6, it can be observed that the MSRCP algorithm outperforms other Retinex algorithms noticeably in enhancing various types of images. This algorithm is capable of effectively enhancing image brightness in low-light environments, while accurately restoring the inherent colors of the target with minimal distortion.

Table 1 documents the average processing times of different algorithms for images under three lighting conditions. The processing time of the SSR algorithm is shorter than that of the three multi-scale algorithms, yet SSR's enhancement performance is unsatisfactory. Among the three multi-scale algorithms, the MSRCP algorithm exhibits the shortest average processing time, significantly outperforming the average processing time of the MSRCR algorithm.

Table 1. Average processing time of different retinex algorithms (s).

Input Image	SSR	MSR	MSRCR	MSRCP
Normal illumination	0.7016	1.5735	2.3035	0.8402
Low illumination	0.6371	1.5313	2.1760	0.8504
Local over-exposure	0.6944	1.6368	2.4340	0.8590

Further evaluation of the enhancement efficacy of the MSRCP algorithm on image quality was conducted based on objective parameters extracted from the images themselves, as depicted in Table 2. Under all three lighting conditions, the average brightness, image standard deviation, contrast, and information entropy of the images were improved after enhancement. Particularly, under low illumination condition, the average brightness increased from 9.90 to 87.02, standard deviation rose from 13.87 to 61.02, image contrast enhanced from 19.62 to 54.53, and information entropy elevated from 5.43 to 7.71, all exhibiting significant improvements. The effectiveness of the algorithm in enhancing image quality, especially in challenging lighting scenarios such as low and dim light conditions, has been convincingly demonstrated.

Table 2. Objective evaluation of image quality enhancement effect by MSRCP algorithm.

Indicators	Normal illumination		Low illumination		Local over-exposure	
	Input	Output	Input	Output	Input	Output
Mean Luminance	116.40	171.32	9.90	87.02	105.41	167.29
Standard Deviation	67.90	77.85	13.87	61.02	52.45	55.50
Contrast	51.59	55.71	19.62	54.53	45.57	49.15
Information Entropy	6.50	7.47	5.43	7.71	6.46	7.59

4.2 Circular Feature Detection Experiment

To validate our ellipse feature detection method, we selected a publicly available satellite image dataset captured by the NEXTSat satellite optical camera under different lighting conditions, along with spacecraft model images taken under different lighting conditions in our laboratory. Comparative methods included random Hough transform (RHT) [14], ELSDc [15], YEAD [16], JiaQi [17], and manually annotated ground truth (GT). The results are presented in Fig. 7 and Table 3.

Figure 7 and Table 3 demonstrate that our ellipse feature detection algorithm achieved favorable results on different images. The number of valid ellipses detected closely approximated the ground truth, significantly outperforming other algorithms, while ensuring the integrity and smoothness of the ellipse contours.

To further validate our approach, we imaged and detected the target from various angles and distances under three different lighting conditions. Partial results are presented below.

Tables 4, 5 and 6 illustrate partial experimental results, representing the recognition of circular feature of the docking ring under different lighting conditions at close distances with varying angles. When the camera is directly facing the docking ring plane, the contour of the docking ring is intact, resulting in a recognition of a perfect circle. However, when there is an angle between the camera and the docking ring plane, the recognition result appears as an ellipse. Due to the angle and partial over-exposure, parts of the docking ring contour may be obscured by the engine nozzle, resulting in incomplete contours. The greater the angle and the stronger the local exposure, the higher the proportion of obscured contour parts. By employing our recognition method, even in cases where the contour is obscured, the complete contour can be relatively accurately

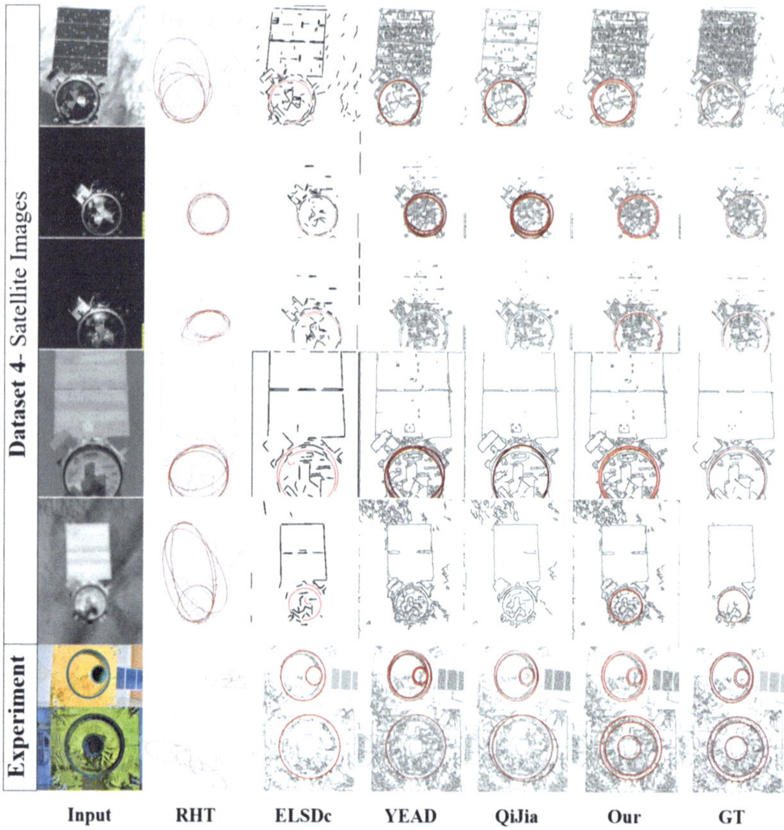

Fig. 7. Comparison of different ellipse detection algorithms on test dataset.

Table 3. Number of valid ellipses detected by different algorithms.

Input Image	RHT	ELSDC	YEAD	QiJia	Our	GT
1	0	0	1	0	**2**	2
2	1	0	1	1	**2**	2
3	0	0	0	0	**1**	1
4	0	0	0	0	**2**	2
6	0	1	0	0	**1**	1
7	0	2	2	3	**4**	4

Table 4. Experimental results under normal illumination condition.

Position	Input	Enhancement	Edge Detection	Contour Fitting	Result
1.0m, 0°	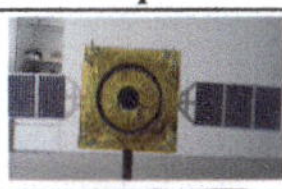		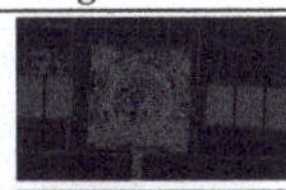	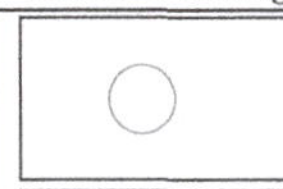	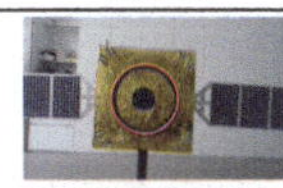
1.0m, 30°				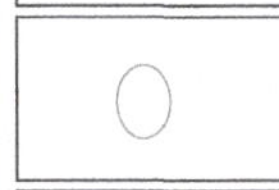	
1.0m, 60°			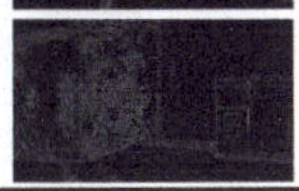		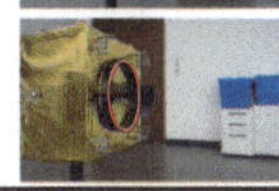

Table 5. Experimental results under low illumination condition.

Position	Input	Enhancement	Edge Detection	Contour Fitting	Result
0.5m, 0°	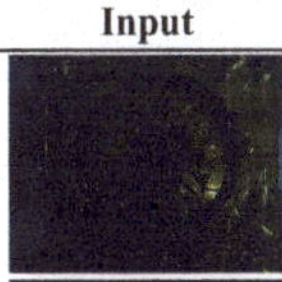		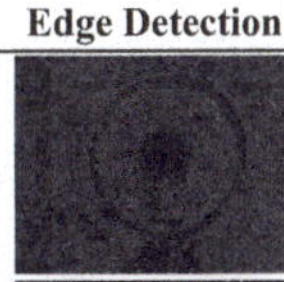	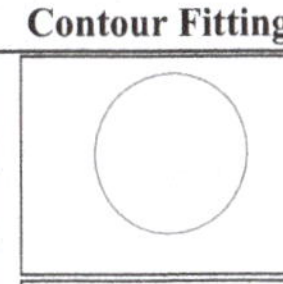	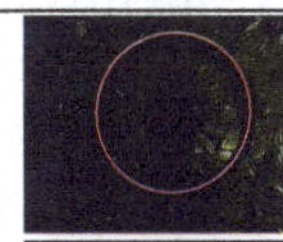
0.5m, 30°	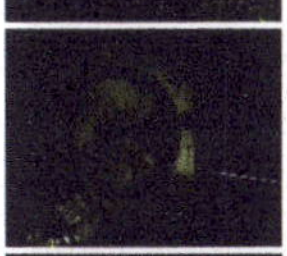			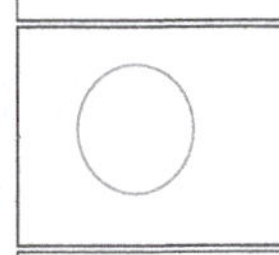	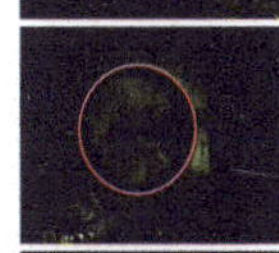
0.5m, 60°	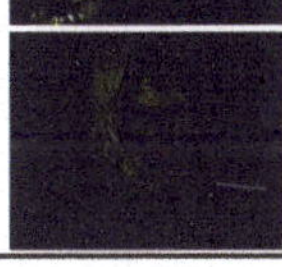		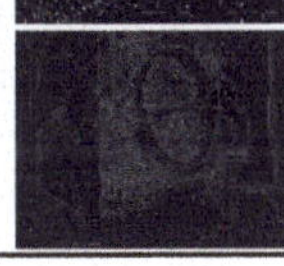	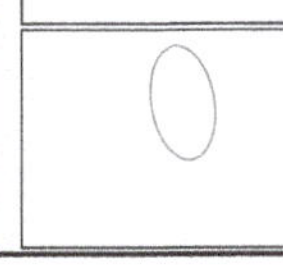	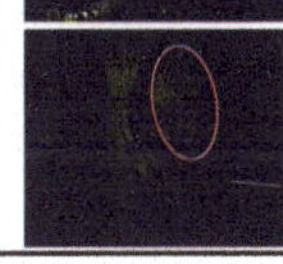

fitted. Furthermore, with the aid of image enhancement algorithms, the image quality under low-light conditions has been significantly improved, enabling accurate recognition of the docking ring, which was previously almost unrecognizable. The experimental results thoroughly validate the effectiveness and robustness of our approach in handling complex lighting conditions.

Table 6. Experimental results under local over-exposure condition.

Position	Input	Enhancement	Edge Detection	Contour Fitting	Result
1.0m, 0°			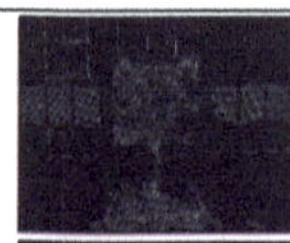	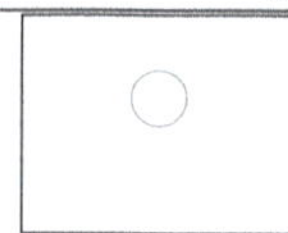	
1.0m, 30°			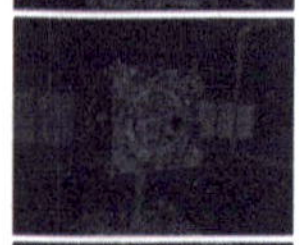		
1.0m, 60°			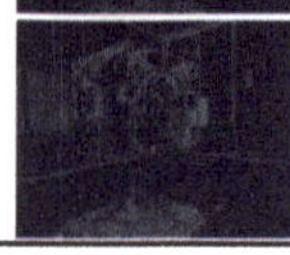		

5 Conclusion and Discussion

In this paper, we propose an adaptive circular feature recognition method under different lighting conditions. Compared to previous research efforts, our study innovatively incorporates the MSRCP enhancement algorithm in the pre-processing step. We conduct experiments using the docking ring of a satellite model as the detection target under different lighting conditions.

The experimental results indicate that our method can effectively enhance image metrics such as average brightness and information entropy, which reflect image quality. Furthermore, when the docking ring is partially obscured by the engine nozzle, causing incomplete contours, our method can accurately fit an ellipse with complete contours. The experimental results under different conditions thoroughly validate the adaptability of our method to different lighting scenarios, providing valuable insights for target recognition in on-orbit service.

However, there are limitations in the interference and deviation observed in the recognition results under extremely low illumination condition and extremely large tilt angle. Future research will focus on investigating the mechanisms behind these issues and improving recognition methods to achieve better results.

References

1. Chen, X.Q., Bai, Y.Z., Zhao, Y., et al.: Space non-cooperative target security approach control technology. Science Press, Beijing (2023)
2. Segal, S., Carmi, A., Gurfil, P.: Stereovision-based estimation of relative dynamics between noncooperative satellites: theory and experiments. IEEE Trans. Control Syst. Technol. **22**, 568–584 (2014)
3. Zhang, L., Pan, W., Ma, X.: Real-time docking ring detection based on the geometrical shape for an on-orbit spacecraft. Sensors **19**, 5243 (2019)
4. Zhang, S., Li, J., Zhang, W., Zhou, W.: Research on docking ring pose estimation method based on point cloud grayscale image. Adv. Space Res. **70**, 3466–3477 (2022)

5. Keipour, A., Pereira, G.A.S., Scherer, S.: Real-time ellipse detection for robotics applications. IEEE Robot. Autom. Lett. **6**, 7009–7016 (2021)
6. Liu, Y., Xie, Z., Liu, H.: An adaptive and robust edge detection method based on edge proportion statistics. IEEE Trans. Image Process. **29**, 5206–5215 (2020)
7. Land, E.H.: The retinex theory of color vision. Sci. Am. **237**, 108–128 (1977)
8. Jobson, D.J., Rahman, Z., Woodell, G.A.: Properties and performance of a center/surround retinex. IEEE Trans. Image Process. **6**, 451–462 (1997)
9. Jobson, D.J., Rahman, Z., Woodell, G.A.: A multiscale retinex for bridging the gap between color images and the human observation of scenes. IEEE Trans. Image Process. **6**, 965–976 (1997)
10. Jobson, D.J.: Retinex processing for automatic image enhancement. J. Electron. Imaging **13**, 100 (2004)
11. Rahman, Z., Jobson, D.J., Woodell, G.A.: Multi-scale retinex for color image enhancement. In: Proceedings of 3rd IEEE International Conference on Image Processing vol. 3, pp. 1003–1006. IEEE (1996)
12. Lu, C., Xia, S., Shao, M., Fu, Y.: Arc-support line segments revisited: an efficient high-quality ellipse detection. IEEE Trans. Image Process. **29**, 768–781 (2020)
13. Fitzgibbon, A., Pilu, M., Fisher, R.B.: Direct least square fitting of ellipses. IEEE Trans. Pattern Anal. Mach. Intell. **21**, 476–480 (1999)
14. McLaughlin, R.A.: Randomized Hough Transform: improved ellipse detection with comparison. Pattern Recognit. Lett. **19**, 299–305 (1998)
15. Pătrăucean, V., Gurdjos, P., von Gioi, R.G.: A parameterless line segment and elliptical arc detector with enhanced ellipse fitting. In: Fitzgibbon, A., Lazebnik, S., Perona, P., Sato, Y., Schmid, C. (eds.) Computer Vision – ECCV 2012, pp. 572–585. Springer, Berlin, Heidelberg (2012). https://doi.org/10.1007/978-3-642-33709-3_41
16. Fornaciari, M., Prati, A., Cucchiara, R.: A fast and effective ellipse detector for embedded vision applications. Pattern Recognit. **47**, 3693–3708 (2014)
17. Jia, Q., Fan, X., Luo, Z., Song, L., Qiu, T.: A fast ellipse detector using projective invariant pruning. IEEE Trans. Image Process. **26**, 3665–3679 (2017)

The Thermal Expansion Properties of Ultra-Low Thermal Expansion Silica Glass Under Temperature Cycles and Irradiation

Chenyang Zhang(✉), Lanjian Nie, Zhufeng Shao, Hui Wang, Hongjie Wang, Yanan Jia, Bo Fu, Lei Wang, Zaikui Xiang, and Sijian Gao

China Building Materials Academy Co., Ltd., Beijing 100024, People's Republic of China
3365620071@qq.com

Abstract. Our study examines the thermal performance and radiation tolerance of ultra-low thermal expansion silica glass made from SiO_2-TiO_2 binary components in a simulated space environment. The low coefficient of thermal expansion, thermal shock resistance, and solderability of this material make it an essential material for the fabrication of high-precision optical components. Thermal expansion coefficients of ultra-low levels allow consistency of imaging quality regardless of temperature fluctuations in fluctuating thermal environments. A gamma radiation environment was simulated to assess the material's resistance to ionizing radiation in space and the thermal expansion coefficient (TEC) of the material after 30 temperatures cycling from −65 °C to 175 °C. After temperature cycling, the TEC of the material did not change significantly, and after radiation exposure of 10^5 Gy the TEC remained stable. It is clear from this result that the ultra-low thermal expansion silica glass performs well under extreme conditions and meets stringent space application reliability standards. Due to these results, a future space optical system could be designed with ultra-low thermal expansion silica glass.

Keywords: Ultra low thermal expansion silica glass · Space service environment · Thermal expansion performance

1 Introduction

Ultra-low thermal expansion silica glass is a binary component glass material composed of SiO_2-TiO_2. This material is based on the CVD gas deposition process of high-purity silica glass. By introducing a specific amount of TiO_2 during the preparation of high-purity silica glass, the thermal expansion coefficient can reach $0 \pm 30 \times 10^{-9}$/°C within the temperature range of 5–35 °C. The material's ultra-low thermal expansion coefficient, high thermal stability, and weldability make it the preferred material for space remote sensors. The characteristic of low thermal expansion coefficient can significantly reduce the requirement of the optical components for the temperature control ability of the observation system, ensuring the image quality of the optical system under complex thermal conditions [1]. Furthermore, the thermal stability and weldability of the glass allow the material to undergo thermal processing methods such as melting, welding,

H. P. Urbach et al. (Eds.): ISSOIA 2023, SPP 191, pp. 54–61, 2024.
https://doi.org/10.1007/978-981-97-6718-2_6

and sealing, achieving the lightweight processing of the honeycomb mirror, reducing the weight of the mirror while improving the rigidity of the mirror body, ensuring the stability of the mirror surface shape. Internationally famous telescopes such as the Hubble Space Telescope, the next-generation Roman Space Telescope, and Geoeye-1, among others, all use this material to prepare the main mirror of the reflector [2]. Beyond its extensive application in the preparation of high-resolution reflector blanks, ultra-low thermal expansion silica glass is also used in the fabrication of precision optical components such as extreme ultraviolet lithography (EUV) photomask substrates [3], micro hemispherical resonators, and Fabry-Perot (F-P) resonant cavities [4].

Since 2014, China Building Materials Academy Co., Ltd has been producing large-scale ultra-low thermal expansion silica glass bulk materials based on the vertical chemical vapor doping deposition process, to meet domestic needs for the localization of glass bulk materials [5]. Concurrently, our company is conducting research to expand product size and improve material stability in order to gradually meet related usage requirements. As the localization of ultra-low thermal expansion silica glass materials continues to progress, it is necessary to research and evaluate the tolerance of localized materials to space environmental conditions. With these goals in mind, this paper primarily investigates the impact of large temperature cycles and space radiation - conditions encountered in space environment applications - on the thermal expansion performance of materials.

2 Materials and Methods

In this study, a low-expansion silica glass material was prepared using the Chemical Vapor Deposition (CVD) method. By conducting a flame hydrolysis reaction of silicon-containing and titanium-containing materials in a hydrogen-oxygen flame, SiO_2-TiO_2 deposition particles are formed [6]. These particles undergo vitrification under high-temperature conditions in the furnace, directly resulting in large-scale ultra-low thermal expansion silica glass bulk material. The experimental sample size was $\Phi 7 \times 20$ mm, which meets the related requirements for thermal expansion measurement.

The large temperature range cycling process and the measurement of the thermal expansion coefficient of the material are primarily based on the Linseis L75/VD/LASER/500LT laser thermal dilatometer (as shown in Fig. 1). This thermal dilatometer is built on the Michelson interference principle, enabling it to meet the measurement requirements for nanoscale length changes. The device is equipped with a low-temperature control system, which allows for precise measurement of the material's thermal expansion coefficient within a temperature range from the liquid nitrogen temperature zone to a maximum of 500 °C. The large temperature range cycling process for the material was achieved through the device's own temperature cycle control system. After completing the temperature cycle treatment, the thermal expansion coefficient of the material can be accurately measured.

For the large temperature range cycling treatment experiment of the material, the article referred to the relevant test methods of GJB 548B-2004, and treated the test samples first with a high and low temperature cycle ranging from −65 to 175 °C. The entire temperature cycle process was based on the laser Thermal dilatometer system.

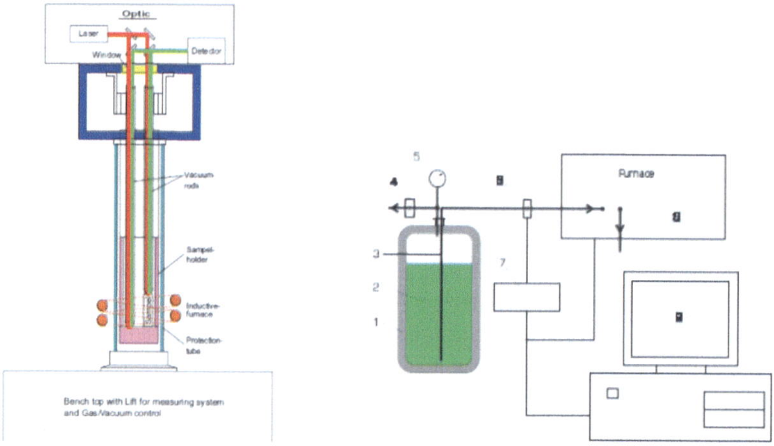

Fig. 1. Schematic diagram of the measurement principle of the Thermal dilatometer and the temperature cycling measurement device.

After completing the set temperature cycle periods (10 times, 20 times, 30 times), the thermal expansion coefficient of the samples was measured using the thermal dilatometer.

For the space irradiation process simulation experiment, the article adheres to the aerospace industry standard QJ20289-2014 "Test Method for Ionizing Radiation Effects of Non-Metallic Materials in Space". A Co-60 γ radiation source was used to simulate the space environment radiation conditions, with a total radiation dose of 10^5Gy. After the total dose irradiation test, the thermal expansion coefficient of the sample was measured, and the changes in the thermal expansion coefficient of the sample before and after irradiation were compared.

3 Results and Discussion

3.1 Analysis of the Effect of Temperature Cycle on Thermal Expansion Performance

In this work, all thermal expansion measurements were carried out at a constant heating rate of 1 °C/min. The linear thermal expansion coefficient of the samples was calculated from the measurement results of length and temperature changes. The linear thermal expansion coefficient is defined as:

$$\bar{\alpha} = \Delta L/(L_0 \Delta T)$$

Here, ΔL and ΔT represent the average change in length and temperature of the sample, respectively. L_0 is the length of the sample at 20 °C.

Through this process, the result of the thermal expansion coefficient of the low-expansion silica glass before the temperature cycle treatment is depicted in Fig. 3 as red hollow dots.

To better quantify the above data, the results are fitted to the following polynomial function [7]:

$$\alpha\left(10^{-6}\,\mathrm{K}^{-1}\right) = a + bT + cT^2 + dT^3 + eT^4 + fT^5 + gT^6 + hT^7$$

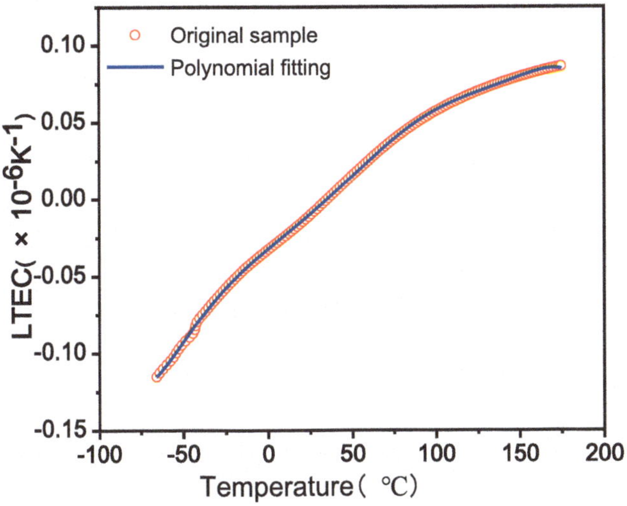

Fig. 2. Thermal expansion coefficient values of the sample before temperature cycling, and the fitting results of the thermal expansion coefficient.

In the fitted polynomial function, the coefficients are as follows: $a = -0.03223 \pm 1.17452 \times 10^{-5}$, $b = 9.05115 \times 10^{-4} \pm 5.32223 \times 10^{-7}$, $c = -2.06979 \times 10^{-6} \pm 1.2912 \times 10^{-8}$, $d = 9.65804 \times 10^{-8} \pm 3.54577 \times 10^{-10}$, $e = -5.06366 \times 10^{-10} \pm 2.74466 \times 10^{-12}$, $f = -1.03521 \times 10^{-11} \pm 6.28172E-14 \times 10^{-14}$, $g = 1.03621 \times 10^{-13} \pm 5.48064 \times 10^{-16}$, $h = -2.5813 \times 10^{-16} \pm 1.41123 \times 10^{-18}$. These parameters were determined using the least squares fitting method. The fitted function is represented as a blue curve in Fig. 2. The standard deviation of the measured results and the fitted data is $0.00059 \times 10^{-6}\mathrm{K}^{-1}$.

Figure 3 shows the deviation between the sample's expansion coefficient and the initial measured thermal expansion coefficient fitting function after 10, 20, and 30 temperature cycles. The calculation results show that the temperature cycle results all fall within the error limit of the fitting curve of $0.0045 \times 10^{-6}\mathrm{K}^{-1}$, confirming that the sample maintains high consistency in linear thermal expansion before and after temperature cycling.

3.2 Analysis of the Impact of γ Irradiation on Thermal Expansion Performance

After subjecting the sample to γ irradiation at a dose of 10^5Gy, the results show that the sample underwent a darkening phenomenon due to irradiation (Fig. 4). To analyze the

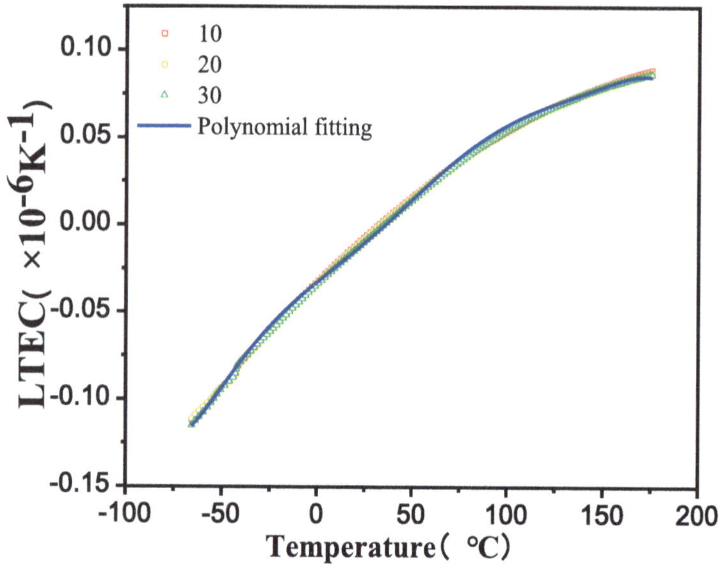

Fig. 3. Measurement results of the thermal expansion coefficient during the temperature cycling process.

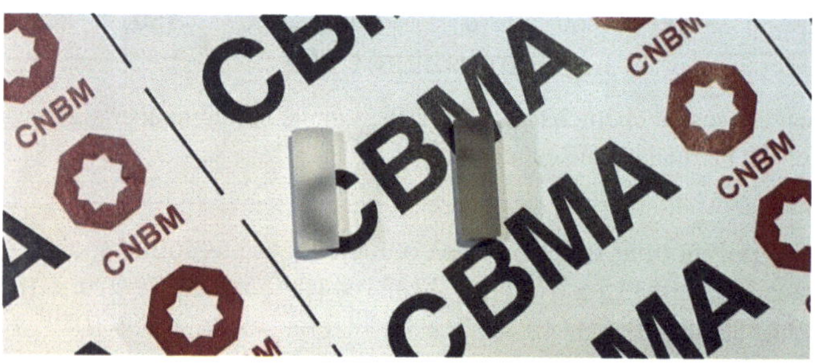

Fig. 4. Comparison of Characteristics Before and After Sample Irradiation.

impact of irradiation on the material's thermal expansion properties, Fig. 5 compares the changes in the thermal expansion coefficient of the sample before and after irradiation.

The test results indicate that after irradiation, the material's thermal expansion coefficient remains essentially consistent with the pre-test results within the temperature range of 10–45 °C. In the temperature range of 45–140 °C, compared to the original sample, its thermal expansion coefficient showed a certain degree of decrease. The standard deviation of the irradiated sample results from the original sample data is 0.00968 $\times 10^{-6}K^{-1}$. Comparing with Corning's related test results, the results show that the material's thermal expansion coefficient essentially maintains a comparable level.

Many scholars have reported the radiation-induced expansion phenomenon of fused silica glass material. Primak and Kampwirth [8] believe that impurities exist inside silica

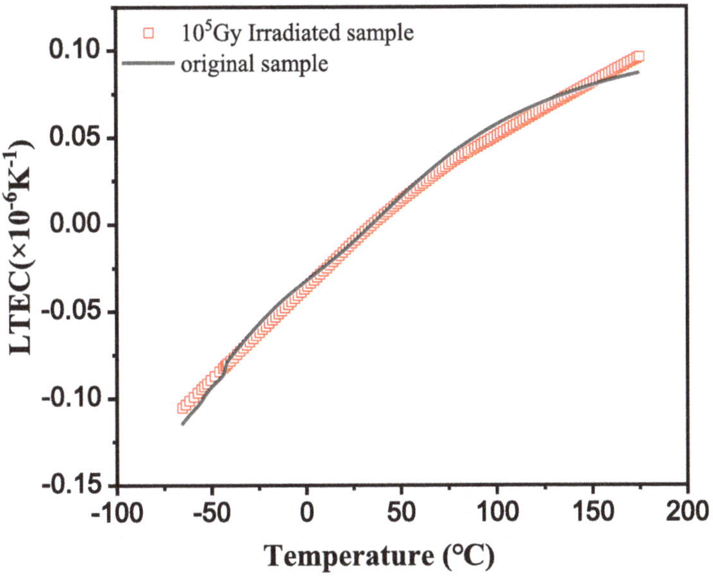

Fig. 5. Comparison of the Thermal Expansion Coefficient of Materials Before and After 10^5Gy γ Irradiation.

glass generated by natural quartz melting. The ionizing electrons produced by the irradiation process are captured by the Al ions within, causing the breakage of Al-O bonds. The destruction of the glass network structure ultimately leads to material expansion. Some scholars believe that in any case, radiation will generate displacements, electron defects, and/or the breakage of Si-O-Si bonds. This leads to the relaxation of the glass network structure, the destruction of the multi-ring network structure and the formation of smaller ring network structures [9]. These structures will fill the relatively large gaps present in the network where silicon and oxygen atoms are interconnected, leading to the compaction effect of the glass network. The expansion effect and compaction effect introduced above will occur synchronously during the irradiation process. The net effect of the two effects will ultimately cause the apparent expansion effect of the material.

Currently, the low expansion characteristics of TiO_2-SiO_2 glass can be attributed to the following: the decrease in mean cation mean square displacement as a result of the increase in Ti^{4+} ions, the increase in free volume available to oxygen atoms allowing them to move more for the same length change, and the increase in low frequency modes by Ti reduce the distortion and thermal expansion of SiO_4. The irradiation environment normally results in compaction effects, recombination effects, and other phenomena related to glass materials. Compression occurs when an increase in density occurs within the glass material as a result of strong ionization. In general, recombination reduces ionization by recombining electron-hole pairs.

Due to the presence of titanium elements in low-expansion silica glass, some titanium ions will form Ti^{3+} during the irradiation process, which is a factor affecting the irradiation response. Glass' thermal expansion coefficient can be affected by the formation of defects like vacancies in its internal structure after irradiation treatment.

Experiments by HIGBY and others have shown that the coefficient of thermal expansion (CTE) of ULE glass material does not change after electron irradiation at a dose of 10^9 rad at room temperature, and only slight changes occur at temperatures below 300 K [10].Ueno's related analysis of the thermal expansion performance of ULE materials also shows that ULE materials do not have significant changes under the conditions of electron beam irradiation and gamma-ray irradiation. They can meet the index requirements for radiation tolerance under the 20-year service cycle of geosynchronous orbit [11]. The results of our sample are essentially consistent with the above test results in terms of thermal expansion performance.

4 Conclusion

Relevant research has been conducted on the thermal expansion characteristics of domestically produced ultra-low thermal expansion silica glass materials under simulated space service environment conditions. The results show that:

1) Under wide temperature test conditions, after 30 cycles of temperature cycling, the standard deviation of the material's thermal expansion coefficient from the original sample remains within the range of $0.0045 \times 10^{-6}K^{-1}$, with no significant changes in thermal expansion coefficient observed.
2) Under the condition of 10^5Gy dose γ irradiation, the thermal expansion coefficient of the material did not significantly change compared to the original material, and it can maintain consistency with the performance of similar imported materials.

The various properties of this material can meet the technical requirements under the simulated space service environment, and can satisfy the reliability requirements for aerospace applications.

References

1. Hu, R., Cheng, Y.T., Chen, X., Zhang, Y.Y., Liu, H.: Study on thermal stability of low expansion glass and SiC lightweight mirror in space environment. In: Proceedings of the Sixth Annual High-resolution Earth Observation Academic Conference (Part 1), vol. 10 (2019)
2. Wells, B.J., Edwards, M.J., VanBrocklin, R.R.: Corning's success with the Discovery Channel 4.3 Meter Telescope. In: Optomechanical Technologies for Astronomy. p. 627320. International Society for Optics and Photonics (2006)
3. Carapella, A.P., Duran, C.A., Hrdina, K.E., Sears, D.J., Tingley, J.E.: ULE® Glass for EUVL applications, a fictive temperature correlation. J. Non-Cryst. Solids **367**, 37–42 (2013)
4. Sanjuan, J., et al.: Long-term stable optical cavity for special relativity tests in space. Opt. Express, OE. **27**, 36206–36220 (2019)
5. Zhufeng, S., Xiang, Z., Rao, C., Wang, J., Sui, M., Wang, Y.: Low expansion large diameter lightweight silica glass reflector. China Build. Mater. Technol. 40–43 (2016)
6. Zhang, C., et al.: Morphology and structural characteristics of TiO_2-SiO_2 binary glass deposition particles. J. Phys.: Conf. Ser. **1759**, 012006 (2021)
7. Watanabe, H., Yamada, N., Okaji, M.: Linear thermal expansion coefficient of silicon from 293 to 1000 K. Int. J. Thermophys. **25**, 221–236 (2004)

8. Rajaram, M., Friebele, E.J.: Effects of radiation on the properties of low thermal expansion coefficient materials: a review. J. Non-Cryst. Solids **108**, 1–17 (1989)
9. Efthimiopoulos, I., et al.: Femtosecond laser-induced transformations in ultra-low expansion glass: microstructure and local density variations by vibrational spectroscopy. J. Appl. Phys. **123**, 233105 (2018)
10. Higby, P.L., et al.: Radiation effects on the physical properties of low-expansion-coefficient glasses and ceramics. J. Am. Ceram. Soc. **71**, 796–802 (1988)
11. Ueno, H., Kamiya, T., Mizutani, T.: Radiation resistance of physical properties of low CTE ceramics and glass in high-dose-rate environments. In: Astronomical Optics: Design, Manufacture, and Test of Space and Ground Systems II. p. 111160S. International Society for Optics and Photonics (2019)

Satellite-Based Operating System Design for New Remote Sensing Platforms

Liang Zhao(✉), Mingdao Xu, and Xutao Sun

Shandong Aerospace Electro-Technology Institute, Yantai 264003, Shandong, China
zxflyr00@163.com

Abstract. With the advancement of space remote sensing technology, the functionality and performance of remote sensing satellites continues to improve, placing higher demands on new remote sensing platforms, such as high real-time capability, high performance, and high reliability. Traditional operating systems are no longer able to meet these requirements. In response to this situation, this paper investigates the design of a satellite-based operating system for new remote sensing platforms. Specifically, this paper proposes a design scheme for a satellite-based operating system based on multi-core processors and partition management. Compared with traditional single-core processors, multi-core processors have higher computing power and parallel processing capabilities, which can better meet the performance requirements of new remote sensing platforms. Meanwhile, through partition management, effective resource allocation and utilization can be achieved, improving the reliability and security of the system. This paper first analyzes and summarizes the operating system requirements of new remote sensing platforms, and clarifies the objectives and requirements of the design scheme. Then, based on the characteristics of multi-core processors, a structure for the satellite-based operating system is designed, using a two-level scheduling method to achieve deterministic scheduling and efficient real-time scheduling for multi-core processors, fully utilizing the high performance of multi-core processors. At the same time, through partition management design, the system resources are divided into different regions, achieving partition isolation and protection to prevent the spread of software problems from affecting other functionalities, thus improving the reliability of the remote sensing platform. Finally, the effectiveness and feasibility of the design scheme are verified through experiments and evaluations. The experimental and evaluation results show that this operating system exhibits high performance and reliability in new remote sensing platforms, and it has significant application value. The research in this paper provides guidance for improving the level of operating system design for new remote sensing platforms, and can better promote the development of remote sensing technology.

Keywords: New Remote Sensing Platform · Satellite-Based Operating System · Multi-Core Processor · Partition Management

H. P. Urbach et al. (Eds.): ISSOIA 2023, SPP 191, pp. 62–76, 2024.
https://doi.org/10.1007/978-981-97-6718-2_7

1 Introduction

In the past several decades, remote sensing technology has been widely used, providing people with a large amount of useful information, from atmospheric environment monitoring to geological exploration, and from agricultural production to urban planning. In recent years, China has deployed several major projects in the field of satellite remote sensing, such as the "Major Special Project on High-Resolution Earth Observation System" and the "National Civil Space Infrastructure", which have given a strong impetus to the development of remote sensing technology in China's aerospace industry. China's remote sensing satellites are gradually developing in the direction of multiple payloads, complex imaging modes, and complex satellite management, which has put forward new demands for the new generation of remote sensing satellites. On the one hand, various types of users have gradually increased their requirements for easy-to-use satellites, which also means that the satellite system needs to break the original mode of use, enhance the intelligence level of the system, and improve the user experience using autonomous task planning, intelligent information processing, and other technologies, to satisfy the application requirements for flexible use. On the other hand, users hope to realize real-time acquisition, transmission, and processing of information through the collaborative work of multiple types of satellites, to enhance the satellite system's quick response-ability to emergency environments, such as natural disasters, environmental emergencies, and emergencies in neighboring hotspots. Therefore, the new generation of remote sensing satellites has put forward higher requirements for the new remote sensing platform, such as high real-time, high performance, and high reliability. The traditional operating system can no longer meet the above requirements, so it is necessary to design and develop a new satellite-based operating system to meet the needs of the new remote sensing platform.

2 Requirements Analysis of New Remote Sensing Platform Operating System

The task scale and complexity of the software system of the new remote sensing platform are constantly rising, and the existing satellite-based operating system can no longer meet the demands of the existing real-time computing and bus management tasks, let alone match the requirements of satellite data intelligent processing, autonomous health monitoring and mission planning. The occupancy rate of computer processors of remote sensing platforms is about to be saturated, and the computing performance needs to be improved. For example, the integral time operation and imaging strip planning tasks have occupied larger processor resources, and the satellite constellation has to process the satellite networking and dynamic routing functions to ensure the real-time interstellar communication. The operation of these tasks has made the processor resource margin less than 20%, which makes it difficult to undertake more tasks. Therefore, based on summarizing and analyzing the main problems faced by the current new remote sensing platform satellite-based operating system in application, it is necessary to upgrade the existing satellite-based operating system to meet the following requirements.

1. Multi-core processor real-time scheduling and management. To solve the problem of a large amount of computation, a large amount of bus data, and high real-time requirements of the new remote sensing platform tasks, we need to unify the management of high-performance multi-core processors to ensure the rapid processing, real-time, and stability of the software system, and to improve the efficiency of the integrated management of multi-tasking of the remote sensing platform software.
2. Requirements for error isolation and reliable operation. If there is no proper isolation means, different software is concentrated together, a single software failure spread to the overall software is almost inevitable. Therefore, there is a need to implement a partition management mechanism for isolation and protection of application software in space to improve the reliability of satellite-based operating system.

According to the above requirements, this paper designs a satellite-based operating system structure based on the characteristics of multi-core processors, adopts a two-level scheduling method to realize the deterministic scheduling and efficient real-time scheduling of the satellite-based multi-core processors, and fully exerts the high-performance of multi-core processors. At the same time, through the partition management design, the system resources are divided into different areas to realize the isolation and protection between the partition areas, to avoid the spread of software problems affecting other functions, and to improve the reliability of the remote sensing platform.

3 Design of Satellite-Based Operating System

3.1 Design of Satellite-Based Multi-core Partitioned Operating System Architecture

The design of the satellite-based multi-core partitioning operating system architecture is shown in Fig. 1, which is divided into three layers.

1. Module support layer. Used to adapt the specific target machine hardware platform processor system, mainly to realize the processor clock configuration, storage space configuration, timer management adaptation, interrupt management adaptation, print serial port adaptation, and other functions.
2. Core operating system layer. The core operating system provides resource management and partition operating system of the processor target platform. The core operating system isolates applications to different partitions and allocates execution time for each partition according to the configuration table to realize partition rotation scheduling. The core operating system manages all shared resources, including interrupts, system time, memory, and I/O services, to ensure that the resources required by a partition are available after a partition switch. The core operating system can support partition-to-partition communication and partition-to-I/O device communication by configuring appropriate communication channels.
3. Partition operating system layer. The partition operating system provides resource management and task scheduling within a single application partition. A partition operating system exists within each partition in the system. The partition operating system runs only within the time window in which the partition is located, and whenever the partition window is switched, the partition operating system will also be

switched. The basic software unit running in the partition is the task. Tasks in the partition are scheduled by priority preemption, while the partition operating system supports time slice scheduling for tasks of the same priority. The partition operating system manages all resources within the partition, including partition memory, time within the partition, interrupts, and all object resources. Because the partition is isolated, the resources managed by the partition operating system are not accessible by other partition operating systems. Information interaction between partitions must be done through the core operating system.

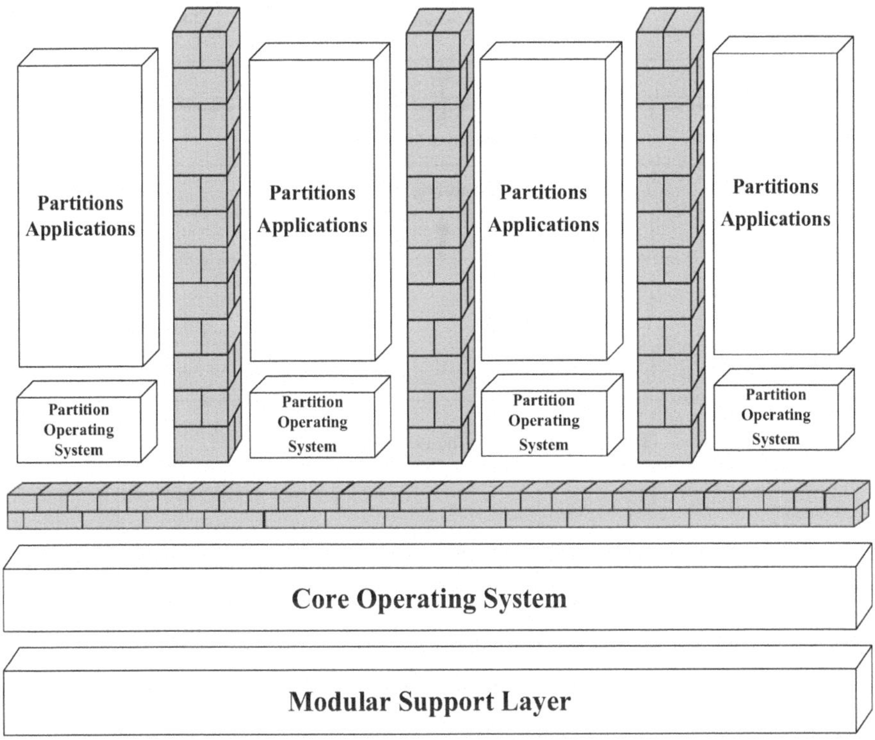

Fig. 1. Satellite-based operating system architecture.

3.2 Design of Multi-core Multi-partition Scheduling

The satellite-based operating system uses a two-level scheduling method that combines inter-partition scheduling and intra-partition scheduling, as shown in Fig. 2. Inter-partition scheduling is time slice rotation scheduling based on statically configured partition time scheduling table, and intra-partition scheduling is priority preemption-based task scheduling. The partition scheduler designed in this paper is based on a statically configured partition time scheduling table, extends the scheduling table to multi-core and multi-partition, multi-core share the same partition scheduling table, and performs partition scheduling in strict accordance with the non-overlapping of partitions, only one partition can be run in each time window, and the tasks within a partition can be SMP tasks and BMP tasks as shown in Fig. 3.

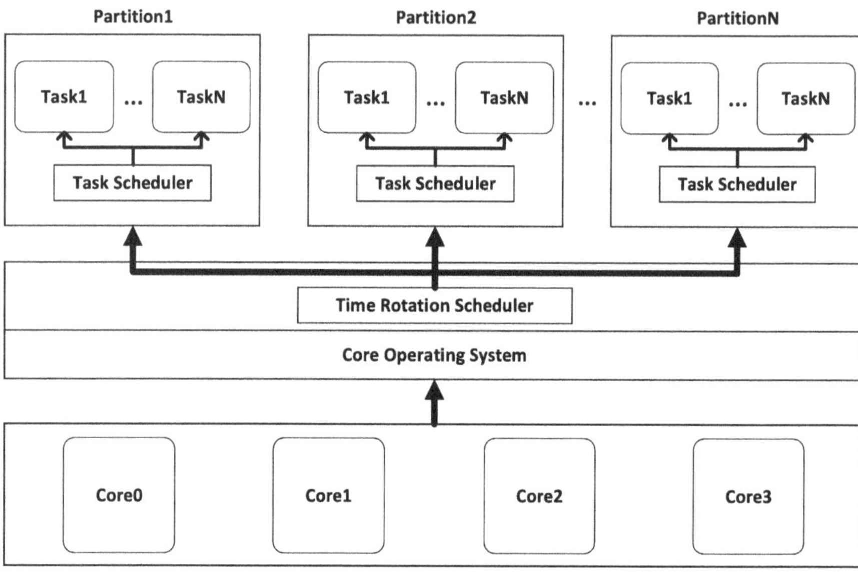

Fig. 2. Diagram of the two-level scheduling.

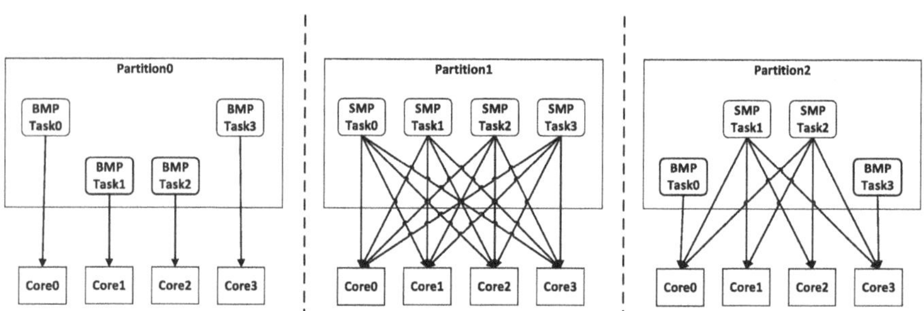

Fig. 3. Diagram of the multi-core partition scheduling.

Inter-partition scheduling takes each partition as the basic object and is implemented by the core operating system to switch between different partitions. The inter-partition scheduling policy is based on a statically configured partition time scheduling table, including the execution time and memory address space of each partition. Each partition is strictly executed according to the time order of the partition time scheduling table, and the next partition is executed immediately after the execution is finished, as shown in Fig. 4. The above partition switching mechanism ensures the temporal isolation of each partition, and enhances the predictability of program operation through the pre-configured partition time scheduling table.

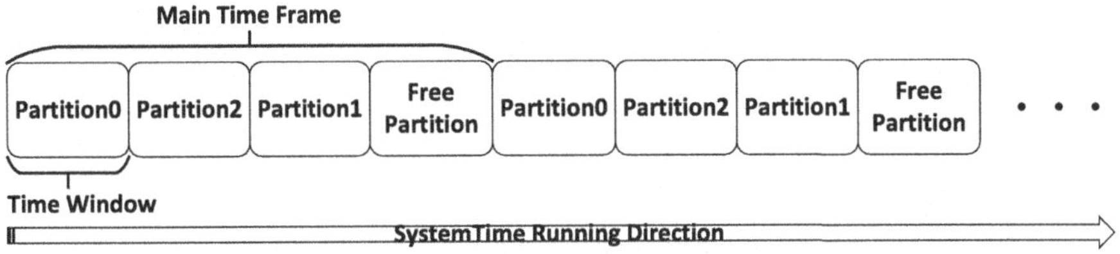

Fig. 4. Diagram of the partition scheduling.

The partition time scheduling tables are maintained by the core operating system. There can be more than one partition time scheduling table in the core operating system, but only one table can be running at the same moment for all processor cores. The partition time scheduling table maintains multiple time windows in the main time frame, which are the execution times allocated to the partition by the system. The core data structure variables of the partition time scheduling table include a count of currently running time slices, a count of executable time slices, and a partition control block. The core operating system uses the system time to meter the running time of the partition and performs inter-partition scheduling at the end of the current partition time window.

Inter-partition scheduling is shown in Fig. 5, and the process is as follows:

1. The processor core holding the system clock (typically core 0) saves the partition context in the core operating system and passes it to the partition operating system of the current partition to be switched;
2. Zero out the running time slice count of the current partition time window and stop the current partition operation utilizing hanging the current partition tasks of all processor cores in the core operating system;
3. Synchronize all other processor cores through inter-core interrupts so that their current partition contexts are saved and enter inter-core interrupt processing to wait for the partition switch to complete;
4. According to the partition time scheduling table, the time window is switched to the next one, and the partition corresponding to this window is acquired as the partition to be scheduled;
5. Activate the current partition tasks of all processor cores in the core operating system, so that the partition to be scheduled into the ready state;
6. Set the partition switching completion flag to notify other processor cores that the partition switching has been completed and can be scheduled;
7. All cores perform core operating system task scheduling, so that the corresponding task of the partition is scheduled into the running state, and then all cores perform intra-partition scheduling, intra-partition tasks continue to be executed, and inter-partition scheduling is complete.

The above intra-partition scheduling method has a time complexity of O(1), which guarantees time-sharing execution of the partition system on a multi-core processor,

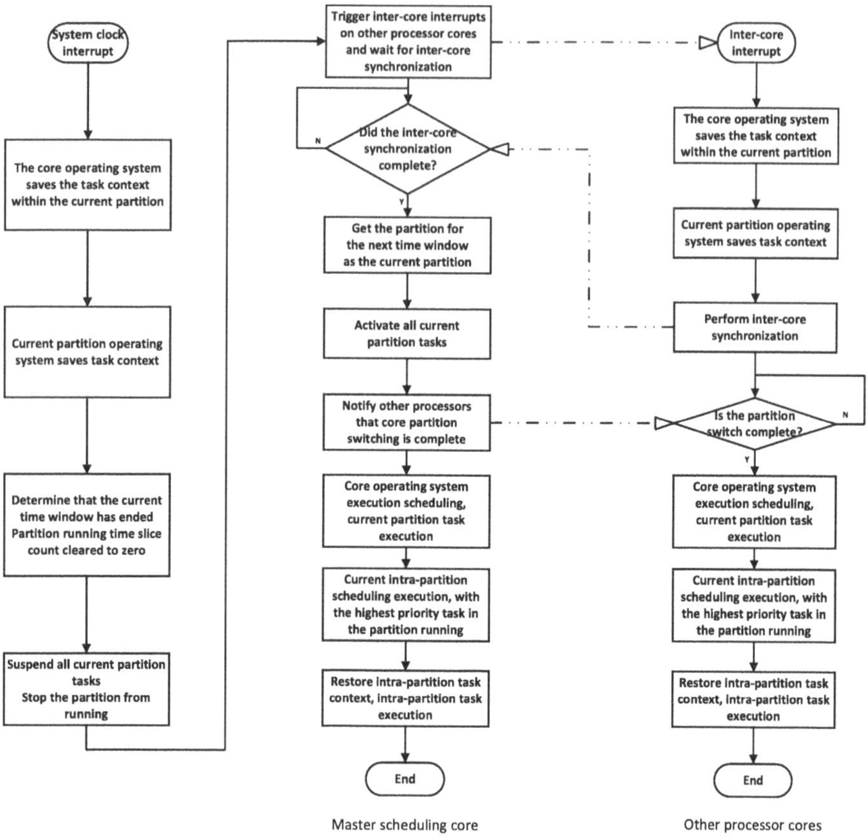

Fig. 5. Multi-core intra-partition scheduling flow.

and the partition running at any moment is fully controllable, ensuring multi-core determinism in intra-partition scheduling.

Intra-partition scheduling takes the task as the basic object and uses a strategy based on priority preemption task scheduling, with a priority range of 0 to 255, the smaller the number the higher the priority. Intra-partition scheduling supports a mixture of SMP and BMP scheduling, and two levels of task-ready queues are designed in the partition, including a global ready queue and each processor core's local ready queue. If there is a higher priority task in the global ready queue that needs to be executed, the tasks running on each core in the partition need to be compared, and the task in the global ready queue will preempt the core that is running with the lowest priority. The local ready queue is the affinity queue, and the tasks in the affinity queue need to specify which core they have an affinity for when they are created so that the tasks are added to the affinity queue on the corresponding core, rather than the global ready queue.

The system uses a bitmap approach for the highest priority task lookups, as shown in Fig. 6. The 256 spaces in the figure represent the 256 bits in the bitmap, and the number in each space characterizes the corresponding priority of that bit. As shown in the figure, when a task of a certain priority is ready, its corresponding bit in the bitmap is set to 1, which is used to quickly determine the current maximum priority. The use of the bitmap makes the finding time of the highest priority task independent of the number of tasks, with a time complexity of O(1), thus guaranteeing the real-time and determinism of the system.

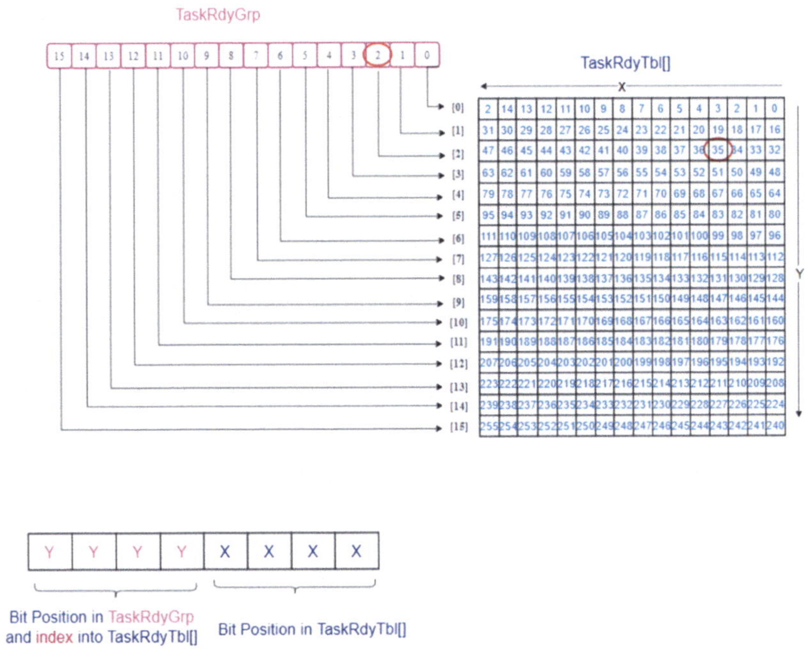

Fig. 6. Bitmap Scheduling Schematic.

3.3 Design of Multi-core Partition Management Design

There are two concepts of partitioning in a multi-core partition operating system, one is spatial partition and the other is time partition. Spatial partition can divide the memory into memory regions that do not overlap each other, so that each partition has its own unique storage space, and the program runs in the unique storage space in the partition, thus reducing the mutual interference between programs in different partitions. Time partition means that the system time is divided into several time windows, and each partition occupies at least one time window. The method of time partitioning makes the running time of different partitions predictable and improves the real-time, accuracy and predictability of the program running in the partition.

In a satellite-based multi-core partition operating system, users can configure the scheduling information of space and time partitions through configuration files to realize dynamic configuration of scheduling of space and time partitions. The user realizes spatial partition by configuring partition configuration information, including ram address information, rom address information, boot address, matching address, and partition initialization address so that the memory address areas used by all the partitions do not overlap with each other; at the same time, the user realizes time partition by configuring the partition time scheduling table, determining the execution order and execution time of each partition, which in turn divides the time window of each partition. The core operating system loads the above partition configuration information, creates and starts each partition, executes partition scheduling, and completes partition management.

In the partition operating system environment, a partition is always in one of the four states: initialized, running, suspended, and deleted:

1. Initialized: the state where the partition has been initialized and is waiting to be started;
2. Running: The state where the partition has finished starting and is running in its time window within the partition time scheduling table;
3. Suspended: The partition is suspended and cannot run, even within its time window, until the partition is resumed running;
4. Deleted: the partition has been deleted and can only enter the running state by reinitializing and restarting.

The partition state changes during the operating system operation due to resource changes or program instructions, and the transitions between the states of each partition are shown in Fig. 7.

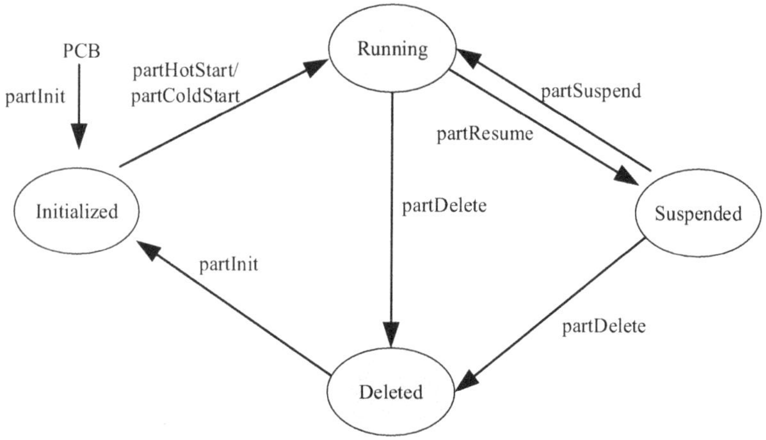

Fig. 7. Partition state transition.

After BSP (Board Support Package) completes the initialization of hardware platform resources (including CPU, memory, clock, interrupt controller, and other peripherals) and related software resources, it starts to initialize the partition operation according to the user partition configuration information. When initializing the partition, to achieve the partition multi-core dynamic control function, it is necessary to create a set of high-priority partition tasks for the dynamic control function of the partition, the number of which is the same as the number of processor cores, and each core is bound to a partition task. After the initialization of the partition is completed, it is in the initial state and can be transferred to the running state through partition start.

Partition start is divided into cold start and hot start. In the cold start process, the partition program is first loaded from ROM to RAM according to the address information configured in the partition; then the partition preparation interface function is called to complete the environment preparation work before establishing the multi-tasking environment in the partition; finally the partition matching interface function is called to bind the partition operating system call function for the core operating system, to enable the partition to enter into the running state with the ability to run. The hot start process is similar to the cold start, the difference is that the hot start does not execute the load partition program.

When a partition or a task in a partition needs to perform a suspend operation on the target partition because of some reason/fault (for instance, error handling, exception handling, schedule switching, etc.). A partition suspend causes the partition to move to the suspended state and not run even during its time window. Performing a partition suspend is similar to the process of switching partitions during intra-partition scheduling by suspending the current partition tasks for all processor cores in the core operating system. When a suspended partition needs to be restored to its original state, a partition resume must be called to transition the partition from the suspended state to the running state. Partition resume only requires the partition state to be switched to the running state, with intra-partition scheduling accomplishing the activation of the current partition tasks for all processor cores in the core operating system at the beginning of the partition time window.

When a partition needs to perform a restart operation on the target partition based on some reason/failure (for instance, error processing activation, scheduling table switching, or setting the partition state by the user, etc.), the partition is first deleted, then reinitialized, and finally restarted by starting the partition via a cold or hot start to complete the restart of the partition. When the partition is deleted, the partition tasks of all processor cores are deleted, all system call bindings are unbound, the partition control block is invalidated, and the partition state is switched to the deleted state.

Using partition management, the allocation and utilization of resources can be effectively realized. The system resources are divided into different parts, and different software runs in different partitions. Even if a software module has a problem, it won't affect the whole system and lead to the failure of other software modules, which avoid the proliferation of software problems affecting other functions. Inter-partition isolation and protection improves the reliability and security of the remote sensing platform.

4 Experimentation and Evaluation

4.1 Experimental Platform and System

This paper uses a project's integrated electronic center machine module as a new remote sensing platform simulation experiment platform with the following interfaces:

1. Two internal CAN buses (500 kbps) connected to other modules in the device through the baseboard;

2. Two external CAN buses (500 kbps) are hooked up to the satellite CAN bus network;
3. Two external CAN buses (500 kbps) communicate with the GNC component to collect status information from the gyro or magnetometer;
4. The center machine module receives RS422 uplink remote control signals from the telemetry remote control module through the base plate; sends RS422 downlink telemetry signals to the remote-control module through the baseboard;
5. One external satellite rocket separation signal is used for the center machine to obtain the separation status of the satellite and the rocket;
6. Four external and two internal high-speed uplink interfaces in the form of asynchronous RS422;
7. One second pulse signal of the GNSS module is received by the baseboard to realize GNC and timekeeping and calibration function;
8. Two OC command input signals (host power-up standby power-off command, standby power-up host power-off command).
9. One external reserved x1 SRIO high-speed interface;
10. One x1 SRIO interface on the baseboard realizes high-speed data interaction with the routing module.

The hardware block diagram of the experiment platform is shown in Fig. 8.The CPU system of the experimental platform is realized by using SHSiP02-2020, which is a high-performance data processing and controlling micro system product for satellite-based computer applications. SHSiP02-2020 can run at a main frequency of up to 1GHz, with fixed-point arithmetic power of not less than 2,200MIPS, and floating-point arithmetic power of not less than 1,200MFLOPS. SHSiP02-2020 can meet the computing needs of satellite affair management, GNC management, mission planning, autonomous health management and other functions, and leave enough arithmetic power to complete the processing of more tasks.

Through system-level packaging technology, high-performance CPU, FPGA, DDR3, NOR FLASH, ADC, and other devices are packaged together to form an integrated chip product. Compared with board-level products with the same function, the size is reduced by more than 60%, which can realize the replacement of the smallest system of a single-board computer. The microsystem is characterized by high integration, reconfigurable, and scalable.

The SHSiP02-2020 microsystem indicators are as follows:

1. Dual-core PowerPC processor, main frequency 1 GHz, fixed-point performance $\geq$ 2000 MIPS, floating-point performance of 1000MFLOPS@1GHz, total dose index $\geq$ 100 K rad (si), SOI process, SEL not less than 75 meV.cm^2/mg, SEU is less than 1×10–4 times error/day;
2. FPGA equivalent logic gate number is not less than 2 million gates, total dose index $\geq$ 100K rad (si), SEL not less than 75 meV.cm^2/mg;
3. Program storage capacity of 512Mbit, TMR three-mode redundancy, ECC checksum, total dose index $\geq$ 30 K rad (si), SEL not less than 60 meV.cm^2/mg;
4. Data storage capacity 512 MB, 64bit + 8bit, ECC calibration, total dose index $\geq$ 100 K rad (si), SEL not less than 75 meV.cm^2/mg;
5. Eight ADC acquisition channels, acquisition precision 12bit, total dose index $\geq$ 100 K rad (si), SEL not less than 60 MeV.cm^2/mg;

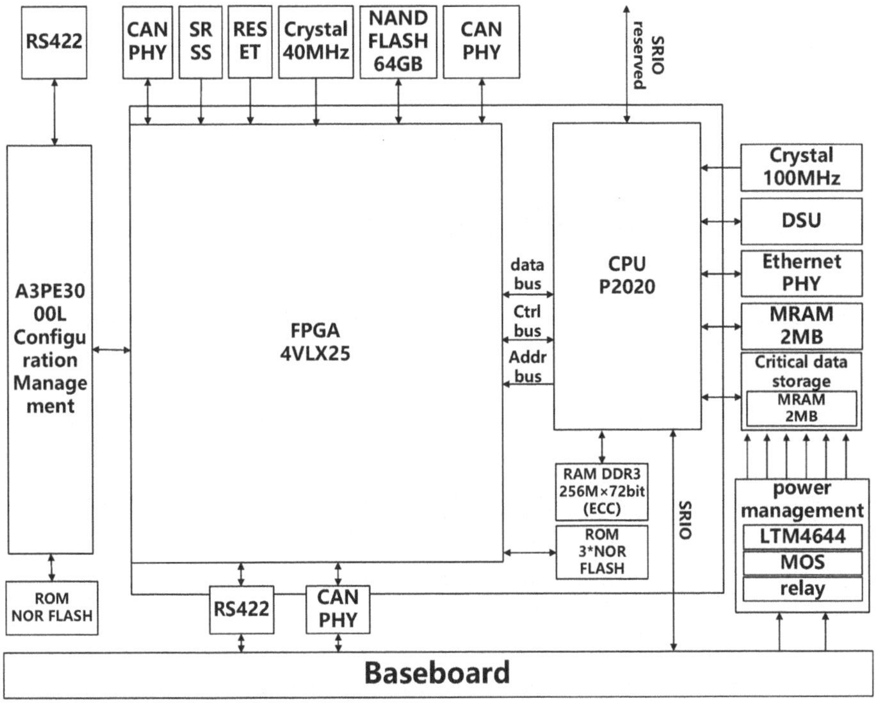

Fig. 8. Experiment platform hardware block diagram.

6. External interfaces: UART, SRIO, PCIE, Ethernet, SPI, I2C, GPIO and so on.

The experimental system simulates the tasks of the new remote sensing platform and completes the functions of satellite affairs management, autonomous mission planning, GNC control, autonomous health management, etc. It consists of satellite affairs software, mission planning software, GNC control software, health management software, and core function software, as shown in Fig. 9.

The partition scheduling timetable of the satellite affairs software, mission planning software, GNC software, and health management software is shown in Fig. 10, with 100ms as the running cycle. Among them, the idle partition is used to run core operating system tasks with lower priority than the partition tasks, which means the core function software runs in the idle partition.

The main functions of the satellite affairs software are telemetry collection and sending, telecommand forwarding, upload command forwarding, thermal control, power management, and load management. The mission planning software mainly completes mission reception, mission planning, command generation, and command execution. GNC software mainly works for inertial group data acquisition, attitude control calculation, solenoid valve output, telemetry transmission, telecommand execution, and injection instruction execution. The health management software realizes telemetry sending, telecommand receiving, injection receiving, and health monitoring management. Partition dynamic management function is the main function of the core function software,

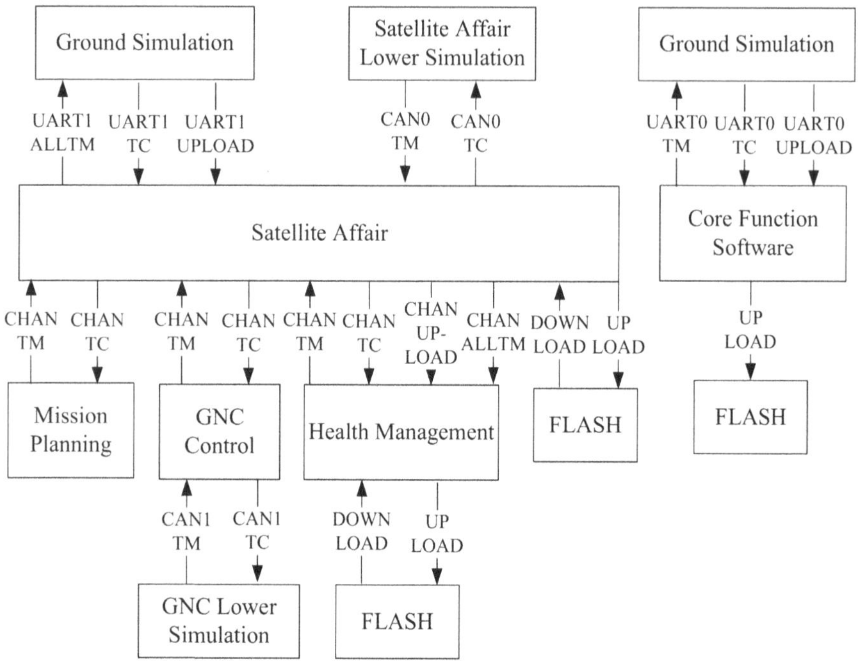

Fig. 9. Experimental system composition.

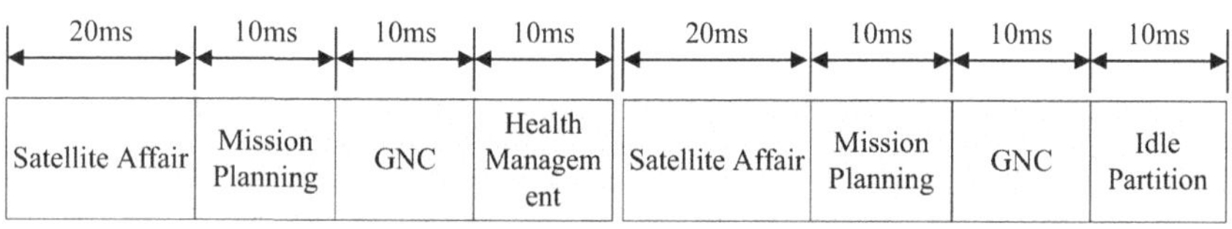

Fig. 10. Experimental system partition scheduling.

including partition burning, partition deletion, partition cold/hot start, partition injection, partition suspending, and partition resuming.

4.2 Experimental Results and Analysis

The experiment is conducted using the new remote sensing platform mission simulation experiment system, which mainly focuses on the satellite-based operating system functions and the functions of each application software partition such as satellite affair, mission planning, GNC control, and health management. The experimental results are shown in Table 1.

The experimental results show that the satellite-based operating system designed in this paper is fully capable of supporting the functions of the new remote sensing platform. The architecture of multi-core and multi-partition realizes the deterministic scheduling and efficient real-time scheduling of the satellite-based multi-core processor, gives full play to the high performance of the multi-core processor, and improves the reliability of the remote sensing platform. The above experimental results verify the effectiveness and feasibility of the satellite-based operating system design for the new remote sensing platform.

Table 1. New remote sensing platform mission simulation system experiment results

Test item	Test content	Test results
Multi-Core Multi-Partition Scheduling	The operation of the partition and the tasks within the partition	Each application partition runs strictly according to the partition table; Tasks within a partition run according to priority preemption
Multi-Core Partition Management	Test the suspend, resume, delete, and restart functions of the satellite affair partition, mission planning partition, GNC partition, and health management partition	The suspend, resume, delete, and restart functions of the satellite affair partition, mission planning partition, GNC partition, and health management partition are all correctly tested by the core function software
Satellite Affair Function Test	Testing the satellite affairs software partition functions	After testing, the functions of the autonomous cutter, command cutter, satellite rocket separation signal acquisition, and CAN bus communication in the satellite affairs software partition are all correct
Mission Planning Function Test	Test the mission planning software partition functions	After testing, the autonomous mission planning function of the mission planning software partition is correct, and it can autonomously plan and execute the visible camera/infrared camera/SAR remote sensing mission at any point under the satellite
GNC Function Test	Test the GNC software partition functions	After testing, the GNC software controls the attitude control engine function correctly based on the inertia attitude data
Health Management Function Test	Test the health management software partition functions	After testing, the diagnostic rate reaches 100% for 10 fault states such as temperature measurement reference fault, thermistor fault, solar cell charging/discharging fault, and battery fault

5 Summary

To meet the high real-time, high-performance, and high-reliability requirements of the new remote sensing platform, this paper studies the satellite-based operating system for the new remote sensing platform. This paper first analyzes and summarizes the operating system requirements of the new remote sensing platform, and clarifies the objectives and requirements of the design scheme. Then, a satellite-based operating system architecture is designed based on the characteristics of multi-core processors, and a two-level scheduling method is adopted to realize the deterministic scheduling and efficient real-time scheduling of the satellite-based multi-core processors, to give full play to the high performance of multi-core-processors. At the same time, through the partition management design, the system resources are divided into different regions to realize the isolation and protection between the partitions, avoiding the proliferation of software problems affecting other functions and improving the reliability of the remote sensing platform. Finally, the effectiveness and feasibility of the design scheme are verified through experiments and evaluations. The experimental and evaluation results show that the operating system has high performance and reliability in the new remote sensing platform and has important application value. The research in this paper has certain guiding significance for improving the design level of the operating system in the new remote sensing platform, which can better promote the development of remote sensing technology.

References

1. Zhan, P., Qi, Z.: Design of onboard partition operating system supporting multi-core processor. Spacecraft Eng. **29**(1), 47–53 (2020)
2. Zhen, X.: Research on integrated avionics for remote sensing satel-lite, Ph.D. thesis. University of Science and Technology of China (2021)
3. Zhou, T., Zhu, X., Ren, X.: A binding group scheduling method for multi-core partitioned operating system, Aeron. Comput. Techn. **52**(3), 85–88 (2020)
4. Wang, J., Yang, Y., Hu, Y.: Analysis on status quo and development trend of optical surveying and mapping satellites. Geom. Inform. Sci. Wuhan Univ. **48**(3), 333–338 (2023)
5. Liu G., Ye H., Li Y., et al: Design of deterministic scheduling method in multicore processor based on partition operating system. Aeron. Comput. Techn. **46**(1), 99–102 (2016).

Research on Nutation Coupling Technology in Near Ground Laser Communication

Baiqiu Zhao[1,2], Shoufeng Tong[1(✉)], Xiaonan Yu[1], Peng Lin[1], and Lei Zhang[1,2]

[1] Changchun University of Science and Technology Institute of Space Optoelectronics Technology, Changchun 7089 Weixing Road130022, China
tsf1998@sina.com

[2] School of Photo-Electronic Engineering, Changchun University of Science and Technology, 7089 Weixing Road, Changchun 130022, China

Abstract. This article proposes a method to reduce the loss of space laser communication links, based on the mutual nutation coupling algorithm of space light to fiber coupling, to improve coupling efficiency. On the basis of coupling theory, we designed a pair of near-Earth laser communication optical transceiver with a communication aperture of 50mm. We verified the mutual nutation coupling calculation method under a 1km space link near the ground, and conducted communication performance testing on a 10Gbps link. Experiments have shown that this algorithm can further improve coupling efficiency and reduce spatial communication link losses on the basis of tracking.

Keywords: Optical communication · Nutation · Coupling

1 Introduction

The commercialization maturity of free space laser communication is gradually increasing. Compared with microwave communication, it has the characteristics of being less susceptible to electromagnetic interference, unrestricted spectrum, and excellent performance in communication capacity and speed. Compared with fiber optic communication, its installation is convenient and flexible, and it can achieve mobile communication. The difficulty of free space laser communication mainly lies in the fact that in order to achieve higher antenna emission gain and transmit longer distances, the emission beam divergence angle of the laser is usually very low. Capturing, tracking, and coupling such a narrow beam divergence angle has certain difficulties [1–4].

Since Swanson first proposed fiber optic nutation technology [5], Multiple research institutions and scholars have conducted research on the coupling technology of space light to single-mode fiber. C. Ruilier et al. investigated the coupling problem between turbulent distorted beams and single-mode waveguides [6]. H. Takenaka et al. measured fiber coupling losses of 10–19 dB in satellite to ground laser communication links under atmospheric turbulence, which is in good agreement with the predicted fiber coupling efficiency of 17 dB [7].

H. P. Urbach et al. (Eds.): ISSOIA 2023, SPP 191, pp. 77–85, 2024.
https://doi.org/10.1007/978-981-97-6718-2_8

To achieve more efficient coupling of spatial light to single-mode fibers, this paper proposes a bidirectional nutation coupling tracking scheme based on the nutation algorithm. Firstly, the definition of coupling between spatial light and optical fibers was interpreted, and the nutation algorithm was elaborated. Then, the end machine equipment used in the bidirectional nutation experiment was introduced in terms of optical, servo, and overall control units. A near ground bidirectional nutation coupling tracking experiment was conducted using this terminal device, and communication transmission performance tests were conducted to verify the effectiveness of the proposed scheme in this paper.

2 Spatial Optical Coupling Mechanism and Nutation Algorithm

2.1 Principles of Spatial Optical Coupling

The Gaussian beam emitted by the optical system, after being transmitted through a long-distance spatial link, can be approximately processed as an ideal plane wave at the antenna end face of the receiving optical system. After convergence, the light field in the focal plane presents an Airy distribution, and the light intensity distribution is:

$$E_O(r_o) = \pi R^2 \frac{\exp(jkf)}{j\lambda f} \exp\left(jk\frac{r_o^2}{2f}\right)\left[2J_1\left(\frac{2\pi R r_o}{\lambda f}\right)/\frac{2\pi R r_o}{\lambda f}\right]$$

where λ denotes the wavelength, 2R represents the coupling lens' aperture stop, f denotes the coupling lens' focal length, k indicates the wavenumber, and J1 represents the first-order Bessel function [8].

The coupling efficiency η is the ratio of the optical power coupled into a single-mode fiber to the optical power received by the upper receiving end's pupil surface [9]. The coupling efficiency calculation formula is as follows

$$\eta = \frac{|\int E_{i.b}{}^* E_{f.b} \mathrm{d}s|^2}{\int |E_{i.b}|^2 \mathrm{d}s \cdot \int |E_{f.b}|^2 \mathrm{d}s}$$

where $E_{i.b}$ is the spot mode field where spatial light converges on the end face of a single-mode fiber, $E_{i.b}^*$ is the complex conjugate of $E_{i.b}$, $E_{f.b}$ is the mode field of a single mode fiber. The schematic diagram of the coupling principle is shown in the figure (Fig. 1):

2.2 Nutation Coupling Algorithm

The laser nutation algorithm is shown in the figure. After laser emission, space light is emitted through a reflector and coupled into the fiber through a coupling lens. After passing through a certain splitting ratio, a portion of the energy in the fiber optic power detector is fed back through the control system, and the other portion is output to the user for information transmission. The nutation algorithm provides a two-dimensional explanation of the fiber coupling end face, using a fast reflector to deflect the converging light spot on the fiber end face for circumferential scanning. The scanning points are N, the circumference radius is AC, and the center of the circumference is point A. At each

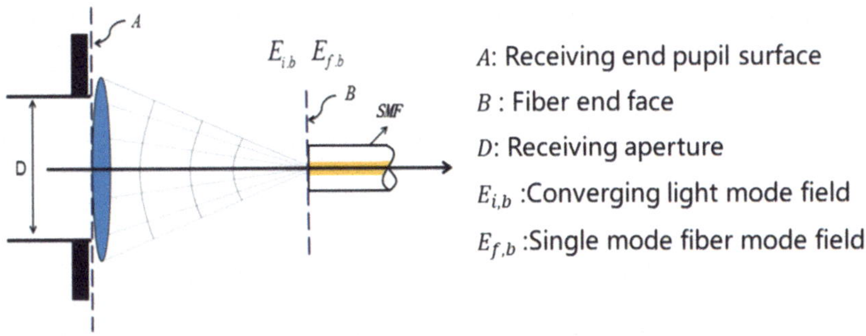

Fig. 1. Schematic diagram of the coupling principle between spatial light and single-mode fiber

scanning point, the energy of the back-end fiber power detector is recorded. When the scanning is completed in one cycle, compare the power values of N points and extract the highest energy value sampling point, which is the direction where the fiber center is located. Control the reflector to converge in this direction, that is, scan the center point A of the circumference and converge in the direction of the fiber center. After convergence, repeat the above operation and perform a circular scan again. After several rounds of convergence, the beam gradually approaches the center of the fiber, improving the coupling efficiency (Fig. 2).

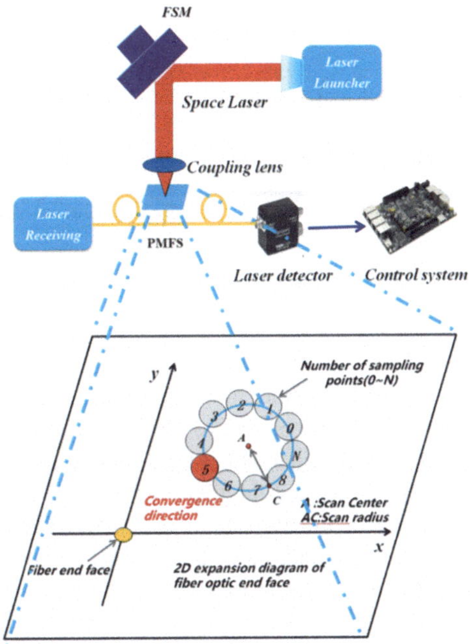

Fig. 2. Schematic diagram of nutation algorithm principle

In the nutation tracking process, the calculation formula for the execution amount of the reflector is as follows

$$\begin{cases} X = A_r \cos(2\pi \frac{i}{N}) + X_{center} + \sum K_p \cos(2\pi \frac{i_{max}}{N}) \\ Y = A_r \sin(2\pi \frac{i}{N}) + Y_{center} + \sum K_p \sin(2\pi \frac{i_{max}}{N}) \end{cases}$$

X is the azimuth execution amount of the reflector, Y is the elevation execution amount of the reflector; A_r is the circumferential scanning radius; i is the scanning point during a single scanning process, N is the total number of scanning points during a single scanning process; $A_r \cos\left(2\pi \frac{i}{N}\right)$ and $A_r \sin\left(2\pi \frac{i}{N}\right)$ synchronously, Constituting nutation circular scanning; X_{center} and Y_{center} are the initial offset points during the scanning process, Scan the center of the circle starting from this point; K_p is the convergence coefficient; i_{max} is the point with the highest power among the number of scanning points in the entire cycle; Decompose the convergence coefficients in azimuth and pitch through $\cos(2\pi \frac{i_{max}}{N})$ and $\sin(2\pi \frac{i_{max}}{N})$ (Fig. 3).

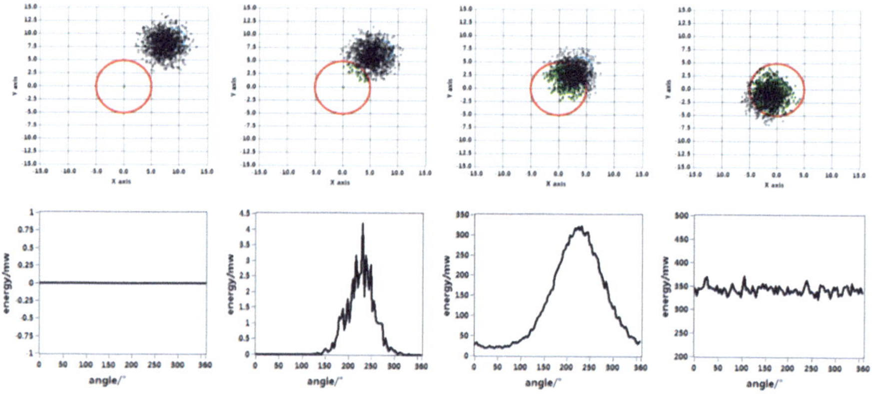

Fig. 3. Four stages of nutation process

The nutation tracking process is divided into four stages. In the first stage, the light spot completely converges outside the fiber mode field, and there is no fluctuation in the fiber power detection curve; In the second stage, during the circular scanning process, the light spot partially converges within the fiber mode field range. At this time, the fiber power curve shows an energy peak, and the highest point of this peak corresponds to the fiber convergence direction; In the third stage, during the circular scanning process, the entire process spot is within the fiber mode field range. At this time, the fiber power curve increases overall compared to the second stage and still has a significant peak; In the fourth stage, the light spot gradually converges to the center of the fiber mode field, and at this point, the fiber power curve reaches its highest and tends to stabilize.

3 Configuration of Experimental Conditions for Nutation Coupling

The overall principle of the near ground nutation experimental equipment is shown in the figure (Fig. 4).

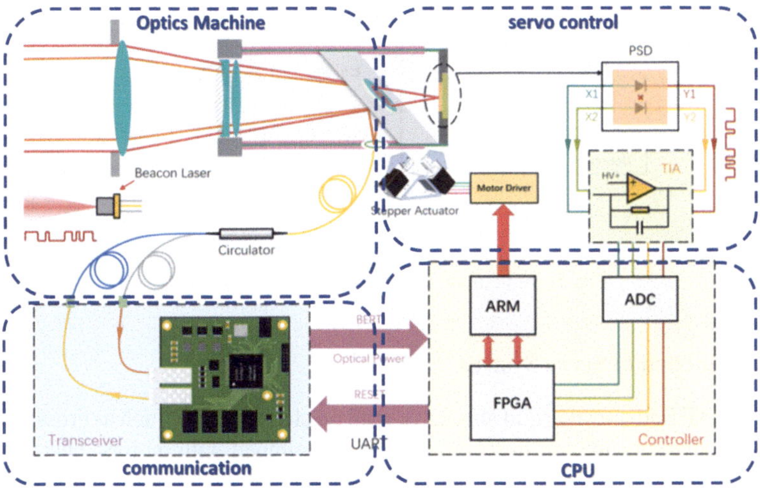

Fig. 4. General schematic diagram of near-ground optical communication equipment

The equipment is generally divided into four parts, namely the optical mechanical part composed of a transmission optical system and a fiber optic circulator, the servo part composed of a stepper motor and a position sensitive detector, the optical transceiver communication part driven by the high-speed GTX interface of FPGA, and the overall control part of ARM+FPGA.

3.1 Introduction to Optics and Mechanics

The overall device adopts a symmetrical design, and the hardware of terminal A and terminal B are identical. The 850nm beacon optical transmission channel has an independent aperture, and the 1550 nm communication optical transmission channel and 850nm beacon reception channel share a common aperture design, with an aperture of 50 mm. After the communication light and beacon light are jointly received by the main optical aperture, they are isolated through wavelength splitters. 850 nm is off target by the position sensitive detector in the horizontal branch, and 1550nm is reflected to the vertical branch, coupled from the spatial light to the fiber optic circulator, and used to achieve communication transceiver isolation. The user input is amplified through EDFA for optical signal amplification and connected to the transmitting end of the fiber optic ringer. The receiving end of the fiber optic ringer undergoes 1:9 splitting processing, of which 10% is used for receiving power detection as feedback. 90% are received as users for business transmission (Fig. 5).

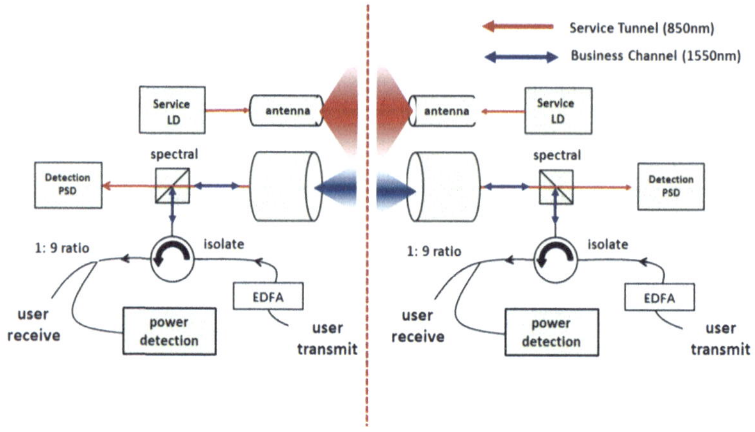

Fig. 5. Schematic diagram of optical principles

3.2 Introduction to Servo System

Traditional APT are mostly rigid structures with mechanical axes, such as cross tracking frames and single pendulum mirrors, which are combined with motors, bearings, and encoders to achieve tracking and pointing of spatial targets. This article adopts a push-pull flexible shaft structure, where the flexible shaft undergoes elastic deformation under the action of force, resulting in angular displacement. The driving source consists of two X-shaped linear motor structures placed vertically at 90°, which achieve complete fit with the driving motor through the decomposition of their own weight. When the angular resolution of the linear motor is fixed, the larger the distance between the motor and the flexible support point, the higher the spatial angular accuracy. This structure has the characteristics of small volume and low cost (Fig. 6).

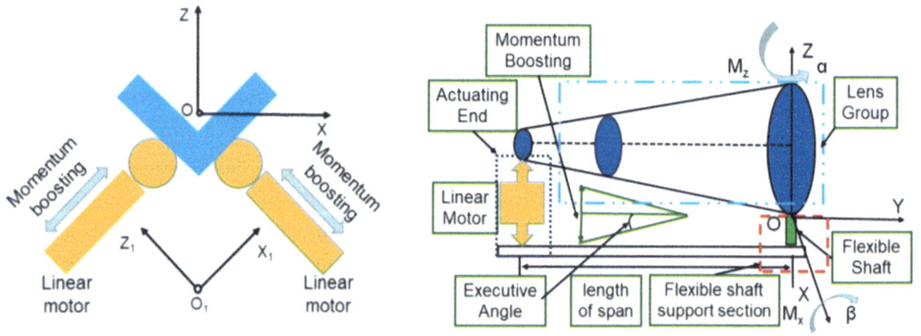

Fig. 6. Principle of servo drive

3.3 Introduction to Communication and Main Control Unit

The main control unit adopts an ARM+FPGA structure, utilizing the high-speed parallel processing capability of FPGA to collect various detection sensors and be responsible for the output of each actuator. Utilize the mathematical computing power of ARM to perform algorithmic analysis on the sensor data collected by FPGA, and send the calculation results to FPGA for execution. The information exchange between the two adopts FSMC bus. Simultaneously utilizing the high-speed GTX interface of FPGA to drive commercial optical models for the emission and detection of optical signals.

4 Near Ground Nutation Experiment and Discussion

After the near ground experimental equipment is installed at a certain distance, the gun aiming device on the equipment is first used to aim at each other and complete the initial pointing. After pointing, enter the scanning capture phase and adopt a scanning skip scheme.

The image below depicts the skip/scan track is shown by the host computer. The transmitting end is in a staring condition, while the receiving end executes raster helical scanning. When the receiving end scans the uncertain area for one week, a flag bit is generated, and the transmitting end executes a skip step. The trigger flag bit of the skip step is transmitted through the baseband information modulated on the beacon. The acquisition process is terminated until the receiving optical power of the detector reaches the acquisition threshold, after which the two-way nutation procedure is started. The execution trajectory of the scanning end and the skipping end are same, with the exception that the scanning end is continually run, and the skipping end executes a point only when it gets the flag bit of the beacon (Fig. 7).

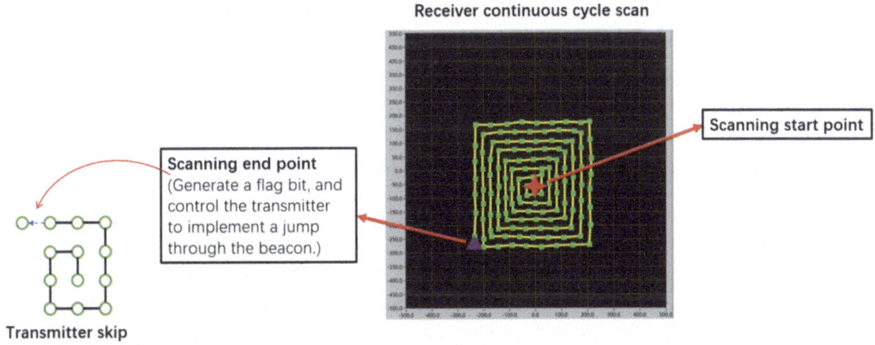

Fig. 7. Schematic diagram of skip scanning principle

The capture is completed, and energy is detected in the receiving end fiber, as shown in point A of the figure. Subsequently, the nutation tracking at the receiving end is turned on, and the energy gradually converges. A peak energy fluctuation represents a scanning period. The nutation tracking at the receiving end is completed, and the energy is increased to point B. Then turn on the nutation tracking at the transmitter, further increasing the energy to point C. At this point, the double ended nutation is completed, and the coupling efficiency is improved to the highest level. After the link was fully aligned and established, a 10Gbps communication rate error rate detection was conducted, and the measured error rate was 2.479E-9 (Fig. 8).

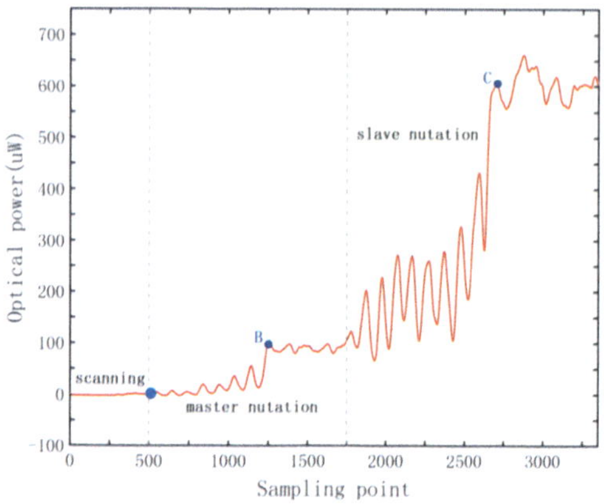

Fig. 8. Scanning capture, single end nutation, and dual end nutation energy changes

5 Conclusion

This article proposes a nutation coupling method based on mirrors, and on this basis, proposes that bidirectional nutation can further reduce the loss of space laser communication links. After single end nutation coupling, although the coupling efficiency has been improved to a certain extent, the beam alignment at the other end still has a certain deviation. The bidirectional nutation simultaneously corrects the alignment deviation between the two ends, further improving the coupling efficiency. This scheme has been validated in near-ground laser communication experiments and plays a certain role in providing coupling efficiency and reducing link losses in space laser communication.

References

1. Toyoshima, M., Fuse, T., Kolev, D.R., et al.: Current status of research and development on space laser communications technologies and future plans in NICT. In: 2015 IEEE International Conference on Space Optical Systems and Applications (ICSOS), October 26–28, 2015, pp. 1–5. New Orleans, LA, USA. New York: IEEE (2015)
2. Wu, W.R., Chen, M., Zhang, Z., et al.: Overview of deep space laser communication. Sci. China Inform. Sci. **61**(4), 040301 (2018)
3. Carrasco-Casado, A., Mata-Calvo, R.: Space optical links for communication networks. Springer Handbook of Optical Networks, pp. 1057–1103. Springer (2020)
4. Cahoy, K., et al.: The CubeSat Laser Infrared CrosslinK mission (CLICK). Proc. SPIE **11180**, 111800Y (2019)

5. Swanson, E.A., Bondurant, R.S.: Using fiber optics to simplify free-space lasercom sys-tems. Proc. SPIE **1218**, 1–13 (1990)
6. Ruilier, C., Cassaing, F.: Coupling of large telescopes and single-mode waveguides: application to stellar interferometry. J. Opt. Soc. Am. A **18**(1), 143 (2001)
7. Takenaka, H., Toyoshima, M., Takayama, Y.: Experimental verification of fiber-coupling efficiency for satellite-toground atmospheric laser downlinks. Opt. Express **20**(14), 15301–15308 (2012)
8. Toselli, I., Agrawal, B., Restaino, S.: Gaussian beam propagation in maritime atmospher-ic turbulence: long term beam spread and beam wander analysis. Proc. SPIE **7814**, 78140R (2010)
9. Winzer, P.J., Leeb, W.R.: Fiber coupling efficiency for random light and its applications to lidar. Opt. Let. **23**(13), 986–988 (1998)

Object-Independent Image Restoration Based on Deep Learning

Hongwei Qi and Bing Dong(✉)

School of Optics and Photonics, Beijing Institute of Technology, Beijing 100081, China
bdong@bit.edu.cn

Abstract. The optical imaging system is affected by both internal factors (such as manufacturing and alignment errors) and external factors (such as atmospheric turbulence), preventing it from reaching diffraction-limited imaging. The blind image restoration algorithms typically require lots of iterative computations, leading to poor real-time performance. Image restoration methods based on neural network have limited application scenario. To solve the above mentioned issues, this paper introduces an object-independent image restoration method based on deep convolutional neural network (DCNN). The DCNN utilizes object-independent three-channel image feature as its input. Without estimating the wavefront aberration modes, the blurred point spread function (PSF) is predicted by the DCNN. Subsequently, a deconvolution operation can be applied to restore the image. Not only the blurred PSF, the restored image is also incorporated in the loss function to enable effective image restoration under complex aberrations. It is proved that our method can achieve good image restoration results even for complex aberrations involving 35 Zernike aberration modes. Compared to traditional blind restoration algorithms, it exhibits advantages in restoration time and better SSIM and PSNR.

Keywords: Deep learning · Object-independent · Image restoration · Jointloss

1 Introduction

The resolution of imaging systems is usually limited by wavefront aberrations. Wavefront aberrations can be caused by various sources such as manufacturing and assembly errors in the optical system, atmospheric turbulence. And they can be corrected by using adaptive optics. However, due to the influence of system errors, wavefront observation errors, noise, and other factors, the image after AO system closed-loop correction may not meet the image quality requirements. In this case, image restoration algorithms are often used to improve image quality [1].

Traditional image restoration algorithms can be classified into two categories based on different known conditions: non-blind restoration algorithms and blind restoration algorithms [2–9]. Non-blind restoration algorithm [10], also known as deconvolution, is a method of estimating clean image from blurred image through mathematical operations when the PSF is known. Non-blind restoration is more focused on how to effectively restore a clean image from noisy blur images. Most of the blurry images in real life

H. P. Urbach et al. (Eds.): ISSOIA 2023, SPP 191, pp. 86–100, 2024.
https://doi.org/10.1007/978-981-97-6718-2_9

situations are unknown PSF, so blind restoration methods are commonly used for processing. The blind restoration algorithm simultaneously solves the clean image and PSF based on some prior constraints. Common algorithms include IBD algorithm, NAS-RIF algorithm [4], MFMD algorithm [5], multi-frame adaptive image deconvolution algorithm [6], TVBD algorithm [7], NSM algorithm [8], and L0 regularization algorithm [9]. Most of these algorithms regularize the sparsity of image based on prior constraints to obtain stable solutions. This prior constraint of smoothness often results in loss of edge information, leading to excessive smoothing of the image and vulnerability to noise, support domain errors, and complex backgrounds. Moreover, these algorithms are typically based on iterative ideas, resulting in poor real-time performance.

In the field of wavefront sensing, Phase Diversity (PD) [11] reconstructs the wavefront through multiple iterations using two images. This algorithm has the advantages of simple structure and the ability to recover continuous and discontinuous phase distributions. It is also applicable to both point and extended target imaging, making it widely used in image restoration and wavefront sensing fields. However, the algorithm is based on the numerical iterative approach, which can lead to low computational efficiency, time-consuming solutions, and issues of local optimality.

Since Angel et al. [12] first used deep learning technology to invert the wavefront phase, neural network has gradually been introduced to the field of wavefront sensing. In recent years, with the development of computer, deep learning has found increasingly widespread applications in intensity-based wavefront sensing and image restoration [13–15]. However, current research mainly focuses on point target imaging scenarios [16], or specific extended target imaging scenarios [17], and once the imaging scene changes, the neural network model may struggle to effectively complete the task. Some network models have attempted to improve generalization ability by increasing the diversity of training scenarios [18]. However, this approach can lead to enormous computational requirements, making it time-consuming and difficult to apply in real-world.

Therefore, it is meaningful to obtain an evaluation function for image aberrations that is independent of the imaging target. Kendrick [19] provided three object-independent feature extraction methods in his article, namely Ratio Metrics, Power Metrics, and Sharpness Metrics. In the field of co-phase error detection [20–22], several researchers have used Power Metrics and Sharpness Metrics to achieve good detection accuracy. Qi Xin [23] et al. proposed a method to obtain a new object-independent feature by taking the modulus of the inverse Fourier transform of the Ratio Metrics, and utilized an LSTM network to achieve wavefront sensing with a dynamic range of 4–9 orders.

This paper proposes an object-independent image restoration method based on deep learning. And a new target-independent feature was used as the input of the network. To avoid the problems of limited order of prediction aberration coefficients and aberration coupling, it is proposed to directly predict the PSF that causes image blurring. JointLoss which include the image restoration quality is proposed as the network loss function. Simulation results shown that our method can effectively restore distorted images.

2 Optical Imaging Model and Feature Extraction

Assuming the imaging target is illuminated by incoherent quasi-monochromatic light, and the imaging system is a linear shift-invariant system. Then the intensity image collected by CCD can be written in the following form:

$$i(x, y) = o(x, y) * PSF(x, y) \tag{1}$$

where $i(x, y)$ is the intensity image collected by the CCD, $o(x, y)$ is the imaging target, $*$ is the convolution operation, $PSF(x, y)$ is the point spread function. The point spread function can be modeled by the following equation:

$$PSF(x, y) = |\mathcal{F}\{P(x, y)\}|^2 \tag{2}$$

where $\mathcal{F}$ is the Fourier transform, $P(x, y)$ is the pupil function. If there is aberration in the optical system, the expression for the generalized pupil function is:

$$P(x, y)) = p(x, y) \cdot e^{[i \cdot \varphi(x, y)]} \tag{3}$$

where $p(x, y)$ is the pupil function of 0–1 distribution, and $\varphi(x, y)$ is the wavefront phase with aberration. In order to completely express the information of wavefront aberration, the weighted combination of Zernike polynomials is generally used to express the wavefront aberration:

$$\varphi(x, y) = \sum_{k=1}^{k} a_k Z_k(x, y) \tag{4}$$

so the point spread function(PSF) of defocus plane is:

$$PSF_{de}(x, y) = |\mathcal{F}\{P_{de}(x, y)\}|^2 = \left|\mathcal{F}\{p(x, y) \cdot e^{[i \cdot \varphi(x, y) + \phi(x, y)]}\}\right|^2 \tag{5}$$

where $\phi(x, y)$ is the defocus phase, According to the Fourier optical principles, in the infocus and defocus plane have:

$$\begin{cases} I_{in}(u, v) = O(u, v) \cdot OTF_{in}(u, v) \\ I_{de}(u, v) = O(u, v) \cdot OTF_{de}(u, v) \end{cases} \tag{6}$$

where $I_{in}(u, v)$, $I_{de}(u, v)$ is the spectrum of infocus and defocus plane, $O(u, v)$ is the spectrum of imaging target, $OTF_{in}(u, v)$, $OTF_{de}(u, v)$ is the optical transfer function of infocus and defocus plane.

Kendrick proposed three methods for extracting object-independent features in his article, namely Ratio Metric, Sharpness Metric and Power Metric. The expressions are as follows:

$$\begin{cases} F_{ratio} = \frac{I_{in}(u,v)}{I_{de}(u,v)} \\ F_{sharpness} = \frac{I_{in}(u,v)I_{in}^*(u,v) - I_{de}^*(u,v)I_{de}(u,v)}{I_{in}(u,v)I_{in}^*(u,v) + I_{de}(u,v)I_{de}^*(u,v)} \\ F_{power} = \frac{I_{in}(u,v)I_{de}^*(u,v) - I_{in}^*(u,v)I_{de}(u,v)}{I_{in}(u,v)I_{in}^*(u,v) + I_{de}(u,v)I_{de}^*(u,v)} \end{cases} \tag{7}$$

In this paper, a new three-channel feature is used to describe object-independent information. That is, the inverse Fourier transform and modulo operation are performed on the Power metric, Sharpness metric and Ratio metric respectively. In order to eliminate the influence of denominator being zero, a regularization term ε is introduced. Therefore, the three-channel feature extraction formula is:

$$\begin{cases} Channel_1 = \left| \mathcal{F}^{-1}\{\frac{I_{in}(u,v)}{I_{de}(u,v)+\varepsilon}\} \right| \\ Channel_2 = \left| \mathcal{F}^{-1}\{\frac{I_{in}(u,v)I_{in}^*(u,v) - I_{de}^*(u,v)I_{de}(u,v)}{I_{in}(u,v)I_{in}^*(u,v) + I_{de}(u,v)I_{de}^*(u,v)+\varepsilon}\} \right| \\ Channel_3 = \left| \mathcal{F}^{-1}\{\frac{I_{in}(u,v)I_{de}^*(u,v) - I_{in}^*(u,v)I_{de}(u,v)}{I_{in}(u,v)I_{in}^*(u,v) + I_{de}(u,v)I_{de}^*(u,v)+\varepsilon}\} \right| \end{cases} \tag{8}$$

where $\varepsilon = 1\text{e} - 10$, $\mathcal{F}^{-1}$represents the inverse Fourier transform. The feature extraction method can obtain richer phase information of object-independent features. And its information is mainly distributed in the central area of image, which helps the neural network extract high-dimensional features and speed up the convergence speed.

Figure 1 shows the extracted three-channel features from infocus and defocus images by replacing two objects under the same phase aberrations.

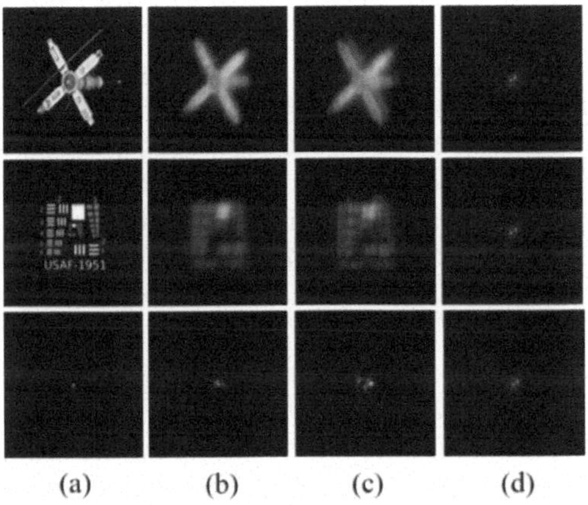

Fig. 1. The feature of different target images, (a–c) is respectively the clean, infocus and defocus image, (d) is the three-channel feature.

3 Method

3.1 Process of Restoration

Specific process of image restoration based on object-independent three-channel feature is illustrated in Fig. 2:

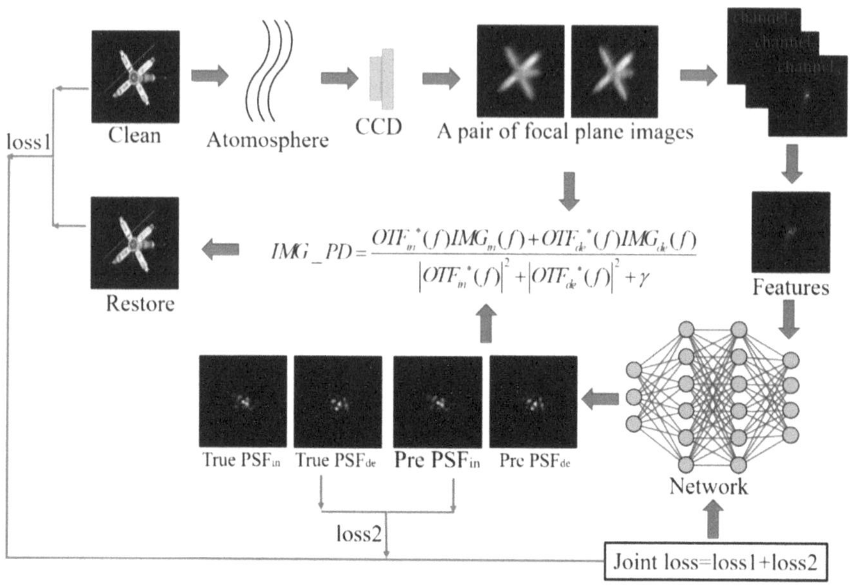

Fig. 2. The strategy of image restoration.

The main steps are as follows: First of all, the clean image adds wavefront aberration to obtain distorted images of the infocus and defocus plane on the CCD. Then according to Eq. (8), the object-independent features are obtained and assembled into a three-channel feature image. Then the three-channel feature image is used as the network input, and infocus and defocus plane PSF are fed into network training as the output. The loss function of network is divided into two parts: the first part is the loss between predicted PSF and real PSF. The second part is the loss between clean image and restored image based on the PD deconvolution formula of Eq. (9):

$$IMG_{PD} = \frac{OTF_{in}^{*}(f)IMG_{in}(f) + OTF_{de}^{*}(f)IMG_{de}(f)}{\left|OTF_{in}^{*}(f)\right|^{2} + \left|OTF_{de}^{*}(f)\right|^{2} + \gamma} \tag{9}$$

Where IMG_{PD} denotes the image restored by deconvolution, $OTF_{in}^{*}(f)$ and $OTF_{de}^{*}(f)$ respectively represent the complex conjugates of infocus and defocus plane OTF(OTF is equal to the Fourier transform of PSF), $IMG_{in}(f)$ and $IMG_{de}(f)$ respectively denote the aberrated images at infocus and defocus plane. Therefore, the joint loss function of network is shown in Eq. (10):

$$JointLoss = MSE(PSF_{true}, PSF_{pre}) + MSE(IMG_{true}, IMG_{pre}) \tag{10}$$

Where PSF_{pre} and PSF_{true} respectively denote the network-predicted PSF and the real PSF, IMG_{pre} and IMG_{true} respectively represent the restored image obtained through deconvolution with network-predicted PSF and clean image.

During testing, you only need to collect the blurred images of infocus and defocus plane, obtain three-channel feature and send them to trained network. Then you will obtain the PSF of infocus and defocus plane. Finally, the restored image can be obtained based on Eq. (9).

3.2 Structure of Model and Dataset

In 2015, Ronneberger [24] et al. first proposed the Unet network structure, which has shown good performance in end-to-end tasks such as computer vision and bioinformatics.we hopes to establish the mapping relationship between three-channel feature and infocus and defocus plane PSF, which is an end-to-end task. Therefore, based on Unet, we improved it to obtain the network OIIR-Net(Object Independent Image Restoration) used in this paper.

The OIIR-Net network structure is shown in Fig. 3. And the overall network is divided into two parts: encoder part and decoder part.

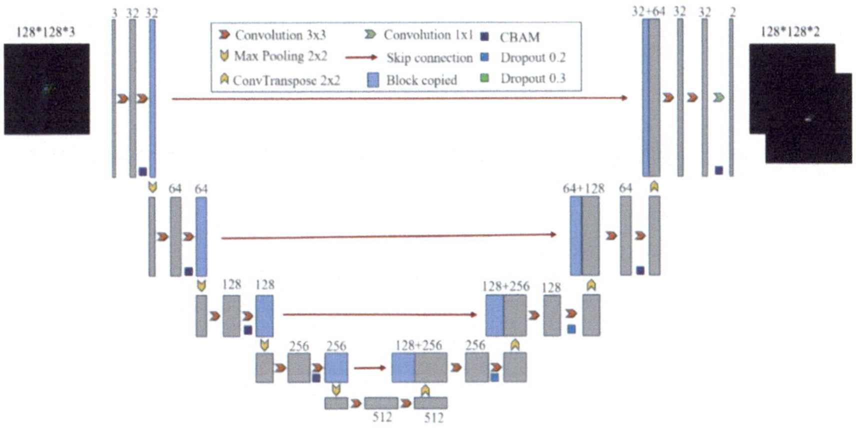

Fig. 3. The network structure of OIIR-Net.

Function of encoder part is to downsample the input three-channel object-independent features and extract image information. In order to improve the feature extraction and expression capabilities of network, a CBAM [25] attention module is added before each downsampling, including channel attention module and spatial attention module. The role of decoder part is to upsample image and connect features of same resolution together through skip connections. Finally, in order to alleviate over-fitting problem that easily occurs in end-to-end tasks, Dropout layers are added before upsampling.

3.3 Dataset and Training

The simulation parameters of optical imaging system are as follows: wavelength $\lambda =$ 635 nm, entrance pupil diameter $D = 6.3$ mm, focal length $F = 250$ mm, CCD pixel size $Pixel = 6.45\ \mu m$, PV of defocused phase $W = 1\lambda$. A total of 20000 sets of feature and PSF data pairs were generated, with 18000 used for training and 2000 for validation. The range of Zernike coefficient is 4 to 35, wavefront RMS is uniformly distributed at [0, 0.3 λ], and noise level is 50 dB.

The input is three-channel feature which image size is 128 * 128 * 3, and output of network are infocus and defocus plane PSF which image size are 128 * 128 * 2.

Training settings of neural network are as follows: initial learning rate is set to 1e–4; batchsize is set to 32; loss function is JointLoss; Adam optimizer optimizes the parameters. And we introduce a learning rate attenuation mechanism to help the network converge; and the maximum number of epochs is set is 100; the learning rate attenuation factor is set to 0.5, patience parameter is set to 10. Also, l2 regularization is introduced. OIIR-Net is trained with a deep learning workstation based on CPU (Intel(R) Core(TM) i7-10700K CPU, 32GB) and GPU (NVIDIA GeForce RTX 3090, 24GB) with Python = 3.8.0 and Pytorch = 1.12.1.

3.4 Result

Network training takes about 5 h. The Loss curve of network after training is shown in Fig. 4. It can be seen that the network converges to a smaller value at 40th epoch.

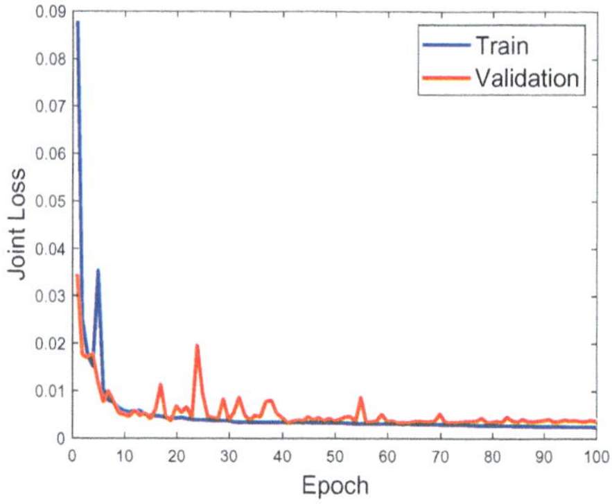

Fig. 4. Loss curve of network training.

In order to test the performance of the network in restoring images, 1000 pairs of data with wavefront RMS uniformly distributed in [0,0.3 λ] were randomly generated as a test set. Figure 5 shows the restoration results of three randomly selected groups of samples.

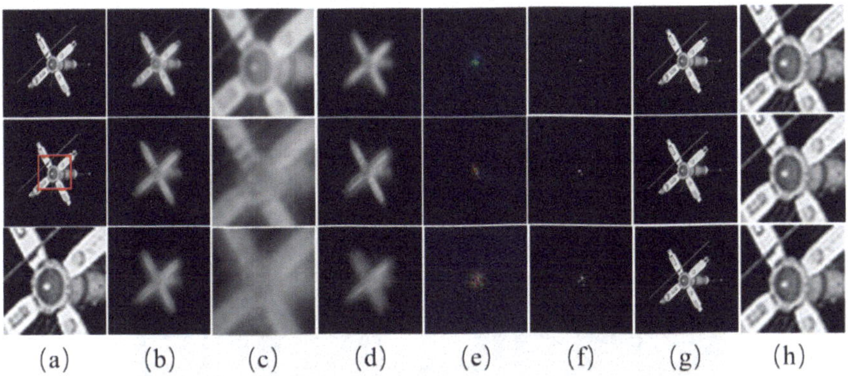

Fig. 5. Final result of network restored image. The first picture in (a) is the clean image, the third picture in (a) is an enlarged picture of Region Of Interest(ROI), (b) is the infocus distortion image, and (c) is the infocus image of ROI, (d) is the defocus plane image, (e) is three-channel feature image, (f) is the infocus PSF predicted by the network, (g) is the restored image, (h) is the restored image of ROI.

Wavefront RMS values for three distorted images in Fig. 5 are respectively 0.1411 λ, 0.2466 λ, and 0.2974 λ. The SSIM between infocus distorted image and clean image are 0.8001, 0.7173, and 0.6551, while the PSNR values are 24.0622 dB, 22.1594 dB, and 19.9984 dB. After restoration, the SSIM values between the reconstructed images and the clear image are 0.9089, 0.8854, and 0.8822, with corresponding PSNR values of 31.8338dB, 29.7050 dB, and 28.8634 dB.

In order to compare the overall restoration performance of the network under different wavefront RMS values, three test sets were generated. Each consisting of 1000 sets of RMS values within the ranges of [0, 0.1 λ], [0.1 λ, 0.2 λ], and [0.2 λ, 0.3 λ]. The average SSIM and PSNR values before and after restoration are shown in Fig. 6.

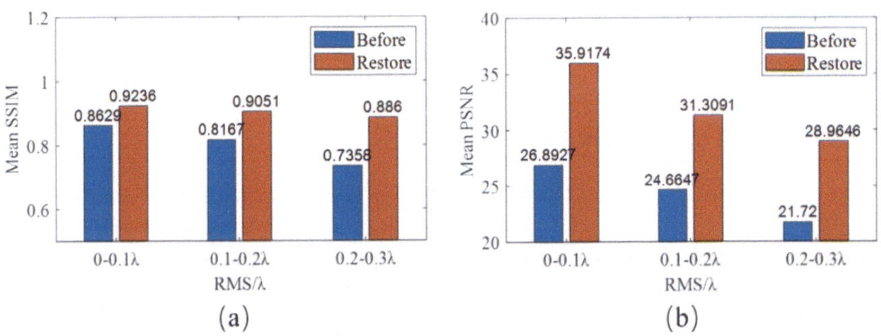

Fig. 6. The average results of 1000 test sets under different magnitudes of aberrations, where (a) represents the average SSIM, and (b) represents the average PSNR.

The results in Figs. 5 and 6 show that the network can have a good restoration effect on distorted images with order 4–35 and RMS in the range of [0, 0.3 λ]. As aberration increases, effect of restoring image becomes worse.

In order to verify the object-independent characteristics of network, 1000 pairs of data were generated using the USAF1951 image as a test set. The result is as Fig. 7 shows:

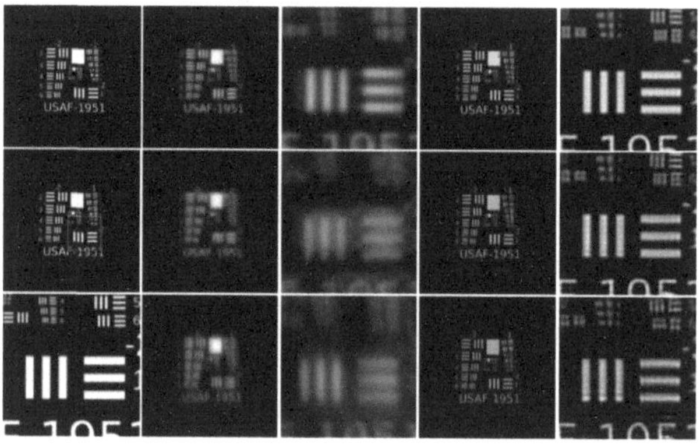

(a) (b) (c) (d) (e)

Fig. 7. Final restored image of USAF1951. The first picture in (a) is the clean image, the third picture in (a) is enlarged picture of ROI, (b) is the infocus plane distortion image, and (c) is the infocus plane image of ROI, (d) is the network restored image, (e) is the network restored image of ROI.

Wavefront RMS values for three distorted images in Fig. 7 are respectively 0.1451 λ, 0.2306 λ, and 0.2887 λ. The SSIM between infocus distorted image and the clear image are 0.8008, 0.7252, and 0.6874, while the PSNR values are 19.7225 dB, 17.3467 dB, and 16.6743 dB. After restoration, the SSIM values between the reconstructed images and the clear image are 0.9499, 0.8975, and 0.8527, with corresponding PSNR values of 24.3417 dB, 23.3238 dB, and 21.9624 dB.

4 Comparison

4.1 Comparison of Different Features

We also compare three-channel feature proposed in this paper with other feature extraction methods. They are sharpness feature and power feature, which are widely used in co-phase error detection. And the ratio feature of inverse fourier transform proposed by Qi Xin in 2019. Similar to the steps in 3.3, use OIIR-Net network for training. The final network Loss curve is shown in Fig. 8.

It can be seen that all four features can converge to a smaller level, but the loss value of three-channel feature image we proposed finally converges to the minimum, and loss curve is smooth.

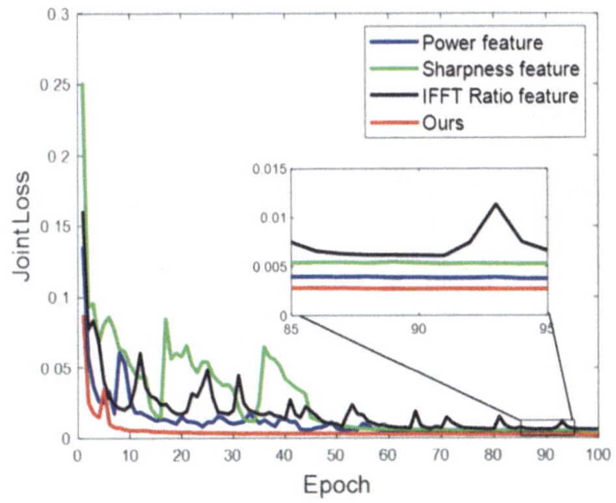

Fig. 8. Loss curves for training with different features

In order to compare the effects of final restored images with different features, 1000 pairs of distorted images with wavefront RMS in the range of [0, 0.3λ] were randomly generated for restoration. The average SSIM and PSNR results of the final restored image are shown in Fig. 9:

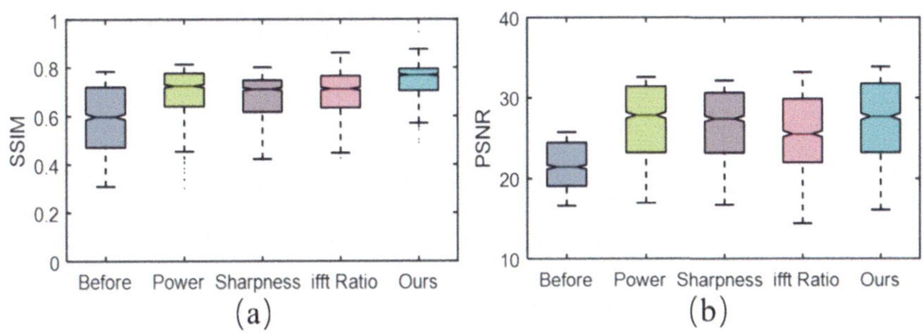

Fig. 9. SSIM and PSNR results of the Final restored image using four features

It can be seen that three-channel feature proposed in this paper has the highest SSIM and PSNR parameters of the restored image among the four features, proving the effectiveness of three-channel feature.

4.2 Comparison of Different Loss Functions

The loss function PSFLoss, which is only related to predicted PSF and real PSF, is effective when only exists low-order aberrations. Figure 10 shows the restoration results when there are only 4–10th Zernike modes aberrations:

It can be seen that image can be restored when only low-order aberrations exist. However, PSFLoss often fails when high-order aberrations exist. The JointLoss can have a better restoration effect even when high-order aberrations exist. Figure 11 respectively

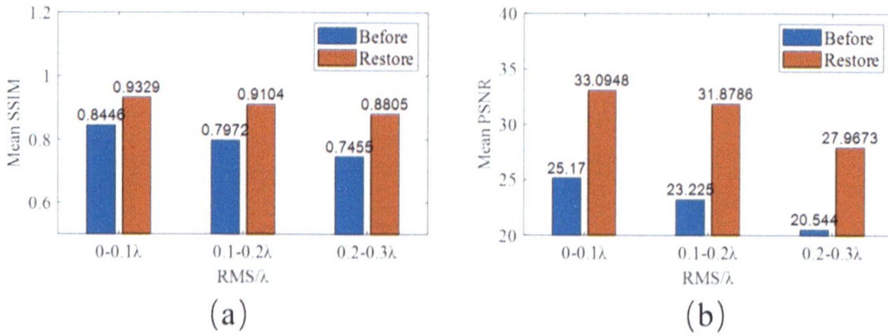

Fig. 10. The average results of 1000 test sets under different magnitudes of aberrations, where (a) represents the average SSIM, and (b) represents the average PSNR.

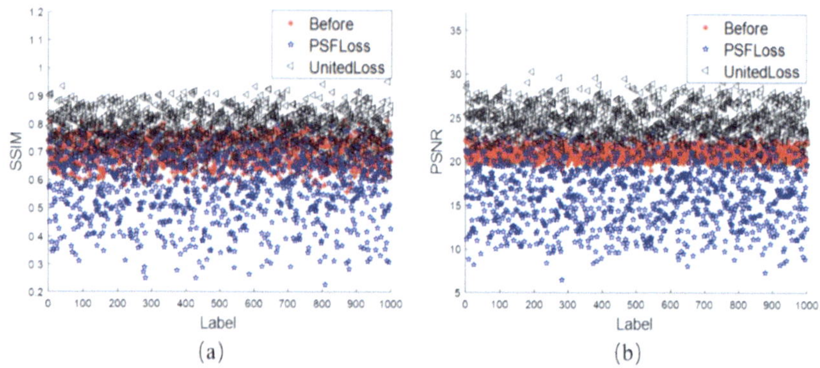

Fig. 11. Result which PSFLoss and as loss function

shows the restoration results with wavefront RMS value in the range of [0.2 λ, 0.3 λ] using PSFLoss and JointLoss.

Among them, the average values of SSIM and PSNR before restoration of 1000 test sets are 0.7358 and 21.7200 dB. When PSFLoss is loss function, the average values of SSIM and PSNR are 0.6983 and 18.2424 dB. We can see that using PSFloss even makes the final image worse. When UnitedLoss is loss function, the average values of SSIM and PSNR are 0.8826 and 28.9646 dB.

The results show that the JointLoss proposed in this paper has a significant effect compared with the PSFLoss in the presence of high-order aberrations. This is because when there are high-order aberrations, the structure of PSF is often complex, and it is extremely difficult to predict the PSF correctly. JointLoss uses final restored image and clean image as part of t loss function, so that even if the network predicts PSF is not accurate enough, image can be restored to a good level.

4.3 Comparison with Traditional Blind Restoration Methods

This section compares the method proposed in this paper with the traditional blind restoration algorithm. The algorithms compared are Total Variation Blind restoration algorithm (TVBD), L0 regularization (L0), and Normalized Sparse Measure (NSM). Three groups of samples with wavefront RMS = 0.0964λ, RMS = 0.1450λ, and RMS = 0.2518λ were selected for restoration respectively. The final results are shown in Tables 1, 2 and 3 and Figs. 12 and 13.

Table 1. Results of RMS = 0.0964 λ.

	SSIM	PSNR/dB	Time
before	0.8367	25.6647	–
TVBD	0.9445	29.4225	15.1640
L0	0.8903	28.6021	1.9760
NSM	0.9025	29.5824	3.0990
Ours	0.9351	33.3091	0.0342

Table 2. Results of RMS = 0.1450 λ.

	SSIM	PSNR/dB	Time/s
before	0.7958	23.7200	–
TVBD	0.8691	25.1583	19.6910
L0	0.8250	23.7089	1.6480
NSM	0.8518	26.0234	4.8162
Ours	0.9260	31.9646	0.0342

Table 3. Results of RMS = 0.2518 λ.

	SSIM	PSNR/dB	Time
before	0.7229	20.7273	–
TVBD	0.8316	25.2656	21.1400
L0	0.7465	20.2802	1.6220
NSM	0.8203	25.4600	5.0745
Ours	0.8959	29.7675	0.0342

The results show that the method proposed in this paper is the best in terms of SSIM, PSNR, and restoration time compared to traditional blind restoration algorithms.

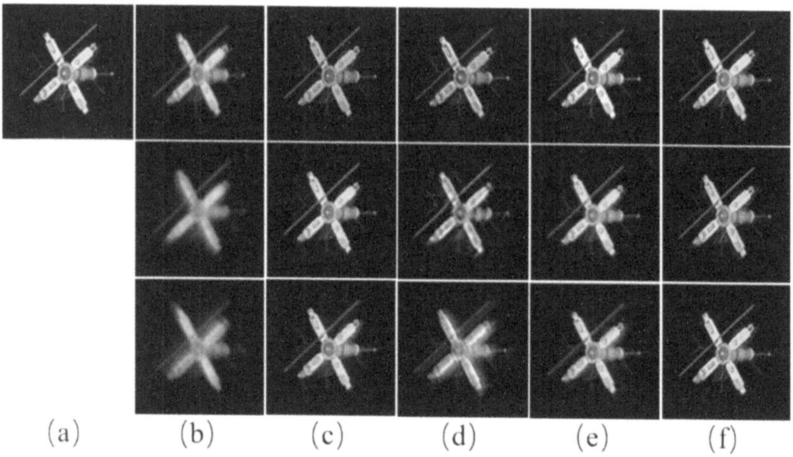

Fig. 12. The results obtained by using different restoration methods. And (a) is the clean image, (b) is the blurred image, (c) is the restored image using TVBD method, (d) is the restored image using L0 method, (e) is the restored image using NSM method, and (f) is the restored image using our method.

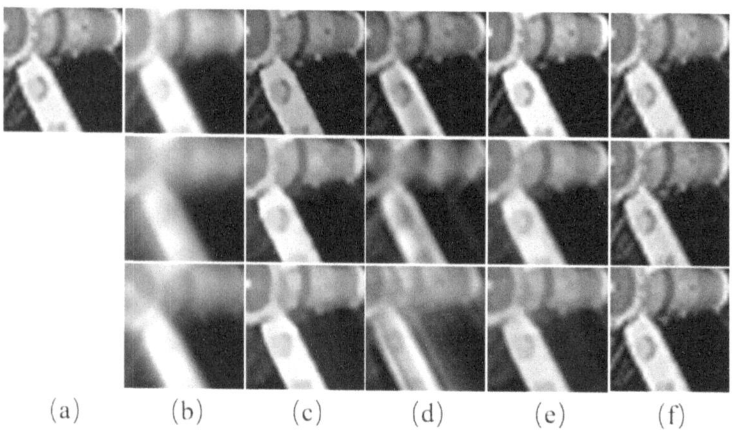

Fig. 13. The results of ROI obtained by using different restoration methods.

5 Conclusion

This paper proposes an image restoration method based on deep learning network, where both the training and application of the network are object-independent. Compared with traditional blind restoration algorithms, our method has better restoration effects and less time consumption. The core innovation of this paper lies in the fact that we extracted three-channel feature image that are associated with aberrations but independent of the extended target. Secondly, we propose to directly predict the PSF, and make the difference between network restored image and clean image as part of loss function, so that the network's prediction of PSF can effectively improve the quality of distorted images. Overall, the method proposed in our paper can serve as a new approach for image restoration.

However, it is important to note that the method proposed in this paper is simulated with known parameters of the optical imaging system. For issues where it is challenging to obtain clean images and ideal PSF, there is no need to calculate the loss function during testing, thus eliminating the requirement for clean images and real PSF. When the simulation parameters align with those of the actual optical imaging system, the network trained through simulation can also improve images distorted by the actual optical imaging system.

In cases where the parameters of optical system are unknown, such as when optical remote sensing satellites operate in orbit and need to collect real-world image data for network training, the method described in this paper may not be applicable. Instead, a self-supervised learning approach can be employed to train the network. This involves collecting multiple pairs of infocus and defocus images to solve for object-independent features for network training. The self-supervised learning network is divided into two parts: the first part is a trainable network encoder that predicts infocus and defocus PSF. The second part is a physics-driven model decoder, which convolves arbitrary imaging targets with the infocus and defocus PSF to obtain distorted images, and then solves for object -independent features. The network is iteratively trained by comparing the decoder's output of object-independent features with the input. This method may serve as a solution to challenges in obtaining clean images and ideal PSF.

References

1. Fienup, J.R.: Phase retrieval algorithms: a comparison (1982)
2. Chan, T., Amizic, B., Molina, R., Babacan, D., Katsaggelos: A blind im-age deconvolution. CRC Press (2007)
3. Gunturk, B.K., Li, X.: Image restoration fundamentals and advances. Digital Imaging & Computer Vision (2012)
4. Kundur, D., Hatzinakos, D.: A novel blind deconvolution scheme for image restoration using recursive filtering. IEEE Trans. Signal Process. **46**, 375–390 (1998)
5. Mugnier, L.M., Conan, J.M., Fusco, T., Michau, V.: In bayesian inference for inverse problems (1998)
6. Zhou, X., Mateos, J., Zhou, F., Molina, R., Katsaggelos, A.K.: Variational dirichlet blur kernel estimation. IEEE Trans. Image Process. *24*, 5127–5139 (2015)
7. Hom, E.F.Y., Marchis, F., Lee, T.K., Haase, S., Sedat, J.W.: AIDA: an adaptive image deconvolution algorithm with application to multi-frame and three-dimensional data. J. Opt. Soc. Am. A **24**, 1580–1600 (2007)
8. Krishnan, D., Tay, T., Fergus, R.: CVPR, pp. 233–240 (2011)
9. Pan, J., Su, Z.: Fast -regularized kernel estimation for robust motion deblurring. IEEE Signal Process. Let. **20**, 841–844 (2013)
10. Hang, Y.: Survey of non-blind image restoration. Chin. Opt. **15**, 954–972 (2022)
11. Yue, D., Li, Y., Nie, H., Jin, G.: Real-time wavefront reconstruction for extended object based on phase diversity. Optik – Int. J. Light Electron Opt. **144** (2017)
12. Angel, J.R.P., Wizinowich, P., Lloyd-Hart, M., Sandler, D.: Adaptive op-tics for array telescopes using neural-network techniques. Nature **348**, 221–224 (1990)
13. Ju, G., Qi, X., Ma, H., Yan, C.: Feature-based phase retrieval wavefront sensing approach using machine learning. Opt. Express **26**, 31767–31783 (2018)

14. Paine, S.W., Fienup, J.R.: Machine learning for improved image-based wavefront sensing. Opt. Lett. **43**, 1235–1238 (2018)
15. Rivenson, Y., Zhang, Y., Günaydın, H., Teng, D., Ozcan, A.: Phase re-covery and holographic image reconstruction using deep learning in neu-ral networks. Light: Sci. Appl. **7**, 17141–17141 (2018)
16. Sinha, A., Lee, J., Li, S., Barbastathis, G.: Lensless computational imag-ing through deep learning. Optica **4**, 1117–1125 (2017)
17. Nishizaki, Y., et al.: Deep learning wavefront sensing. Opt. Express **27**, 240–251 (2019)
18. Kupyn, O., Budzan, V., Mykhailych, M., Mishkin, D., Matas, J.: (2017)
19. Kendrick, R.L., Acton, D.S., Duncan, A.L.: Phase-diversity wave-front sensor for imaging systems. Appl. Opt. (1994)
20. Hui, M., et al. Object-independent piston diagnosing approach for seg-mented optical mirrors via deep convolutional neural network. Appl. Opt. **59**, 771–778 (2020)
21. Tang, J., et al.: Object-independent tilt detection for optical sparse aper-ture system with large-scale piston error via deep convolution neural net-work. Opt. Express **29**, 41670–41684 (2021)
22. Wang, Y., Zhang, C., Guo, L., Xu, S., Ju, G. Decoupled object-independent image features for fine phasing of segmented mirrors using deep learning. Remote Sens. **14** (2022)
23. Xin, Q., Ju, G., Zhang, C., Xu, S.: Object-independent image-based wavefront sensing approach using phase diversity images and deep learn-ing. Opt. Express **27**, 26102–26119 (2019)
24. Ronneberger, O., Fischer, P., Brox, T.: U-Net: convolutional networks for biomedical image segmentation. Springer International Publishing (2015)
25. Woo, S., Park, J., Lee, J.Y., Kweon, I.S.: CBAM: convolutional block attention module. Springer, Cham (2018)

A Novel Star Identification Method Assisted by Stellar Brightness for Star Tracker

Xiang Li[1,2], Junfeng Xie[2(✉)], Fan Mo[2], Chaopeng Xu[2], and Junye Shen[3]

[1] School of Remote Sensing and Information Engineering, Wuhan University, Wuhan 430070, China

[2] Land Satellite Remote Sensing Application Center, Ministry of Natural Resources of China, Beijing 100048, China

idealeexiang@163.com

[3] School of Surveying, Mapping and Geographical Sciences, Liaoning University of Engineering and Technology, Fuxin, Liaoning, China

Abstract. In modern deep space exploration missions, star trackers are crucial attitude determination equipment, which establish the positioning of spacecraft by capturing stars. With the advancement of multi-heads star tracker technology, its designs are increasingly geared towards achieving higher pixel resolution, smaller fields of view (FOV), and the ability to detect fainter stars. However, this development trend inevitably leads to an increase in the number of navigation stars in the field of view, thus challenging the applicability of existing star identification methods. For this, this paper develops a novel star identification method assisted by stellar brightness. This method first establishes a set of magnitude screening criteria and establishes the corresponding relationship between star brightness and apparent magnitude in the star image. Subsequently, stars within the filtering range are prioritized for identification based on stellar brightness and angular distance information. On this basis, the identification process is extended to the stars outside the screening range, and both stages are verified with the help of the directed-loop method to improve the accuracy. Taking the dual-head star tracker of the Gaofen-7 satellite as an example, this method's identification accuracy and attitude determination precision were analyzed by processing its multi-orbit star image data. The results demonstrate that the method not only has a high identification rate but also improves the attitude precision by 0.05 to 0.1 arc seconds compared to telemetry downlinked data. The method proposed in this study provides strong theoretical support for the technological development of future multi-heads star trackers and has potential for practical application.

Keywords: Star identification · Star tracker · GF-7

1 Introduction

Star trackers, as high-precision optical sensors, play a crucial role in the advancement of deep space exploration and astronomical observation by providing the absolute three-axis attitude of spacecraft through the observation of stars [1]. Stars possess distinct

H. P. Urbach et al. (Eds.): ISSOIA 2023, SPP 191, pp. 101–113, 2024.
https://doi.org/10.1007/978-981-97-6718-2_10

characteristics such as small proper motion, stable position, and extensive distribution. Astronomers have meticulously compiled star catalogs through long-term observation and study, mainly recording the position, parallaxes, brightness (usually described by magnitude), and other information of stars. Star trackers calculate attitude by matching the observed stars with the stars recorded in the star catalog, a process known as star identification. Generally, the position information and magnitude of stars are their most significant features, and the main methods can be divided into three categories: subgraph isomorphism [2, 3], pattern recognition [4, 5], and artificial intelligence [6, 7]. Most of these methods mainly use the position information of stars to construct matching features. With the development of the small field of view (FOV) and multi-FOV star trackers, to ensure a sufficient number of navigation stars in the FOV, the limiting magnitude of the star tracker is generally increased, which leads to a denser distribution of stars in the small FOV. Traditional star identification methods lack uniqueness in matching features, making it difficult to meet the requirements of efficient and accurate recognition, and it is also hard to adapt to the development of star tracker hardware.

The reason why most methods do not consider star magnitude is that the brightness information provided in the star catalog is the apparent magnitude, color index, and spectral type. For the same star, the instrument magnitude detected by the star tracker can differ from its apparent magnitude by 2 to 3 magnitudes. However, current research still focuses on finding new matching features, and most experiments are performed at magnitudes less than 6.5, which presents challenges in directly extending to higher limit magnitude detection. For star identification at higher limit magnitudes, this paper proposes a magnitude-assisted star identification method. This method initially utilizes range magnitudes to filter and identify observed stars, and then, based on successful identification, recognizes the remaining stars within the field of view of the star image.

2 Method

The number of stars in a star catalog is closely related to their magnitude. As the magnitude increases, the number of stars in the catalog rises sharply. When the magnitude is 8, the total number of stars in the celestial sphere is approximately 64,172. The capture of a large number of stars by cameras due to high limit magnitude detection can lead to identification difficulties. Hence, this paper advocates for prioritizing stars based on magnitude for initial identification, necessitating a statistical analysis aligned with the star tracker's field of view to establish the required magnitude filtering range. To ensure normal star identification, there should be at least four or more navigation stars in a single field of view pointing randomly within the magnitude range of a specific sky area. Generally speaking, the larger the FOV angle of the star tracker, the smaller the magnitude filter range. Design 10,000 random pointing experiments with different FOV angles and extreme detection amplitudes of the star tracker. A success is considered if there are four or more navigation stars within the specified field of view. The success rate is then calculated to determine the optimal amplitude filtering range for each FOV angle (see Fig. 1).

It can be seen that when the limit magnitude for star detection is 6 and the angle of FOV is 4 degrees, the probability of having more than four navigation stars within the

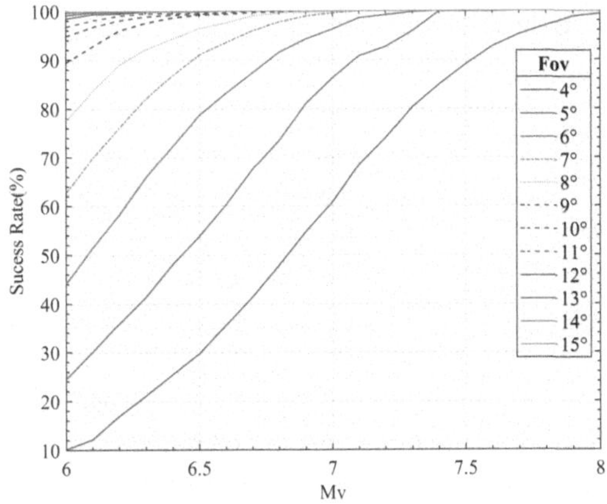

Fig. 1. The probability that the number of stars within the field of view exceeds four under different magnitudes for various FOV angles

FOV is only 10.2%. However, when the angle of FOV is 15 degrees, this probability reaches 99.9%. The optimal magnitude upper limit (OMUL) for each FOV is shown in the Table 1.

Table 1. Optimal magnitude upper limit for each FOV

FOV (degrees)	4	5	6	7	8	9	10	11	12	13	14	15
OMUL (Mv)	8.2	7.7	7.4	7.1	7.1	6.9	6.7	6.6	6.6	6.3	6.1	6.1

In practical applications, since the size of the field of view angle is not always an integer, to determine the optimal magnitude upper limit, one can perform probability statistical analysis based on the actual field of view angle parameters set, or round down to find the corresponding optimal magnitude upper limit.

The features used for identifying the main star include the stellar inter-star angle between the main star and companion stars, as well as their magnitudes. When establishing the navigation star catalog, each star within the magnitude range is considered as a main star. The companion stars that fall within its FOV radius and meet the magnitude requirements are calculated and their numbers recorded. The ID number and magnitude of each companion star are then recorded in sequence, and their inter-star angle from the main star is calculated, as Eq. 1. The data is described in the Table 2.

$$\theta_{ij} = \arccos(V_i^{\mathrm{T}} V_j) \tag{1}$$

Where, $\mathbf{V}_i = \begin{pmatrix} \cos\alpha_i \cos\delta_i \\ \sin\alpha_i \cos\delta_i \\ \sin\delta_i \end{pmatrix}$. α_i represents the right ascension of star i, δ_i represents the declination of star j.

Table 2. Format of the navigation star catalog

Star ID	Star Mag	Num of companion stars	Star List		
1	Mag_{main}	N	Star ID_1; Mag_1; θ_{11};	Star ID_2; Mag_2; θ_{12};	...
2	Mag_{main}	N	Star ID_1; Mag_1; θ_{21};	Star ID_2; Mag_2; θ_{22};	...
......					
i	Mag_{main}	N	Star ID_1; Mag_1; θ_{i1};	Star ID_1; Mag_1; θ_{i2};	...

In Table 2, Star ID_n, Mag_n, θ_n represents the ID number, magnitude, and inter-star angle of the n-th companion star relative to the current main star.

It should be noted that before adopting the method of this paper, it is necessary to first use other star identification methods for magnitude training. This involves statistically analyzing star points from multiple orbits and multiple star images, training and fitting the relationship between the sum of the grayscale of defocused image pixels of stars with different brightness and their apparent magnitude. Then, using the completed navigation star catalog, the original star image transmitted by the star tracker is identified. The main process of star identification is as follows:

1. Determine the optimal upper magnitude limit based on the FOV of the star tracker, and calculate the apparent magnitude using the sum of the grayscale of star points. Then, use the optimal upper magnitude limit to filter the stars.
2. Calculate the distance of the filtered stars to the principal point. The star closest to the center is designated as the main star.
3. Identify preliminary candidate main stars. Mark the navigation stars whose magnitude difference from the main star in the image is less than ε_{Mag} in the catalog. (ε_{Mag} is the maximum magnitude error).
4. Match the main star by calculating the inter-star angle between the candidate stars and the main star, and arrange them in ascending order (see Fig. 2a) to form an array of inter-star angle. Define two pointers, where Pointer 1 points to the fifth element of the candidate stars in the navigation star catalog, and Pointer 2 points to the first element of the inter-star angle array. If the inter-star angle of the two elements satisfies Eq. 2 and their magnitudes satisfies Eq. 3, then a successful match is confirmed, and a record is made. Pointer 1 then moves back three times, and Pointer 2 moves back once. If Eq. 3 is not satisfied, the pointer with the smaller value is moved.

$$\mathrm{d}_{ij} = \arccos(W_i^{\mathrm{T}} W_j) \tag{2}$$

Where, $W_i = \frac{1}{\sqrt{(x_i^2+y_i^2+f^2)}}\begin{pmatrix} -x_i \\ -y_i \\ f \end{pmatrix}$, f represents focal length of star tracker camera.

$$\left|d_{ij} - \theta_{ij}\right| \le \varepsilon_{\mathrm{Distance}} \tag{3}$$

Where, $\varepsilon_{\mathrm{Distance}}$ is an error window depending on both the star tracker measurement error and the star catalog position uncertainly. If $\varepsilon_{\mathrm{Distance}}$ is too large, the possibility of a misidentification increases. If it is too small, a misidentification is unlikely, but the number of identified stars is apt to decrease.

5. Record the candidate star that appears most frequently, along with the other companion stars, and note their ID numbers. Look up their right ascension and declination information. Using that star as the origin, re-establish the image coordinate system, where the new system's x and y axes are parallel to the x and y axes in the image center coordinate system. Connect each neighboring star to the origin and sequentially calculate the angle each line makes with the x-axis of the new coordinate system (see Fig. 2b), connecting them in a counterclockwise direction in order of increasing angle. Record the inter-star angle between adjacent stars and the IDs of the stars at both ends of each star image point on the image, as shown in the Table 3.

Table 3. Circular star image structure

Inter-star angle	Star ID1	Star ID_2
$d_{(4,5)}$	4	5
$d_{(5,3)}$	5	3
$d_{(3,1)}$	3	1
$d_{(1,2)}$	1	2
$d_{(2,6)}$	2	6
d(6,4)	6	4

6. In the first connection sequence: 4 -> 5 -> 3 -> 1 -> 2 -> 6 -> 4, calculate whether the inter-star angle satisfy Eq. 3. If within the error range, retain them and record their ID numbers. If not, cancel the connection. For example, in the Fig. 2b, $d_{(2,6)}$ does not satisfy the conditions. If no other stars are available, then calculate $d_{(2,4)}$. After connecting the first and last stars, the final loop is: 4 -> 5 -> 3 -> 1 -> 2 -> 4. If the number of nodes in the directed loop is less than half of the number of navigation stars within the magnitude range in the FOV, then sequentially change the candidate stars and continue matching the directed loop until the requirements are met.
7. Find all the stars outside the magnitude range of the primary star in the FOV, calculate the inter-star angle of the primary star and a randomly selected confirmed companion star (companion star 1) in the navigation star catalog to form a database of navigation star diagonal distances. With the main star at the center, establish a main star measurement coordinate system. The new system's x and y axes are parallel to the x and y axes in the image center coordinate system. Connect each neighboring star to the origin and sequentially calculate the angle each line makes with the x-axis of the new coordinate system. Connect them in a counterclockwise direction in order of increasing angle, and assign them numbers (see Fig. 2c).
8. Calculate the inter-star angle in the image coordinate system between all unidentified stars and both the main star and companion star 1, forming an image inter-star angle database. If for any star *i*, both diagonal distances are smaller than the stellar diagonal distance deviation, then record its ID number and perform a circular verification. Start

from star point 1 and conduct a sequential verification, that is, from 1- > 2- > 3- >…- > 10. If a star fails the verification, it is removed, and finally, the largest search loop is returned.

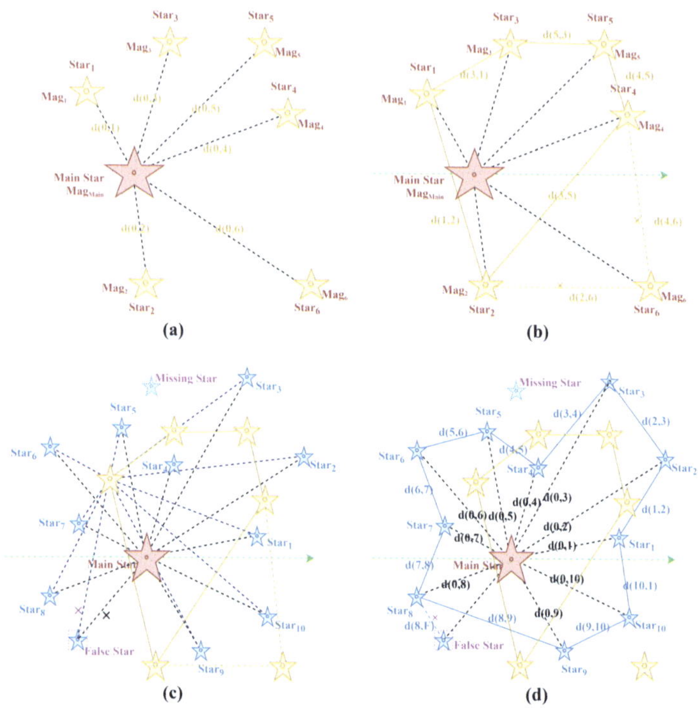

Fig. 2. Identification steps

It is important to note that in the image, if a star point do not meet the requirements for inter-star angle from both the main star and companion stars, they are marked as unidentified stars. There are two cases for these unidentified stars: one is that the star is not present in the navigation star catalog, and the other is that the star image point represents another point light source in space, known as a false star. When a star exists in the navigation star catalog but is not captured, it is referred to as a missing star. The feature of this paper is the use of inter-star angle, and construct circular verification. Neither false stars nor missing stars affect the method's matching. Also, in step 7, one can choose to prepare a navigation star catalog for recognized companion stars in advance, thus increasing the capacity of the navigation star catalog to enhance the efficiency of the method's operation.

3 Experiment

3.1 Data and Result

This experiment focused on the Gaofen-7 satellite, which is the first Chinese satellite equipped with a dual-head star tracker (APS-01). It achieves sub-arcsecond level attitude measurement accuracy through its high-precision star trackers and gyroscopes. This star tracker is based on a CMOS APS detector and, during its orbital period, provides real-time telemetry attitude data at 8 Hz to meet the constraints of the telemetry channel, along with the downlink of original star chart data, ensuring the feasibility of attitude post-processing. At the same time, the attitude derived from the original star chart processing (post-processing attitude) can serve as a check and backup for the telemetry attitude data. If there are wrong with the telemetry attitude data, the system switches to the post-processing attitude for production. This star tracker is characterized by its high precision, high stability, high update rate, and high reliability.

The Gaofen-7 satellite was launched in 2019 and has been in operation for nearly four years. APS-01 onboard has consistently outputted stable quaternions throughout its operational period. It has two FOV designated as "a" head and "b" head, with identical parameters for both heads as outlined in Table 4.

Table 4. Parameters of APS-01

parameters	specifications
Focal length	13164.646770 pixels
Principle point	(1024, 1024) pixel
APS size	2048 × 2048 pixels
FOV	8.9° × 8.9°
Pixel Size	15 μm × 15 μm
Rate	1Hz(default)

The APS-01 has a detection limit magnitude of 8.5. The number of navigation stars was analyzed using a uniformized Hipparcos star catalog, as shown in Fig. 3.

As can be seen from the figure, when the limit magnitude is 8.5, the number of navigation stars in the entire sky is 60,298, and approximately 90 navigation stars will appear in a single FOV. It will be difficult to identify star features using traditional star map recognition algorithms. It can be concluded from Table 1 that its optimal magnitude upper limit should be 6.9. By matching and analyzing the stars in the GF-7 multi-orbit image, an exponential function is used to fit the relationship between the discrete pixel brightness sum of the star and apparent magnitude. The equation is:

$$y = 2.301 \times e^{-0.000478x} + 5.357 \times e^{-1.879 \times 10^{-5}x} \tag{4}$$

In this experiment, the magnitude fitting error is 0.4Mv, and the APS-01 original star image data (see Fig. 4) from the first frame of the 012326th orbit, transmitted on August 3, 2022, was used as an example.

Table 5. Result of star identification

		ID	Coordinate	Sum of grayscale	magnitude	Inter-star angle of navigation (Rad)	Inter-star angle of obversion (Rad)
Main star		21512	(1226.35583, 1228.25989)	1366	6.41		
Companion stars	1	21526	(1267.67664, 1342.14465)	968	6.71	0.025734	0.025727
	2	21532	(1365.33887, 1431.25562)	4745	5.13	0.035732	0.035718
	3	21536	(1494.36316, 1505.5675)	837	6.81	0.046711	0.046689
	4	21534	(1622.57239, 1540.58289)	3141	5.56	0.055896	0.055872
	5	21574	(930.68994, 1695.82239)	4944	5.10	0.045844	0.045844
	……						
Remaining stars		21502	(1227.04822, 1103.76843)	311	7.30	0.014863	0.014887
		21496	(1275.96265, 1043.80371)	107	7.53	0.019043	0.018984
		21501	(1394.12769, 1133.73755)	141	7.49	0.027665	0.027614
		21497	(1129.11145, 994.22913)	323	7.29	0.010880	0.010875
		21505	(1501.03284, 1212.85449)	213	7.41	0.036662	0.036625
		21499	(1050.32153, 996.18945)	227	7.39	0.007896	0.007919
		……					

A navigation star catalog was constructed for the APS-01 carried by Gaofen-7 to homogenize the stars recorded in the Hipparcos catalog. The navigation star catalog was constructed according to Table 2. The star image was denoised to extract star points and the sum of grayscale of defocused image pixels was calculated. Finally, the partial identification results were (Table 5):

3.2 Accuracy Verification

For the APS-01, it's "a" head and "b" head are fixedly mounted to each other, and their angular separation can be considered constant over a short period. The triaxial accuracy

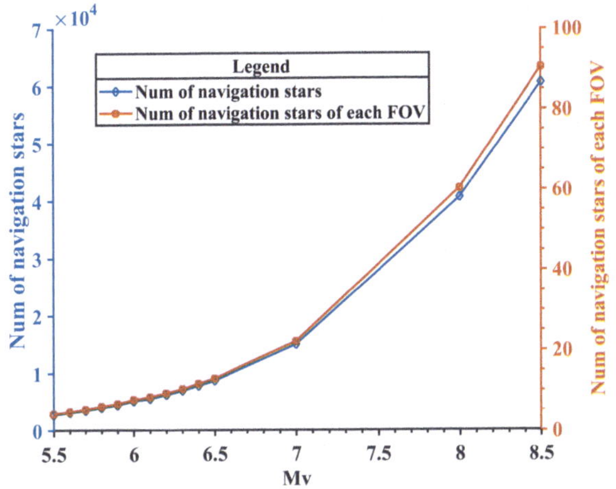

Fig. 3. The number of navigation stars in the entire celestial region and in a single FOV with magnitude varies

Fig. 4. A star image of 012326th orbit

of the attitude is highest with respect to the optical axis (see Eq. 5). Therefore, calculating the standard deviation of the angle between the two optical axes provides an intuitive analysis of its accuracy. After star identification, the QUEST algorithm [8] is used for attitude determination (see Table 6), resulting in the attitude quaternion q = [q0, q1, q2, q3]. By identifying the original star image data of the APS-01's "a" head and "b" head for the entire orbit, calculating the standard deviation of the angle between the two optical axes, and comparing it with the standard deviation calculated from the star's telemetry data, a comparative analysis can be performed.

Table 6. Attitude of "a" head and "b" head

TimeCode	Field of View	
	The quaternion of "a" head	The quaternion of "b" head
271227736	[0.63655, −0.02491, 0.29441, 0.71239]	[0.23884, −0.01841, 0.57174, 0.78467]
271227737	[0.63665, −0.02537, 0.29461, 0.71221]	[0.23905, −0.01867, 0.57207, 0.78437]
271227738	[0.63677, −0.02583, 0.29479 0.71200]	[0.23924, −0.01893, 0.57239, 0.78407]
271227739	[0.63689, −0.02630, 0.29498, 0.71179]	[0.23945, −0.0192, 0.57271, 0.78376]
271227740	[0.63701, −0.02676, 0.29517, 0.71159]	[0.23967, −0.01948, 0.57303, 0.78345]
271227741	[0.63714, −0.02723, 0.29536, 0.71138]	[0.23985, −0.01972, 0.57336, 0.78315]
……		

$$\mathrm{H} = \begin{bmatrix} 2(q_1q_3 + q_0q_2) \\ 2(q_2q_3 - q_0q_1) \\ 2\left(q_0^2 + q_3^2\right) - 1 \end{bmatrix} \tag{5}$$

To determine the optical axis $\mathrm{H_a}$ of the "a" head and $\mathrm{H_b}$ of the "b" head, the angle θ (in units of arcseconds) between these two vectors is calculated as Eq. 6:

$$\theta = \arccos\left(\frac{H_a \times H_b}{|H_a| \times |H_b|}\right) \times 3600 \tag{6}$$

Accuracy is determined by calculating the standard deviation of the angular separation between the optical axes (Fig. 5 and 6).

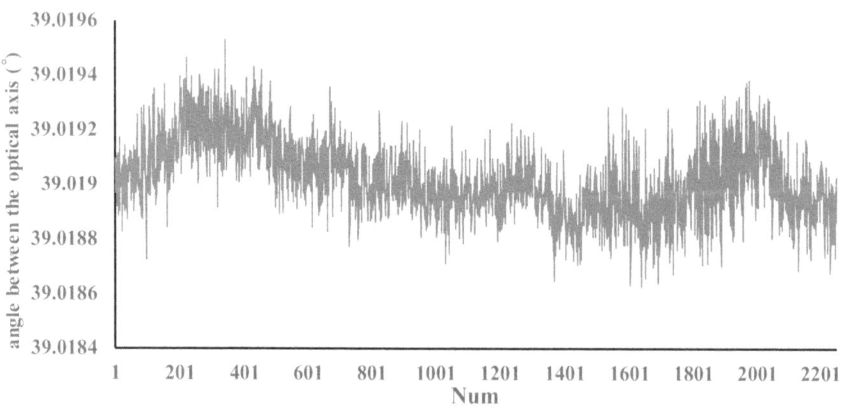

Fig. 5. Angle between optical axis calculated by telemetry data

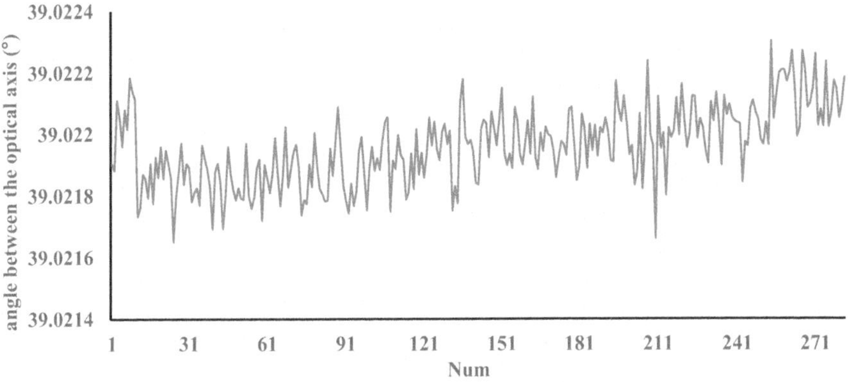

Fig. 6. Angle between optical axis calculated by the method proposed in this paper

The results show that the standard deviation of the optical axis angle for this algorithm is 0.49 arcseconds, which is higher than the 0.51 arcseconds obtained from the star's telemetry data. To enhance the credibility of the experimental results, multiple orbital experiments were conducted and calculate the star image identification rate, as shown in Table 7.

Table 7. Comparison of accuracy between telemetry data and the method proposed in this paper

Track	Num of image	Accuracy calculated by telemetry data	Accuracy calculated by method of this paper	Identification Rate
20220117-012326	197	0.5010687731	0.4500051975	99.49%
20220116-012310	203	0.5031774034	0.4527908929	99.01%
20220115-012281	147	0.3876431206	0.3322905406	100%
20220115-012281	138	0.3876431206	0.3322905406	100%
20211231-012067	105	0.3955927005	0.3946716428	100%
20211212-011778	160	0.4067259122	0.3900040379	99.37%
20211126-011535	188	0.573617884	0.5715670694	100%
20211016-010912	173	0.4692667027	0.4543230643	100%
20210920-010508	169	0.4443173985	0.4334284843	100%

It can be observed that problems occurred in the identification of some star images. Generally, the issue with problematic star images is primarily due to the failure in matching the main star. This can be resolved by adjusting the threshold values for inter-star angle and magnitude deviation. However, excessively high threshold values can affect the matching time, while too low values may lead to matching errors. The threshold needs to be adjusted based on experience. Analysis reveals that the method proposed in this paper slightly higher the accuracy of attitude telemetry data. The main reason for this difference is that the on-orbit method favors real-time and are limited by hardware [9]. Compared to ground-based method, the completeness of the on-orbit star catalog is somewhat lacking. Additionally, due to the need for real-time attitude determination, the usage of on-orbit computational resources must be kept within reasonable limits, typically resulting in the matching of 6–30 stars. In contrast, ground-based environments allow the use of a more comprehensive navigation star catalog and afford ample time for star identification, resulting in a higher number of correctly identified navigation stars than on-orbit. Nevertheless, since the number of navigation stars recognized on-orbit already far exceeds the requirements for attitude determination, simply increasing the number of identified stars does not significantly enhance attitude accuracy.

4 Conclusion

This study focuses on high limit magnitude star trackers and proposes an autonomous, all-sky star image identification algorithm that assists with star brightness. The method uses star magnitude as the basis for star selection and proceeds in two steps. Initially, it performs preliminary identification using the diagonal distance and magnitude information of main star and companion stars. Subsequently, based on the identified main star and companion stars, it further identifies remaining stars in the field of view. During the identification process, a circle model is constructed for detection, significantly enhancing the accuracy rate of the matching results. This method cleverly circumvents the issue of feature ambiguity caused by the increased limit magnitude of star trackers. Additionally, the attitude accuracy derived from the method is slightly higher than that obtained from attitude telemetry data. Notably, this study's use of the Hipparcos star catalog could be substituted with the Gaia catalog to further improve attitude accuracy.

Acknowledgments. This study was supported in part by National Natural Science Foundation of China (42371391, 42001416).

Disclosure of Interests. The authors have no competing interests to declare that are relevant to the content of this article.

References

1. Wang, Z.: Global field-of-view imaging model and parameter optimiza-tion for high dynamic star tracker. Opt. Express **26**, 33314–33332 (2018)
2. Mortari, D.: The pyramid star identification technique. J. Navig. **51**(3), 171–183 (2004)
3. Guangjun, Z.: Star map identification based on a modified triangle algo-rithm. Acta Aeronautica et Astronautica Sinica **27**(6), 1150–1154 (2006)
4. Jian, L.: Improved grid algorithm based on star pair pattern and two-dimensional angular distances for full-sky star identification. IEEE Access **8**, 1010–1020 (2019)
5. Mehta, D.S.: A rotation-invariant additive vector sequence based star pat-tern recognition. IEEE Trans. Aerospace Electron. Syst. **55**(2), 689–705 (2019)

6. Jiang, J.: Star identification based on spider-web image and hierarchical CNN. IEEE Trans. Aerospace Electron. Syst. **56**(4), 3055–3062 (2019)
7. Shaofei, Y.L.: Robust and Efficient star identification technique based on 1D convolutional neural network. IEEE Trans. Aerospace Electron. Syst. **58**(5) (2022)
8. Shuster, M.D.: In quest of better attitudes. In: 11th Annual AAS/AIAA Space Flight Mechanics Meeting, pp. 2089–2117. Santa Barbara, CA (2001)
9. Cheng, H.: A fast algorithm of all-sky star map automated identification. Aerospace Control Appl. **43**(4) (2017)

SBRDF Hyperspectral Ray Tracing Rendering for 3D Representation Based on Target Surface Material

Jia Yi[1,3], Huilin Jiang[4,5,6](✉), Xiaoyong Wang[3], and Du Ying[2]

[1] Graduate School, Changchun University of Science and Technology, Changchun 130022, China

[2] School of Physics, Changchun University of Science and Technology, Changchun 130022, China

[3] Beijing Institute of Space Mechanics and Electricity, Beijing 100190, China

[4] National and Local Joint Engineering Research Center of Space Optoelectronics Technology, Changchun University of Science and Technology, Changchun 130022, China

15754350608@163.com

[5] Fundammontal Science On Space-Ground Laser CommunicationTechnology Laboratory, Changchun University of Science and Technology, Changchun 130022, China

[6] Key Laboratory of Education Ministry Optoelectronics Measurement and Control and Optical, Shapingba, China

Abstract. Spectral technology is one of the key means for long-distance non-contact detection and identification of materials. It is an important force for ground-based detection satellites which can play a role in monitoring satellite operation status and safety, and identifying satellite types except for radar. However, for complex targets detected at ultra long distances, the detection object is equivalent to a point source, and the distribution of surface materials is not clear. Usually, material spectral unmixing can only focus on the distribution of planar materials, making it difficult to analyze the target structure. This article uses the spectral BRDF function to construct a hyperspectral ray tracing rendering 3D representation method based on the relationship between illumination, target surface normals, and detection points. It obtains a 3D rendering of a single material on a composite target, laying the foundation for analyzing the structural features of long-distance point target detection.

Keywords: SBRDF · hyperspectral ray · material spectral unmixing · tracing rendering 3D representation

1 Introduction

Spectral technology is one of the key means for long-distance non-contact detection and identification of materials, and the spectral bidirectional reflection distribution function is also a recent development focus in this field [1]. With the development of computer graphics, the spectral bidirectional reflectance distribution function also plays an important role in the field of global illumination research. The field of global illumination

H. P. Urbach et al. (Eds.): ISSOIA 2023, SPP 191, pp. 114–122, 2024.
https://doi.org/10.1007/978-981-97-6718-2_11

research in graphics involves complex optical processes, involving physical factors such as the type, intensity, direction of illumination, and the refraction and reflection relationships between objects [2]. In the real world, the scene seen by the human eye is caused by the scattering or radiation of light from different structural surfaces of objects, therefore, people have thc mapping and perception of colors and structures; In the world of computer graphics, computer simulated scenes are constructed using both mathematical and physical knowledge. How to truly simulate the colorful world seen by the human eye in a computer, and how to make the scene simulated more realistic in a computer, requires a combination of physical knowledge and computer algorithm models and other interdisciplinary knowledge, such as lighting models [3] and rendering rendering techniques [4].

2 Theory of Rendering Technology

2.1 SBRDF Hyperspectral Theory

In the visible light band, the interaction between scene and light is essentially the absorption and scattering process of light generated by photons of different wavelengths colliding with each solid surface in the scene, as shown in Fig. 1. Assuming a beam of photons is directed towards a flat plane in the scene, the micro elements of that plane are dA, select the corresponding coordinate system and its spatial relationship is as follows: $\vec{l}$ is the direction perpendicular to the plane and also the normal direction, $\vec{l} = (cos\theta_1, \sin\theta_1\cos\varphi_1, \sin\theta_1\sin\varphi_1)$, $\vec{n}$ represents the normal perpendicular to the horizontal direction, which is also the z direction, $\vec{n} = (0, 0, 1)$, $\vec{s}$ is the radiation direction of vertical reflection, assuming its radiation follows the Oxy plane and the z-direction is 0, $\vec{s} = (\cos\theta_s, \sin\theta_s, 0)$, The direction of the signal received by the $\vec{\zeta}$ sensor, assuming that the angle between the sensor reception and the normal direction is ψ, then $\vec{o} = (\cos\theta_0, sin\theta_0\cos\psi, \sin\theta_0\sin\psi)$。

Assuming that the sunlight is projected along the $\vec{s}$ direction and its irradiance is E_s, then the irradiance of the plane in the $\vec{l}$ direction is E_+, then

$$\frac{E_+}{E_s} = \frac{(\vec{s}\cdot\vec{l})}{(\vec{s}\cdot\vec{n})} = I_+,\ E_+ = I_+E_s \tag{1}$$

$$\vec{s}\cdot\vec{l} = \cos\theta_s\cos\theta_1 + \sin\theta_s\sin\theta_1\cos\varphi,\ \vec{s}\cdot\vec{n} = \cos\theta_s \tag{2}$$

$$\text{then } I_+ = \cos\theta_1(1 + \tan\theta_s\tan\theta_1\cos\varphi) \tag{3}$$

Assuming that the sunlight is projected along the $\vec{o}$ direction and its irradiance is E_o, then the irradiance of the plane in the $\vec{l}$ direction is E_-, then.

$$\frac{E_-}{E_o} = \frac{(\vec{o}\cdot\vec{l})}{(\vec{o}\cdot\vec{n})} = I_-,\ E_- = I_-E_o \tag{4}$$

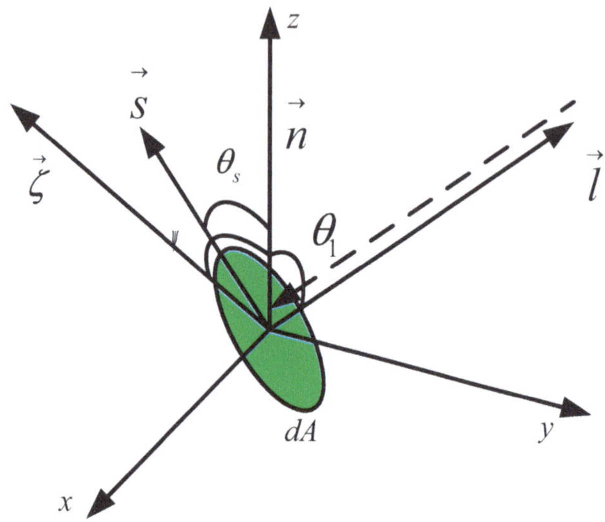

Fig. 1. Schematic diagram of the relationship between a single plane and lighting in the scene

$$\vec{o} \cdot \vec{l} = \cos\theta_o \cos\theta_1 + \sin\theta_o \sin\theta_1 \cos(\varphi - \psi),\ \vec{o} \cdot \vec{n} = \cos\theta_o \tag{5}$$

$$I_- = \cos\theta_1[1 + \tan\theta_o \tan\theta_1 \cos(\varphi - \psi)] \tag{6}$$

For I_+, when $\theta_s + \theta_1 < \frac{\pi}{2}$, then $\sin\theta_1 < \cos\theta_s$, $\sin\theta_s < \cos\theta_1$,

$$\tan\theta_s \tan\theta_1 = \frac{\sin\theta_s}{\cos\theta_s} \cdot \frac{\sin\theta_1}{\cos\theta_1} \tag{7}$$

then $\tan\theta_s \tan\theta_1 < 1$, but $\cos\varphi$ value in [−1,1], so $1 + \tan\theta_s \tan\theta_1 \cos\varphi > 0$, and $0° \leq \theta_1 < 90°$, then $\cos\theta_1 > 0$, $I_+ > 0$, at this point, it indicates that the solar radiation falls on the plane.

When $\theta_s + \theta_1 = \frac{\pi}{2}$, then $\tan\theta_s \tan\theta_1 = 1$,when $\varphi = \pi$, $\cos\varphi = -1$, so $1 + \tan\theta_s \tan\theta_1 \cos\varphi = 0$ then $I_+ = 0$, at this point, it indicates that the direct sunlight is parallel to the illuminated surface.

When $\theta_s + \theta_1 > \frac{\pi}{2}$, $\tan\theta_s \tan\theta_1 > 1$, let $1 + \tan\theta_s \tan\theta_1 \cos\varphi = 0$, Assuming acritical angle is α,

$$\alpha = \cos\varphi = \frac{-1}{\tan\theta_s \tan\theta_1} \tag{8}$$

When $\varphi = \alpha$, the direct sunlight rays on the surface are parallel to the plane; when $\varphi < \alpha$, The sunlight falls on a flat surface, when $\alpha < \varphi \leq \pi$ The sunlight falls on the back of the plane.

The same principle can also be applied to I_-, when $\theta_o + \theta_1 < \frac{\pi}{2}$,the sensor always observes the upper surface of the plane, when $\theta_o + \theta_1 = \frac{\pi}{2}$, and $\varphi = \pi$, The sensor cannot observe the plane, and when other values of φ are taken, the sensor can observe the upper surface of the plane, when $\theta_o + \theta_1 > \frac{\pi}{2}$, and $\varphi = \alpha$ the sensor cannot observe the plane,$\alpha < \varphi \leq \pi - \varphi$, The sensor observes the opposite side of the plane, and when φ takes other values, the sensor observes the upper surface of the plane.

The above observation is for a single wavelength, usually the response wavelength of the human eye to the visible wavelength is 380nm-780nm. What the human eye sees is the superposition of this wavelength, and the visual color is represented by the tristimulus value, which is the number of red, green, and blue primary colors corresponding to the isokinetic spectral color of each wavelength in the spectrum, called the spectral tristimulus value, denoted as $r(\lambda)$, $g(\lambda)$, $b(\lambda)$.Therefore, the color equation for equipotential spectral color λ with a matching wavelength of $C(\lambda)$ is:

$$C(\lambda) = \overline{r}(\lambda) + \overline{g}(\lambda) + \overline{b}(\lambda) \tag{9}$$

Correspondingly, the column vector V in the model of array detection sensors can be represented as [5]:

$$V = \begin{bmatrix} \overline{r} \\ \overline{g} \\ \overline{b} \end{bmatrix} = \int_{\lambda_{\min}}^{\lambda_{\max}} \rho(\lambda)D(\lambda)\text{o}(\lambda)r_{BRDF}(\lambda)d\lambda + \varepsilon \tag{10}$$

In the formula, $\rho(\lambda)$ is the spectral power distribution of the lighting source, and $r_{BRDF}(\lambda)$ is the reflectance of the spectral bidirectional reflection distribution function of the imaged scene in the detection direction,$\text{o}(\lambda)$ is the spectral transmittance of the array detection sensor filter,$D(\lambda)$ is the spectral sensitivity function of the device, and ε is noise.

2.2 Ray Tracing Optimization Algorithm

Ray tracing algorithm is one of the most commonly used algorithms in realistic graphics, which is used to simulate each ray and produce effects such as reflection, soft shadows, and indirect lighting. In an ideal situation, a diffuse material will reflect the light it receives from various angles, and the reflection effect of objects made of different materials varies greatly under the influence of material characteristics; The diffuse reflection material has a uniform and consistent reflection effect; The reflection of metal materials is manifested as a hard specular reflection; The transparent glass material moves towards the extreme on the other side, presenting as light refraction. In order to achieve better rendering results for objects with different material properties and optimize the realism of ray tracing, the contribution factor of objects to indirect light is calculated through a lighting model to improve the realism of rendering. The formula for optimizing the lighting model is as follows:

$$I = \sum_{i=1}^{M} \sum_{\lambda=380nm}^{780nm} I_{light(i)}(\lambda) + \left(r_{reflex} \cdot I_s \cdot k_s + r_{refraction} \cdot I_s \cdot k_s\right)_{indirect-light} \tag{11}$$

In the above equation, I_{light} is established based on the Blinn phong model [6], calculate the contribution value of direct light (number M) that can be counted, with the latter two being the contribution of indirect light reflection and refraction.

3 Implementation of SBRDF Hyperspectral Ray Tracing and Rendering

3.1 Experimental Equipment

Scattered light from asymmetric materials irradiated at any angle will exhibit polarization characteristics, and physical properties such as material thickness and surface smoothness correspond to different polarization information of scattered light. BRDF is a physical quantity used to represent the surface reflection characteristics of objects and is the basis for studying the scattering characteristics of test objects. The advantage of BRDF and spectral BRDF lies in their ability to provide detailed surface reflection characteristics information, including scattered light information at different wavelengths. By measuring and analyzing BRDF and spectral BRDF, it is possible to more accurately understand the optical properties of the target surface, thereby achieving analysis and identification of the target structure. This method is of great significance in long-distance detection and identification of complex targets, which can help improve the accuracy and reliability of detection.

Therefore, the advantage of BRDF and spectral BRDF lies in their provision of more comprehensive and detailed surface reflection characteristics information, which helps to achieve accurate analysis and identification of target structures. This is of great significance for long-distance detection and monitoring of targets, as it can improve the efficiency and accuracy of detection. The following is a schematic diagram of polarization information detection from different angles.

3.2 Experimental Results

Through experiments, the incident zenith angles of the gold coating were measured to be 20°, 36°, 44°, and 60°, the incident and reflection azimuth angles were 180°, and the reflection zenith angle was 0–80°. The data was sampled at intervals of 2°, with a wavelength range of 380 nm–780 nm.

Using the measured spectral data of golden insulation film, white paint coating, and silicon based solar cell panels, the overall surface scattering spectrum of the object is rendered, and the geometric model of the object is established and surface mesh is divided; Extract the material information of grid cells and calculate the normal angle of grid cells; Grid area calculation and coordinate system conversion; Calculate the BRDF of each grid by weighted superposition. If the geometric dimensions and surface material type of the object are set, as shown in Table 1 (Table 2).

Using 3D software to mesh the target object, the accuracy of the model mesh division is shown in Fig. 2. It can be seen that the accuracy analysis error varies greatly in the short wavelength range. Therefore, the smaller the mesh size, the higher the accuracy, but the computational complexity is also greatly improved. Grid size selection: around 100000 times the minimum wavelength (Fig. 3 and 4).

Based on the additivity of the spectrum [1], the BRDF of the entire object surface is calculated using the BRDF of each grid unit, and the image rendering is shown in Fig. 5.

In terms of image realism, ray tracing models are based on ideal models that simulate ray paths. Due to the energy loss caused by light refraction in real scenes, there

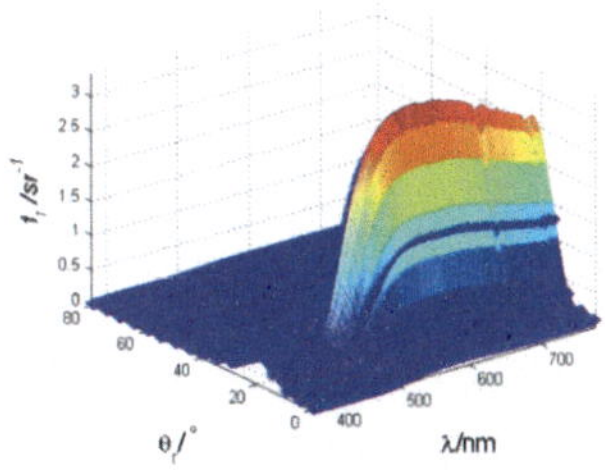

(a) The incident zenith angle is 20 °

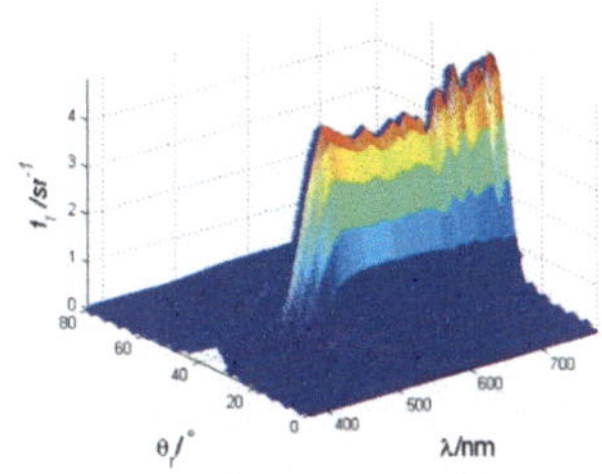

(b) The incident zenith angle is 36°

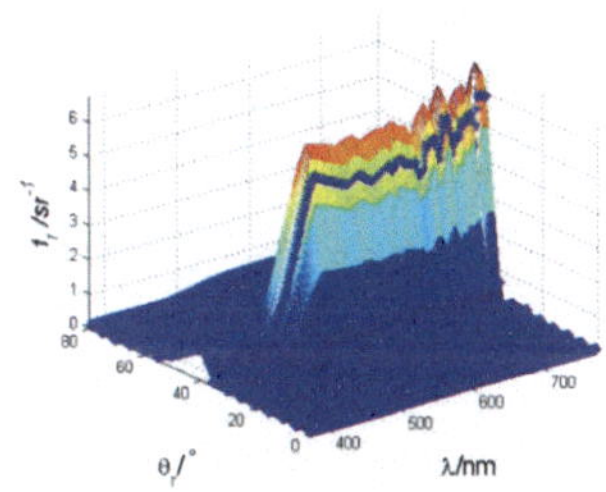

(c) The incident zenith angle is 44°

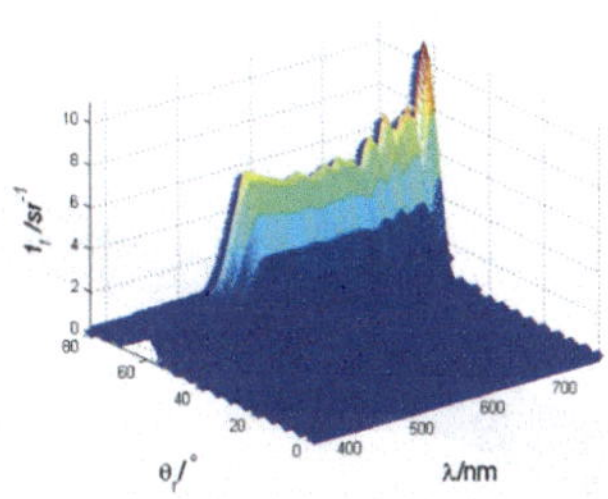

(d) The incident zenith angle is 60°

Fig. 2. BRDF measurement values of golden insulation film with a relative azimuth angle of 180° for incident and reflection and a change in zenith angle

is still a certain gap from fully depicting 3D scenes. If we want to further simulate the transmission of light, we also need to consider the content of radiation measurement.

Table 1. Geometric dimensions and surface material types of object models

	Length, Width, and Height (m)	Surface material of length
Satellite body	1 * 1 * 1	Insulation film
Sailboard	3 * 0.05 * 1	Solar cell
Antenna	1 * 0.05 * 1	White paint

Table 2. Grid partitioning parameters

Name	Content
Grid type	Square grid
Grid size	0.05 m * 0.05 m
Number of grids	6864

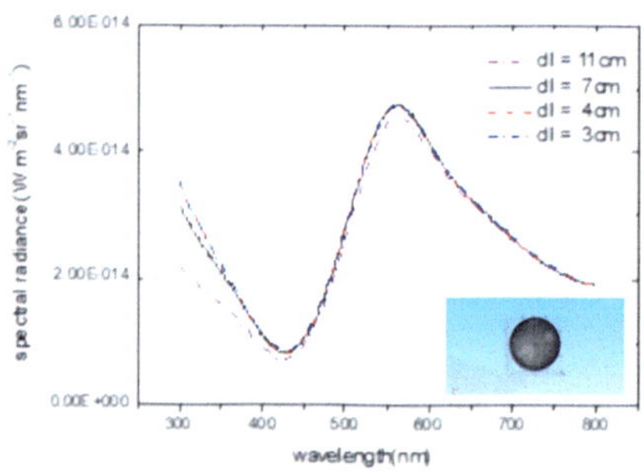

Fig. 3. Precision analysis of grid partitioning

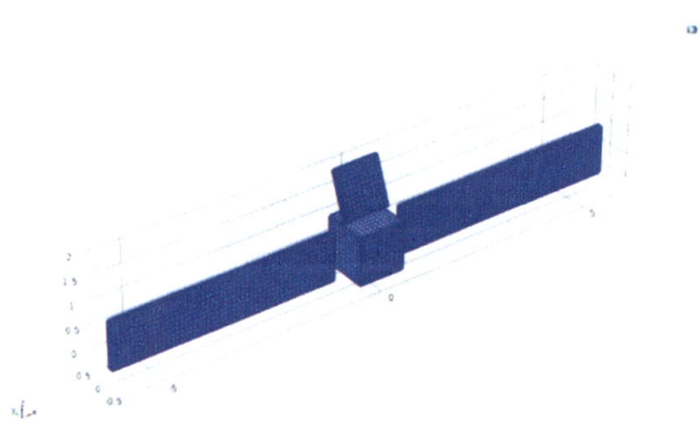

Fig. 4. Grid division diagram of the target object

Radiation measurement method is a method used to measure the surface reflection characteristics of objects, usually used to study the optical properties of optical materials. By using radiation measurement, optical parameters such as reflectivity, transmittance,

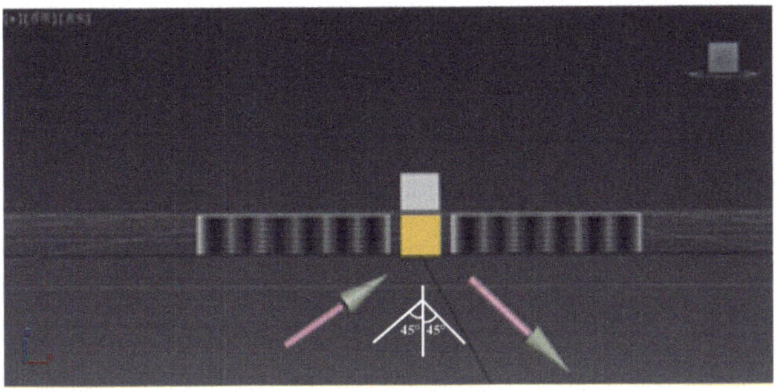

Fig. 5. Rendered image with an incident zenith angle and an outgoing zenith angle of 45°

and absorption of objects at different wavelengths or angles can be obtained, thereby understanding the interaction between objects and light. Radiation measurement methods typically involve using a light source to illuminate the surface of an object, and then measuring optical parameters such as reflected and transmitted light. Common radiation measurement methods include reflectance measurement, transmittance measurement, absorption measurement, etc. These measurement results can be used to analyze the optical properties of materials and understand their absorption, reflection, and transmission of light of different wavelengths. To verify its effectiveness, Fig. 6 shows multi angle rendered images using different insulation films.

Fig. 6. Rendering images of different gold insulating films

4 Conclusion

At present, physical rendering based on ray tracing is considered the most effective method to improve visual experience. However, with the development of algorithm research and hardware technology, it has reached a turning point from traditional raster rendering methods to real-time ray tracing on consumer platforms. SBRDF hyperspectral ray tracing rendering based on target surface material is a key method for physical rendering. This article constructs a hyperspectral ray tracing rendering 3D representation method based on the relationship between illumination, target surface normals, and detection points, which will lay the foundation for analyzing the structural features of long-distance point target detection.

References

1. Song, W., et al.: Research on space debris classification based on scattering spectra, spectroscopy and spectral analysis **35**(6), 1464–1468 (2015)
2. Philip, D., Philippe, B., Kavita, B.: Advanced Global Illumination, Second Edition. CRC Press 04–08 (2016)
3. Li, Z.: Research and Implementation of Realistic Illumination Technology Chengdu. University of Electronic Science and Technology (2011)
4. Kajiya, T.: The rendering equation. ACM **20**(4), 143–150 (1986)
5. Liang, J.X., Wan, X.: Spectral reconstruction from single RGB image of trichromatic digital camera. ACTA Opt. Sin. **37**(9), 370–377 (2017)
6. Blinn J F. Models of light reflection for computer synthesized pictures. Seminal Graph. ACM, 192–198 (1977)

Typical Failure Cases and Analysis of COTS Polarization Maintaining Fiber Couplers Used in Spacecraft

Jia Qiuyang(✉), Ding Xiaoyan, Li Jiang, Wang Yue, and Chen Yan

China Aerospace Components Engineering Center, China Academy of Space Technology, Beijing, China
jiaqy091@163.com

Abstract. Polarization maintaining fiber coupler is a special type of fiber coupler with low additional loss, low polarization crosstalk, and the ability to maintain the polarization state of linearly polarized light in transmission. They are mainly used in optical communication equipment or fiber optic gyroscopes that require high precision and performance. With the increasing emphasis on cost control in the aerospace industry, COTS polarization maintaining fiber couplers have also been widely used in the aerospace field. This article lists two typical failure cases of COTS polarization maintaining fiber couplers during the evaluation and verification process. After analysis, it is believed that both issues are related to the materials and processes used in polarization maintaining fiber couplers. The COTS polarization maintaining fiber coupler is not designed and manufactured for space applications. In order to improve performance and reduce costs, the materials and processes used are different from similar products used in high reliability fields. However, due to its low cost and good performance parameters, spacecraft still choose a portion of COTS polarization maintaining fiber couplers. This article analyzes the mechanism for the failure of COTS polarization maintaining fiber couplers, and proposes suggestions for selecting COTS polarization maintaining fiber couplers for future spacecraft based on aerospace application requirements.

Keywords: Polarization Maintaining Fiber Couplers · COTS components · Failure Analysis

1 Working Principle, Structure and Parameters of Optical Fiber Coupler

1.1 Working Principle of Fiber Optic Couplers

Fiber optic couplers are passive devices that divide light waves from one fiber into multiple fibers or combine multiple fibers into one fiber. They are widely used in fields such as fiber optic communication and fiber optic sensing [1–4]. There are three manufacturing methods [4–6]: melt taper, micro-optical method, and optical waveguide method. For those with fewer channels (channel number $\leq$ 4), the melt taper method is generally

H. P. Urbach et al. (Eds.): ISSOIA 2023, SPP 191, pp. 123–132, 2024.
https://doi.org/10.1007/978-981-97-6718-2_12

used for manufacturing; Large channel numbers (channel numbers $\geq$ 8) are generally manufactured using the optical waveguide method. The melt taper method is to burn and stretch multiple fiber optic with coating removed together to bring the fiber cores together, based on the principle of mutual coupling of evanescent fields between single-mode optical fibers, in order to achieve optical wave coupling. There are various methods of melting, such as flame heating [7], electric heating [8], laser heating [9, 10], etc. Currently, the commonly used method is hydrogen oxygen flame heating, and the process parameters are controlled by the mechanical parameters (i.e. process parameters) of the manufacturing equipment, including flame width and height, hydrogen flow rate, stretching speed, and heating time. The stopping condition of the cone pulling is based on the preset coupling ratio.

To manufacture 2 $\times$ 2 optical fiber couplers, the coating layer is first peeled off from two optical fibers, and then they are leaned against each other. Then, they are heated and stretched towards both sides to evenly melt and form a double cone within a certain length distance. In the coupling zone, the diameter of the fiber core decreases, and the normalized frequency decreases. The smaller the value, the larger the mode field diameter, that is, the more part of the mode field diameter exceeds the fiber diameter. In this way, more parts of a mode propagate in the cladding part of the coupling zone and are then coupled into the core of another fiber.

1.2 Structure and Main Parameters of Polarization Maintaining Fiber Coupler

The product involved in the case in this article is a 2 $\times$ 2 polarization maintaining fiber coupler suitable for 1550 nm optical wavelength, which has low additional loss, return loss, and extinction ratio. The main performance indicators of this product are shown in Table 1.

Table 1. Main Parameter Indicators of Polarization Maintaining Fiber Coupler

Serial number	parameter	unit	index
1	Additional loss	dB	$\leq$1
2	extinction ratio	dB	$\geq$16
3	Return loss	dB	$\geq$45
4	working temperature	°C	−5 −+ 70
5	Storage temperature	°C	−40 −+ 85

This product is packaged in a stainless steel shell, with both ends fixed with stainless steel buckles. The internal optical fibers are fixed in the glass shell using organic adhesive. The coupler is made using the melting cone pulling process, and the two optical fibers are fused and tapered by heating to achieve coupling of optical power. Then, the polarization maintaining optical fibers are separated into two channels, The tail is connected using four fiber optic connectors (see Fig. 1 for structural schematic and Fig. 2 for cross-sectional photos).

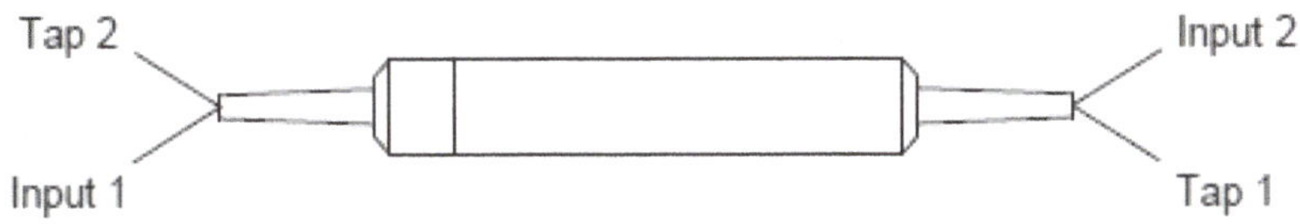

Fig. 1. Schematic diagram of fiber optic coupler structure

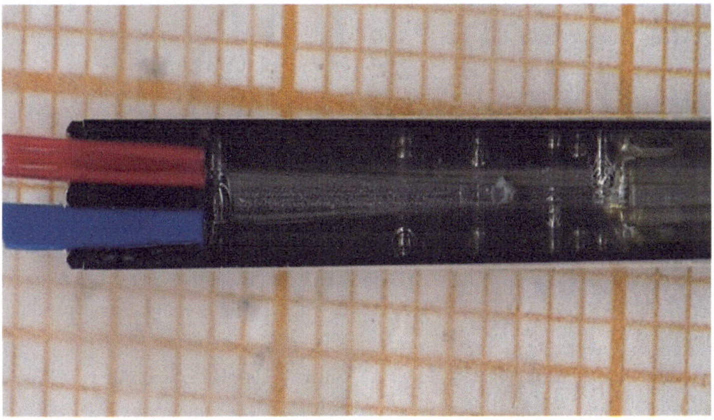

Fig. 2. Cross section photo of fiber optic coupler

2 Evaluation and Verification Status

To verify the aerospace applicability and quality reliability of this product, some products were selected for assessment testing. All samples conducting assessment tests are selected as qualified products. The setting of the assessment test items and conditions is mainly compatible with the mechanics, environment, life requirements of the spacecraft, and the parameter characteristics of the COTS coupler itself, referring to relevant military standards. The main test items and conditions are shown in Table 2.

Table 2. Assessment Test Items and Conditions

Group	Test	Condition
Group 1	High Temperature Storage Test	Temperature: 85 °C (± 2 °C) Humidity: <40% RH Test Duration: 2000 h
	Tensile load	Maximum load: 100N Load application acceleration rate: 2000N/min Maximum load holding time: 1 min

(*continued*)

Table 2. (*continued*)

Group	Test	Condition
	Input power	Power: 2W
Group 2	Temperature Cycling	Temperature: −40°C to 85°C (± 2 °C) Dwell Time at Extremes: ≥15 min Number of Cycles: 500
	Low Temperature Storage	Temperature: −40 °C (±5 °C) Humidity: Uncontrolled Test Duration: 1000 h
	Vibration	Vibration Type: Sinusoid Frequency: 10 to 2000 Hz Acceleration (or Amplitude): 20 G maximum acceleration Sweep Time: 10–2000, 10 Hz in 20 min Duration: Sweep cycle performed 12 times in each of the three mutually perpendicular directions (total of 36 times)
	Mechanical Shock	Shock Type: Half Sine Pulse Number of Shocks: 3 times per direction for 6 directions (on 3 axes), (18 shocks, total) Shock Level: 500 g Duration: 1 mS
Group 3	Fiber Integrity Straight Pull	Load: 450 g Test Duration: 1 min
Group 4	Fiber Integrity Sise Pull	Load: 230 g, 2 × 90° Test Duration: 5 s
	Fiber Integrity Twist	Load: 450 g, 0 to 90 to −90° Test Duration: 10 cycles

3 Test Results and Failure Analysis

3.1 No Signal on the Input1 End of Polarization Maintaining Fiber Coupler After Temperature Cycling

After undergoing 500 temperature cycling tests in group 2 in Table 2, a polarization maintaining fiber coupler product was tested at room temperature and found no signal on the Input1 end of the fiber coupler. It was also found that the black fiber protection sleeve of the fiber coupler was loose in the fiber connector and could be pulled out of the connector. The black protection sleeve was partially pulled out with residual adhesive, and the surface of the adhesive was smooth, which was a typical debonding phenomenon, The optical fiber in the connector did not break, and the morphology is shown in Fig. 3.

X-ray examination revealed that the optical fiber inside the black protective sleeve fractured at the coupler, as shown in Fig. 4.

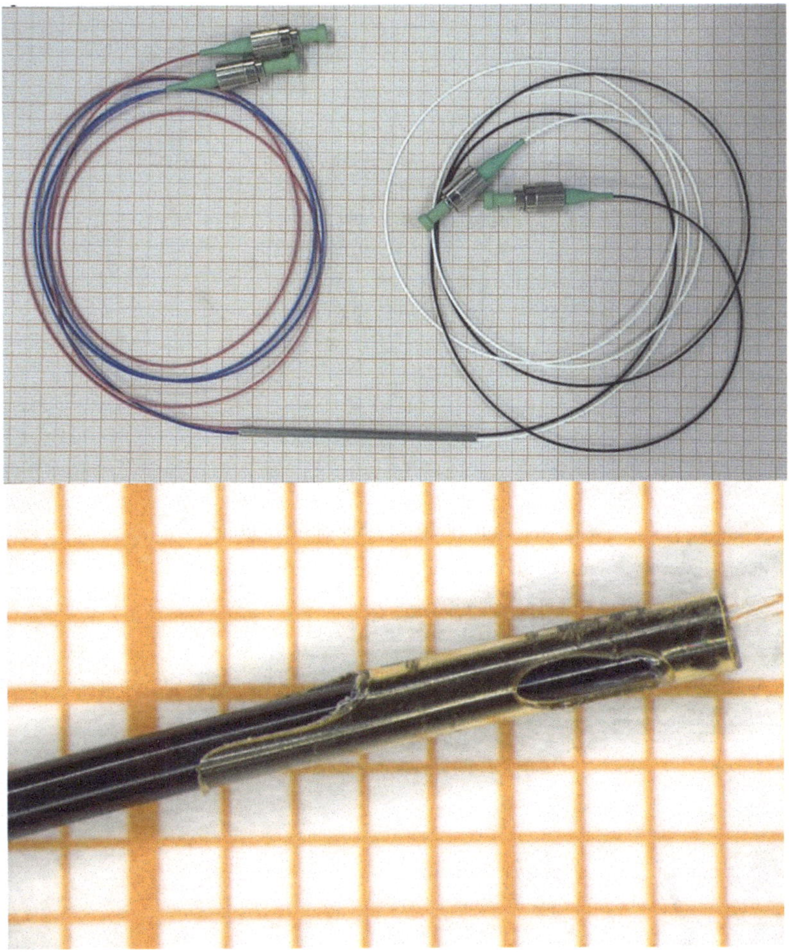

Fig. 3. Appearance of Failure Polarization Maintaining Fiber Coupler

Use mechanical method to unseal the fiber optic coupler and expose the fiber optic fracture. Upon inspection, it was found that the fiber optic fracture was located inside the coupler, approximately 4 mm away from the fiber optic fusion position. The inspection also found that the area approximately 5.5 mm away from the fiber optic fusion position was a bare fiber, with the coating layer removed. The other fiber at the input end and the two fibers at the output end of the coupler were all wrapped in the coating layer, without any bare fiber phenomenon. The morphology is shown in Fig. 5.

Scanning electron microscopy was used to examine the fracture surface of the optical fiber. Some areas were covered with organic adhesive, and cleavage patterns were visible on the fiber cross-section. The fracture surface exhibited a typical brittle fracture morphology, as shown in Fig. 6.

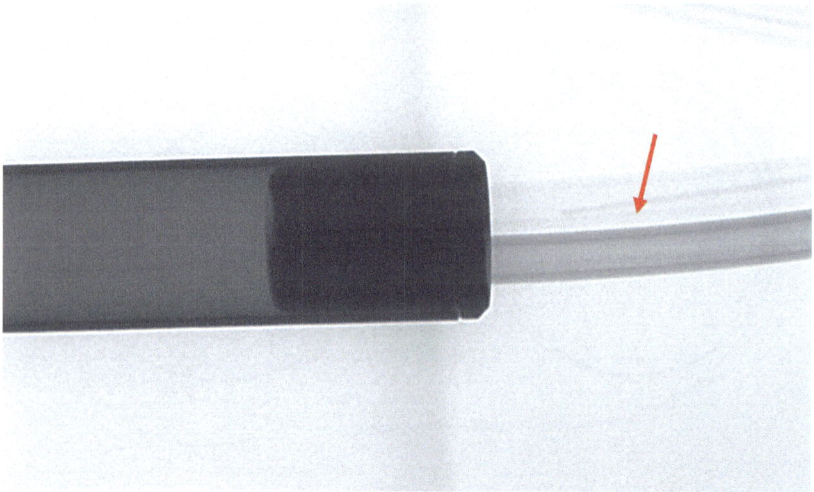

Fig. 4. X-ray photo of fiber optic fracture

Fig. 5. Morphology of Broken Fiber

Based on the inspection results of the failed polarization maintaining fiber coupler mentioned above, it can be concluded that the problem of no signal in the Input1 end test of the polarization maintaining fiber coupler after temperature cycling is due to the detachment of the protective sleeve of the black optical cable from the fiber connector. The loss of the protective sleeve of the fiber optic cable leads to a decrease in its load-bearing capacity and its fracture under external force. The fiber optic fracture is located

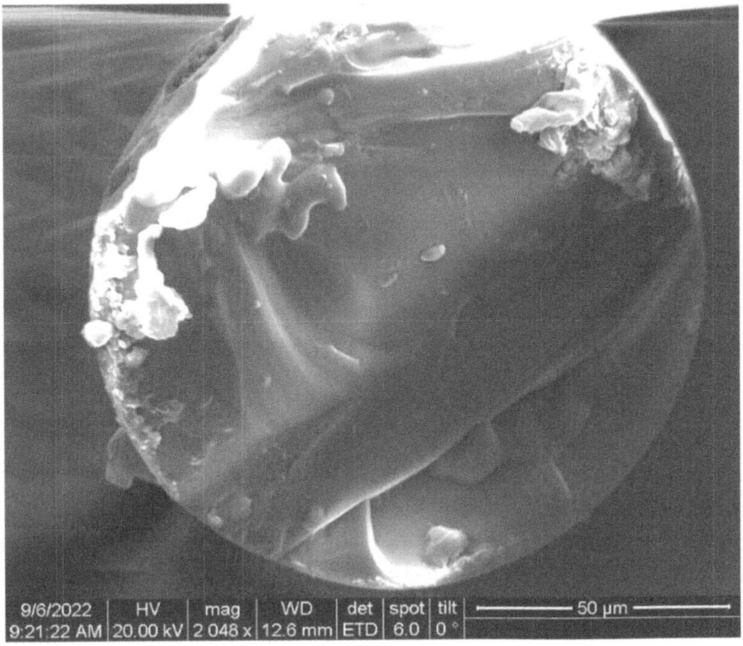

Fig. 6. SEM morphology of fiber optic fracture surface

in the bare fiber segment, where there is a lack of flexible coating protection, further reducing its load-bearing capacity. In addition, it is not ruled out that the fiber optic may be damaged during the manufacturing process of the coupler by peeling off the coating layer, resulting in the fiber optic breaking at the damaged point under normal stress.

3.2 The Leading-Out End of Polarization Maintaining Fiber Coupler is Loose After Fiber Integrity-Sise Pull

During the Fiber Integrity Sise Pull test of group 4 in Table 2, one polarization maintaining fiber coupler failed to achieve the predetermined load of 230 g. After removing the fiber optic coupler, an appearance inspection revealed that the blue end of the fiber optic coupler was loose and the red end tension curve was normal. Further inspection revealed that the blue protective sleeve could be pulled out of the fiber optic coupler for about 3 mm in length, and no obvious adhesive was found on the pulled out part of the blue protective sleeve, indicating a typical debonding phenomenon. X-ray inspection did not reveal any fracture of the optical fiber, as shown in Fig. 7.

Anatomy of fiber optic couplers using mechanical methods. Anatomical examination revealed that the blue protective sleeve had no adhesive force with the coupler sleeve and could easily be detached. There were no obvious adhesive marks on the surface of the detached part of the blue protective sleeve, and there were no obvious adhesive marks on the inner wall of the coupler sleeve. The red protective sleeve was firmly adhered to the sleeve. The internal optical fibers of the coupler and the optical fibers inside the blue protective sleeve are not broken, and the morphology is shown in Fig. 8 and Fig. 9.

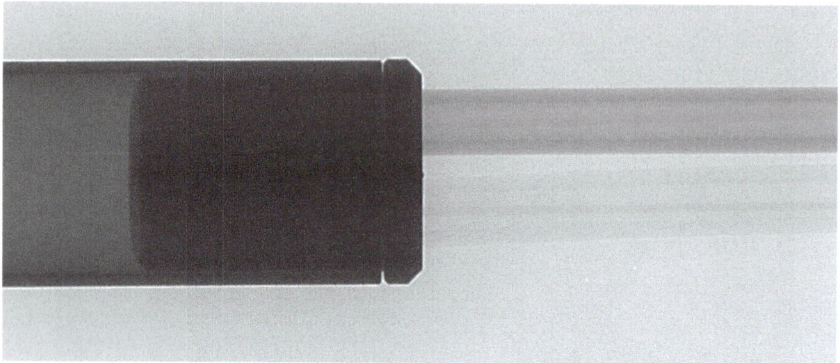

Fig. 7. X-ray inspection morphology

Based on the inspection results of the failed polarization maintaining fiber coupler mentioned above, it can be concluded that the problem of loose lead end after the Fiber Integrity-Sise Pull test of the polarization maintaining fiber coupler is due to a small amount of adhesive, and the blue protective sleeve and coupler sleeve did not form an effective bond, resulting in the blue optical cable becoming loose during the test and unable to reach the target load.

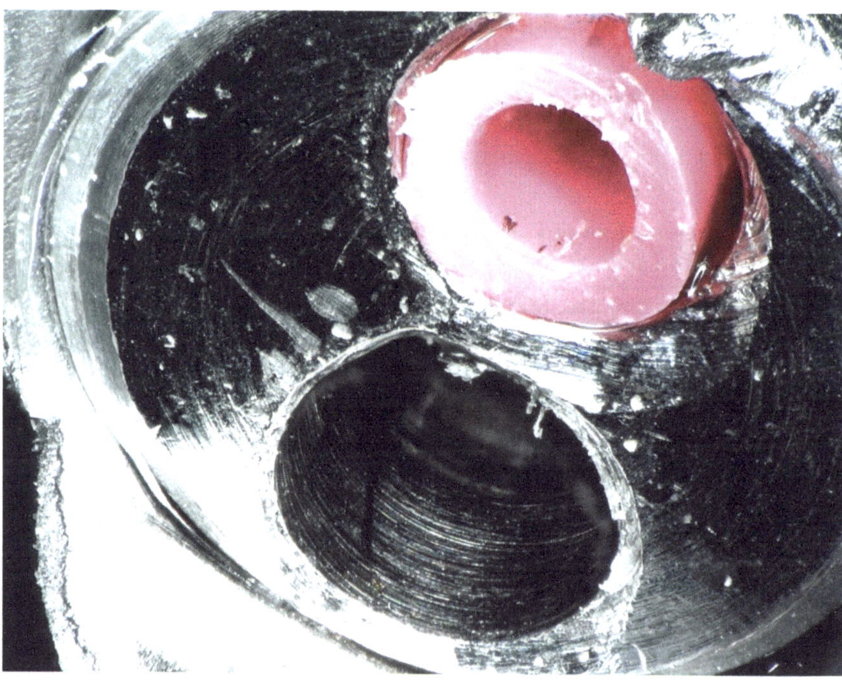

Fig. 8. Internal Morphology of Fiber Coupler Sleeve

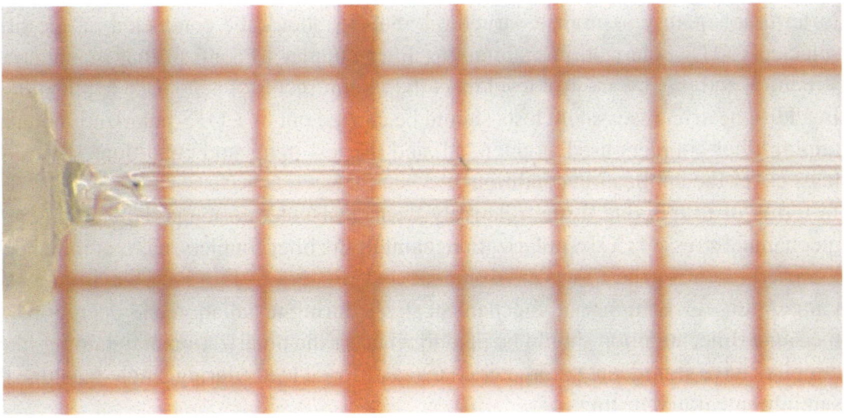

Fig. 9. Morphology of Optical Fibers in the Fiber Coupler

4 Suggestions

Based on the dissection of polarization maintaining fiber coupler samples, we found that the manufacturing method of this polarization maintaining fiber coupler is currently the most mainstream melt taper method. It appeared in 1985, using the melt taper process to bring two or more bare fibers (excluding the coating layer) together in a specific way, and achieve high-temperature melt stretching to achieve coupling of transmitted optical power within the fiber. Until the set splitting ratio is reached, heating stops, and the production of the fiber coupler is completed. The manufacturing of polarization maintaining fiber couplers using the melt taper method generally follows the steps of "alignment, melt taper, packaging, and testing", where the packaging process includes specific steps such as installing a quartz substrate, dispensing glue, installing a heat shrink sleeve, and sealing a stainless steel shell.

From the above failure cases, it can be found that the main cause of failure is a problem in the packaging process. In the packaging process, dispensing is the process of using epoxy resin adhesive to bond and fix the fiber coupling area with the quartz substrate. The quality of the dispensing process will have a significant direct impact on the performance indicators and future stability of the coupler. Due to considerations such as cost and process difficulty, the adhesive materials and bonding processes used in COTS polarization maintaining fiber couplers are often difficult to adapt to the wide temperature range and high mechanical requirements of aerospace applications.

In response to the above failure cases and the design and process factors behind them, the following suggestions are proposed for the subsequent application of COTS polarization maintaining fiber couplers in spacecraft:

1) In terms of selection, priority should be given to mature suppliers and their products, preferably polarization maintaining fiber couplers with certain flight experience, or products that can prove that their materials and processes can meet the temperature, mechanics, and irradiation environments of aerospace applications;

2) In terms of quality assurance, structural analysis should be conducted on the first selected COTS polarization maintaining fiber coupler to confirm that its material, structure, and process control level meet the requirements of aerospace applications; In addition, strict assessment tests should be carried out on COTS polarization maintaining fiber couplers used as intended, and the test items and conditions should at least cover the application conditions;
3) In terms of usage, due to the relatively weak resistance to temperature stress and mechanical stress of COTS polarization maintaining fiber couplers, it is recommended to strictly control the usage environment and process to minimize environmental temperature fluctuations and mechanical stress during installation and operation. At the same time, attention should be paid to whether the polarization maintaining fiber couplers have significant parameter changes or loose leads during single machine or satellite integrated testing.

References

1. Li, H., Lou, S.: Experimental study on the loss of fused tapered optical fiber coil. J. Optoelectron. Laser **21**(10), 1459–1461 (2010)
2. Ren, J., Hu, Y.: A unified model of fused single mode fiber coupler. Semicond. Optoelectron. **32**(1), 18–23 2011 (2011)
3. Pal, B.P., Chaudhuri, P.R., Shenoy, M.R.: Fabric and modeling of fused biconical tapered fiber couplers. Fiber Integr. Opt. **22**(2), 97–117 (2003)
4. Wang, J., Zheng, Y., Wu, Y., et al.: Coupling coefficient analysis of fused polarization maintaining fiber couplers. J. Natl. Univ. Defense Technol. **29**(2), 292–295 (2007)
5. Hsu, C.M., Su, C.T., Liao, D.: A new approach for optimizing the optical performance of the broadband tap coupler. Int. J. Syst. Sci. **34**(3), 215–220 (2003)
6. Wong, W.Y., Choy, L.K.: The manufacturing of an optical fiber coupler by the fusion elongation method. J. Mater. Process. Technol. **63**(1), 301–310 (1997)
7. Yokohama, I., Okamoto, K.: Fiber coupler fabric with automatic fusion elongation processes for low excess loss and high coupling ratio accuracy. J. Lightwave Technol. **5**(7), 910–915 (1987)
8. Takeuchi, Y., Noda, J.: Novell fiber coupling process using a microheater. Photonics Technol. Lett. IEEE **4**(5), 465–467 (1992)
9. Timothy, E.D., George, K., Timothy, A.B.: Carbon dioxide laser fabric of fused fiber couplers and tags. Appl. Opt. **33**(38), 6845–6848 (1999)
10. Bayle, F., Meunier, J.P.: Efficient fabric of fused fiber biological tap structures by a scanned CO2 laser beam technique. Appl. Opt. **44**(30), 6405–6411 (2005)
11. Li, H., Lou, S.: Experimental study on the relationship between profile and loss of fused tapered optical fiber coupler. Infrared Laser Eng. **38**(S11), 292–295 (2009)

MTF Automatic Measurement Method Based on Line Support Area

Zhang Bingxian(✉)

Beijing Institute of Space Mechanics and Electronics, Beijing 100094, China
409313966@qq.com

Abstract. The degradation function of satellite images is a function that changes with time and location, however, due to the limitation of ground facilities, the degradation function measured by commonly methods can only be a fixed value, which does not conform to the time-varying characteristics of the degradation function, therefore, we propose an adaptive MTF measurement method based on line support area. The method automatically identifies ground reference objects and calculate the MTF value in real-time, provides high-precision degradation functions for subsequent restoration. Firstly, we use gradient detection operators to extract points with high contrast, and calculate the gradient direction of these points as the phase direction to construct a "line support area", then we use optimized linear fitting method to fit the line support area and locate the line edge in order to avoid the phenomenon of edge points not being on a straight line due to grid display issues, secondly, considering the uniform gray distribution on both sides of the linear edge that meets the MTF calculation, we introduce the KURTOSIS value to determine the candidate area, and then calculate the MTF value from candidate area. Finally, we verifies the accuracy of the proposed algorithm by using the three-line array target of ZY3. it can be found that the measurement accuracy of the MTF curve obtained by the adaptive edge extraction method at the Nyquist frequency is 96.7%, while the measurement accuracy of the MTF curve obtained by manual methods at the Nyquist frequency is 62%. It can be seen that manually selected methods are prone to introducing noise, leading to a decrease in measurement accuracy, which is not conducive to the construction of degraded models, The MTF curve obtained through adaptive methods not only better suppresses the influence of random noise, but also has better measurement accuracy than the MTF curve obtained manually.

Keywords: MTF value · Line support area · KURTOSIS value

1 Overview

The information of an image mainly includes background information and edge information, while, edge information is defined as areas with significant changes. The edge target that need to be extracted in this article belong to step like edges. For step like edge detection, the following points should be achieved as much as possible [1]:

1) Accurately locate the position of sub-pixel edges;

H. P. Urbach et al. (Eds.): ISSOIA 2023, SPP 191, pp. 133–146, 2024.
https://doi.org/10.1007/978-981-97-6718-2_13

2) Avoid bilateral phenomena in the detection results;
3) Being able to identify edges with the same grayscale feature under different energies;
4) Being able to identify edges with different grayscale features that meet specific requirements under the same energy;
5) Effectively suppressing the impact of noise on edge detection;
6) It can effectively detect edge features with different edge slopes.

If the above requirements can be well met, it can be ensured that the detected edge lines can serve as candidate edge targets for computer automatic extraction. However, in reality, due to the wide variety and complex distribution of ground objects in satellite images, it is difficult to find a method that fully meets the above detection requirements. At present, general edge detection mainly includes the following steps: smoothing processing, image enhancement, edge detection, and edge localization. In general edge detection steps, although image filtering and image enhancement can improve the accuracy of edge detection to a certain extent, image filtering and image enhancement can increase the computational complexity of edge detection and reduce the efficiency of edge detection. Especially for large high-resolution satellite images (such as the Resource 3 satellite), the computational complexity leads to the most significant reduction in computational speed, This does not meet the time requirements of the ground preprocessing system, therefore it is not conducive to automatic MTF calculation.

2 MTF Calculation Method Based on Line Support Area

Based on the characteristics of MTF calculation, it is necessary to construct a linear feature extraction algorithm with low computational complexity, continuous detected edge points, and sub pixel level edge localization to meet the needs of production. This article analyzes the characteristics of existing algorithms, extracts points with high contrast based on gradient detection operators, and calculates the gradient direction of these points as the phase direction to construct a "line support area". The "line support area" can effectively avoid the phenomenon of broken lines in the image. By linear fitting of the line support area, the line where the edge is located can be well located, Avoiding the phenomenon of edge points not being in a straight line due to grid display issues. This article takes into account the fact that the grayscale distribution on both sides of the linear edge of the ground reference object that meets the MTF calculation should be relatively uniform, and introduces the KURTOSIS value to determine the candidate area of the ground reference object. Then, based on the grayscale distribution of the ground objects on both sides of the linear edge of the candidate area, it is ultimately determined whether it is a ground reference object that meets the MTF calculation requirements. The specific calculation process is shown in Fig. 1:

2.1 Detection of Pixel Grayscale Changes

Edge contour refers to the collection of areas in an image where all grayscale changes are the most severe [2]. Therefore, the grayscale change information of the image, i.e. gradient value, is a key factor in edge extraction, and its specific meaning is shown in the following figure:

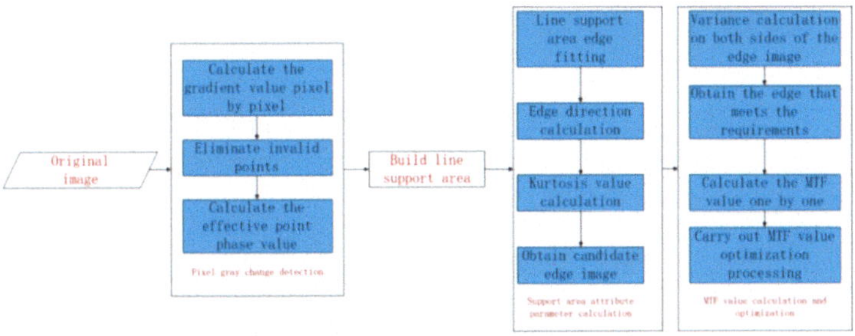

Fig. 1. The flowchart of adaptive MTF calculation

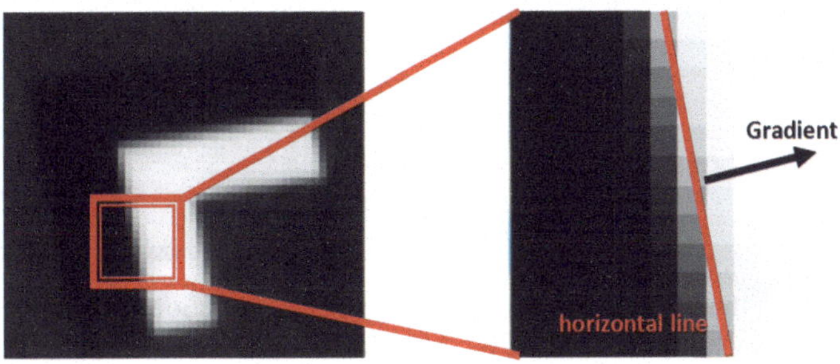

Fig. 2. Schematic diagram of image gradient changes and horizontal lines

The direction of the horizontal line determines the direction of the fastest gradient change. From Fig. 2, it can be seen that the direction of change of all edge pixels together constitutes the direction of the horizontal line. In other words, it can be considered that for pixels with the same horizontal line change direction, if their spatial distance is small, they should be pixels belonging to the same edge. Therefore, the direction of the horizontal line is an important factor in detecting edge straight lines, So it is necessary to extract the horizontal line before locating edges. The simplest horizontal direction is the gradient direction.

The function of pixel grayscale change detection is to calculate the gradient amplitude of each pixel in the image. By setting an appropriate threshold, values with smaller grayscale changes are eliminated, while values with larger grayscale changes are retained. Due to the requirement for grayscale differences on both sides of the edge during MTF calculation, this threshold is easy to set, and generally 30% to 50% of the radiation quantification index is chosen as the gradient threshold. Calculate the horizontal line direction of each point for the remaining points, laying the foundation for the construction of subsequent line support areas. The method for calculating gradients in this article is to use a 2 * 2 template, as shown in Fig. 3:

Where $i(x, y)$ represents the grayscale value of a pixel point with pixel coordinates (x, y), the gradient calculation formula for the image is:

$$g_x(x, y) = \frac{i(x+1, y) + i(x+1, y+1) - i(x, y) - i(x, y+1)}{2} \tag{1}$$

$\ddots$	$\vdots$	$\vdots$	$\cdots$
$\cdots$	$i(x, y)$	$i(x+1, y)$	$\cdots$
$\cdots$	$i(x, y+1)$	$i(x+1, y+1)$	$\cdots$
$\cdots$	$\vdots$	$\vdots$	$\ddots$

Fig. 3. Template of gradient calculation

$$g_y(x, y) = \frac{i(x, y+1) + i(x+1, y+1) - i(x, y) - i(x+1, y)}{2} \tag{2}$$

The calculation formula for the horizontal direction is:

$$angle = \arctan(\frac{g_x(x, y)}{-g_y(x, y)}) \tag{3}$$

The calculation formula for gradient amplitude is:

$$G(x, y) = \sqrt{g_x(x, y)^2 + g_y(x, y)^2} \tag{4}$$

After obtaining the gradient amplitude and horizontal direction of each pixel in the image, it is necessary to encode each pixel in the image. Based on the encoding, invalid points are removed and similar points are merged. The coding rules are as follows:

(1) For points with gradient amplitude less than 30% of the quantification index, assign a value of 0, otherwise assign a value of 1;
(2) Divide the flat space evenly into 8 parts, namely –180°~–135°, –135°~–90°, –90°~–45°, –45°–0°, 0°–45°, 45°–90°, 90°–135°, 135°–180°, and represent each direction range as 1, 2, 3, 4, 5, 6, 7, and 8. If the horizontal line of a pixel falls within the corresponding direction range, it is assigned a corresponding encoding for that range, and the encoding value is multiplied by the encoding result in step (1) to obtain a comprehensive encoding result.
(3) Calculate the angle formed between a point with an encoding greater than 1 and adjacent two points, and increase the encoding value by 0.5 for points with an angle equal to 180° to highlight their importance among similar points. The calculation method for the angle is shown in Fig. 4:

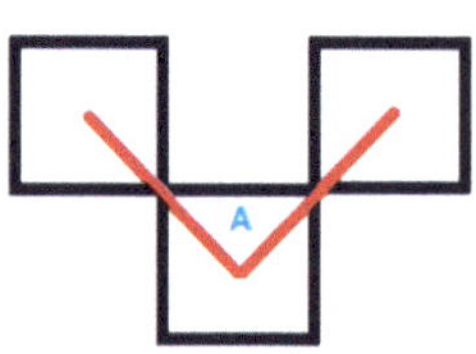

Fig. 4. Schematic diagram of pixel angle calculation

The black square represents a single pixel, the red line represents the line connecting the centers of two pixel points, and the angle A formed by the two red lines is the angle formed between adjacent pixel points. For a point with an angle equal to 180°, it forms the smoothest edge with the surrounding two points, making it the best edge point.

This article takes the artificially deployed target graphics in domestic high-resolution satellites as an example to detect and encode grayscale changes. The encoding results are shown in Fig. 5. In order to better display the encoding results, this article multiplies the encoding values by 25 and uses grayscale for representation.

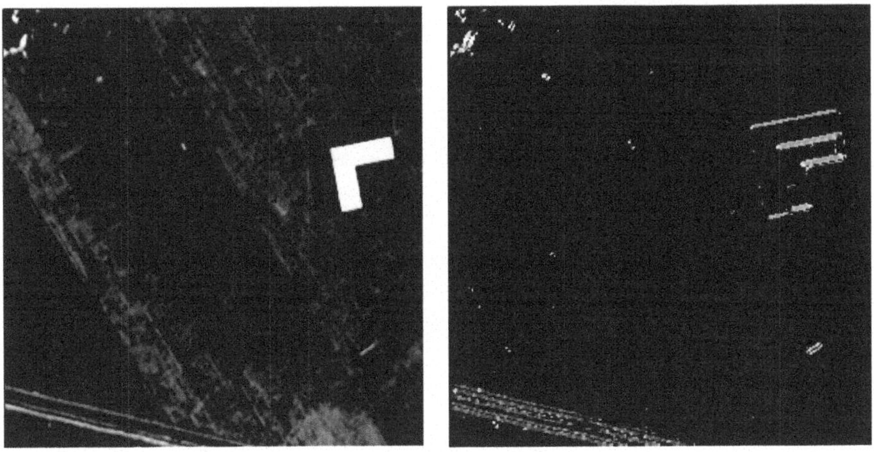

(a) original image (b) Graph of encoding result

Fig. 5. Schematic diagram of pixel angle calculation

2.2 Construction Line Support Area

The line support area refers to a collection of pixels with adjacent geometric positions. Due to the continuity of the edge of the object, it can be preliminarily considered that these points with encoding values less than 1 and gradient amplitudes greater than 30% of the maximum quantization belong to the same edge. The schematic diagram of the construction of the line support area is shown in Fig. 6:

As shown in Fig. 6, the colored regions represent different line support areas. It can be observed from the figure that pixels within the same line support area have similar horizontal line directions, which means they have similar encoding. As mentioned earlier, this article sets the threshold of this encoding to 1, which means that points with encoding less than 1 belong to the same type of line support area. However, this will cause a problem. Due to the increase in the encoding value of points with a pixel angle of 180° by 0.5 during encoding, So, points with codes 5.5 and 6.5 will be divided into the same line support area, causing errors in subsequent edge positioning. Therefore, it is necessary to optimize the line support area and eliminate coarse points. The specific processing method is as follows (Fig. 7):

The edge distribution of terrain features is continuous, as shown in Fig. 6. Without considering intersections, the encoding of most points in the edge area should tend

(a) original image (b) Horizontal Line Direction Map (c) Line support area

Fig. 6. Schematic diagram of line support area construction

Fig. 7. Schematic diagram of line support area construction

towards one direction. However, it is not ruled out that noise interference or calculation errors may lead to the introduction of pixels in another direction during the detection process of edge point grayscale changes. Considering that the number of such points is relatively small, a method of removing gross errors can be adopted to process them. The specific method is to use the least squares method to perform linear fitting on all pixel points, and the fitting formula is as follows:

$$\begin{aligned} M_1 &= N_1 * a + b \\ M_2 &= N_2 * a + b \\ &\cdots\cdots\cdots \\ M_k &= N_k * a + b \end{aligned} \tag{5}$$

M represents the pixel coordinates in the Y direction of the pixel, N represents the pixel coordinates in the X direction of the pixel, k represents the number of pixels in the line support area, and a and b represent the slope and intercept of the fitted line, respectively. Write the above formula in the form of a matrix as follows:

$$[M_1, M_2 ..., M_k] = \begin{bmatrix} N_1, 1 \\ N_2, 1 \\ .... \\ N_k, 1 \end{bmatrix} \begin{bmatrix} a \\ b \end{bmatrix} \tag{6}$$

That is:

$$Y = XA \tag{7}$$

where Y represents the matrix of the coordinate set in the Y direction of the pixel point, X represents the extended matrix of the coordinate set in the X direction of the pixel point, and A represents the parameter matrix of the linear fitting result, then the error equation is:

$$V = XA - Y \tag{8}$$

The corresponding normal equation is:

$$X^T PXA = X^T PY \tag{9}$$

The solution of the normal equation is:

$$A = (X^T PX)^{-1} X^T PY \tag{10}$$

Find the orthogonal difference between all points in the line support area and the solved line equation, and remove the points with differences exceeding the threshold. Assuming the line equation is: $ax + by + c = 0$, then the orthogonal difference between the pixel points and the solved line equation is:

$$d = \frac{|ax + b\tilde{y} + c|}{\sqrt{a^2 + b^2}} \tag{11}$$

Perform least squares linear fitting on the point set after removing gross errors, and then calculate the orthogonal difference between the remaining points and the new fitted line. Repeat this iteration until all pixel points have an orthogonal difference less than the threshold with the fitted line. The final fitted line obtained can be initially considered as an edge line, which is used to determine the edge image and therefore has low accuracy. Subsequent MTF calculations will introduce polynomial fitting to further improve the fitting degree of the edge line.

2.3 Calculation of Support Area Attribute Parameters

After edge linear positioning of the support area, it is necessary to calculate some attribute characteristics within the support area to determine whether there are edge images that comply with MTF calculation in the support area. In order to reduce computational complexity, preliminary screening is conducted on all support areas. The principle of screening is to use an expandable rectangular box to calculate the Kurtosis value and Skewness value of the image within the rectangular box, The calculation formula is as follows:

$$K = \frac{\sum_{i=1}^{k} (x_i - \overline{x})^4}{(n-1)sd^4} - 3 \tag{12}$$

$$SK = \frac{n\sum (x_i - \overline{x})^3}{(n-1)(n-2)sd^3} \tag{13}$$

Among them, K represents the Kurtosis value, SK represents the Skewness value, x_i represents the value of the sample size, $\overline{x}$ represents the mean of the sample size, sd represents the standard deviation of the sample size, and n represents the number of samples.

Kurtosis is a statistical measure that describes the degree of steepness in the distribution of all values in a population. Usually represented by bk, the normal distribution has bk = 0, bk < 3 indicates that the distribution has insufficient kurtosis, and bk > 3 indicates that the distribution has excessive kurtosis. The edge image that meets the MTF calculation requirements should meet the bimodal distribution characteristics of the grayscale histogram, that is, have excessive kurtosis, as shown below:

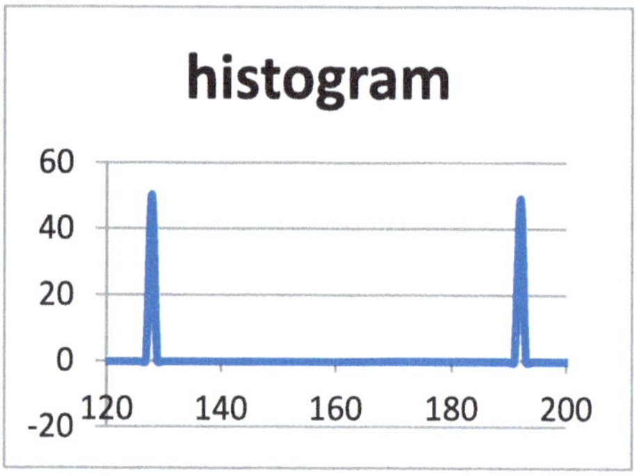

(a) Simulated edge target image (b)Gray histogram of simulated edge target image

Fig. 8. Schematic diagram of line support area construction

In Fig. 8 (b), the value of the horizontal axis represents the size of the grayscale value, and the value of the vertical axis represents the number of pixels. The kurtosis value of the edge target image simulated above is 122.5387, indicating that the Kurtosis value of the multi peak image is relatively high.

Skewness, similar to kurtosis, is a statistical measure that describes the distribution of data, describing the symmetry of the distribution of values in a population. Skewness is represented by bs, where bs approaches 0, indicating that the distribution is symmetric. It can be observed that for edge images, their distribution should be close to 0.

Due to the use of an extensible rectangular box to calculate Kurtosis and Skewness values, it is necessary to limit the size of the rectangular box. Cai Xinming demonstrated through experiments that the size of the edge image has an impact on the calculation of MTF [3]. The experimental results show that when the height of the blade image is less than 15, the stability of MTF values is poor. However, when the height of the blade image is greater than 17, the MTF calculation value is relatively stable and does not change much. Therefore, this article selects an expandable rectangular box with a minimum size of 17 * 17 and a maximum size of the line support area. When the size of the line support area is less than 17 * 17, it can be directly removed, while removing line support areas that do not meet the Kurtosis and Skewness threshold requirements.

Cai Xinmin et al. further verified the influence of edge inclination factor [3], and the experimental results showed that the stability of the calculation results was maintained well when the inclination angle of the blade was between 3° and 20°, and the calculation results were relatively ideal. Generally, the optimal inclination angle of the blade is around 8°. When using the blade method to calculate MTF in practice, it is generally not recommended to choose blade images with an inclination angle less than 3° or greater than 20°. Therefore, by obtaining the horizontal line direction of the edge line, it is possible to further eliminate non compliant line support areas and obtain the optimal dialog box for compliant line support areas.

In the candidate line support area, based on the optimal dialog obtained in the above steps, calculate the standard deviation of the grayscale of the features on both sides of the edge separately. Since the grayscale distribution on both sides of the edge should be relatively uniform, the standard deviation of the grayscale distribution on both sides should be relatively small. Here, we select a sample standard deviation threshold of 6.0 by analyzing the distribution characteristics of the actual uniform features. Considering that the size of the optimal dialog box may not be accurate due to the previous calculation of attribute parameters, the size of the dialog box can be appropriately adjusted to obtain the dialog box size with the smallest standard deviation of gray distribution on both sides of the edge, while meeting the minimum range of 17 * 17. The image in this dialog box is the selected edge image.

Finally, based on the conventional MTF calculation method, the corresponding MTF value can be calculated using the selected edge image.

3 Experiment Results

The target image obtained by ZY3 was selected for this experiment, and the experimental results are as follows:

It can be clearly observed that the gradient distribution map is a grayscale value, while for the horizontal direction distribution map, different grayscale values will be displayed according to different directions. The construction of the support area is based on different grayscale directions in the horizontal direction distribution map. By establishing a removal rule, the edge edges that meet the requirements are generated, and the results are as follows:

Due to the small edge area and the fact that most of the background images in Fig. 10 are black, which is not conducive to display, in order to illustrate the effectiveness of

(a) original image (b) Gradient distribution map of candidate points

(c) Horizontal Line Direction Distribution Map of Candidate Points

Fig. 9. Schematic diagram of the results of pixel grayscale changes in the target image

Fig. 10. Edge localization results in target images that meet MTF calculation

the algorithm extraction in this paper, the red rectangular box area in Fig. 10 is enlarged and overlaid with the original image for display. The display results are as follows:

Fig. 11. Edge positioning results with target images that meet MTF calculation (local magnification + overlay display)

Due to the current measurement of MTF values in the cross orbit direction, the extracted edge images should be in the same direction, as shown in Fig. 11. The red lines represent the edge position of the edge image, where the edges are all along the cross orbit direction. Whether it is a target image or a conventional image, the extraction of the edge visually meets the requirements of MTF calculation for the characteristics of the edge image. In order to verify the impact of the edge images extracted by this algorithm on the accuracy of MTF calculation, this paper uses the edge targets and three-line array targets in Fig. 9 to determine the effectiveness of the algorithm in this paper. As shown in Fig. 11, adopting the adaptive edge extraction method will result in three different MTF curve distribution results, as shown below:

Take the average of the three curves obtained in Fig. 12 as the final MTF distribution curve, and compare this curve with the manually selected edge image MTF curve. Considering that there is a three-line array target laid in the figure, the laboratory calculation formula can be used to obtain the MTF value of the three-line array target at the Nyquist frequency as the true value to compare the accuracy between the adaptive acquisition method and the manual method, The specific results are shown in Fig. 13:

From Fig. 13, it can be observed that the adaptively obtained MTF curve distribution is smoother. In addition, the MTF value at the Nyquist frequency obtained using the three-line array target is:

$$Ny = \frac{\left(\frac{1230-927}{1230+927}\right)}{1.0} * \frac{\pi}{4} = 0.1103 \tag{14}$$

Taking this MTF value as the true value, it can be found that the measurement accuracy of the MTF curve obtained by the adaptive edge extraction method at the Nyquist frequency is 96.7%, while the measurement accuracy of the MTF curve obtained by manual methods at the Nyquist frequency is 62%. It can be seen that manually selected methods are prone to introducing noise, leading to a decrease in measurement accuracy, which is not conducive to the construction of degraded models.

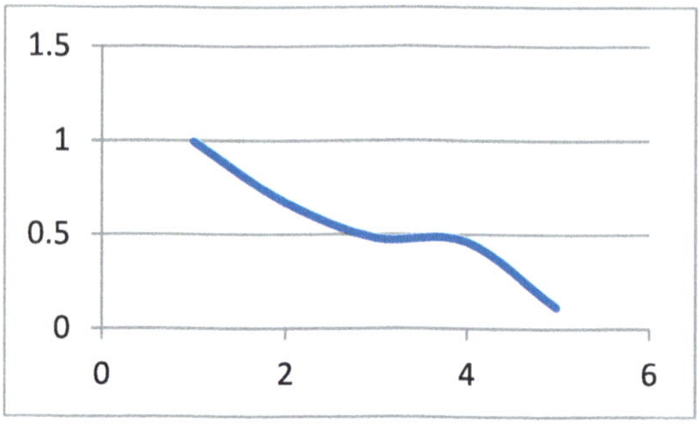

(a) MTF curve obtained from the left edge image(Nyquist=0.117886)

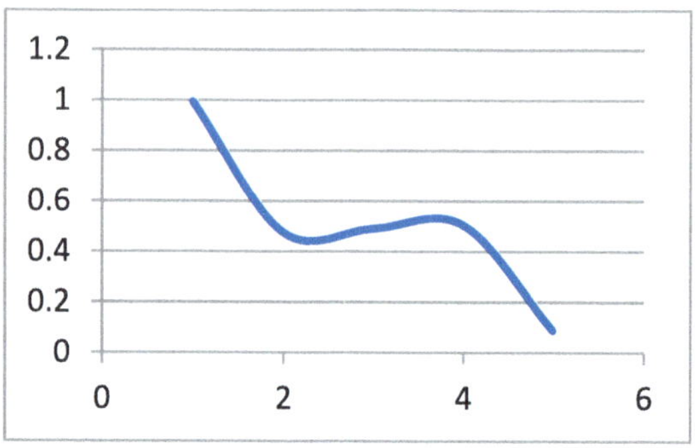

(b) MTF curve obtained from the middle edge image(Nyquist=0.091443)

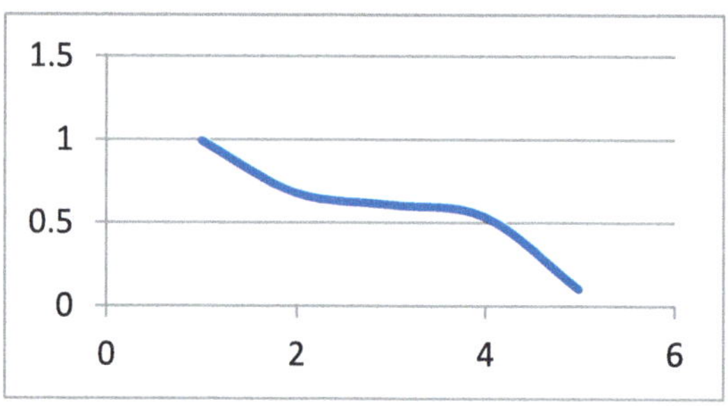

(c) MTF curve obtained from the right edge image(Nyquist=0.106639)

Fig. 12. Schematic diagram of MTF curve for adaptive edge image acquisition (including target image)

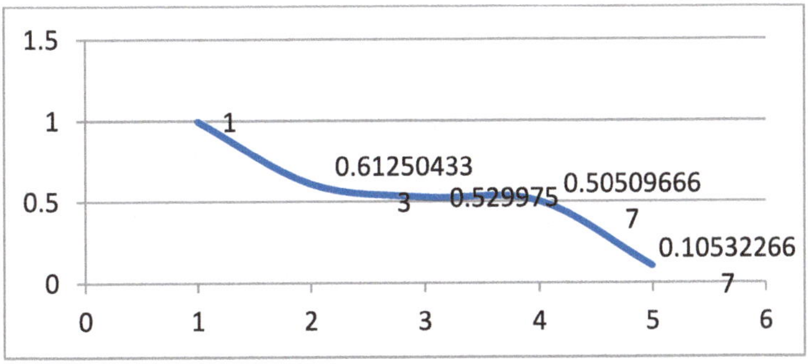

(a) MTF curve distribution obtained by adaptive edge extraction (Nyquist=0.106639)

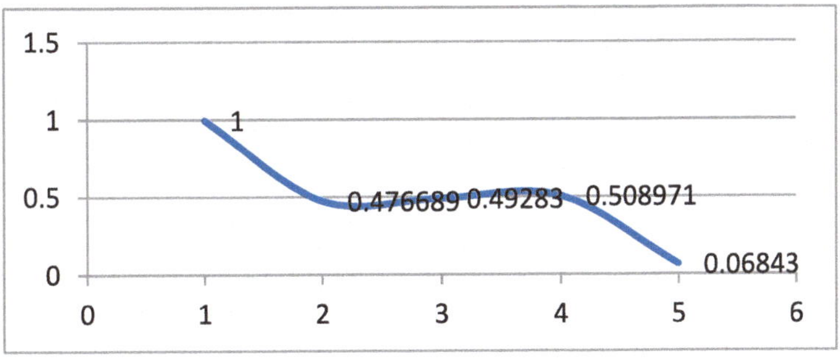

(b) MTF curve distribution obtained by manual edge extraction (Nyquist=0.06843)

Fig. 13. Comparison diagram of MTF curves obtained from adaptive edge images and manually captured edge images

4 Conclusion

The MTF curve obtained through adaptive methods not only better suppresses the influence of random noise, but also has better measurement accuracy than the MTF curve obtained manually. This experiment confirms the effectiveness of our algorithm and the universality of our algorithm through verification with conventional images. However, the calculation time of the method in this article is still relatively long, and in the future, GPU acceleration can be considered to improve the efficiency of the algorithm.

References

1. Duan, R.: Overview of research on image edge detection algorithms. Optics **31**(3), 415–419 (2005)
2. Von Gioi, R.G., Jakubowicz, J., Morel, J.M., Randall, G.: LSD: A line segment detector. Image Process. Line 1–19 (2012)
3. Cai, X.: MTF Calculation and Analysis Based on Satellite Remote Sensing Images. Nanjing University of Science and Technology, NanJing (2007)

4. Helder, D.L., Choi, T.: IKONOS Satellite in Orbit Modulation Transfer Funciton (MTF) Measurement using Edge and Pulse Method. South Dakota State University, South Dakota State (2001)
5. Choi, T.: IKONOS Satellite on Orbit Modulation Transfer Function Measurement using Edge and Pulse Method. South Dakota State University, South Dakota State (2002)
6. Ge, P., Wang, M., Pan, J.: Research on adaptive MTF image restoration processing for high resolution TDI-CCD imaging data. Land Resour. Remote Sens. **4**, 23–28 (2010)
7. Gu, X., Li, X., Min, X.: CBERS 02B satellite CCD camera MTF in orbit measurement and image MTF compensation. Chin. Sci. **35**(B12), 26–40 (2005)
8. Mu, X.: Remote sensing image restoration based on spatial optical remote sensor MTF compensation. Jilin: Changchun Institute of Optics, Precision Machinery and Physics, Chinese Academy of Sciences (2012)
9. Pang, T.: Fuzzy Kernel Estimation and Image Restoration for Motion Blurred Images. Southwest University, Chongqing (2011)
10. Qiu, X.: Research on Remote Sensing Image Restoration Technology Based on MTF. Nanjing University of Science and Technology, NanJing (2006)
11. Qureshi, A.G.: Constrained kalman filtering for image restoration. In: Proceedings of the Conference on ICASSP- 89, pp. 1405–1408 (1989)
12. Bensebaa, K., Banon, G.J.F., Fonseca, L.M.G.: On-orbit spatial resolution estimation of CBERS-2 imaging system using ideal edge target. In: Proceedings of SITIS, pp. 466–475 (2006)
13. Johnson, C.B.: Point Spread Functions, Line Spread Functions, and Edge Response Functions Associated With MTFs of the Form EXP = [-(OMEGA-OMEGA-C)N]. WASHINGTON: 1031–1033(1973)

Thermal Balance Test for Optical Remote Sensing Platform Based on Specialized Dual Strategy External Heat Flux Simulation Method

Yan Yinxue[1(✉)], Kong Qingle[1], Zhang Yue[1], and Fu Chunyu[2]

[1] Beijing Institute of Space Mechanics and Electricity, No. 104 Youyi Road, Haidian Beijing, China
yanyinxue@126.com

[2] Beijing Institute of Spacecraft Environment Engineering, No. 104 Youyi Road, Haidian Beijing, China

Abstract. Thermal balance test is one of the important tests for verifying the thermal model of the satellite and assessing the design of the thermal control system. This paper is oriented to the thermal balance test problems of China's optical remote sensing platform satellites, and based on the comparison of domestic and foreign thermal balance test methods, systematically analyzes the influence of the remote sensing satellite's complicated configuration layout, variable external heat flow and other elements on the test, and puts forward a dual external heat flow simulation strategy for a certain model of optical remote sensing satellites. The infrared heating cage realizes the simulation of external heat flow on the surface of satellite cabin plate, and the thin-film electric heater realizes the simulation of external heat flow on the exposed and complex surface of satellite, and combines with the active tracking and temperature control method to reduce the heat leakage from the docking flange of the satellite and other parts of the satellite. The test results show that the method can accurately simulate the external heat flow of the satellite under the working conditions of low-temperature initial period and high-temperature final period, effectively assess the thermal control design, provide a basis for the optimization of the satellite transient power dissipation design, and provide support for the test and verification of China's future optical remote sensing satellites.

Keywords: Optical Remote Sensing Platform · Thermal Design · Thermal Balance Test · Star Sensor

1 Introduction

Satellite thermal control technology is the process of controlling the thermal exchange between the internal and external environments of a satellite, ensuring that different parts of the satellite and the onboard instruments and equipment operate within the specified temperature range throughout the mission. It is a crucial assurance system within the various systems of a satellite, maintaining thermal balance within the required

H. P. Urbach et al. (Eds.): ISSOIA 2023, SPP 191, pp. 147–155, 2024.
https://doi.org/10.1007/978-981-97-6718-2_14

temperature range. The performance and reliability of the satellite thermal control system directly impact the working state of other systems and the operational lifespan of the satellite [1]. Thermal balance testing is conducted under vacuum and thermal radiation conditions in a space environment simulator to examine the temperature distribution during spacecraft orbital flight and validate the accuracy of thermal analysis models [2]. Both domestic and foreign satellites can experience malfunctions or failures due to excessively high or low temperatures. Therefore, it is necessary to conduct thermal balance testing before launching a satellite. In the United States and Europe, external heat flux simulation devices are used in spacecraft thermal balance testing, which involve the use of sun simulators and motion simulators for simulation purposes. The thermal balance test of spacecraft in China mainly uses infrared heating devices for simulation [3–6]. This paper presents a dual strategy external heat flux simulation method for thermal balance testing of a specific optical remote sensing satellite platform. The external heat flux simulation for the satellite panel surface is achieved by using an infrared heating cage, while the external heat flux simulation for the complex exposed surfaces of the satellite is achieved through thin film electric heaters. Active tracking and temperature control methods are employed to reduce heat loss from parts such as the docking flange of the satellite. This method can provide support for the testing and verification of future optical remote sensing satellites.

2 Characteristics of High-Resolution Remote Sensing Satellite Thermal Control

The thermal control system of a remote sensing satellite primarily consists of thermal control coatings, multi-layer insulation components, heat pipes, temperature sensors, electric heaters, thermal fillers, heat spreaders, and temperature controllers [7–10]. As an essential support system, it needs to meet the temperature requirements of all onboard instruments and equipment under specific orbits, attitudes, and operational modes. Satellite thermal control design typically involves a combination of passive thermal control and active thermal control methods. Passive thermal control stabilizes the overall satellite temperature within the normal operating range by effectively utilizing insulation and heat conduction materials. Active thermal control, on the other hand, uses heating circuits to ensure that individual components maintain temperatures above the lower limit requirements.

The thermal control of high-resolution remote sensing satellites imposes strict temperature control accuracy requirements on the star sensor components. This is due to the increasing demands for control precision and image positioning accuracy of high-resolution sensors. The attitude determination accuracy of the satellite plays a significant role as a source of control and imaging errors, making it a crucial technology for achieving high-precision and stable control. The accuracy of high-precision attitude determination is influenced by factors such as sensor layout design and in-orbit thermal deformation of attitude sensors [11]. In terms of sensor layout design, an improvement can be made by changing the traditional approach of installing star sensors on the satellite platform to directly installing them on the main structure of the optical remote sensing system

[12]. Given that the main source of error for star sensor measurements in orbit is thermal deformation caused by temperature changes, it is necessary to tightly constrain the temperature of the star sensor components within a narrow range [13].

3 Thermal Balance Test Plan

3.1 Thermal Flow Simulation Methods

The thermal flow simulation method proposed in this paper, which adopts a dual-strategy approach, mainly utilizes an infrared heating cage to simulate the external heat flow on the surface of satellite panels and a thin film heater to simulate the external heat flow on the complex surfaces exposed to space. This allows for precise simulation of the external heat flow on satellites.

The design of the infrared heating cage primarily involves determining parameters such as the maximum heating power and the heating belt area. First, the maximum power required for each zone of the infrared heating cage, denoted as P_i, is calculated according to Eq. (1):

$$Q_{i\max} = P_i K_1 K_2 K_3 \tag{1}$$

where, Q_{imax} is the maximum heat flux in the i-th heating zone, W/m^2; i is the zone number; K_1 is the coefficient considering the contribution of radiation in both directions of the infrared cage belt; K_2 is the angle coefficient; K_3 is the safety coefficient.

The angle coefficient K_2 is an important parameter for calculating the maximum power of the heating cage. As shown in Fig. 1, a simplified representation of two planes (length x, width y, spacing z) is considered. Two non-dimensional lengths, X = x/z and Y = y/z, are defined. The angle coefficient can be calculated using Eq. (2) to determine the coefficient between the emitting surface and the receiving surface.

$$K_2 = \frac{2}{\pi XY}\left\{\begin{array}{l} \ln\left[\frac{(1+X^2)(1+Y^2)}{1+X^2+Y^2}\right]^{1/2} + X\sqrt{1+Y^2}arctan\frac{X}{\sqrt{1+Y^2}} \\ +Y\sqrt{1+X^2}arctan\frac{Y}{\sqrt{1+X^2}} - X arctan X - Y arctan Y \end{array}\right\} \tag{2}$$

According to Eq. (3), the resistance value of each zone in the heating cage, R_i, is calculated. Then, using Eq. (4), the heating belt area for each zone, A_i, can be determined.

$$R_i = \frac{P_i}{I^2} \tag{3}$$

$$A_i = \frac{R_i}{1.667}W \tag{4}$$

where, I is the designed current; W is the width of the heating belt.

The design of thin film heaters primarily involves calculating their resistance values. First, the required maximum power, P_j, is calculated according to Eq. (5). Then, the resistance value of the heating element is determined using Eq. (6).

$$Q_{j\max} = P_j K_5 K_6 \tag{5}$$

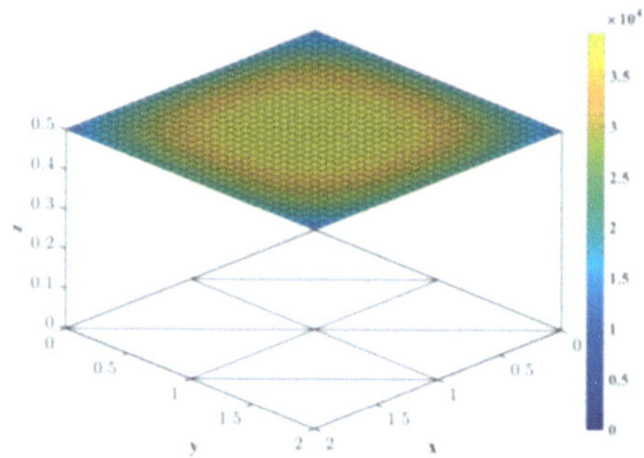

Fig. 1. Distribution of angle coefficient for parallel plates

$$R_j = \frac{P_j}{I^2} \tag{6}$$

where, Q_{jmax} is the maximum heat flux in the j-th heating zone, W/m^2; K_5 is the surface emissivity of the heating element; K_6 is the safety coefficient.

3.2 Satellite Test Technical Status

The remote sensing satellite used in the experiment has a mass of approximately 220 kg and adopts a box-type panel structure. The ± Z panels, ± Y panels, and central cabin panel are made of composite panel aluminum honeycomb, while the ± X panel is made of aluminum alloy. The star sensor is installed on the main body structure of the optical remote sensing instrument, as shown in Fig. 2.

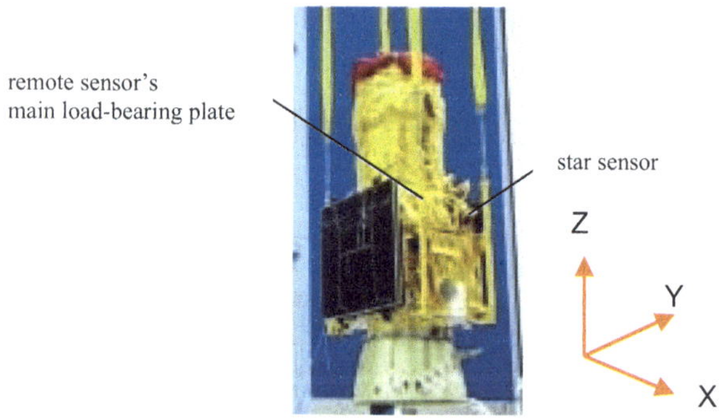

Fig. 2. Satellite diagram.

The satellite and the infrared heating cage are vertically installed on a parking platform, which is then inserted into a vacuum chamber. The parking platform is secured to the bottom of the chamber using leveling devices. The satellite's orientation is as follows: The + X axis points towards the vacuum chamber door, and the + Z axis points towards the top of the vacuum chamber. The infrared heating cage and the test fixture are shown in Fig. 3.

Fig. 3. Photograph of the experimental setup

3.3 Test Operating Conditions

Determine the high and low temperature operating conditions for the thermal balance test based on parameters such as the satellite's attitude, thermal dissipation, operational modes, coating performance, and solar constant during on-orbit flight. By applying high and low temperature external heat flux to the satellite in a vacuum cryogenic environment, the temperature levels and variation patterns of the satellite and its star sensors are obtained. This information is then used to refine the thermal analysis model and predict the in-orbit temperature of the satellite.

1) Low temperature test conditions: The external heat flux simulation is based on the early stage of the satellite's life, with a beta angle of 26.978° (corresponding to maximum solar incident heat flux). The satellite's attitude is set to sun-pointing cruise mode, and the internal heat source power consumption is loaded according to the operation of sun-pointing cruise mode.
2) High temperature test conditions: The external heat flux simulation is based on the late stage of the satellite's life. When the beta angle is 17.142° (corresponding to minimum solar incident heat flux), the satellite's attitude is set to imaging side-swing –30°. When the beta angle is 26.978° (corresponding to maximum solar incident heat flux), the satellite's attitude is set to imaging side-swing + 30°. The components operate in a "capture before transmit" mode.

3.4 Test Environmental Conditions

For all operating conditions, the vacuum level meets the requirement of not exceeding 1.3×10^{-3} Pa. The temperature of the heat sink does not exceed 100K (–173°C). The inner wall of the vacuum chamber is coated with black paint, with a surface emissivity exceeding 0.9, in accordance with the requirements of the thermal balance test outline.

4 Test Analysis

For the low temperature early stage conditions, active temperature control is only performed on the battery, fiber optic gyroscope, star tracker cold plate, and brackets. The duty cycle for active temperature control of other devices is zero. The battery pack has a duty cycle of 42.21%, fiber optic gyroscope A has a duty cycle of 36.65%, star tracker cold plate A has a duty cycle of 33.05%, and star tracker bracket A has a duty cycle of 69.11%.

For the high temperature late stage conditions, active temperature control is only performed on the battery, fiber optic gyroscope, star tracker cold plate, and brackets. The duty cycle for active temperature control of other devices is zero. The battery pack has a duty cycle of 33.75%, fiber optic gyroscope A has a duty cycle of 19.73%, star tracker cold plate A has a duty cycle of 19.73%, and star tracker bracket A has a duty cycle of 67.5%.

The temperature data of star sensors are shown in Fig. 4. Table 1 presents a summary of the temperature data for the thermal balance test.

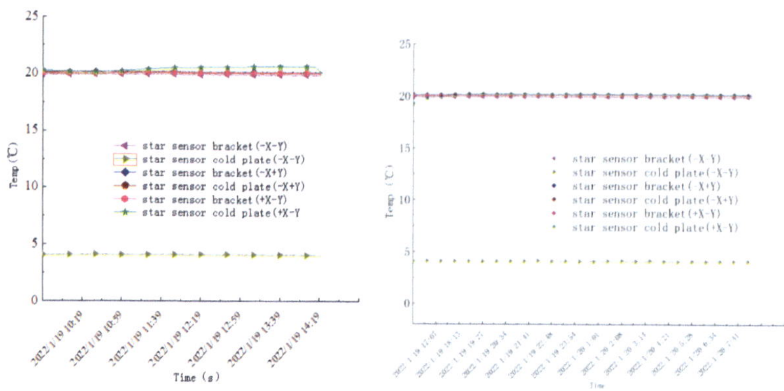

a. initial condition at low temperatures b. final condition at high temperatures

Fig. 4. Satellite panel and star sensor temperature data

Table 1. Presents a summary of the temperature data for the thermal balance test

Equipment Name	Performance Requirements(°C)	Test Data		Whether the indicators are met
		Early low-temperature stage	Late high-temperature stage	
Star sensor A	–40~+40 (0~+10 best)	4.05–4.16	4.05–4.20	yes
Star sensor A bracket	20 ± 3	19.8–19.98	19.84–19.98	yes
Star sensor B	–20~+40 (20 ± 2 best)	20.0–20.17	20.05–20.17	yes

(*continued*)

Table 1. *(continued)*

Equipment Name	Performance Requirements(°C)	Test Data		Whether the indicators are met
		Early low-temperature stage	Late high-temperature stage	
Star sensor B bracket	20 ± 3	20.07–20.24	20.04–20.27	yes
Star sensor C	–20~+40 (20 ± 2 best)	16.07–16.69	15.47–16.74	yes
Star sensor C bracket	20 ± 3	20.04–20.09	20.01–20.12	yes
Three-axis magnetometer	–15~+50	7.09–9.96	7.37–14.56	yes
Three-axis fiber optic gyroscope	–20~+45 (+10~+25 best)	13.66–15.91 13.63–15.95 (backups)	13.78–15.93 13.78–15.97 (backups)	yes
Reaction wheel Y	–15~+50	8.80–11.73	11.62–19.94	yes
Reaction wheel Z	–15~+50	11.60–15.63	11.02–21.45	yes
Reaction wheel S	–15~+50	34.85–37.91	29.27–41.52	yes
Magnetic torque converter Y	–10~+45	7.32–9.87	1.77–18.61	yes
Magnetic torque converter Z	–10~+45	–1.40–1.70	–1.34–8.50	yes
Power controller	–15~+50	10.66–13.25	11.75–18.35	yes
Storage battery	+ 10~+30	16.97–21.12	18.96–22.94	yes
Stack	–15~+50	0.32–2.76	0.36–7.34	yes
Network box	–15~+50	2.38–5.10	6.46–15.90	yes
Digital transmission integrated machine A	–15~+50 (work) –25~+60 (storage)	1.43–4.76	1.43–24.40	yes

(continued)

Table 1. (*continued*)

Equipment Name	Performance Requirements(°C)	Test Data		Whether the indicators are met
		Early low-temperature stage	Late high-temperature stage	
Digital transmission integrated machine B	−15~+50 (work) −25~+60 (storage)	3.79–7.68	3.62–24.81	yes
Dual channel amplifier	−25~+55 (work) −25~+65 (storage)	1.98–5.24	2.15–29.19	yes
Data transmission antenna	−90~+90	−8.58~−12.61	−10.24–9.05	yes
Camera integrated electronic unit	−15~+50	6.62–10.37	6.36–29.63	yes
Compressed storage unit	−25~+60	−1.15–0.79	−1.13–27.47	yes
Secondary power supply	−15~+50	−1.48–2.49	−2.02–9.07	yes
Focal plane unit two circuit	−15~+50	−0.87~−0.03	−0.91–3.01	yes

5 Experimental Conclusion

The results of the thermal balance experiment conducted on the optical remote sensing satellite indicate that:

(1) The test temperatures of all satellite subsystems and star trackers are within the specified temperature range, with duty cycles below 70%.
(2) The thermal balance experiment identified a lower-than-expected internal heat source power consumption, which has significant guidance implications for correcting the thermal control status of the satellite after the experiment.
(3) The method of combining infrared heating enclosure and thin film electric heaters for simulating external heat flux is deemed reasonable and feasible, and can be applied to the thermal balance testing of optical remote sensing satellites.

References

1. Guirong, M.: Satellite Thermal Control Technology. China Astronautics Press, Beijing (1991)
2. Huang, B., Ma, Y.: Space Environment Test Technology of Spacecraft. 1st edn. National Defense Industry Press, Beijing (2002)

3. Test requirements for launch, upper-stage and space vehicles: SMC-S-016 (2014)
4. Test requirements for launch, upper-stage and space vehicles: MIL-STD-1540E (2006)
5. Space engineering thermal control general requirements:ECSS-E-ST-31C (2008)
6. Thermal Balance Test Method for Spacecraft: GB/T 34515-2017 (2017)
7. Meng, Q., Yang, T., Yu, Z., et al.: Transient numerical simulation and orbit verification of pipeline heat pipe for space remote sensor. J. Beijing Univ. Aeronaut. Astronaut. **46**(11), 2045–2055 (2020)
8. Cong, S., Zhang, D., Chang, M., et al.: Design of low temperature environment device for space borne CCD test based on semiconductor refrigeration. Spacecraft Environ. Eng. **36**(1), 95–99 (2019)
9. Fan, S., Guan, M., Zheng, D., et al.: Refrigerated CCD imaging system for fluorescent cell observation. Spacecraft Environ. Eng. **27**(2), 205 (2010)
10. Yan, X., Li, Y., Zhang, C., et al.: Simulation and analysis of thermal state for micro—nano satellite. Chin. Space Sci. Technol. **43**(3), 35–42 (2023)
11. Bian, Z., Cai, C., Wang, L.V., et al.: High accuracy and high stability control technology of remote sensing satellite. Aerosp. Shanghai **31**(3), 24–38 (2014)
12. Fan, J., Qingwen, W., Ju, L., et al.: Thermal design of lightweight space remote sensor integrated with satellite in low earth orbit. Chin. Opt. **6**(2), 237–243 (2013)
13. Qiao, P., He, X., Wei, Z., et al.: Calibration of high-accuracy star sensor. Infrared Laser Eng. **41**(10), 2779–2784 (2012)

Structure Design of Remote Sensing Camera Based on Silicon Carbide 3D Printing

Li Qinglin[1,2](✉), Zhang Nan[1], and Chang Junlei[1]

[1] Beijing Institute of Space Mechanics and Electricity, Beijing 100094, China
liql1982@163.com

[2] Beijing Key Laboratory of Advanced Optical Remote Sensing Technology, Beijing 100094, China

Abstract. With the continuous maturity of 3D printing technology, it is gradually being applied in the aerospace field. The 3D printing process breaks through the constraints of traditional mechanical processing and provides a new design approach for the structure design of space remote sensing cameras. While meeting the structural strength and stiffness of remote sensing cameras, it can achieve integrated design of multiple parts, high lightweighting rate of parts, and high stability of camera. The article first introduces the characteristics of 3D printing technology and structure design methods for 3D printing; Afterwards, the integrated structure design of the space remote sensing camera was completed using silicon carbide as the raw material; Finally, finite element simulation analysis of remote sensing cameras was carried out under various working conditions. The first order frequency of the remote sensing camera was 72.8 Hz, and the maximum translation of the mirror under gravity and 10 °C temperature fields was 0.012 mm, and the maximum deflection angle was 0.8″. This met the requirements of optical system design indicators and verified that the remote sensing camera has good mechanical and thermal stability. After physical mechanical vibration test verification, it meets the design results.

Keywords: 3D printing · silicon carbide · remote sensing camera · structure design

1 Introduction

Space remote sensing cameras not only need to meet the requirements of the space environment, but also need to adapt to the mechanical environment of the launch segment. The design complexity is high, involving multiple components, traditional manufacturing methods have long cycles, high costs, and complex camera structures are difficult to process. With the development of additive manufacturing technology (also known as 3D printing technology), the complexity of product manufacturing has been reduced, the reliability of products has been improved, and it has gradually been widely applied in fields such as aviation and aerospace [1–2].

3D printing technology is a rapid prototyping technology that uses powdered metal or non-metallic adhesive materials to construct objects through layer by layer printing

H. P. Urbach et al. (Eds.): ISSOIA 2023, SPP 191, pp. 156–166, 2024.
https://doi.org/10.1007/978-981-97-6718-2_15

[3]. By utilizing the technological advantages of 3D printing, the integration of spatial remote sensing camera structure has been achieved, reducing the number of components and the difficulty of product assembly, and reducing the weight of spatial remote sensing cameras [4].

The article proposes an optical mechanical structure design based on silicon carbide 3D printing, which improves the adaptability of space remote sensing cameras to space temperature environments, reduces the weight of space remote sensing cameras, and achieves lightweight and high stability of space remote sensing cameras.

2 3D Printing Design Methods

2.1 Characteristics of 3D Printing Technology

3D printing technology is completely opposite to traditional material reduction manufacturing processes. In terms of processing, 3D printing technology is not affected by traditional machining tools and equipment, and directly realizes the manufacturing and forming of parts with complex features such as hollow, spatial torsion, and lattice, providing a breakthrough in traditional design space. In recent years, 3D printing technology has been promoted and applied in the aerospace field. For example, the shell and battery pack of the Tomsk TPU-120 satellite in Russia are formed using 3D printing, achieving integrated design of key components. In the design of space remote sensing cameras, multiple constraints such as product strength and stiffness, gravity deformation, temperature thermal stress, assembly stress, etc. need to be considered, making the design complex; The product department has a large number of components and multiple assembly relationships. In the face of these problems, 3D printing technology can be used to achieve breakthroughs in traditional design: (1) achieve integrated structure design, reduce the number of parts, and reduce assembly processes; (2) 3D printing can achieve micro lattice and hollow structures, improving the lightweight rate of mechanical structures; (3) 3D printing can achieve the use of the same material for optical and structural components, reducing thermal mismatch caused by different materials, and improving the temperature adaptability of optical and mechanical structures.

Materials are the key to 3D printing technology and also affect the performance of the final product. At present, both metals and ceramics can be formed through 3D printing. The typical materials for space remote sensing camera mirrors to be 3D printed are silicon carbide, while the typical materials for load-bearing structures to be 3D printed are silicon carbide and metal materials. In order to fully leverage the advantages of 3D printing and achieve the use of the same material for the reflector and load-bearing structure of space remote sensing cameras, silicon carbide material is selected, and the drawbacks of high hardness and poor processability of silicon carbide material are solved, fully leveraging the good material properties of silicon carbide.

2.2 Structural Design Method for 3D Printing

The main methods and processes for designing the structure of spatial remote sensing cameras for 3D printing are as follows: (1) Based on the characteristics of 3D printing technology, integrate the design of various components of remote sensing cameras to minimize the number of parts and assembly relationships as much as possible; (2) Based on the layout of the optical system, starting from the interface, design space, and constraint conditions, topological optimization of the structural configuration of each component is carried out; (3) Revise the design results of the parts based on the 3D printing process to ensure that the structure conforms to the 3D printing process; (4) For silicon carbide brittle materials, add flexible link designs at interfaces with connection relationships; (5) Conduct finite element simulation analysis on the entire remote sensing camera to ensure that the design structure meets the usage requirements; (6) Based on the tolerance allocation relationship of the optical system, reserve machining allowances for parts with high precision requirements to ensure the feasibility of subsequent machining.

3 Structure Design and Simulation Analysis of Remote Sensing Cameras

Optical systems can be divided into refractive, reflective, and reflective types. Reflective optical systems do not have chromatic aberration, and their aperture can also be relatively large. Especially vertical axis reflective optical systems can achieve a relatively large field of view without central obstruction, with high optical efficiency and MTF, and are widely used in space remote sensing cameras [5]. In this project, the space remote sensing camera adopts a off-axis TMA optical system. The incident light passes through the primary mirror, secondary mirror, and tertiary mirror to reach the focal plane, and two imaging channels are achieved at the focal plane through a planar reflector. The optical system configuration is shown in Fig. 1.

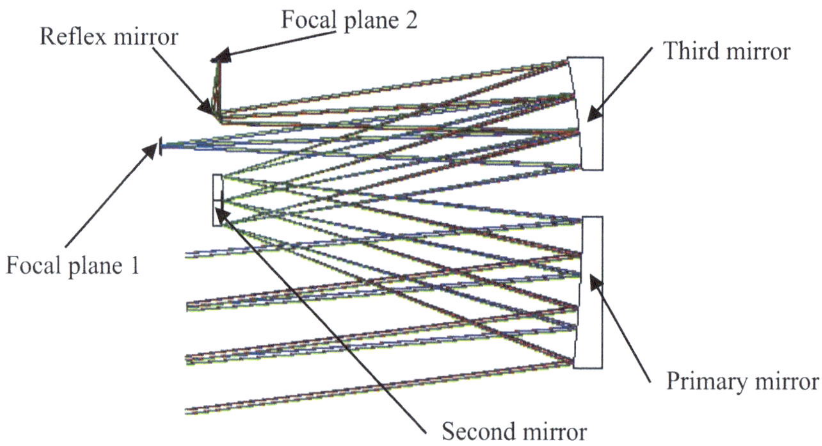

Fig. 1. Off-axis TMA optical system

Based on the design results of the optical system, the structural design of the remote sensing camera is carried out with the overall weight and stiffness as the objectives, and the environmental temperature and mechanical environment of the launch section as constraints. The main material of the camera is 3D printed silicon carbide material.

3.1 Mirror Design

At present, the commonly used forms of support for mirrors mainly include back center support, back three-point support, and side support. The traditional support methods are all separated from the mirror and support structure [6–7], as shown in Fig. 2. The use of 3D printing technology can achieve integrated design of the reflector and its support structure. The reflector support adopts a design form of integrated mirror body and frame, and adopts a three-point fixing method to achieve static support of the reflector. The weight of the reflector component is 1.1 kg, which is about 35% lighter than traditional support schemes.

(a) mirror (b) mirror frame

Fig. 2. Traditional support scheme for reflective mirrors

The mirror of a space remote sensing camera undergoes changes in surface figure due to factors such as temperature changes and gravity deformation during use. A flexible unloading link needs to be installed between the mirror and the load-bearing structure. At the same time, the mirror adopts an integrated design of the mirror body and frame made of silicon carbide material, which needs to overcome the brittleness drawbacks caused by silicon carbide material. The flexible hinge structure can effectively reduce the influence of external mechanics and temperature environment on the reflector. However, due to its low stiffness and easy deformation, it is necessary to optimize the structural parameters of the flexible hinge based on the stiffness of the reflector component and the requirements of the reflector surface figure. Through simulation and comparison, it is determined that the flexible hinge structure adopts an orthogonal structure form, and its three-dimensional model is shown in Fig. 3. The flexible hinge used is made of invar steel material, which has a linear expansion coefficient similar to that of silicon carbide material, reducing material mismatch [8]. The final model of the reflector design is shown in Fig. 4.

A simulation analysis was conducted on the constrained mode, gravity, and temperature field conditions of the reflector. The first-order mode of the reflector is 452 Hz, which meets the requirements of components greater than 200 Hz. The modal shape is shown in Fig. 5. Under the conditions of gravity and temperature fields, the RMS accuracy of the reflector surface is less than 0.01 λ(wavelength $\lambda = 632.8$ nm). The simulation analysis results are shown in Table 1.

Fig. 3. The configuration of flexible hinge

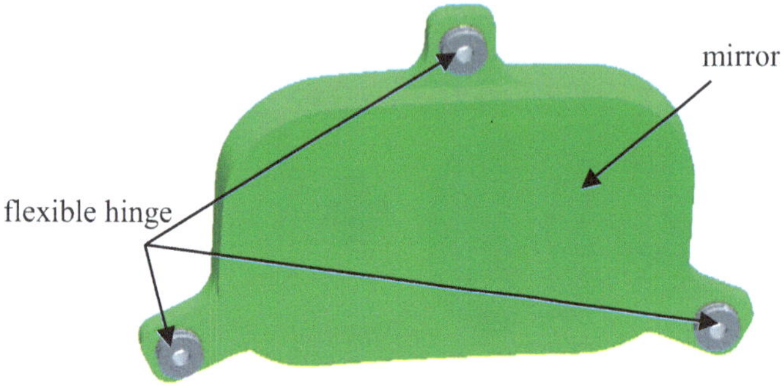

Fig. 4. The configuration of mirror

Fig. 5. The first order modal of mirror

Table 1. Analysis results under different working conditions

Constrained working condition	Design requirement	Result
Fundamental frequency analysis/Hz	>200	452
Gravity analysis/nm	<6.328	4.8
Analysis of temperature rise at 10°C/nm	<6.328	1.9
Forced displacement analysis of 0.01 mm installation surface/nm	<6.328	1.3

3.2 Main Load-Bearing Structure Design

As the supporting structure of the entire remote sensing camera, the main load-bearing structure should not only meet the layout requirements of the optical system, but also have sufficient stiffness and strength to meet the requirements of the mechanical environment of the reflection section [9]. Combining the characteristics of 3D printing technology, the main load-bearing structure in the project adopts a frame structure, with silicon carbide as the material used. According to the design results of the optical system, the secondary mirror and focal plane are arranged on the front face of the main loadbearing structure, while the primary mirror and tertiary mirror are arranged on the back face [10]. The topology optimization method is used to optimize the design of the main load-bearing structure, taking into account the special process of 3D printing and the characteristics of high brittleness and low fracture toughness of silicon carbide materials [11]. During the structural optimization process, the main load-bearing structure should also meet the following requirements: (1) while meeting the optical interface requirements, minimize the number of through holes as much as possible to reduce the risk of local low strength; (2) The minimum wall thickness and lightweight form of the structure should meet the requirements of 3D printing process to ensure the yield of 3D printing; (3) Add metal nesting at the connection gap to enhance the strength of the stressed area. Through optimization design, the final model of the main load-bearing structure is shown in Fig. 6. The weight of the main load-bearing structure is 4.2 kg, which is about 40% lighter than traditional material manufacturing solutions. The first modal frequency of the main load-bearing structure is 639.9 Hz, and the vibration mode diagram is shown in Fig. 7.

Fig. 6. The configuration of main bearing structure

3.3 Simulation Analysis

The remote sensing camera mainly consists of a main load-bearing structure, a primary mirror, a secondary mirror, tertiary mirrors, a flat mirror, and baffle. In order to reduce the mechanical response of key parts of the remote sensing camera in the launch phase, vibration isolation measures are adopted to connect it to the satellite platform. The overall size of camera is less than 400 mm × 400 mm × 450 mm, and the weight is less

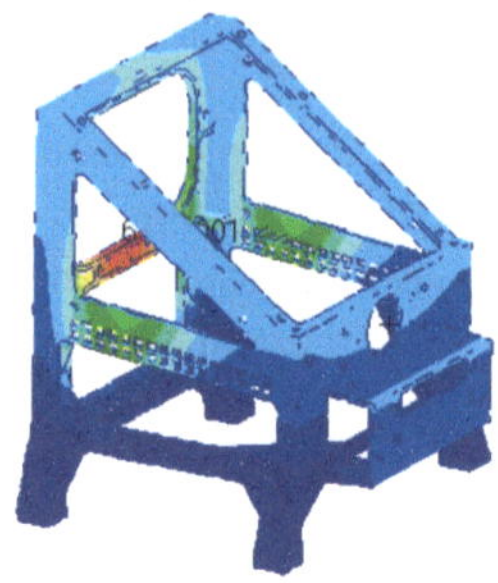

Fig. 7. The first order modal of main bearing structure

than 10kg, with a weight reduction ratio of about 30%. The design results of the remote sensing camera are shown in Fig. 8.

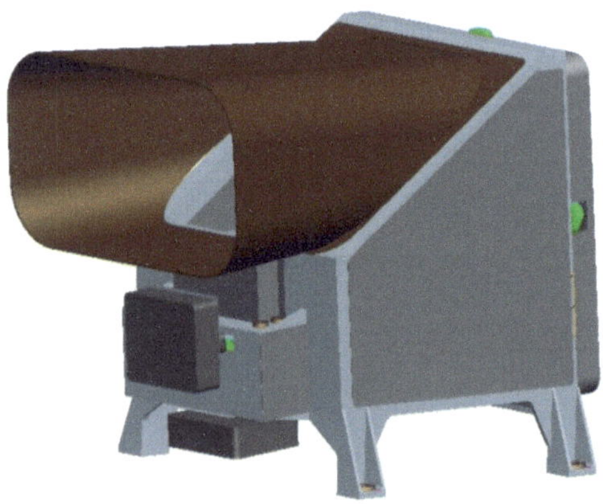

Fig. 8. The overall configuration of remote sensing camera

To evaluate the performance of remote sensing cameras, finite element simulation analysis was conducted. The main analysis content includes constrained modal analysis, gravity stability analysis, and temperature field stability analysis. Through simulation analysis, the first order frequency of the remote sensing camera is caused by the absence of vibration isolation, with a frequency of 72.8 Hz, which meets the requirement that it's greater than 60 Hz. The vibration mode diagram is shown in Fig. 9.

The gravity stability analysis mainly assesses the imaging performance changes of the camera caused by the inconsistency between heaven and earth. Through the simulation analysis results of applying one time of gravity to the remote sensing camera, it can be seen that the maximum translation of the mirror component is 0.002 mm, and the maximum deflection angle is 0.8″, meeting the requirements of optical system translation less than 0.04 mm and deflection angle less than 6″. The gravity deformation diagram of the remote sensing camera is shown in Fig. 10.

Fig. 9. The first order modal of camera

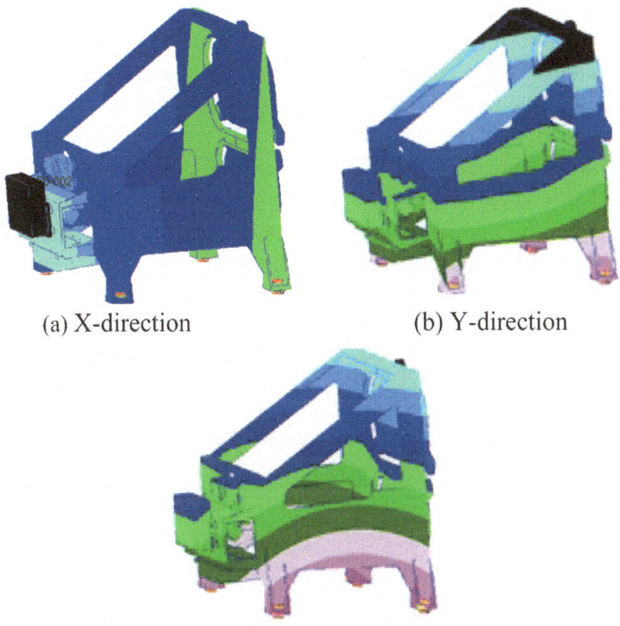

(a) X-direction (b) Y-direction

(c) Z-direction

Fig. 10. The deformation of camera under gravity

The temperature stability analysis mainly simulates the imaging performance changes of the camera caused by temperature changes in orbit. Through the simulation analysis results of the remote sensing camera under a temperature change of 10 °C, it can be seen that the maximum translation of the mirror component is 0.012 mm, and the maximum deflection angle is 0.8″, meeting the requirements of optical system translation less than 0.04 mm and deflection angle less than 6″. The temperature deformation diagram of the remote sensing camera is shown in Fig. 11.

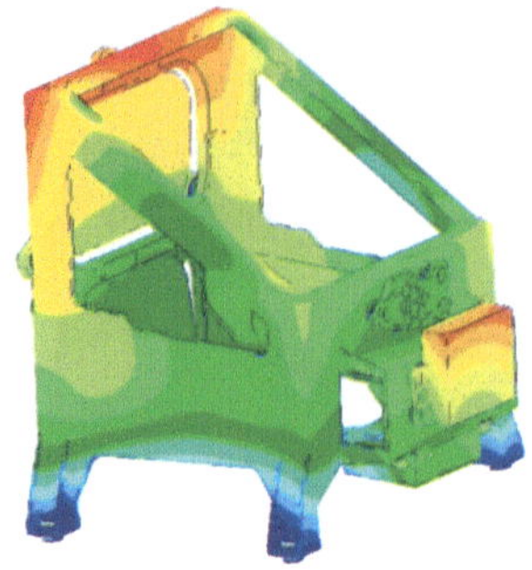

Fig. 11. The deformation of camera under temperature field

To verify the temperature adaptability of the remote sensing camera, the structural deformation data analyzed under temperature load was further subjected to optical system simulation analysis. Through analysis, it was found that the changes in the principal point of the optical system were 0.002 mm, the principal distance was 0.001 mm, the focal length was 0.001 mm, and the MTF was 0.001. The changes were all very small, and the impact on imaging quality could be ignored. It can be seen that the adaptability of 3D printed silicon carbide remote sensing cameras to the environment has been improved from the traditional 2 °C to 10 °C, which can meet the imaging requirements under large temperature changes in orbit. The changes in optical system parameters are shown in Table 2.

Table 2. Variation of optical system under temperature condition

Items	principal point/mm	principal distance/mm	focal length/mm	MTF
Variation of items	0.002	0.001	0.001	0.001

3.4 Test Verification

(1) Mechanical vibration test
Based on the design results, the processing and manufacturing of the remote sensing camera were completed, and mechanical environment tests were conducted on the camera. The frequencies of the remote sensing camera in the X, Y, and Z directions were 116.3 Hz, 82.6 Hz, and 69.9 Hz, meeting the requirement of a first-order frequency greater than 60 Hz. The consistency with the simulation results reached 96%. The test site is shown in Fig. 12.

Through the angle testing of the reference cubic mirror installed on the remote sensing camera, the maximum change in the position angle of the reflector before and after mechanical vibration is less than 5″, meeting the design index requirements. This further verifies that the remote sensing camera structure developed using silicon carbide

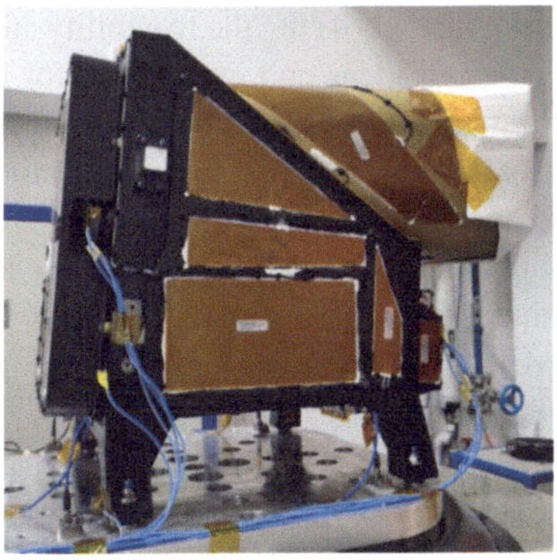

Fig. 12. Mechanic vibration test

3D printing has high stability, and its design ideas and methods are reasonable and feasible.

(2) Thermal vacuum test
During the thermal vacuum test, the remote sensing camera controlled the temperature at 20 °C ± 5 °C and conducted MTF testing under extreme conditions. The test results of MTF were 0.170–0.172, with a maximum variation of 0.002. The defocus distance of the remote sensing camera's focal plane is less than 0.003 mm under different extreme working conditions, which is much smaller than the camera's half focal depth of 0.048 mm. According to the experimental results, it has been proven that the remote sensing camera has high stability in imaging quality in a thermal environment within the range of 10 °C and has good adaptability to temperature environments. The experimental site is shown in Fig. 13.

Fig. 13. Thermal vacuum test

4 Conclusion

The article introduces the characteristics of 3D printing technology and the application of silicon carbide materials in 3D printing technology. It summarizes the design method for developing remote sensing camera structures using silicon carbide 3D printing technology, and uses a off-axis TMA remote sensing camera as an example to design the reflector and main bearing structure, verifying the feasibility and correctness of this technology method, it provides design ideas and experience for the application of silicon carbide 3D printing technology in space remote sensing cameras.

The remote sensing camera using silicon carbide 3D printing has achieved the design of the same material optical and mechanical structure, improved the temperature adaptability of the remote sensing camera, proposed a design scheme of integrated mirror body and frame, reduced the assembly process of the mirror, improved the reliability of the mirror, and provided technical support for the development of remote sensing cameras towards high stability and lightweight.

References

1. Fang, X., Xiao, H., Sun, D.: The application of 3D printing technology in space compo-sites manufacturing. Spacecraft Recov. Remote Sens. **36**(2), 77–82 (2015). (in Chinese)
2. Liu, J., Shi, Y., Cui, Y., et al.: Research progress and prospect of on-orbit additive manufacturing technology. Chin. Space Sci. Technol. **42**(6), 23–34 (2022)
3. Zhang, X., Tang, S., Zhao, H., et al.: Research status and key technologies of 3D printing. J. Mater. Eng. **44**(2), 122–128 (2016). (in Chinese)
4. Wang, C., Zhang, N., Wang, Y., et al.: Structure design of space optical camera based on 3D printing. Spacecraft Recov. Remote Sens. **41**(1), 85–90 (2020). (in Chinese)
5. Liu, F., Cheng, Z., Shi, L., et al.: Design and analysis of supporting structure for rectangu-lar mirror. Infrared Laser Eng. **44**(5), 1512–1517 (2015). (in Chinese)
6. Wang, K., Xin, H., Xu, H., et al.: Lightweight design of fast steering mirror for space cam-eras. Infrared Laser Eng. **48**(4), 0418001 (2019). (in Chinese)
7. Zhang, N., Pang, S., Chang, J., et al.: Design and analysis of four-point spherical for mirror. Spacecraft Recov. Remote Sens. **39**(6), 64–71 (2018). (in Chinese)
8. Yang, L., He, B.: Summarization of joining technologies of SiC mirror of space remote sensor. OME Inf. **27**(12), 56–61 (2010). (in Chinese)
9. Cai, W., Fan, B., Zhang, F., et al.: High opto-mechanical stability design of multi-spectral camera. Spacecraft Recov. Remote Sens. **33**(3), 85–92 (2012). (in Chinese)
10. Nan, Z., Qinglin, L., Junlei, C.: Design and verification of main bearing structure of re-mote sensing camera based on SiC 3D printing. Infrared Laser Eng. **51**(3), 20210267 (2022). (in Chinese)
11. Chuanqi, H., Hailin, L., Xiaoting, H., et al.: Microstructure and mechanical properties of reaction sintered silicon carbide. China Ceramic Ind. **27**(5), 7–11 (2020). (in Chinese)

Rolling Attitude Effect on High Resolution Space Camera Temperature

Henghui Meng[1,2(✉)], Dongyang Han[1,2], Chuanqiang Zhang[1,2], and Liyin Geng[1,2]

[1] National Key Laboratory of Spacecraft Thermal Control, Beijing 100094, China
menghenghui@126.com

[2] Beijing Institute of Spacecraft System Engineering, Beijing 100094, China

Abstract. With the development of aerospace technology, spacecraft have a fast and good application demand, and there is an increasing number of sudden missions in orbit. Imaging of key areas is completed through attitude adjustment. The attitude adjustment disrupts the stable thermal environment of the spacecraft and affects the imaging accuracy of the spacecraft's optical camera. High precision optical cameras need to have the ability to adapt to complex spatial thermal environments, which is the trend of future development. An optical camera was researched, and for the first time systematically analyzed the variation law of spatial heat flux, providing guidance for thermal design schemes. The optical camera conducted thermal analysis and thermal balance tests simultaneously, and the simulation results were consistent with the experimental data, verifying the correctness and effectiveness of the thermal design. The results indicate that the side sway has a relatively small impact on the temperature field inside the camera body, but a greater impact on the external light shield, and active thermal control can weaken the influence of the side sway on the external heat flux. To further reduce the impact of lateral sway on high-precision cameras, improvement measures have been proposed to guide the development of subsequent spacecraft: increasing the length of the light shield, such as a flexible light shield, to weaken the influence of external heat flow on the star's internal light shield; Perform high-precision temperature control on key components to reduce the impact of complex heat flow.

Keywords: Camera · Roll · Temperature field · Thermal control design · Thermal experiment

1 Introduction

Pressure and temperature are the mainly two aspects impact of space environment on the camera imaging for spacecraft in near-Earth orbit. For pressure, it changes from an atmospheric pressure to zero quickly when spacecraft is launched, which should bring mirror deformed, then produce focal plane out of focus, and reduce the quality of image. Pressure impact can be corrected by the in-orbit calibration, for the change of pressure is a one-time process, and the change is stable. While for temperature, its change is complex, as for suffering to out heat flux and surrounding environment, and

H. P. Urbach et al. (Eds.): ISSOIA 2023, SPP 191, pp. 167–178, 2024.
https://doi.org/10.1007/978-981-97-6718-2_16

it is known that different temperature distributions bring different effects to the camera optical system [1–4]. Root-Mean-Square (in short RMS), is used for tolerance allocation on several high-resolution space telescope, and the tolerance of thermal control system is usually up to half of the total budget of tolerance [5–7]. For thermal design project, the technology of optical instruments elimination heat will minimize temperature changes [7–8], and high-precision thermal control measures are used to achieve temperature requirement, and ensure the camera's image quality and reliability. Based on the basic theory of optical wave-front aberration, thermal design for high-resolution cameras are studied, and environmental adaptability for optical remote sensor are carried out [9–11], So, there are many methods for thermal design of cameras in a single attitude.

In order to meet the demand for imaging current hot spots mobility, it is achieved by satellite attitude adjustment, which will bring space heat flux differences between rolling attitude and normal attitude, and also effect temperature field for optical remote sensor, finally there will be an impact on the image quality. It has not seen in the open literature for thermal design for optical remote sensor in different attitudes. In this paper, a high-resolution CCD camera as for an example, study the impact of rolling attitude for the camera temperature field, and provides a solution to this effect, which is used to guide the high-precision thermal design of subsequent spacecraft.

2 Thermal Design Difficulties

The high-resolution CCD camera is composed of optical lens, focal plane assembly, hood, barrel and frame. Optical lens comprises a main reflector mirror (in short main mirror), a secondary reflector mirror (in short secondary mirror), and a third reflector mirror (in short third mirror), after lens barrel tube, mirror frames and so on. The focal plane assembly comprises CCD, focal plane electronic circuit boards and motor. Hood and tube are made of aluminum honeycomb, three mirror are made of glass-ceramic, mirror support frames are made of titanium, the secondary mirror-support rods are made of Invar, after lens barrels are made of carbon fiber composite material, focal plane assembly are made of aluminum alloy.

High-resolution CCD camera is mounted on the bottom of payload module, except for the hood, the rest of the camera body are located inside the satellite. Figure 1 is the structure of optical remote sensor, it is seen that the camera is located in X direction, and hood in Z direction. For the satellite flying on the orbit, the X direction is flying direction and the Z direction points to Earth. Heat dissipation is produced by focal plane assembly while the camera is working, and Table 1 is specific heat dissipation distribution. According to optical performance analysis, thermal control requirements are shown in Table 2.

Rolling attitude is carried for obtaining the function of focus on key areas, which will bring another space environment. So, for thermal control design, except for considering the normal working conditions, it is also need to take into account the demand for working in rolling attitude, which require the design a better applicability, and bring thermal design difficulties. As there is difference between normal attitude and rolling attitude in space heat flux, effectively reduce the influence of heat flux changes, which is the direction of thermal design. In this paper, starting from analysis the rule of heat

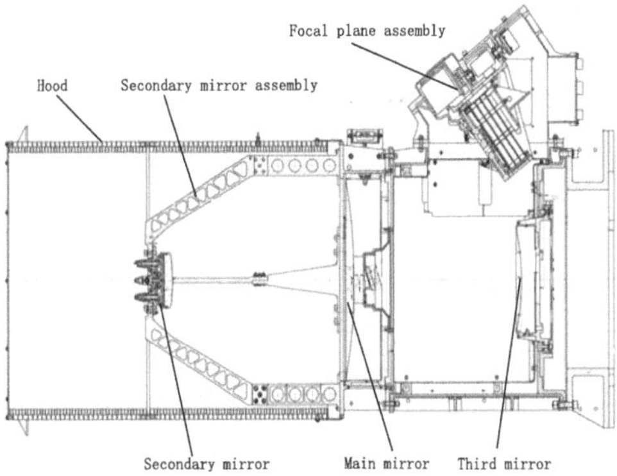

Fig. 1. High resolution CCD camera configuration

Table 1. Heat dissipation distribution

No	Dissipation parts	Dissipation	Work mode
1	Focal plate of CCD circuit board	7.8 W	15 min/orbit
2	CCD drive circuit board	1.4 W	15 min/orbit
3	Data transmission circuit board	5 W	15 min/orbit
4	Timing circuit board	1.3 W	15 min/orbit

Table 2. Temperature target /°C

Parts	Temperature requirement
Main mirror	18.5–21.5
Secondary mirror	18.5–21.5
Third mirror	18.5–22.5
Back mirror body	18.5–22.5
Knightheads	14–21.5
CCD	12–26

flux, and verify the correctness of the design through numerical analysis and thermal balance test.

3 Heat Flux Analysis

Space heat flux of near-Earth orbit spacecraft is composed by solar, albedo, and planet shine, and the absorbed heat flux is demanded by optical properties of the spacecraft, orbit position and attitude changes. As camera hood and radiator directly face to the cold space, the two parts are affected significantly by space heat flux. As the above mention, the two parts are located in the +Z direction of the satellite, and +Z direction of heat flux has more impact to camera temperature, so theoretical analysis is used. Figure 2 is the orientation schematic of spacecraft's orbital plane, sun, and Earth.

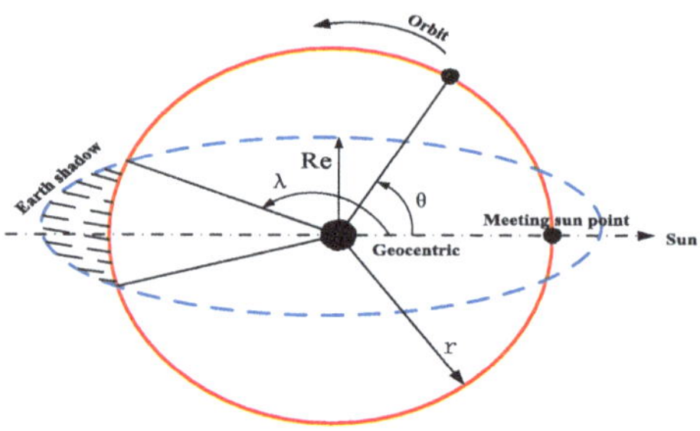

Fig. 2. Earth shade projection in orbit

According to paper [12], + Z surface absorption of solar is expressed in formula 1:

$$Q_{+Z} = S\delta_{+Z}\alpha_{+Z}A_{+Z}\cos\beta|\cos\theta| \tag{1}$$

In formula 1, Q_{+Z} represents + Z surface absorption of solar(W), S represents the arrival solar radiation flux(W/m^2), α_{+Z} represents solar absorption rate of + Z surface, A_{+Z} represents the area of radiator(m^2), βrepresents the angle between sun and orbital plane, θ represents the angle between satellite and meeting sun point, the value of δ_{+Z} has a function of θ,can be expressed in formula 2:

$$\delta_{+Z} = \begin{cases} 0 \; \theta \in [0, \pi/2] \cup [3/2, 2\pi] \\ 1 \qquad \theta \in (\pi/2, 3\pi/2) \end{cases} \tag{2}$$

For normal attitude, +Z axis of satellite points to earth, +X axis is flight direction, +Y axis, X axis, and Z axis build up a right-hand coordinate system. For rolling attitude, as shown in Fig. 3, +X axis is the axis of rotation, the rolling angle is positive by right-hand rules, otherwise it is negative.

Consider rolling case, +Z surface absorption of solar is expressed in formula 3:

$$Q_{+Z} = S\delta_{+Z}\alpha_{+Z}A_{+Z}\cos(\beta - P)|\cos\theta| \tag{3}$$

In formula 3, P represents the rolling angle. For a sun-synchronous orbit, β ranges from 16 to 27°, rolling angle P varies from –32°to 32°, so two specific rolling angles are selected, according to the law of cosines, it is known that.

cos (β–32) > cosβ > cos(β + 32),which means the absorption of solar is maximum in positive attitude, and the absorption of solar is minimum in negative attitude.

For albedo and planet shine, an angle is produced by rolling attitude between +Z direction and earth, which makes less absorbing heat flux than the normal posture, the relationship can be simplified in formula 4:

$$Q_{Rooling} \approx Q_{normal} \cdot \cos P \tag{4}$$

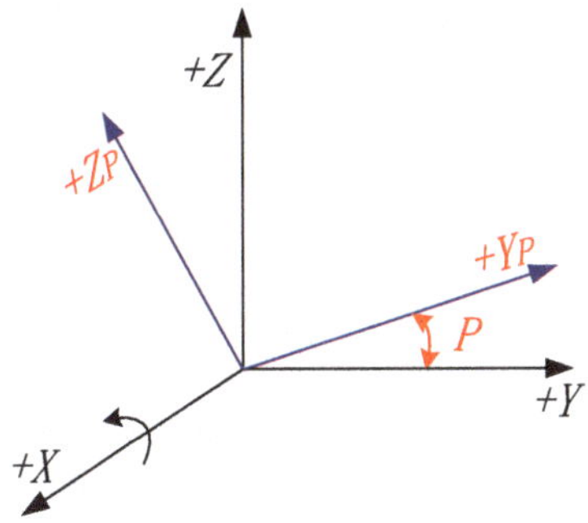

Fig. 3. Rolling attitude sketch

According to formula 3, solar is larger in positive rolling attitude, so albedo is larger in positive rolling attitude than in negative rolling situation.

Winter solstice is selected as an example, the arriving heat flux of + Z direction is analyzed in normal attitude and rolling attitude. Figure 4 shows the heat flux distribution, Table 3 is the average heat flux of + Z direction in normal and rolling attitudes. It can be seen that theoretical analysis is correct in rolling attitude heat flux.

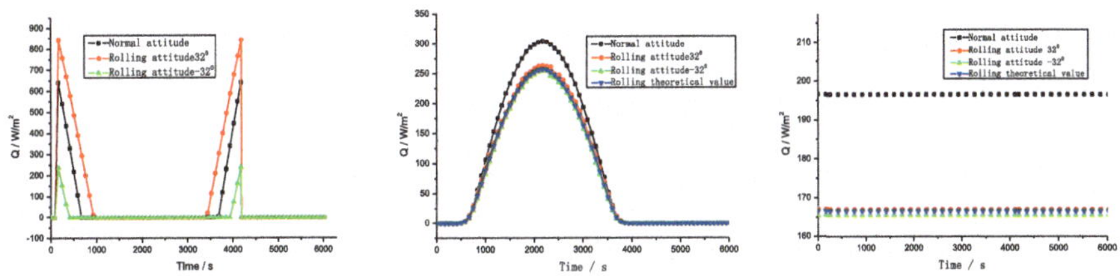

(a) Solar (b) Albedo (c) Planet shine

Fig. 4. Heat flux distribution

Table 3. Average heat flux of +Z direction (W/m^2)

Satellite position	Solar	Albedo	Planet shine
Winter solstice, Normal	54.3	97.3	196.6
Winter solstice, Rolling −32°	9.5	80.0	165.6
Winter solstice, Rolling +32°	109.9	85.3	167.2
Summer solstice, Normal	48.2	95.6	183.3
Summer solstice, Rolling −32°	15.3	79.3	154.5
Summer solstice, Rolling +32°	78.9	82.6	157.6

It is seen that there is little external heat flux changes on summer solstice or the winter solstice in normal attitude; while in rolling attitude, albedo and planet shine decrease.

For rolling angle reaches to −32°, the arrival solar decreases obviously, while increases significantly at rolling attitude of +32°.

4 Thermal Design

For thermal design project, it needs to meet the required for imaging in normal attitude, positive attitude, and negative attitude. Optical-mechanical-electrical-thermal integrated design is used, and reasonable optical materials, structural materials and assembly techniques are carried to ensure temperature uniformity; a combination of active and passive measures of thermal control design is adopted in the scheme to use the thermal control resources reasonably [13]. In order to increase the reliability of thermal design, necessary redundancy and allowance is kept.

1) MLI are installed on the camera outer surface

   In order to minimize the impact of surrounding wall and space heat flux, and keep temperature uniformity and stability, 15 layers MLI are installed on the camera outer surface, which has superior insulation capability.
2) Blackened for internal camera surface

   In order to flatten barrel temperature level between the various metal parts and improve temperature uniformity inside the camera, while meet the requirements of optical extinction stray light, the surface of main metal parts in the light path are blackened, which make sure Infrared emissivity $\varepsilon h \geq 0.85$.
3) Reduce thermal coupling between camera and spacecraft

   In order to reduce the influence of satellite boundary, 10 mm thick thermal insulation pads are installed on the mounting surface, 5 mm thick insulation pads are installed between the camera lens hood and satellite panels. At the same time, 10 layers MLI are installed on the inside surface of camera small cabin, reducing the radiation heat transfer with the camera body.
4) Cooling for CCD

   According to design requirements, the temperature of CCD device needs to be kept within 12–26°C. To meet this requirement, copper conductive strip is pasted behind the CCD device, with one end connected to heat pipes, located in the focal plane box panel(+Z side) by a conducting cable (copper braid). Four heat pipes connect conducting cable and radiating surface, which are located in the +Z panel of the satellite payload module. At the same time, a certain heating power is designed to prevent the temperature of CCD device from becoming too low in rest working mode. Uses cerium glass silver second surface mirror is adopted as the radiating surface, with a size of $400 \times 300\ mm^2$.
5) High-precision temperature control

   In order to meet temperature requirement of camera body, especially the optical component, chip heaters are pasted on the appropriate location of the main parts in the camera and an active heating method is adopted to control the temperature. 14 heating circuits are arranged in the hood, primary mirror, CCD and other parts. The temperature controller collects temperature through thermistors, complete the comparison with the set values and control heaters using an on/off and proportion control strategy. The temperature control precision is up to 0.1°C.

6) Rolling attitude design

Even there is a huge space heat flux change between normal attitude and rolling attitude, compared with the period of 100 min, rolling working time is just only up to 10 min, which is so little time and bring limited impact. In order to reduce the influence of side-sway on the camera's temperature field, the choice of radiating surface is a key factor. Considering the camera layout restrictions on the satellite, the + Z surface radiating surface which is not sensitive to the side-sway is chosen. The exposed part of the camera is coated by MLI to reduce the effect of space heat flux variation on the camera body temperature field. As the radiating surface and entrance optics are the only parts exposed outside of the satellite, the influence of space external heat flux is weakened by adopting reasonable heaters to reduce the temperature fluctuation.

5 Thermal Analysis

Thermal analysis is an important method for thermal design. The main method for solving temperature field of spacecraft is node thermal network analysis techniques, which is based on the finite difference method and lumped parameter method, and it is also adapt for camera thermal analysis.

5.1 Thermal Analysis Method [14–15]

As for micro unit (node), the thermal network model can be expressed as:

$$(cm)_j \frac{dT_j}{d\tau} = \sum_i R_{i,j}\sigma(T_j^4 - T_i^4) + \sum_i D_{i,j}(T_j - T_i) + Q_{sj} + Q_{pj} \tag{5}$$

In formula 5, $(cm)_j$ represents heat capacity of node j; $\tau$is time; Q_{sj} represents absorb outer heat flux of node j; Q_{pj} is thermal power of node j; $R_{i,j}$ is thermal radiation heat transfer coefficient; $D_{i,j}$ is thermal conductivity transfer coefficient; $\sigma$is Stefan-Boltzmann constant.

5.2 Thermal Analysis Mode

According to the state of thermal design, thermal analysis mode is built by using analysis software and the analysis model is divided into 926 nodes. For CCD focal plane assembly, encryption grid is carried to analyze the effect of the surrounding environment.

5.3 Thermal Analysis Case

According to Table 3, it is known that in the normal attitude, the minimal outer heat flux occurs in summer solstice, while the maximum heat flux comes in winter solstice. Taking into account the satellite side swing, and thermal coating degradation, so two special analysis cases are defined: one is the low case, with the rolling angle of $-32°$in summer solstice; the other is hot case, with the rolling angle of $+32°$in winter solstice. Detailed cases are shown in Table 4 (Table 5)

Table 4. Thermal analysis cases

No	Orbit moment	Satellite attitude	Thermal-optical property	Work mode	Boundary condition	Note
Case 1	Summer solstice	Normal	Beginning of life	No operation	−5 °C	Low case
Case 2	Winter solstice	Normal	End of life	20min /orbit	30 °C	Hot case
Case 3	Summer solstice	Rolling − 32°	Beginning of life	No operation	−5 °C	Low case
Case 4	Winter solstice	Rolling + 32°	End of life	20min /orbit	30 °C	Hot case

Table 5. Thermal analysis results t/°C

Parts	Temperature requirement	Case 1	Case 2	Case 3 (Rolling)	Case4 (Rolling)	Note
Main mirror	18.5–21.5	19.07–19.32	20.66–20.69	19.08–19.32	20.40–20.43	Meet requirement
Secondary mirror	18.5–21.5	19.02–19.46	19.02–19.47	19.02–19.45	19.01–19.48	Meet requirement
Third mirror	18.5–22.5	19.00–19.25	22.30–22.37	19.00–19.24	22.02–22.09	Meet requirement
Back mirror body	18.5–22.5	18.47–19.15	21.17–21.80	18.41–19.09	20.88–21.49	Meet requirement
Inner knighthead	14–21.5	18.18–18.56	18.47–19.05	18.08–18.45	18.26–18.74	Meet requirement
Outer knighthead	14–21.5	16.30–17.58	16.78–18.86	16.11–17.14	16.47–18.13	Meet requirement
CCD	12–26	14.43–16.39	14.87–23.98	14.37–16.33	14.82–22.93	Meet requirement

5.4 Analysis Result

Analysis result is present in the following table.

6 Thermal Test

For further study the impact of rolling attitude, thermal balance test is carried out, including hot and low cases in normal attitude and rolling attitude.

6.1 Test Result

Test system is shown in Fig. 5, and test conditions are consistent with Table 4. All results are present in Table 6.

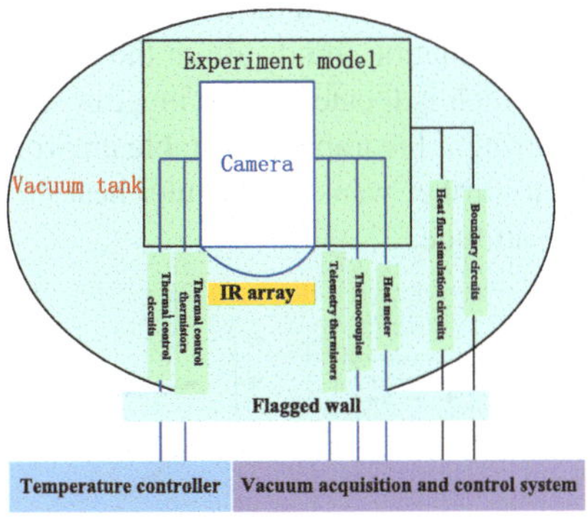

Fig. 5. Experiment system

Table 6. Thermal balance experiment results t/°C

Parts	Temperature requirement	Case 1	Case 2	Case 3(Rolling)	Case (Rolling)	Note
Main mirror	18.5–21.5	18..99–19.01	19.9–19.96	19.01–19.02	19.99–20.08	Meet requirement
Secondary mirror	18.5–21.5	19.09–19.25	19.11–19.31	19.09–19.25	19.14–19.31	Meet requirement
Third mirror	18.5–22.5	18.81–18.83	20.6–20.68	18.83–18.86	21.12–21.22	Meet requirement
Back mirror body	18.5–22.5	19.45–19.62	20.65–20.79	19.48–19.62	21.11–21.26	Meet requirement
Inner knighthead	14–21.5	18.23–18.58	18.64–18.99	18.28–18.59	19.19–19.51	Meet requirement
Outer knighthead	14–21.5	17.23–17.74	17.6–18.07	17.18–17.71	17.91–18.43	Meet requirement
CCD	12–26	13.49–13.75	17.81–22.23	13.46–13.75	18.07–22.40	Meet requirement

It is shown that all results meet the design requirements, and there is only fluctuation of 0.5°C between normal attitude and rolling attitude. Compared with thermal analysis result, there is only deviation of 1°C, which means analysis is consistent with the experimental results.

In rolling attitude, outer heat flux will change, and bring an effect on temperature distribution. Low case as an example, hood and radiator are analyzed to study the impact of rolling attitude.

6.2 Hood

Figure 6 is temperature distribution of hood. Four temperature points are distribution from the outside to the inside, from the figure, it can be seen that hood temperatures are affected by external heat flux, alternating light and shadow: the closer to outer space, the more temperature fluctuation; and farther to outer space, the less temperature fluctuation. All temperature fluctuation trends are in the same shape, especially in the hood temperature point IV, which is located in the innermost of hood, the temperature fluctuations stabilized, and bring a favorable and stable environment for the secondary mirror, the primary mirror, to further weaken the outer heat flux changing between the normal attitudes and rolling attitude.

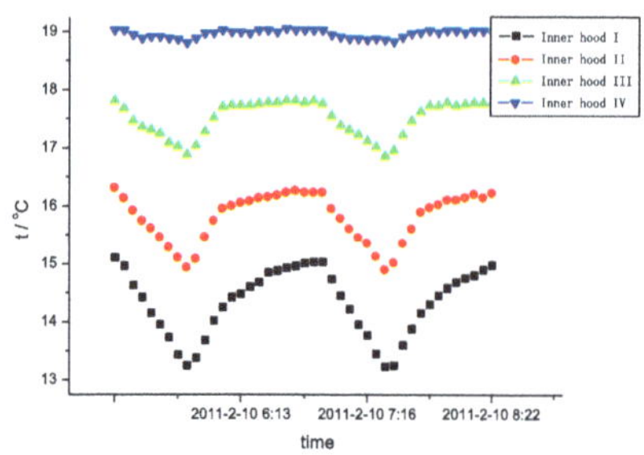

(a) Normal attitude

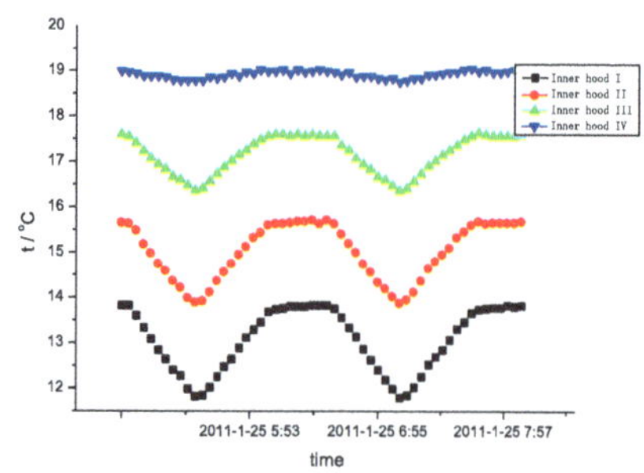

(b) Rolling attitude

Fig. 6. Inner hood temperature distribution

6.3 CCD

Figure 7 is temperature distribution of CCD. CCD is connected with cooling surface by heat pipes, so it should be affected by out heat flux. While the temperature of CCD is maintained around13.6 °C, which proves that it has the ability of avoid of being affected

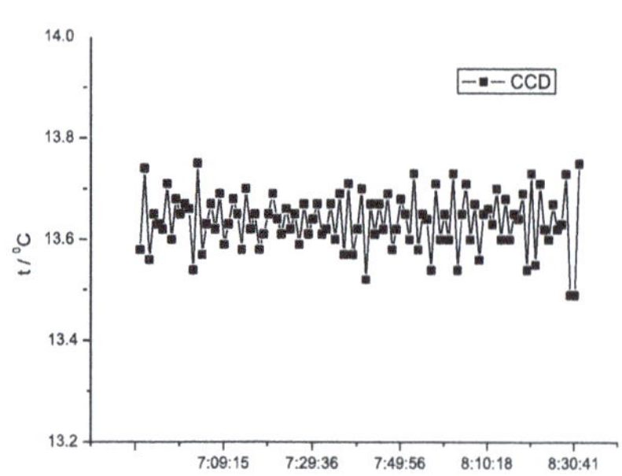

(a) Normal attitude

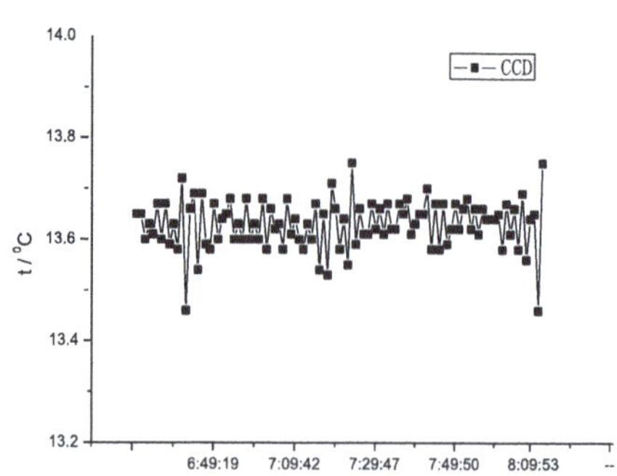

(b) Rolling attitude

Fig. 7. CCD temperature distribution

by out heat flux. On the one hand, for outer heat flux, the difference between normal attitude and rolling attitude is $5-10W/m^2$;on the other hand, a heater is carried on CCD, which decrease heat flux differences by proportional heater control technology. It proves that precision temperature control heaters can further weaken the influence of the outer heat flux.

7 Conclusion

The first systematic analysis of spatial heat flux theory is conducted, providing guidance for high-precision camera thermal design. Simultaneously conduct extreme thermal analysis and corresponding thermal balance tests on the camera under rolling conditions, and the test results are consistent with the thermal analysis results. Short term rolling has little effect on the temperature field of the camera body, and only a greater impact on the hood; precision temperature control can be effectively reduced rolling effects.

In order to adapt to the fast and good application scenarios of future spacecraft and weaken the influence of complex heat flux on camera temperature field, further consideration can be given from the following two aspects:

1) Increase the length of the hood, such as using techniques such as unfolding the hood, to reduce the influence of external heat flux on the inside hood;
2) Perform high-precision temperature control on key components, such as CCD, to reduce the influence of external heat flux.

References

1. Hong, W., Changyuan, H.: Study on the thermal effects of the optical system in a aerospace camera. Optic. Tech. **29**(4), 452–457 (2003)
2. Fujian, D., Yingcai, Li.: An effective optimum design method of the structure of cassegrain reflective system. Acta Photonica Sin. **29**(8), 756–762 (1999)
3. Dun, G.: The thermal effect on image quality of the remote sensor optical system. Changchun Insti. Opt. Fine Mech. Phys. Chinese Academy Sciences (2010)
4. Guoqiang, L., Hong, J., Qiang, L.: Analysis for some factors affecting the tem-perature distribution of a CCD camera. Chin. Space Sci. Technol. *21*(5), 62–70 (2001)
5. Lixin, Z., Ying, S.: Summary of thermal control and thermal optical analysis for space optical system. Spacecraft Recov. Remote Sens. **22**(2), 13–19 (2001)
6. Paul, R.: Opto-Mechanical Systems Design. CRC Press Taylor and Francis Group, California (2006)
7. Giesen, P., Folgering, E.,: Design guidelines for thermal stability in opto-mechanical instruments. Optomechanics 2003, San Diego (2003).
8. Wu, L., Zhang,C.M., Zhao, B.C.: Analysis of the lateral displacement and optical path difference in wide-field-view polarization interference imaging spectrometer. Opt. Commun. **273**(1), 67–73 (2007)
9. Rongli, C., Zhen, M., Wengang, Y.: Thermal-optical analysis and tests for high resolution space camera. Acta Photonica Sin. **39**(11), 2068–2072 (2010)
10. Wenggang, Y., Lei, Y., Rongli, C.: Precise thermal control design and validation for high resolution space camera. ACTA Photonica Sin. **38**(9), 2363–2367(2009)

11. Changxiang, E.L., Qingwen, Y.W.: Research on adaptability of optical re-mote sensors in mechanical and space thermal environments. Chinese J. Opt. Appl. Opt. **2**(5), 364–376(2009)
12. Xianwen, N., Jiaxun, Z., Hai, J.: Extreme external heat flux analytical model for inclined-orbit hexahedral satellite. J. Astronaut. **29**(3), 754–759 (2008)
13. Zengqi, H, Jingang. H.: Themal Control Technology of Spacecraft. Chinese Science and Technology Press (2007)
14. David, G., Gilore.: Spacecraft Thermal Control Handbook. 2nd edn. The Aero-space Press EI Segundo, California (2002)
15. Guirong, M., Shun, G.: Spacecraft Thermal Control. Science Press, Beijing (1998)

Design and Simulation of the Baffle of Low Light Level Imaging Sensor

Shengbo Mu(✉), Yang Li, and Tingcheng Zhang

Beijing Institute of Space Mechanics and Electricity, Beijing 100094, China
msb_20050390@126.com

Abstract. Low light level imaging sensor is used to detect faint visible light, such as the moonlight, starlight, airglow and so on. The sensor has had great development with its high sensitivity, large dynamic range, therefore, stray light can cause imaging failure directly. A two-class truncated cylinder baffle was proposed and baffle design method is discussed in detail. The point source transmittance(PST) is used to evaluate the stray light suppression of the system. The system is used in field bias to avoid ghost in transmission optical system. In the end, the simulation results are analyzed and calculated with Lighttools software. Finally the PST curves were given at the off-axis angles between 18°and 40°. The analysis results show that the two-class truncated cylinder baffle could stop the stray light from the veiling glare rejection angle and reduce the scattered rays reflected by the baffle. For stray light sources with small off-axis angles, the PST is reduced to the order of −7. It is capable of meeting usage requirements.

Keywords: low light level imaging sensor · Baffle · PST

1 Introduction

With the development of high-technology war, the special optics technology and the multiple types of photoelectric detection devices technology have made a leap in development. The high sensitivity and low detection threshold low light level detection remote sensors have emerged one after another. For such weak target detection systems, the stray light can directly cause the failure of the imaging. So the suppression and analysis of stray light has become an important issue. Remote sensing devices for detection weak signal targets at home and abroad mainly use low light imaging systems (visible spectral range) and infrared thermal imaging systems [1–4]. The low light imaging system refers to an optical system that obtains visible light images under low illumination conditions such as night and dawn. Its principle is basically the same as that of visible light imaging. It only has higher requirements for the sensitivity and dynamic range of detection instruments. Compared with infrared thermal imaging system, low light imaging systems are easier to distinguish information that is similar to the characteristics of radiation sources which is the important information related to combat such as heavy fog and surface conditions. So, it is also a technology which countries pay more attention to in their development [5–8].

H. P. Urbach et al. (Eds.): ISSOIA 2023, SPP 191, pp. 179–185, 2024.
https://doi.org/10.1007/978-981-97-6718-2_17

For general optical remote sensors, the design of the light baffle is mainly used to block stray light outside the field of view. It plays a role in protection for field of view. For high sensitivity low light detection remote sensors, its observation target is weak visible light, which is visible light at night such as moonlight, starlight and airglow. Compared to visible light detection, its detection target is much darker and the detector has a larger dynamic range which can generally reach the order of 10^{-6} to 10^{-9} [9, 10]. The stray light generated by the scattering of stray light sources outside the field of view through the surface of the light baffle can be recognized by the detector. It is possible to interfere with the load and even directly cause the imaging failure. High performance external light baffles are required to ensure that intense light sources outside the edge of the field of view do not interfere with the load. Taking a high sensitivity low light imaging remote sensor with a transmission optical system structure as an example, the stray light suppression methods for high sensitivity low light imaging remote sensors were mainly researching. The source of the stray light in the system was analyzed. The design method of the external light baffle was detailed discussion.

2 Baffle Design

2.1 The Principles of Baffle Design

The principle of baffle design of the high sensitivity low light level detection remote sensor is the following three points. First, ensure the imaging light was not blocked by the baffle. Second, avoid large off-axis angle stray light sources directly entering the outer baffle of the remote sensor. At last, reduce stray light scattered multiple times into the remote sensor through the outer baffle at a small off-axis angle. A method which the external and internal baffles were combining together with an asymmetric inclined entrance structure has been proposed in this paper. A field diaphragm was designed used in field bias which can reduce stray light entering the camera caused by the multiple scattering by the baffle. It also can avoid the ghost images can also be avoided and the requirements of the camera can be meet.

2.2 The Design of the Asymmetric Baffle

For low light level imaging remote sensors, the strongest stray light source is the sunlight. So the design of the external baffle needs to avoid direct sunlight entering the baffle. According to the orbital characteristics of the remote sensor, when the remote sensor is working on orbit, if the sun is always on one side of the remote sensor, the baffle can be designed in an inclined entrance form. The design principle of the inclination angle is shown as Fig. 1. The formula 1 is the calculation method. The inclination angle α of the front end surface of the baffle is the key technical indicator that determines whether the sunlight can enter the baffle. The larger the angle α, the larger the angle between the front surface normal and the optical axis, and the less sunlight can enter the baffle. As shown in Fig. 1, for the baffle with any inclination angle α. When the angle θ between the sunlight and the normal of the front surface is less than 90°, the sunlight will not enter the large baffle. Otherwise, when it is larger than 90°, it will enter.

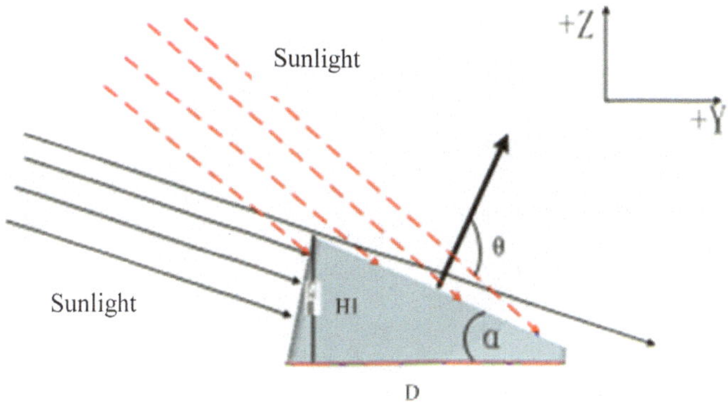

Fig. 1. Design principle of inclination angle of the baffle

$$\tan \alpha = \frac{H_1}{D} \tag{1}$$

Based on the operational characteristics of the remote sensor on orbit, the sunlight is always on one side of the remote sensor. So the design of the external baffle can choose an asymmetric inclined entrance structure form. After analysis and calculation, when the inclined angle α of the baffle is designed larger than 60°, the angle θ between the sunlight and the normal of the front surface is less than 90° all the year. It can completely block direct sunlight. As shown in Fig. 2, the tilt angle of the baffle is taken as 60°.

2.3 Calculation of the Size of the Baffle

The calculation method for the length of a two section asymmetric entrance baffle is shown as the formula (2–5) [11].

$$L_1 = D \times \tan \alpha \tag{2}$$

$$L_1 + L_2 = L \tag{3}$$

$$L_2 = \beta L \tag{4}$$

$$D = d + 2(L_1 + L_2) \times \tan \theta \tag{5}$$

L is the total length of the baffle. L1 is the length of the first level baffle. L2 is the length of the second baffle. d is the aperture of the first surface of the optical system. Θ is the half field angle of the view. ω is the suppression angle of the stray light. D is the maximum aperture of the optical system. Among them, $\beta \geq 0.5$ [12]. According to the above method, the baffle is designed. The structure of the baffle is shown in Fig. 3.

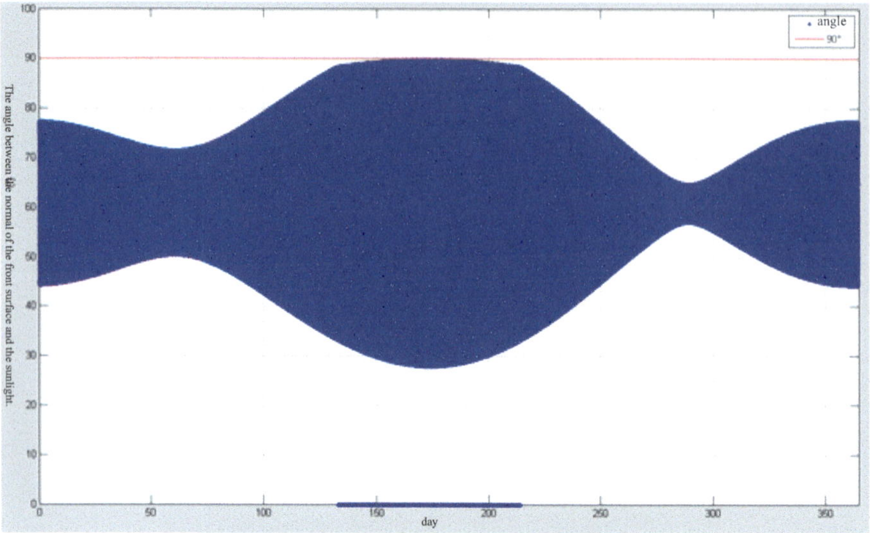

Fig. 2. The angle θ between the normal of the inclined surface of the baffle and the light

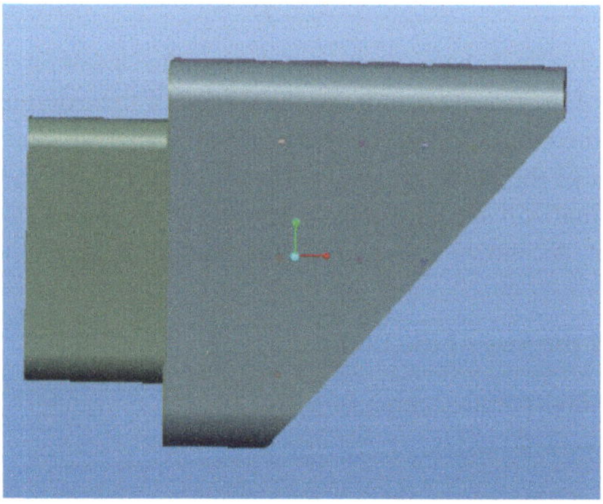

Fig. 3. The structure of the baffle

2.4 The Design of the Vanes

The ghost image of a transmissive optical system is also its main stray light source. The most energetic ghost image is called a focused ghost image. At present, the transmittance of the lens after being coated with an antireflective film is about 99% and the reflectivity is approximately 1%. The ratio of the ghost image energy generated by secondary reflection between mirrors to the energy of imaging light is 1/10000. If the ghost image is a focused ghost image, it can be detected by a high sensitivity low light imaging remote sensor. It will impact the imaging. Because the imaging field of view is rectangular, the baffle vanes are designed as rectangle. To reduce scattering from the surface and the end face of the vane, a certain inclination angle is designed at the end face of the vane. The design results of the baffle vanes are shown in Fig. 4.

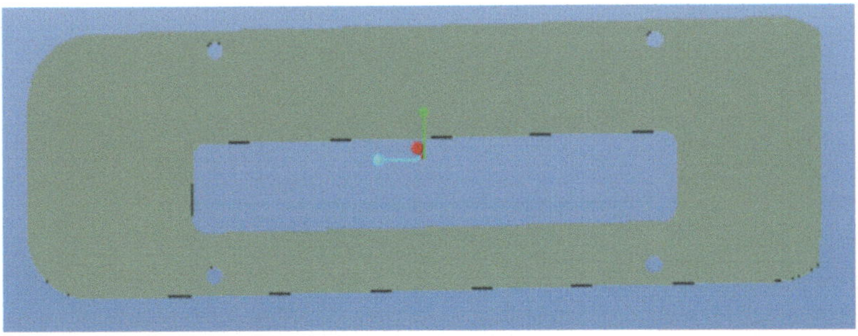

Fig. 4. The schematic diagram of the structure of the baffle vane

3 Simulation and Analysis Results of Baffle

According to the structure of the optical system and the external baffle, the optical mechanical structural models are modeled and analyzed using lighttools software. According to the existing processes, the surface characteristics of the system components are set. The surface absorption rate of the baffle is set to 95%, and the reflectivity of the surface which is the lamb surface is set to 1%. The surface transmittance of the lens is set to 98%, and the reflectivity is set to 1%, the absorption rate is set to 1%. The remaining mechanical structure surfaces are set as blackened surfaces. The parameters adopt the standard values from the surface attribute library in the software. The point source transmittance (PST) is selected to evaluate index. The suppression of stray light under different angles of incidence of sunlight as shown in Table 1 is calculated. The PST curve is shown as Fig. 5.

From the above analysis results, it can be seen that the baffle can prevent stray light sources with an off-axis angle larger than 36.8° from entering the baffle. For stray light sources with small off-axis angles, the PST is reduced to the order of −7. It is capable of meeting usage requirements.

Table 1. The calculation result

Sunlight incident angle (°)	PST
18°	8.85e-7
20°	6.47e-7
22°	5.42e-7
24°	3.91e-7
26°	2.88e-7
28°	1.29e-7
30°	8.69e-8
32°	6.37e-8
34°	4.09e-8
35°	1.12e-8
36.8°	0
40°	0

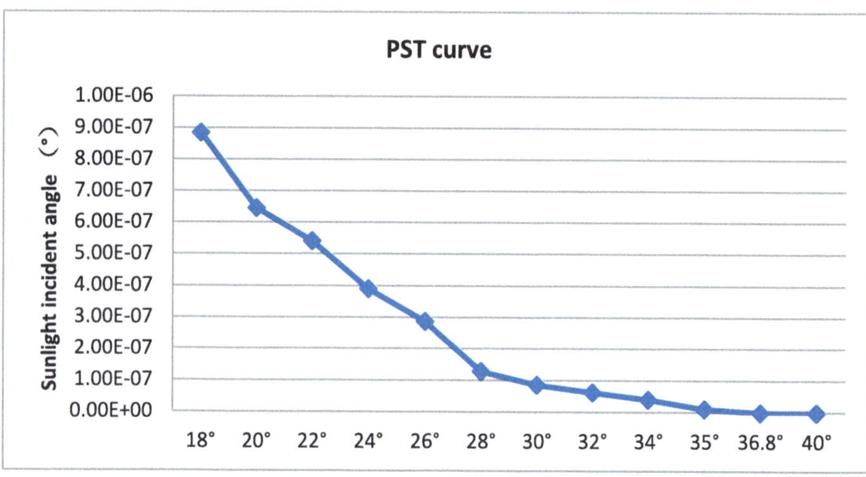

Fig. 5. The PST curve

4 Conclusion

This paper discusses the design method of the external baffle of a high sensitivity low light imaging remote sensor. The design principles for the external baffle of a high sensitivity low light imaging remote sensor have been derived. The design method of a two section inclined entrance type baffle is detailed introduction. Taking a high sensitivity low light imaging remote sensor as an example, an external baffle was designed. Finally, the optical system was simulated and analyzed in the lighttools simulation software.

The PST was calculated. The analysis and calculation results indicate that the baffle can effectively prevent stray light from entering the baffle outside the suppression angle. For stray light sources with small off-axis angles, the PST is reduced to the order of −7. It is capable of meeting the requirements for the use of high sensitivity low light imaging remote sensors.

References

1. Huang, Y., Xu, J., Bai, S., et al.: Satellite infrared remote sensing technology and its application in disaster prevention and relief. Spacecraft Recov. Remote Sens. **41**(5), 118–126 (2020)
2. Wang, H., Li, Q., Bo, S., et al.: Research on the analog front end processing technology of foreign space infrared remote sensors. Spacecraft Recov. Remote Sens. **41**(5), 83–94 (2020)
3. An, M., Liang, D., Nie, Y., et al.: Research on infrared detector imaging test methods for ZY-1 02E satellite. Spacecraft Recov. Remote Sens. **42**(4), 37–45 (2021)
4. Qiao, M., Tan, J., Liu, Y., et al.: Infrared dim small flying target recognition algorithm for space-based surveillance. Chin. Space Sci. Technol. **42**(5), 125–132 (2022)
5. Liu, Y.: Research on Optimization Theory and High Performance System Technology of Low Light Imaging Detection. Nanjing University of Science and technology (2009)
6. Zhou, L.: Progress and prospects of low light imaging technology. In: Progress in Modern Optics and Photonics-Celebrating Academician Wang Daheng's 65th Anniversary of Scientific Activities Album. Tianjin: Tianjin Science and Technology Press, pp. 316−339 (2003)
7. King, N.S.P., Albright, K.L., Jaramillo, S.A., et al.: High-frame-rate CCD cameras with fast optical shutters for military and medical imaging applications. In: Ultrahigh-and High-Speed Photography, Videography, and Photonics 94. Proceedings SPIE **2273**, pp. 56–60 (1994)
8. Loktionov, V.: High technology of night vision system with image intensifier manufacture: state and summary of year 2003. In: SPIE, vol. 5834, pp. 197−202 (2005)
9. Li, H., Li, Y.: Optimization design method for elimination stray light in spaceborne optical system baffle. J. Photonics **25**(10), 914−919 (1996)
10. Chen, S.: The effects on romote sensing image quality from scenes and imaging condition. Spacecraft Recov. Remote Sens. **31**(1), 1−7 (2012)
11. Li, Y., Xiang, L., Li, L.: Design and simulation of the baffle of a new type space camera. Opto-Electronic Eng. **37**(7), 41–44 (2010)
12. Lu, W., Zhan, L., Zhang, J., et al.: Design of lens hood for star sensor and result simulation. Opto-Electronic Eng. **28**(3), 12–15 (2001)

Using Remote Sensing Technologies for Monitoring Urban Heat Islands

Stanislav Sakhniuk[1](✉), Vitaliy Zatserkovnyi[1], Mauro De Donatis[2], Iryna Tsiupa[1], and Anzhelina Mienasova[1]

[1] Taras Shevchenko National University of Kyiv, Institute of Geology, Kyiv 03022, Ukraine
sakhnyuk2003@ukr.net

[2] Università Degli Studi Di Urbino Carlo Bo, 61029 Urbino, Italy

Abstract. The process of urbanization is accelerating day by day, so this rapid development turns the natural landscape into an artificially created one [3]. In turn, the growth of urbanization causes significant changes in the natural landscape. This leads to microclimatic changes, air pollution, heat island effect etc. The UHI is an urban area whose air temperature exceeds the temperature of the suburbs by 4° or more. Due to air pollution by man-made emissions in urbanized areas, the thermal regime is changing, the concentration of carbon dioxide and water vapor has now reached 90% of the total amount of pollutants [8]. Given the potential danger of elevated values of air temperature, caused by UHI, that affect the lives of residents, the method for determining surface temperature and the location of heat islands should be determined.

The purpose of this paper is to analyze the problem of urban heat islands. Using the processing and analysis of data from Google Earth Engine, to analyze changes in the number and condition of green spaces, as well as to analyze areas of interest for changes in surface temperatures over time. To identify the relationship between green spaces in megacities and the UHI. To propose an effective method of monitoring the UHI for further prevention of environmental disasters in megacities.

Keywords: Remote sensing · GIS · UHI · Google Earth Engine · LST · UTVFI

1 Introduction

In recent years, human activities, mainly caused by anthropogenic factors, have resulted in unprecedented climate change and environmental degradation, making global warming and energy shortage extremely apparent. As a result, there has been a significant shift in people's attitudes and behaviors, urging them to take strict measures to reduce carbon emissions. Specifically, in the building sector, which is a significant contributor to energy consumption and greenhouse gas emission, several actions and best practices have been implemented to promote energy efficiency and reduce carbon emissions [7].

The process of urbanization is accelerating day by day, so this rapid development turns the natural landscape into an artificially created one [3]. In turn, the growth of

H. P. Urbach et al. (Eds.): ISSOIA 2023, SPP 191, pp. 186–196, 2024.
https://doi.org/10.1007/978-981-97-6718-2_18

urbanization causes significant changes in the natural landscape. This leads to microclimatic changes, air pollution, heat is-land effect etc. The UHI is an urban area whose air temperature exceeds the temperature of the suburbs by 4 degrees or more [10]. Due to air pollution by man-made emissions in urbanized areas, the thermal regime is changing, the concentration of carbon dioxide and water vapor has now reached 90% of the total amount of pollutants. As a result, there is another problem that contributes to global warming – the "greenhouse effect". Elevated values of air temperature affect human health, lead to breathing problems, convulsions, heat and sunstroke, heat stress and increased mortality [8].

Given the potential danger of elevated values of air temperature, caused by UHI, that affect the lives of residents, the method for determining surface temperature and the location of heat islands should be determined.

The purpose of this paper is to analyze the problem of urban heat islands. Using the processing and analysis of data from Google Earth Engine, to analyze changes in the number and condition of green spaces, as well as to analyze areas of interest for changes in surface temperatures over time. To identify the relationship between green spaces in megacities and the UHI. To propose an effective method of monitoring the UHI for further prevention of environmental disasters in megacities.

The objects of research in this paper are surface temperatures and green spaces of such megacities as Oslo, Kyiv and Rome. The subject of the study is monitoring of urban heat islands, their impact on residents and finding effective preventive solutions.

2 Method and Theory

2.1 Platform and Database Overview

Data from the Sentinel 2 MSI satellite was selected and analyzed for this study: multispectral instrument, level-1C and sensor MODIS: MOD11A2.061 Terra Land Surface Temperature and Emissivity 8-Day Global 1 km.

MODIS (Moderate Resolution Imaging Spectroradiometer) is a spectroradiometer with 36 spectral bands and a resolution of 250–1000 m. It can measure surface temperatures with high separation over a wide area, thanks to which it can help identify regions of cities where the air temperature is higher than in the surrounding area.

Using MODIS data, it is possible to identify urban regions where efforts should be focused to reduce the impact of urban heat islands, for example, by increasing green space, etc. [4].

Using Google Earth Engine (GEE) platform, that uses a combination of satellite imagery, Earth observation data, and machine learning algorithms to allow users to identify and measure changes in land use, ecosystems, and climate patterns at a global scale, it became possible to identify urban regions where efforts should be focused to reduce the impact of urban heat islands. The Google Earth Engine relies on multiple technologies within the Google data center, such as Flume Java framework, Bigtable, Spanner distributed databases [2], Borg cluster management system, Colossus, and Google Fusion Tables [5]. GEE provides accessibility via its API and web-based interface, enabling swift prototyping and visualization of outcomes. The GEE Data Catalogue presents a wide assortment of geospatial data, comprising satellite images that are pre-processed, optimized, and available in a format that mitigates data management issues, leading to amplified computing efficiency [7]. Additionally, the data is accessible to the masses.

2.2 Study Areas

Since the best representatives of the UHI, as mentioned earlier, are megacities, the territories of interest in this work were three European capitals – Oslo, Kyiv and Rome.

The reason for choosing Rome as a territory for analysis was also the fact of the great historical status of the city, which, together with European standards, is the basis for serious restrictions on clearing, and given the geographical location of this city, due to which residents suffer from heat in summer, we would like to identify possible trends in temperature changes, and in case of their absence, confirm the theory that the total preservation of plantations is an effective solution to reduce the impact of heat islands.

As for Kyiv, due to the irresponsibility of the authorities, the city is experiencing problems with massive illegal felling of trees, respectively, serves as an example of a city that is experiencing changes in the number of green spaces, unlike Rome. In addition, both cities tend to have temperature differences with the suburbs, which proves the expediency of their choice as areas of interest.

Oslo is known as the greenest European capital, therefore, it is important to analyze the temperature regime in the city and compare the state of other capitals with the leader.

All three capitals are located in different latitudes and form an imaginary triangle. Each of the cities also demonstrates an example of different climatic conditions, which also affect the condition of vegetation, residents and surface temperatures (Fig. 1).

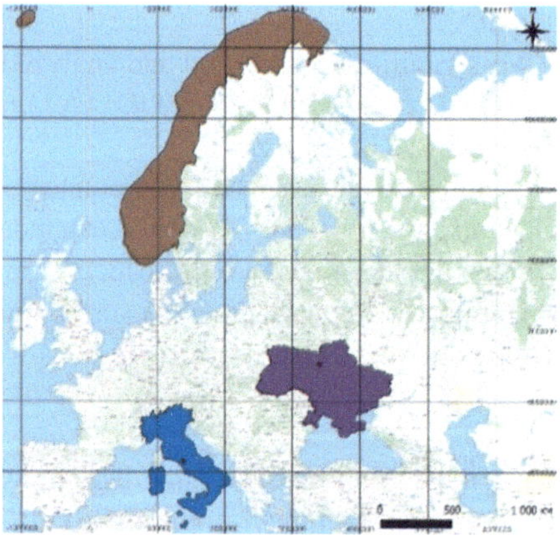

Fig. 1. Location of the study areas – Oslo, Kyiv and Rome, based on OpenStreetMap data

2.3 Methodology

Among the main factors of UHI are changes in the surface of the earth and lack of evaporation due to the lack of vegetation in urban area [6]. It is green areas that have cooling effect for megacities, and carbon dioxide absorption. Therefore, since green areas play such an important role in the temperature regime of the city, we would like to analyze their possible relationship with the UHI in more detail.

In most cases, the vegetation map (NDVI, or Normalized Difference Vegetation Index) accurately reflects the heat islands, as the NDVI provides a way to know the amount of photosynthetic active biomass in plants [1]. Therefore, the calculation of this index is the next step in the analysis of urban heat islands and their interaction with green spaces. This index is a measure of the difference in reflectivity between the red band (RED) and the near-infrared (NIR) band of images. NDVI ranges from -1 to $+1$, where positive values indicate areas covered with vegetation and negative values indicate areas without vegetation.

It is calculated as follows:

$$\mathrm{NDVI} = \frac{\mathrm{NIR} - \mathrm{RED}}{\mathrm{NIR} + \mathrm{RED}} \tag{1}$$

where NIR is the reflection coefficient in the near-infrared zone; RED is the reflectance in the red spectral zone (Fig. 2).

```
var NDVI = image.expression(
    "(NIR - RED) / (NIR + RED)",
    {
      RED: image.select("B4"),    //  RED
      NIR: image.select("B8"),    // NIR
      BLUE: image.select("B2")    // BLUE
    });
Map.addLayer(NDVI, {min: 0, max: 1}, "Oslo NDVI");
Map.addLayer(NDVI, {min: 0, max: 1, palette: ['brown', 'yellow', 'green']}, "NDVI colour");
```

Fig. 2. Code snippet to compute NDVI in Google Earth Engine

When working with spectroradiometer data, it is enough to select the necessary channel that will display the desired data, in the case of urban heat islands, it is usually a surface temperature channel, depending on the data set (Fig. 3).

```
var modis = ee.ImageCollection('MODIS/006/MOD11A2');
var start = ee.Date('2022-06-01');
var dateRange = ee.DateRange(start, start.advance(1, 'months'));
var modLSTday = modis.filterDate(dateRange).select('LST_Day_1km');
var modC = modLSTday.map(function(image) {
  return image
    .multiply(0.02)
    .subtract(273.15)
    .copyProperties(image, ['system:time_start']);
});
var temp_trend = ui.Chart.image.series({
  imageCollection: modC,
  region: roi,
  reducer: ee.Reducer.median(),
  scale: 1000,
  xProperty: 'system:time_start'})
  .setOptions({
    lineWidth: 1,
    pointSize: 3,
    trendlines: {0: {
        color: 'CC0000'
      }},
     title: 'LST  Time Series Rome',
     vAxis: {title: 'LST Celsius'}});
print(temp_trend);
```

Fig. 3. Code snippet to visualize land surface temperature (LST) using MODIS data in Google Earth Engine

Thus, to confirm the relationship between green areas and urban heat islands, maps taken from images of the Sentinel-2 MSI satellite, which have a time interval of seven years, and temperature maps of the capitals surface, made according to data obtained at the same time sensor MOD11A2.061 were constructed (the maps on the top represent data of June 2015, on the bottom – June 2022) (Fig. 4).

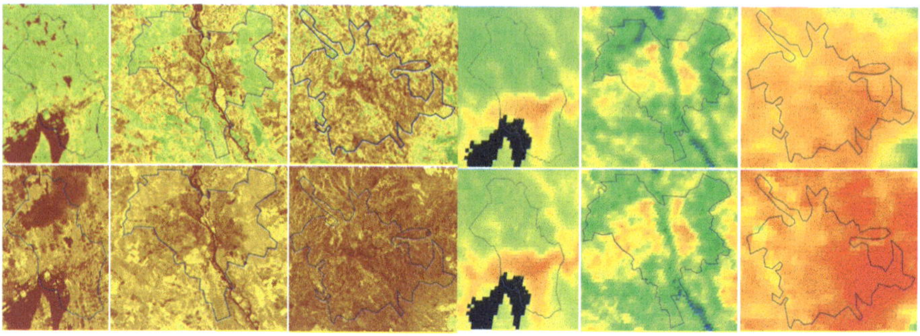

Fig. 4. Vegetation and surface temperature maps of Oslo, Kyiv and Rome from left to right (2015 – on the top; 2022 – on the bottom) from Sentinel-2 and the MODIS sensor. Application date: 20.10.2023

The maps show that the number of yellow and orange pixels (responsible for elevated temperatures) has increased significantly in Kyiv and Rome. At the same time, the temperature map of Oslo has hardly undergone any visual changes, which may indicate the stability of the temperature regime.

The values of the vegetation index of pixels in different part of the cities were compared.

Thus, the average value of NDVI was decreased almost in a half in each of the cities. Given that the condition of the vegetation can be determined by the NDVI indicator, we can say that indicators >0.3 units demonstrate a satisfactory condition of the greenery. However, the difference in latitude should still be considered, so vegetation with an NDVI value of 0.3 units in Rome is likely not identical to vegetation with the same NDVI value in Oslo.

For quantitative analysis and additional confirmation of changes in surface temperatures and impact of greenery on urban cooling, temperature graphs were constructed. The graphs demonstrate temperature changes in June 2015 and June 2022 in each of the earlier mentioned city (Fig. 5).

From the graphs it is noticeable that the average temperature in Kyiv is slightly, but increased, the average monthly temperature in Rome has increased by more than 4 degrees Celsius, and the temperature graph of Oslo stayed almost the same. These graphs could serve as a confirmation of temperature maps demonstrated earlier.

The last step in the search for urban heat islands is the calculation of the Urban Thermal Field Variance Index (UTVFI), which is calculated using the formula below, and the distribution of the obtained values into six categories according to the environmental

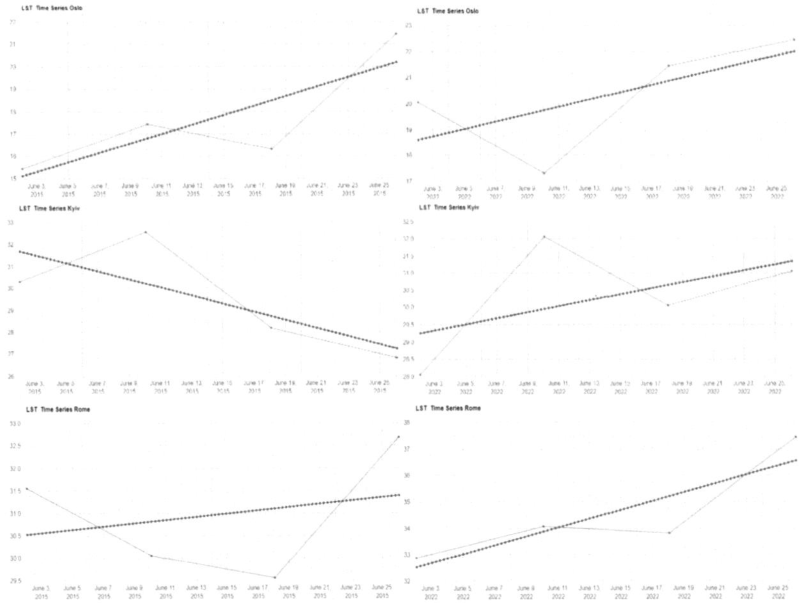

Fig. 5. Temperature change graphs in the capitals in June 2015 (on the left) and 2022 (on the right), plotted from MOD11A2.061. Application date: 20.10.2023

assessment index (Table 2).

$$\mathrm{UTVFI} = \frac{\mathrm{Ts} - \mathrm{Tm}}{\mathrm{Ts}} \tag{2}$$

where Ts is the temperature of the earth's surface; Tm is the average temperature of the earth's surface (Fig. 6) (Table 1).

```
//UTVFI
var lst_mean = ee.Number(LST.clip(geometry).reduceRegion({
reducer: ee.Reducer.mean(),
geometry: geometry,
scale: 30,
maxPixels: 1e9
}).values().get(0));
print(lst_mean, 'lst_mean');

var lst_std = ee.Number(LST.clip(geometry).reduceRegion({
reducer: ee.Reducer.stdDev(),
geometry: geometry,
scale: 30,
maxPixels: 1e9
}).values().get(0));
print(lst_std, 'lst_stdDev');

var utvfi = LST.subtract(lst_mean).divide(LST).rename('UTVFI')
var utvfi_vis = {
  min: -1,
  max: 0.3,
  palette: ['313695', '74add1', 'fed976', 'feb24c', 'fd8d3c', 'fc4e2a', 'e31a1c',
    'b10026']
}
Map.addLayer(utvfi.clip(geometry), utvfi_vis, 'UTVFI AOI')
```

Fig. 6. Code snippet to compute UTVFI in Google Earth Engine

The next step was to divide the UTVFI data into six classes according to the level of environmental assessment discussed earlier (Table 2), which in the future will also make it possible to calculate changes in the size of the UHI. Based on the classification

Table 1. Table of NDVI values of pixels with time interval of seven years

	August of 2015	August of 2022
Oslo (the territory of Ekebergparken, Kongsveien 23)	0,673	0,377
Kyiv (the territory of Shevchenko Park, Volodymyrska Street, 60)	0,671	0,473
Rome (the green areas around the Colosseum, Piazza del Colosseo, 1)	0,695	0,335

Table 2. Table of UTFVI and Environmental Assessment Index thresholds

UTFVI	Urban Heat Island	Environmental Assessment Index
<0,000	None	Perfect
0,000–0,005	Weak	Good
0,005–0,010	Medium	Satisfactory
0,010–0,015	Strong	Bad
0,015–0,020	Stronger	Worse
>0,020	The strongest	The worst

of the values of the urban heat field dispersion index, maps of the distribution of heat islands in each of the capitals for 2015 and 2022 were created (Fig. 7).

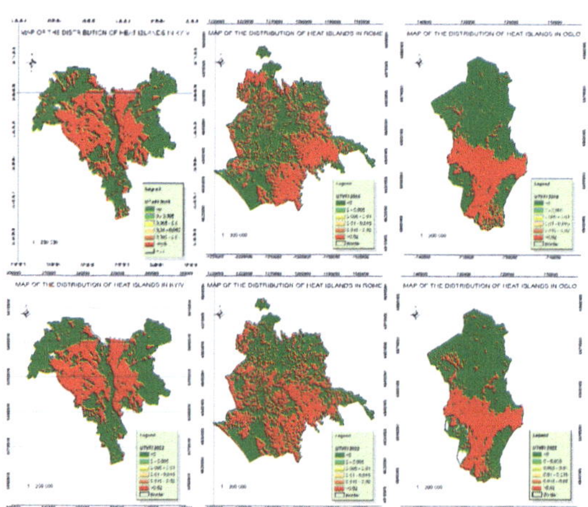

Fig. 7. Maps of the distribution of heat islands in selected cities based on MODIS data. Date of appeal: 20.10.2023

Once the urban heat islands have been found, visualized and grouped according to the environmental assessment, it is possible to quantitatively and qualitatively analyze the changes that may have occurred over time. Thus, the main criteria for the analysis will be the values of vegetation indices, including the previously mentioned NDVI and the equally important leaf area index (LAI).

The leaf area index (LAI) is also quite important during monitoring of green spaces because it shows the ratio of total leaf area to land area. In turn, the amount of intercepted light, accumulated nitrogen, respectively, surface temperature, UHI, etc. depends on the leaf area [1]. Maps of leaf area index values for Oslo, Kyiv and Rome were constructed according to July 2015 and 2022 data and values for the same pixel in certain parts of cities were calculated (the territory of Ekebergparken – Oslo, the territory of Holosiivskyi Park – Kyiv, the territory of the Villa Doria Pamphili Landscape Park – Rome) (Fig. 8).

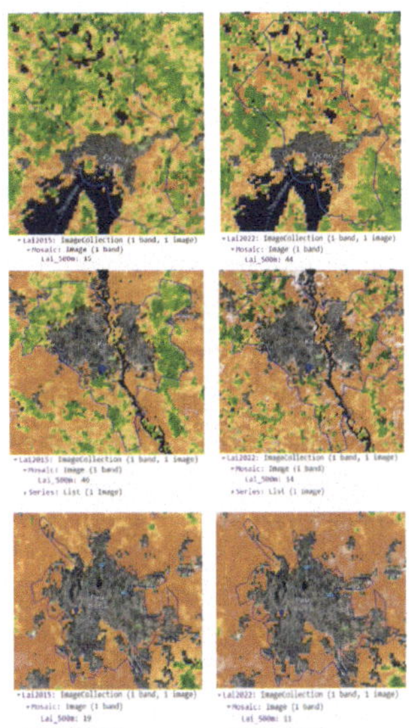

Fig. 8. Leaf surface index maps of the capitals, made using the product MOD15A2H.061: Terra LAI 500 m. Application date: 05.02.2023 and 20.10.2023

Kyiv and Rome have both experienced a decline in LAI values over the past seven years. In Oslo, on the contrary, the value of the leaf area index increased significantly, which may indicate an improvement in the condition or increase in vegetation in the analysis area.

Geodatabases that contain information about the vegetation cover of cities, are also very useful tool to monitor urban heat islands, since analyzing changes in it, it becomes possible to predict the spread of heat islands. Global forest change data in this work was provided by the Hansen Global Forest Change (2000–2023) product made by the Global Land Analysis and Discovery (GLAD) laboratory. Using maps of tree loss and growth from 2000 to 2022, the situation in the capitals were compared (Fig. 9).

Looking at the maps, we can say that Oslo and Kyiv have suffered the largest number of trees cut down (shown in red pixels), but unlike Kyiv, Oslo also has a large number of

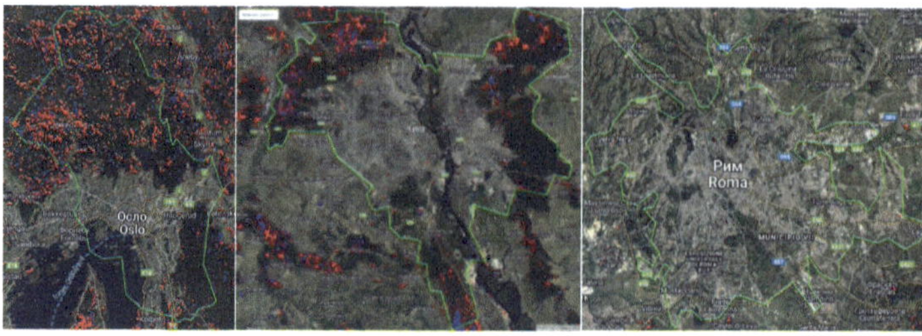

Fig. 9. Information from the Hansen Global Forest Change database (2000–2023). Application date: 05.02.2023 and 20.10.2023

newly planted trees (shown in blue pixels). As for Rome, there were almost no changes in the number of plantations.

3 Results

Analyzing the territories of the cities of Oslo, Kyiv, and Rome, a detailed analysis of changes in surface temperatures between 2015 and 2022 was conducted. Thus, temperature changes in each city were different.

Analyzing the results obtained (Table 3), it is possible to notice an increase in the area of urban heat islands in Rome and Oslo. Thus, the worst trend is in Rome, where the area of the strongest and worst heat islands in terms of environmental assessment has increased by as much as 30 km^2.

Table 3. Table of values of the dispersion index of the urban thermal field

KYIV					
UTVFI 2015			UTVFI 2022		
Class	Sum of Area	%	Class	Sum of Area	%
1 - None	422,32	51,01	1 - None	457,86	55,31
2 - Weak	5,83	0,70	2 - Weak	4,80	0,58
3 - Medium	6,10	0,74	3 - Medium	4,85	0,59
4 - Strong	6,15	0,74	4 - Strong	5,03	0,61
5 - Stronger	6,28	0,76	5 - Stronger	4,82	0,58
6 - The strongest	381,18	46,04	6 - The strongest	350,43	42,33
Total	828	100	Total	828	100
ROME					
1 - None	601,10	46,75	1 - None	589,99	45,88
2 - Weak	29,39	2,29	2 - Weak	24,14	1,88
3 - Medium	31,10	2,42	3 - Medium	26,14	2,03
4 - Strong	31,65	2,46	4 - Strong	26,60	2,07
5 - Stronger	30,34	2,36	5 - Stronger	26,42	2,05
6 - The strongest	562,10	43,72	6 - The strongest	592,60	46,08
Total	1286	100	Total	1286	100
OSLO					
1 - None	292,95	60,97	1 - None	285,73	59,47
2 - Weak	4,24	0,88	2 - Weak	2,64	0,55
3 - Medium	4,25	0,89	3 - Medium	2,68	0,56
4 - Strong	4,02	0,84	4 - Strong	2,63	0,55
5 - Stronger	3,75	0,78	5 - Stronger	2,48	0,52
6 - The strongest	171,24	35,64	6 - The strongest	176,87	36,81
Total	480	100	Total	480	100

Despite the global increase in temperatures, some cities still manage to avoid increasing the UHI area, including the capital of Ukraine (Fig. 10).

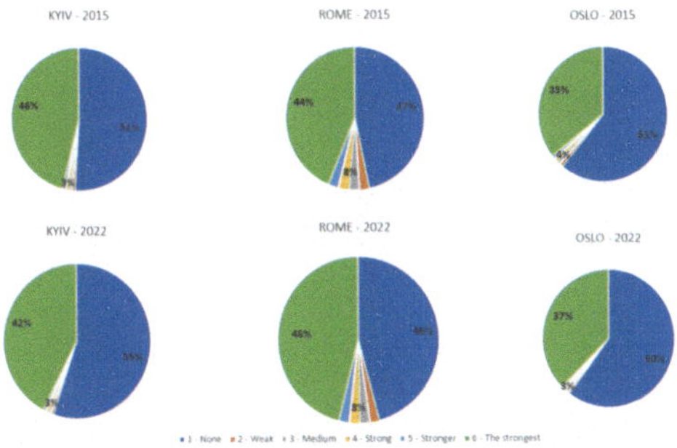

Fig. 10. Percentage charts of change in area of urban heat islands

So, the territory of Kyiv shows a good trend of reducing the number of urban heat islands (30 km^2 less in seven years).

In the case of Rome, the changes in the city are similar to those that occurred in Oslo, namely a slight increase in the area of the UHI.

As for the values of the leaf surface index, it has undergone a significant increase, which may indicate a partial improvement in the vegetation condition in the Oslo. Changes in the vegetation cover of these cities during this time were also analyzed, and the relationship between the increase in city temperature and the decrease in the amount of vegetation in the city was clearly identified. Thus, the territory of Kyiv has experienced a decrease in the LAI value, which may be partly due to the massive tree felling demonstrated by the Hansen Global Forest Change database (2000–2023). As for Rome, the situation in the city remained almost unchanged, except for a slight increase in the average monthly temperature, possibly partly caused by a decrease in vegetation indices.

As a result, the direct dependence of the spread of UHI on green areas was identified and proved. Effective methods have been identified that allow for high-quality and fast monitoring of the city's temperature regime. This makes it possible to create a unified geographic information system that can help local authorities ensure the comfort and safety of residents.

4 Conclusions

Given the potential danger posed by massive deforestation and urban heat islands to residents of modern megacities, it is important to calculate, develop, and implement a number of preventive measures that will allow for the anticipation and avoidance of the negative impacts of these processes. This study demonstrates the need for constant and ongoing monitoring of the UHI in cities, as even in a short period of time certain changes can occur, which then provoke significant negative consequences, as demonstrated in the examples of the areas of interest.

Using the example of European capitals such as Oslo, Kyiv, and Rome, the impact of one of the factors - green spaces on heat islands - was analyzed in detail, namely, maps of surface temperatures were constructed, which confirmed the assumption that surface temperatures depend on vegetation cover, which, in turn, was analyzed by calculating the vegetation index and leaf surface index. We also analyzed areas of interest in terms of quantitative changes in tree cover using high-resolution global maps of forest cover changes and proved the connection with the spread of elevated temperatures in these areas. The result was an understanding of the significance of the impact of green spaces on the temperature regime of the city and confirmation of the need to solve this issue.

References

1. Agrii Company, Radar indices and models. Part 2 (2020). https://agrii.com.ua/radarni-indeks
2. Chang, F., et al.: OSDI 2006 - 7th USENIX Symposium on Operating Systems Design and Implementation, pp. 205–218 (2006)
3. Ghosh, I.: The Dramatic Global Rise of Urbanization (1950–2020). World Economic Forum (2019)
4. GIS Geography. MODIS: Moderate Resolution Imaging Spectroradiometer (2022). https://gisgeography.com/modis-satellite/
5. Gonzalez, H., et al.: Proceedings of the ACM SIGMOD International Conference on Management of Data, pp. 1061–1066 (2010)
6. Li, Y., Zhao, X.: An empirical study of the impact of human activity on long-term temperature change in China: a perspective from energy consumption. J. Geophys. Res. (2012)
7. Le, M.T., Bakaeva N.: Method of identifying urban heat islands by remote sensing based on big data. ESCP-2023 (2023). https://doi.org/10.1051/e3sconf/202340305007
8. Sakhniuk, S., Tovstonoh, D., Monastyrova O., Zatserkovnyi, V.: Monitoring of urban heat islands using remote sensing technologies. In: 16th International Conference Monitoring of Geological Processes and Ecological Condition of the Environment (2022). https://doi.org/10.3997/2214-4609.2022580055
9. Sakhniuk, S., Zatserkovnyi, V., Polyakova, N., Martyniuk, L., Shovkoplias T.: Monitoring of green areas using remote sensing technologies. In: Interna-tional Conference of Young Professionals «GeoTerrace-2022» (2022). https://doi.org/10.3997/2214-4609.2022590075
10. Solecki, W.D., Rosenzweig, C.: Mitigation of the heat island effect in urban New Jersey. Glob. Environ. Change Part 6 Environ. Hazards, 39–49 (2011)

A Scheme for High Spatial Resolution Imaging Spectroscopy Technology

Zhe Zhao[1,2], Jiacheng Zhu[1,2], and Weimin Shen[1,2(✉)]

[1] Key Laboratory of Modern Optical Technologies of Education Ministry of China, School of Optoelectronic Science and Engineering, Soochow University, Suzhou 215006, Jiangsu, China
alleria_zz@163.com, {zjc,swm}@suda.edu.cn

[2] Key Laboratory of Advanced Optical Manufacturing Technologies of Jiangsu Province, School of Optoelectronic Science and Engineering, Soochow University, Suzhou 215006, Jiangsu, China

Abstract. Imaging spectrometers with high spatial resolution and high spectral resolution have low signal to noises ratio(SNR). In order to improve SNR, satellite scanning-back scheme is adopted to improve the integration time of the system and reduce the requirement for large aperture optical system with high spatial resolution. The imaging spectral technology scheme with orbital height of 500 km, swath of 10 km, spatial resolution(GSD) of 1 m, spectral resolution 5 nm, wavelength range of 0.4–1.0 um and SNR of no less than 100 is proposed, and the optical system is designed. The front telescopic system consists of a folded axis three mirror structure a focal system and an inverse Cassegrain structure objective lens system. The 4 m aperture beam is compressed into 250 mm for imaging, with an F/1.375, a field of view (FOV) of 1.146°, and an image slit length of 110 mm. The spectroscopic system consists of two Dyson optical structures, with a single slit length of 55mm, and two splice covering the front system's slit of 110 mm.

Keywords: Optical design · Satellite scanning-back · High spatial resolution · imaging spectrometer · signal to noise ratio

1 Introduction

1.1 A Subsection Sample

In terms of urban environmental remote sensing applications, it mainly involves land classification and utilization, impermeable surface monitoring, etc. However, low spatial resolution can lead to issues such as loss of details and mixed pixels, affecting its application [1]. The spatial variation of the coastal environment is at a fine scale, below 1m, requiring high resolution imaging spectrometers [2, 3].

In 2001, the European Space Agency's PROBA(Project for On Board Autonomous) was equipped with the CHRIS(Compact High Resolution Imaging Spectrometer) imaging spectrometer [4]. The front-end system adopts a dual mirror structure, and the splitter system adopts a curved prism Offner architecture. The swath is 13−15 km. Spectral

H. P. Urbach et al. (Eds.): ISSOIA 2023, SPP 191, pp. 197–216, 2024.
https://doi.org/10.1007/978-981-97-6718-2_19

range: 400−1050 nm. It possesses the ability to adjust the attitude along and through rails. The spatial resolution is 17–40 m. The spectral resolution is 1.25–11.25 nm. The attitude adjustment along the orbit direction of CHRIS can achieve an improvement in SNR, but the spatial resolution is still relatively low, making it difficult to meet the requirements of high spatial resolution scene applications.

In 2001, the US small imaging satellite OrbView-4 was equipped with the OHIS(OrbView Hyperspectral Imaging System) [5]. The width is 5 km, GSD = 8, and the spectral resolution is 11.4 nm. Spectral range is 0.45–2.49 um. The spatial resolution of OrbView-4 has been improved to within 10m, which can achieve high resolution scene applications to a certain extent. However, the width is too narrow to achieve precise detection of large areas.

In 2009, the TacSat-3 satellite in the United States was equipped with ARTEMIS(Advanced Response Tactically Effective Military Imaging Spectrometer) [6].The diameter of the pupil is 350 mm, with a spatial resolution of 4/5m. The spectral range is 0.4–2.5 um, with a spectral resolution of 5 nm. The front telescopic system adopts a dual mirror structure, and the splitter system adopts an Offer structure. The characteristic is low cost, tactical level, and short term development and launch. Although ARTEMIS has a high spatial resolution, its SNR decreases due to the decrease in diffraction efficiency in the purple band, resulting in low image quality.

In 2011, the China Tiangong-1 hyperspectral imager [7] adopted composite prism splitting technology to improve the utilization of light energy. The use of directional mirror motion compensation technology has improved integration time. Visible near infrared(VNIR) band, GSD = 10 m, spectral resolution of about 10 nm, and width of 10km. Short wave infrared(SWIR)band, GSD = 20 m, spectral resolution of about 23 nm, and width of 10 km. Due to the improved utilization of light energy and integration time of the Tiangong-1 optical system, the SNR has been improved, but the spectral resolution is relatively low, failing to achieve a 5 nm level.

The Spaceborne Hyperspectral Applicative Land and Ocean Mission(SHALOM) [8] utilizes an agile satellite platform, with an integration time increased by four times, and uses a large aperture of 650 mm, small F number optical system. GSD <10 m, width> 10 km, spectral range 0.4–2.5 um, spectral resolution 10 nm. By using a small F number system and increasing the integration time, SNR was improved, but the spectral resolution did not reach the 5 nm level.

The French Hyperspectral Imager (HYPXIM) [9] uses an agile satellite platform, a front telescope system with a three mirror structure, and a prism beam splitter system. Aperture of 450 mm, width of 16 km, GSD = 8, spectral range is 0.4–2.5 um, spectral resolution of 10 nm. The utilization rate of light energy has been improved, with a spatial resolution of less than 10 m and a swath greater than 10 km, indicating a high indicator. However, the spectral resolution has not reached the 5 nm level.

The above in orbit or in research high spatial resolution imaging spectrometers, except for ARTEMIS, have a spectral resolution of 10 nm, and fail to meet the technical specification of 5 nm when obtaining high spatial resolution. Additionally, some spectral bands are discrete and discontinuous. High spatial resolution and hyperspectral resolution cannot be achieved simultaneously, and the highest spatial resolution of the above mentioned high resolution imaging spectrometer is at the level of 5 m, yet it has not achieved the level of 1 m. The technical solution proposed in this article can achieve both high spatial and high spectral resolution simultaneously.

This article first introduces the difficulties of high resolution imaging spectrometers and the indicators of typical high resolution instruments in orbit and in research, and points out some of their existing problems. Next, we will conduct system indicator analysis, signal to noise ratio analysis, and introduce satellite scanning back scheme. Then, we will analyze the aperture and scanning magnification. Next, we will select optical systems, design optical systems, and evaluate image quality. Finally, we will summarize this article.

2 Scheme Analysis

According to the requirements of the system's indicators, relevant technical parameters were derived and signal-to-noise ratio analysis was conducted. In the conventional satellite scanning back mode, a super large aperture optical system is required to meet the indicator requirements. To this end, a satellite scanning scheme was proposed and the principle was introduced. Based on the principle of backscattering and combined with the aperture parameters of the optical system, a new round of signal-to-noise ratio analysis is conducted to obtain a set of reasonable and feasible technical parameters, followed by optical system selection analysis. Based on the selection results, the optical system design and image quality evaluation were carried out. Finally, the advantages and disadvantages of this scheme were commented on, and prospects were proposed.

The imaging spectrometer in this scheme is divided into two parts: a front telescopic system and a spectroscopic system. The front telescope system is divided into two parts: an unfocused telescope system and an imaging objective lens. The former compresses a large aperture beam of light, and the imaging objective lens then images the ground object scene onto a narrow image plane. The spectroscopic system consists of two Dyson systems, each covering half of the slit of the front system, and the two splices covering the entire long slit. Finally, the detector records spectral information.

2.1 System Indicator Analysis

The purpose of this article is to achieve an imaging spectral scheme with a width of 10 km, a spatial resolution of 1 m, and a spectral resolution of 5 nm at an orbital height of 500 km. The technical specifications of the system are as follows (Table 1):

The calculated parameters are shown in the table below (Table 2):

The aperture of the optical system needs to be determined based on the SNR index.

2.2 Signal to Noise Ratio Analysis

SNR is defined as the ratio of the effective signal electron number to the noise electron number generated by radiation from the target object on the photosensitive surface of the detector [10], expressed as follows:

$$R_{SN} = \frac{S_{signal}}{N_{noise}}, \tag{1}$$

Table 1. Parameters of Imaging Spectrometer

Parameters	Numerical values
Orbital altitude/km	500
Spatial resolution/m	1
Spectral resolution/nm	5
Wavelength range/μm	0.4–1
Swath width/km	10
SNR	>100
Pixel size/um	11

Table 2. Derived parameters

Parameters	Numerical values
Focal length/mm	5500
FOV/°	1.146
Slit length/mm	110
Satellite point velocity/m/s	7060
Integration time/ms	0.142

In the formula: R_{SN} is the signal-to-noise ratio; S_{signal} Is the number of effective signal electrons; N_{noise} Is the number of noisy electrons.

$$S_{signal} = B(\lambda)\tau(\lambda)Q(\lambda)\frac{\lambda}{hc}t\frac{\pi ab}{4F^2}m\Delta\lambda, \tag{2}$$

In the formula: $B(\lambda)$ is the radiance of the pupil, calculated by MODTRAN; $\tau(\lambda)$ is the system transmittance; $Q(\lambda)$ is the quantum efficiency of the detector; λ is the wavelength; h is the Planck constant; c is the speed of light in a vacuum; t is the integration time; π is a constant; ab is the pixel size of the detector; F is the $F\#$ number of systems; m is the number of pixel merges; $\Delta\lambda$ is the sampling interval for a single pixel spectrum.

$$N_{noise} = \sqrt{\left(\sqrt{N_{photon}}\right)^2 + \left(\sqrt{N_{dark}}\right)^2 + \left(\sqrt{N_{read}^2}\right)^2}, \tag{3}$$

In the formula: $\sqrt{N_{photon}}$ is photon shot noise; $\sqrt{N_{dark}}$ is the dark current of the detector; $\sqrt{N_{read}^2}$ is read out noise for the detector.

The radiance is calculated by MODTRAN and the parameters are as follows: the atmospheric model is selected as 1976 US Standard; The aerosol model is selected as Rural atmosphere VIS = 23 km; Select Slant Path for atmospheric path type; The albedo is 0.3.

The reflectivity of the reflective surface of the optical system is above 0.95 in the 400–450 nm wavelength range; In the 400–900 nm band, the average reflectance is above 0.98 [11]. The transmittance of the refractive surface, in the 400–450 nm wavelength range, is 0.985; 450–1000 nm, with a value of 0.99. The peak diffraction efficiency of the grating is 0.8.

The calculation of SNR requires information such as the quantum efficiency and pixel size of the detector, and the relevant parameters are shown in the table [12] (Table 3):

Table 3. Parameters of detector

Parameters	Numerical values
Pixel size/um	11
Fullwell capacity/e^-	36000
Quantum efficiency	Up to 50%
Spectral range/nm	290–1100
Readout noise/e^-	22(typical)
Dark current/e^-/pixel/s@ +20 °C	530
Dynamic range (A/D)/bit	12
Frame rate/fps	1102–10782

In the conventional push scan mode, calculate the system aperture value that meets the SNR requirements.

According to the SNR indicator requirements, when the system aperture is 12.5 m, i.e. F/0.44, the signal-to-noise ratio curve is shown in the following figure, meeting the requirement of a mean greater than 100. The horizontal line in the figure represents the mean SNR of different solar zenith angles. The mean SNR corresponding to a zenith angle of 30° or less is above 100, and the peak SNR of a zenith angle of 60° is 100.

When the system aperture is greater than 12.5 m, the SNR will continue to increase, but an excessively large aperture makes it extremely difficult to process, adjust, and launch. Next, the aperture of the optical system will be reduced through satellite scanning to make it engineering feasible (Fig. 1).

2.3 Principle of Satellite Scanning Back

As shown in the following figure, during satellite observation of ground targets, the attitude of the satellite platform is adjusted during flight to ensure that the optical system's line of sight is always aligned with the ground target scenery. The target object is imaged on the detector, reducing the image shift rate on the image plane, thereby increasing integration time and improving SNR (Fig. 2).

Now we will compare and analyze the SNR of push scan mode and scan back mode. Equation (4) represents the SNR of push scan mode, and Eq. (5) represents the SNR of

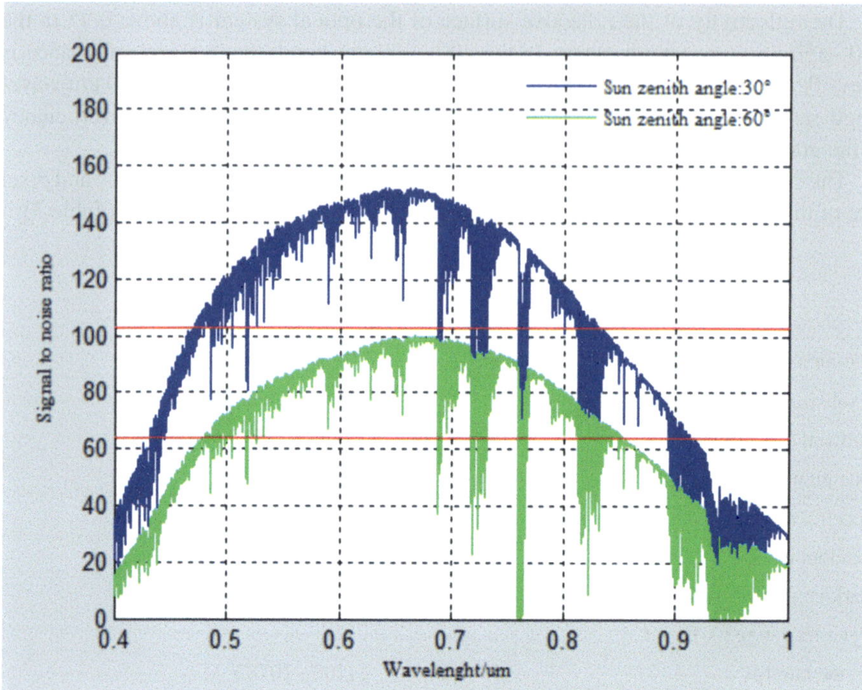

Fig. 1. SNR curve with peak diffraction efficiency of 0.5 in conventional push-scan mode

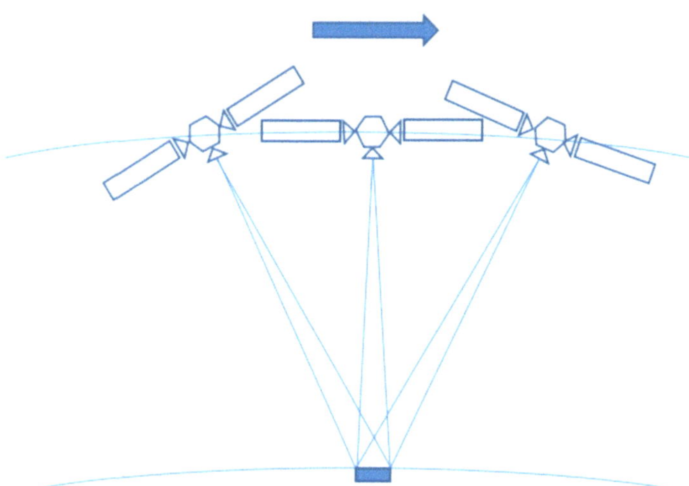

Fig. 2. Principle of satellite scanning back

scan back mode. It can be seen that the denominator of Eq. (5) decreases and the SNR increases by $\sqrt{G}$ times, indicating that the satellite flyback mode can improve the SNR.

$$R_{\mathrm{SN}} = \frac{S}{N} = \frac{\sqrt{G}S}{\sqrt{G\left(\sqrt{S}\right)^2 + G\left(\sqrt{Q}\right)^2}}, \tag{4}$$

$$\begin{aligned} R_{\mathrm{SN}} &= \frac{GS}{\sqrt{\left(\sqrt{GS}\right)^2 + \left(\sqrt{GQ}\right)^2}} \\ &= \frac{\sqrt{G}S}{\sqrt{\left(\sqrt{S}\right)^2 + \left(\sqrt{Q}\right)^2}}, \end{aligned} \tag{5}$$

In the formula: R_{SN} is the signal-to-noise ratio; S is the number of effective signal electrons; N is the number of noisy electrons; G is the satellite flyback multiple; Q is the number of dark noise charges.

The satellite scanning back mode improves SNR by increasing the integration time, even if the image shift rate on the detector image plane is reduced by G times. Compared to the push scan mode, the actual ground scanning path length in the back scan mode will become shorter, becoming 1/G of the push scan mode, causing the actual ground scanning path to become block by block instead of the continuous and uninterrupted ground path in the push scan mode. According to the compensation angular rate formula [13] and the satellite scanning mode parameters, the path length and corresponding time of ground scanning in the scanning mode were calculated, as shown in the table below (Table 4).

Table 4. Time and length corresponding to different multiples

Parameters	Numerical values						
Multiple	8	7	6	5	4	3	2
Time/s	96.5581	98.8900	102.1832	107.1873	115.7067	133.4765	194.2191
Distance/km	96.3827	112.8458	136.0959	171.4248	231.5717	357.0268	785.8692

From this, it can be seen that the larger the scanning magnification, the shorter the corresponding time, the shorter the ground distance scanned, and the higher the control accuracy requirements for scanning.

2.4 Analysis of Aperture and Scanning Magnification

According to the previous analysis, the larger the aperture of the optical system, i.e. the smaller the F-number, the higher the mean SNR. At the same time, the aperture of the optical system needs to consider engineering practice, and it needs to be able to be processed, adjusted, and launched by a carrier rocket, so it cannot be too large.

When the aperture of the optical system remains constant, the larger the backscatter factor, the corresponding integration time increases exponentially, and the higher the mean SNR. However, considering engineering practice, the larger the backscatter factor, the higher the accuracy requirement for the relative error of the satellite platform's angular rate, and the more difficult it is to control. Therefore, the scanning back factor should not be too high.

Taking into account the requirement for SNR mean under a spatial resolution of 1m and the mutual constraints between the optical system aperture and the backscatter ratio, a set of reasonable and feasible parameter indicators were calculated: the optical system aperture is 4m, which is F/1.375, with 8 backscatters. The SNR curve at this time is shown in the following figure.

The horizontal line in the figure represents the mean SNR of different solar zenith angles. The mean SNR corresponding to zenith angles of 15° and below is above 100, meeting the indicator requirements. The SNR peak corresponding to the zenith angle within 45° is above 100 (Fig. 3).

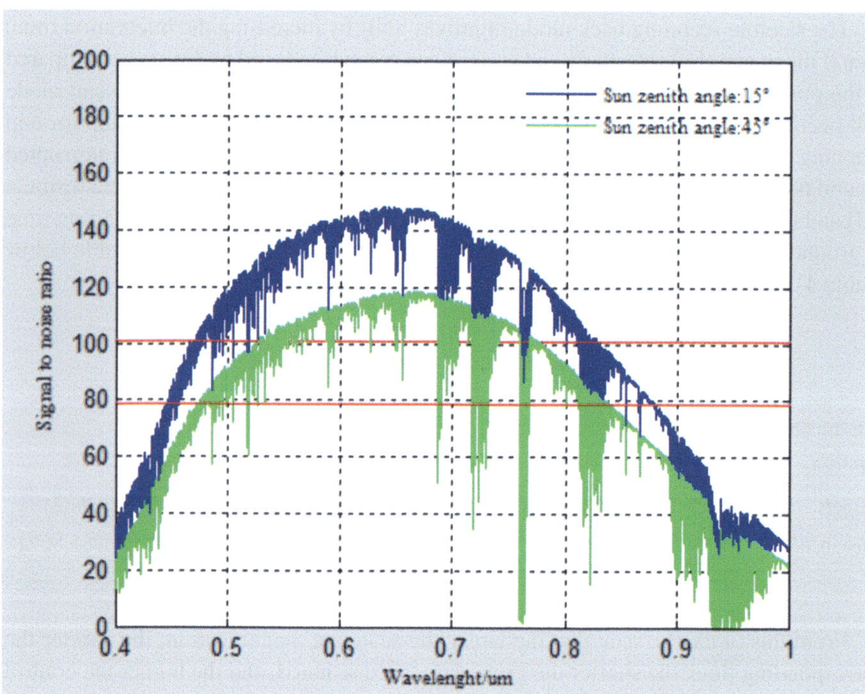

Fig. 3. SNR curve with peak diffraction efficiency of 0.5 in 8 satellite scanning back mode

From the above analysis, it can be seen that when achieving the technical indicators of ground spatial resolution of 1 m and spectral resolution of 5 nm, the satellite echo mode can reduce the system aperture of the conventional push scan mode to within 4 m, while ensuring that the average SNR is above 100, making the meter level resolution imaging spectrometer feasible. The specific indicators are as follows (Table 5):

Table 5. Parameters

Parameters	Numerical values
F#	1.375
Entrance pupil diameter/m	4
Number of fold	8

2.5 Optical System Selection

According to the analysis results of the conventional push scan mode indicators, it can be seen that in order to achieve the SNR indicator requirement of spatial resolution of 1 m, the aperture of the optical system will be large. After adopting the satellite scanning scheme, the system aperture can be reduced, which is feasible in engineering, but still slightly larger.

The overall scheme of the optical system is as follows: the optical system divides the pre telescope system into an unfocused telescope system and an imaging objective system. Using the unfocused telescope system, the large aperture beam is compressed at high magnification, and then imaged into a slit by a small F-number imaging objective. The slit is then divided by the spectroscopic system, and finally a spectral image plane of the slit is formed on the detector. And due to the slit length reaching 110 mm, it is difficult for a single beam splitter system to cover such a wide slit field of view. Therefore, two beam splitters are used for field of view splicing to cover the entire slit.

As shown in the following figure, the large-aperture incident beam from the target object is compressed by the unfocused system to become a parallel beam with a small aperture, which is then imaged by the imaging objective to the slit. After the beam emitted from the slit passes through the spectroscopic system, the spectral image is imaged on the detector (Fig. 4).

The selection of optical systems is divided into three parts: 1. The selection of unfocused systems, 2. The selection of imaging objective lens structures, and 3. The selection of spectroscopic systems. Below, we will discuss each of them one by one.

1. Selection of non focal system: Due to the large aperture of the system, the front telescope system adopts a reflection system. Reflection systems can be divided into two types: non intermediate real images and intermediate real images.

The system without intermediate real images has problems such as secondary occlusion and difficulty in extracting image planes. It is often used in off axis structures, with the advantages of wide field of view, large volume, and no real exit pupil. Due to the meter level caliber of the system, considering the actual engineering situation, off-axis structures cannot be used.

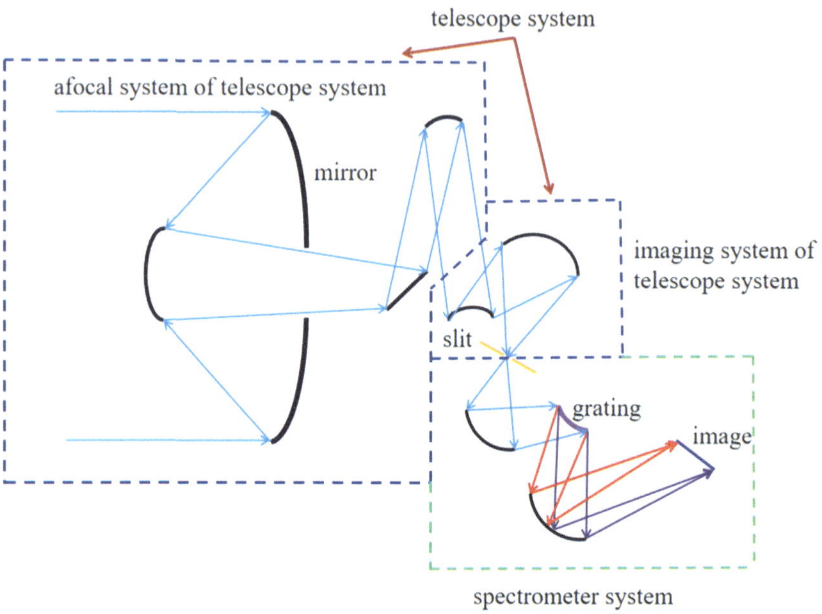

Fig. 4. Schematic diagram of optical system

Systems with intermediate real images have a relatively small field of view and a real exit pupil [13].

Systems with intermediate images can be divided into two anti systems and three anti systems. The two mirror system is suitable for small field of view and low magnification situations [14], while the three mirror system can achieve large field of view and high magnification compression. The unfocused telescope part of the front-end system in this article requires high magnification compression of the meter wide beam to the millimeter level, with a compression magnification of up to 20 times. Therefore, a three mirror structure is adopted, as shown in Fig. 5 (a).

Non off axis three mirror systems with intermediate real images can be divided into two structures: coaxial and foldable. The folded axis three mirror structure is based on the coaxial three mirror structure, with slight eccentricity and tilt of the reflector, which can avoid secondary obstruction, as shown in Fig. 5 (a).

Based on the above analysis, this article adopts a folded axis three mirror structure with intermediate real images as the unfocused system of the front telescope system .

2. Selection of imaging objective structure: The imaging objective is connected to the defocused system, and its entrance pupil matches the exit pupil of the defocused system, while maintaining the telecentric image. It can be divided into two types for selection: refractive system and reflective system.

The refractive system can introduce issues such as color difference, color distortion, and secondary spectroscopy, and even after the beam is compressed, it still has an aperture

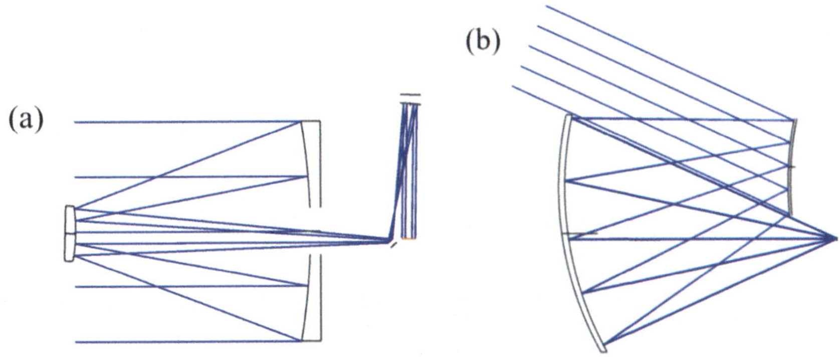

Fig. 5. Structure selection of spectrometer system. (a) Folded axis three mirror system with intermediate real image; (b) Off axis two mirror system without intermediate real image

of several hundred millimeters. Large aperture refractive materials can cause the system to become too heavy.

Reflective objective lenses are divided into two types: coaxial structure and off axis structure. The coaxial structure has a small field of view angle, while beams compressed up to 20 times have a large field of view angle, so coaxial structure is not suitable.

Based on the above analysis, this article adopts an off axis reflective dual mirror structure as the imaging objective of the front-end system, as shown in Fig. 5 (b).

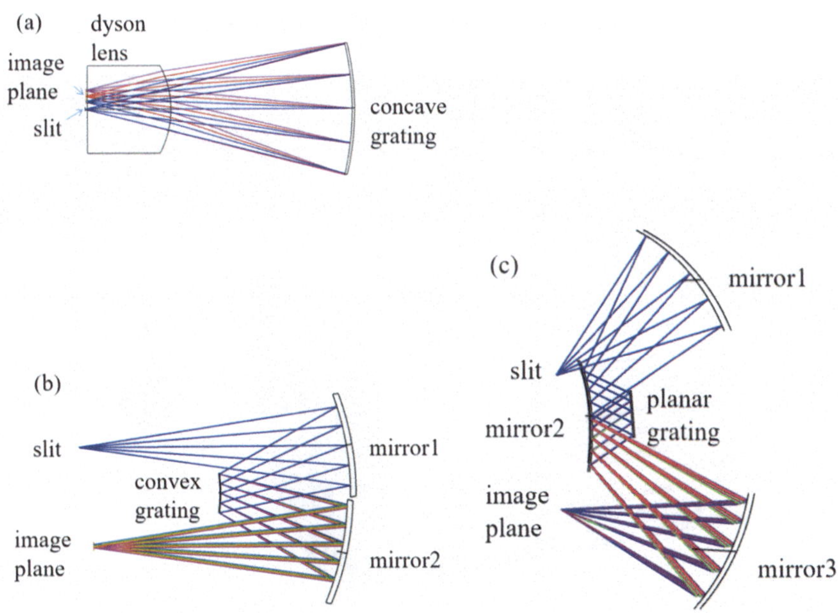

Fig. 6. Structure selection of spectrometer system. (a) Dyson spectrometer system; (b) Offner spectrometer system; (c) Schwarzschild spectrometer system

3. Selection of beam splitting system structure: Due to the overall plan using a small F-number front telescope system, the beam splitting system needs to choose an optical structure that can achieve large numerical aperture to match the front system.

The numerical aperture of the Offner structure cannot be enlarged, making it difficult to match with the front telescope system, as shown in Fig. 6 (b).

The Dyson structure, as shown in Fig. 6 (a), and the Schwarzschild structure, as shown in Fig. 6 (c), can achieve large numerical apertures, thereby matching with the front-end system. Next, optical system simulation analysis was conducted on both Dyson structure and Schwarzschild structure. Based on the simulation results, analyze the size and spatial layout of the two structures to determine whether they can achieve splicing in a narrow field of view. Based on the analysis results, determine the final selection of the splitter system.

3 Optical System and Image Quality Evaluation

3.1 Front End System

The front telescope system consists of an unfocused system and an imaging objective system, as shown in the following figure. The unfocused system adopts a folded axis three mirror structure, and the imaging objective adopts an inverse Cassegrain structure. The exit pupil of the unfocused system is connected and matched with the entrance pupil of the imaging objective. As shown in the following figure, the folding mirror folds the optical axis, saving space and facilitating the connection of the back-end spectrometer.

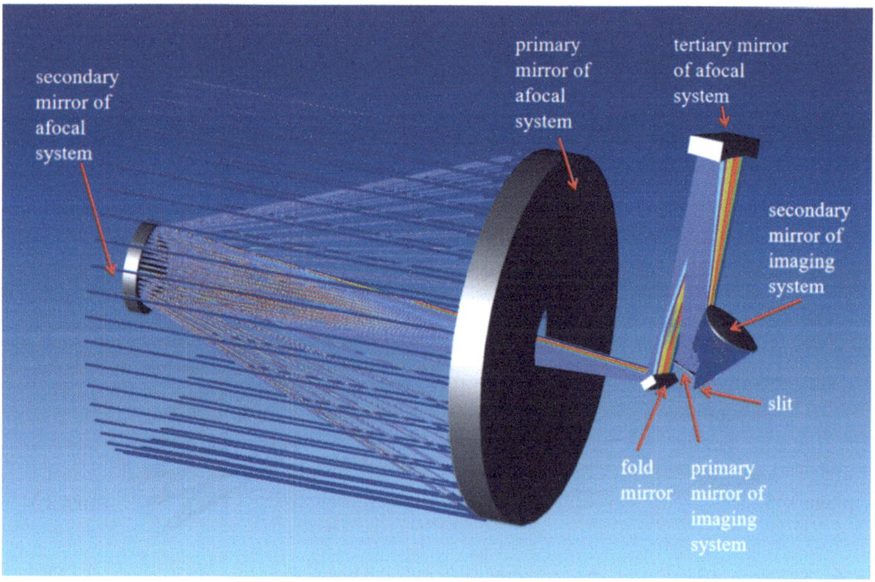

Fig. 7. Telescope system

The main mirror aperture is 4 m, and the beam is compressed to 250 mm before being imaged by two off axis reverse imaging objective lenses, with a slit length of 110 mm. The front-end system F# = 1.375, field of view angle FOV = 1.146°, and corresponding ground clipping width of 10 km at a track height of 500 km (Fig. 7).

The MTF curves at the center wavelength of 0.7 um and edge wavelengths of 0.4 um and 1.0 um are close to the diffraction limit, and at the Nyquist frequency of 45.51 p/mm, they are close to or above 0.8, as shown in the following Fig. 8:

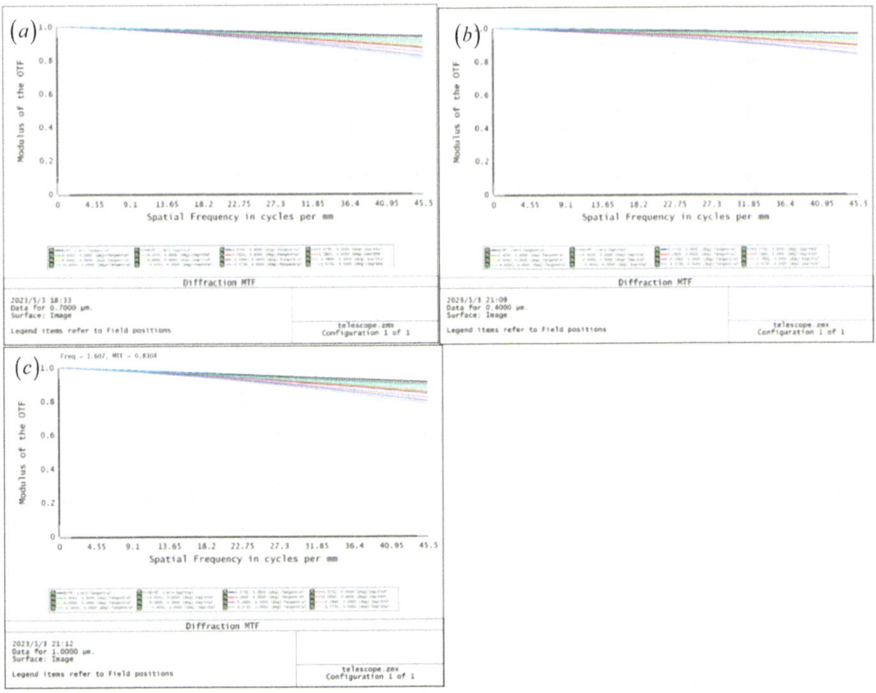

Fig. 8. MTF curve of telescope system. (a) 0.7 um; (b) 0.4 um; (c) 1.0 um

The point charts for different wavelengths and field of view positions are shown below, with the leftmost column showing the point charts for each field of view at a wavelength of 0.4 um. The far right column is a point plot of a 1.0 um wavelength in the field of view. Among them, the small circle represents Airy patches, and the rectangular small square represents a single pixel. The maximum RMS of the point plot is 3.122 um, located within 11 um of one pixel (Fig. 9).

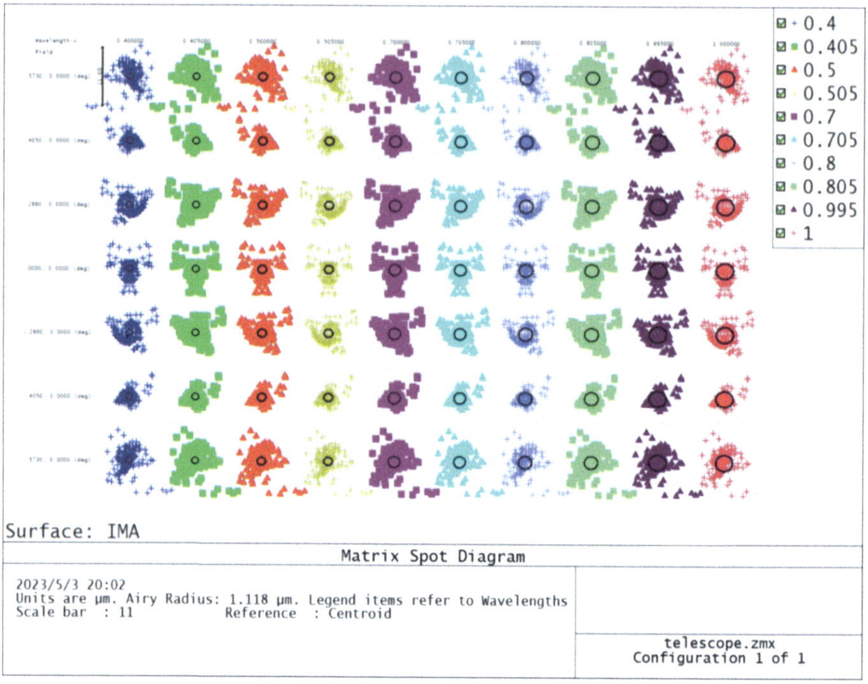

Fig. 9. Spot diagram of telescope system

3.2 Spectroscopic System

The spectroscopic system is shown below, consisting of two Dyson systems for field of view stitching, with a numerical aperture of NA = 0.342. A single Dyson slit is 55 mm long, and the two splices cover the entire 110 mm slit. Dyson lenses are composed of calcium fluoride and fused quartz. The distance between the incident slit and the first Dyson lens is 5 mm, the distance between the outgoing spectral plane and the Dyson lens is greater than 5 mm, and the distance between the spectral plane and the slit is greater than 60 mm. After turning, the space left is convenient for the installation of the slit and detector (Fig. 10).

The point diagram of the spectroscopic system is shown below, where the small circle represents the Airy spot and the rectangular block represents a single pixel. The maximum RMS of the point plot is 2.287 um, which is less than 11 um for one pixel (Fig. 11).

The diffraction energy concentration curve of the spectroscopic system is shown below, with a single pixel concentrating over 90% of the energy (Fig. 12).

The MTF curve of the spectroscopic system is shown in the following figure, which is above 0.8 at the Nyquist frequency of 45.51 p/mm (Fig. 13).

The bending of the spectral lines in the spectroscopic system is shown below, and the bending of the spectral lines does not exceed 5% of a single pixel (Fig. 14).

The color distortion of the spectroscopic system is shown below, with a color distortion not exceeding 5% of a single pixel (Fig. 15).

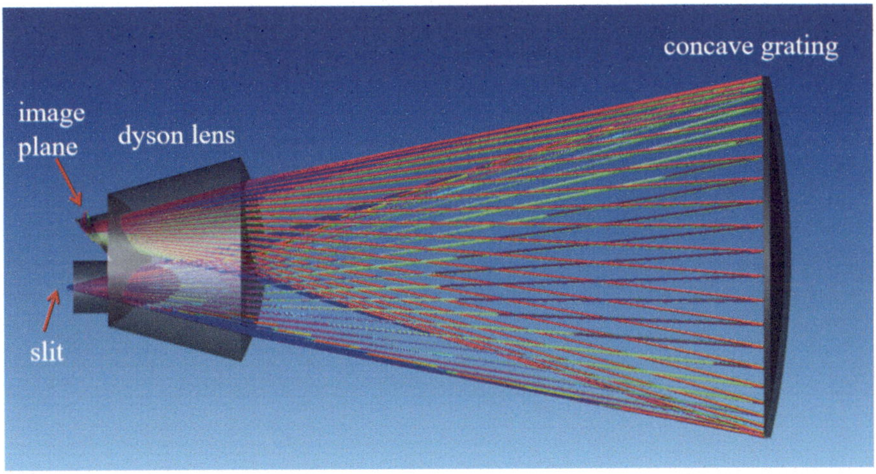

Fig. 10. Dyson spectrometer system

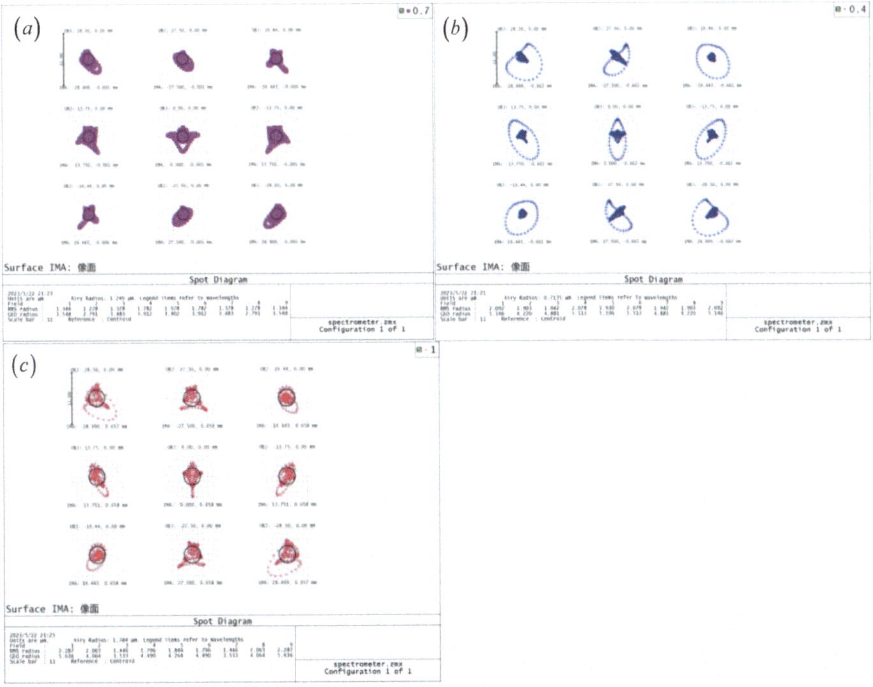

Fig. 11. Spot diagram of spectrometer system. (a) 0.7 um; (b) 0.4 um; (c) 1.0 um

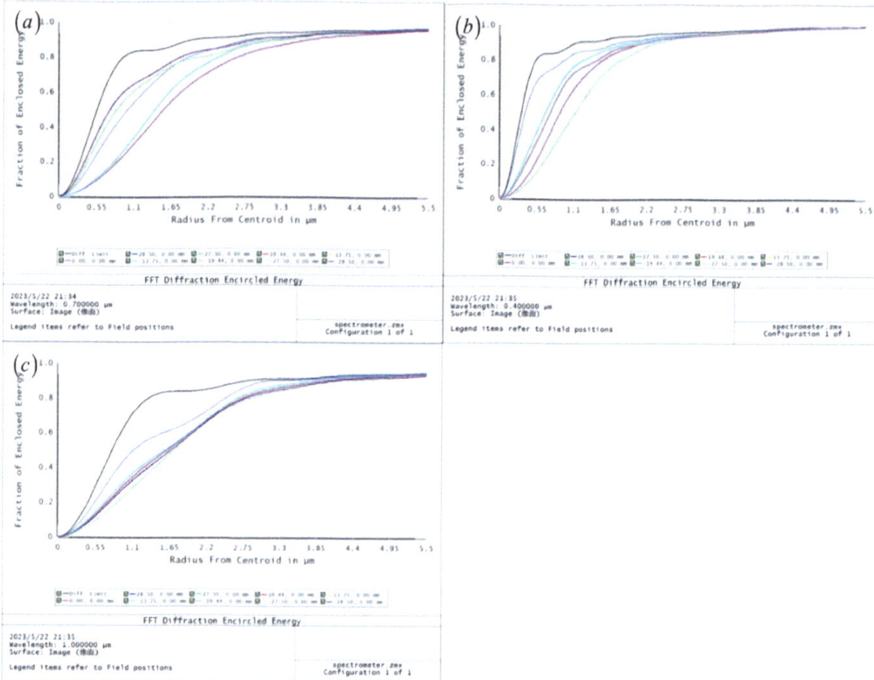

Fig. 12. Curve of diffraction encircled energy. (a) 0.7 um; (b) 0.4 um; (c) 1.0 um.

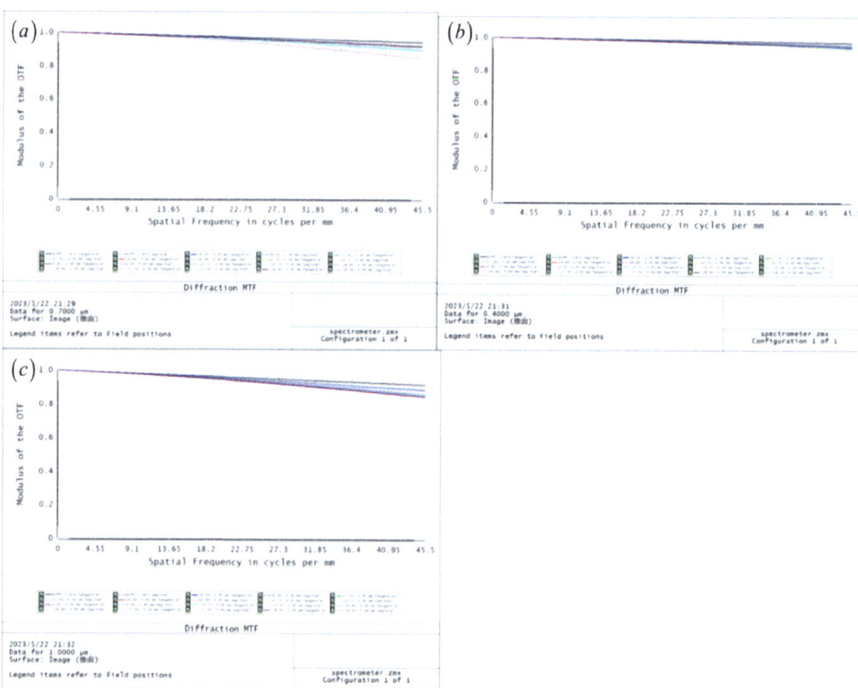

Fig. 13. MTF curve of spectrometer system. (a) 0.7 um; (b) 0.4 um; (c) 1.0 um

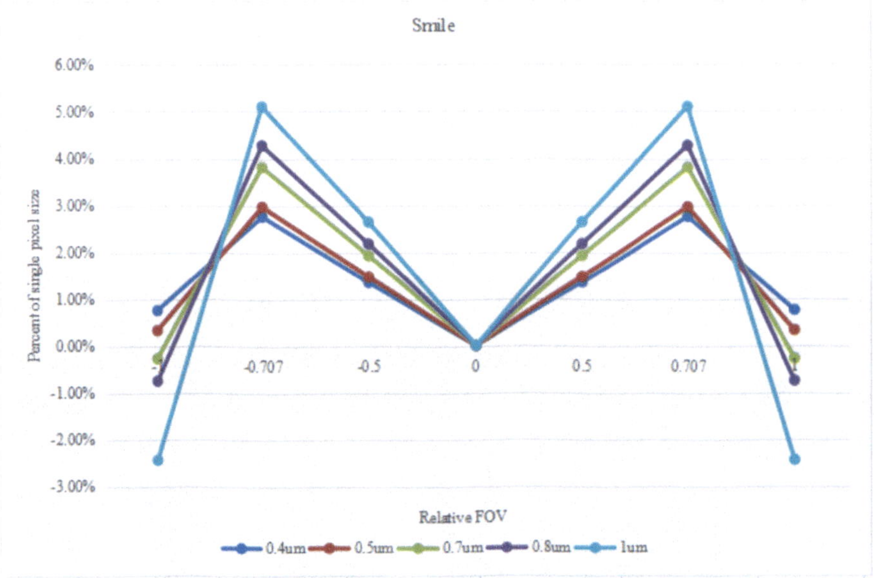

Fig. 14. Smile of spectrometer system

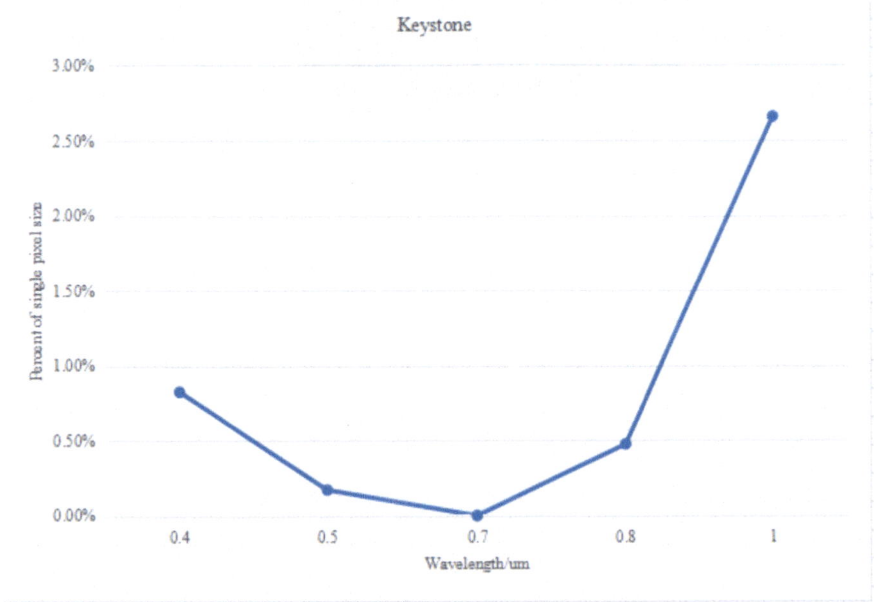

Fig. 15. Keystone of spectrometer system

3.3 Overall System

The optical system as a whole consists of a front telescope system and a beam splitter system, as shown in the following figure. The front telescope system compresses the beam of light by the unfocused telescope system, and then the imaging objective connected is imaged to the slit. Two splitting systems are spliced to cover the long slit of the front system, and their respective slit surfaces are split. The spectral image surface is recorded by CCD (Figs. 16 and 17).

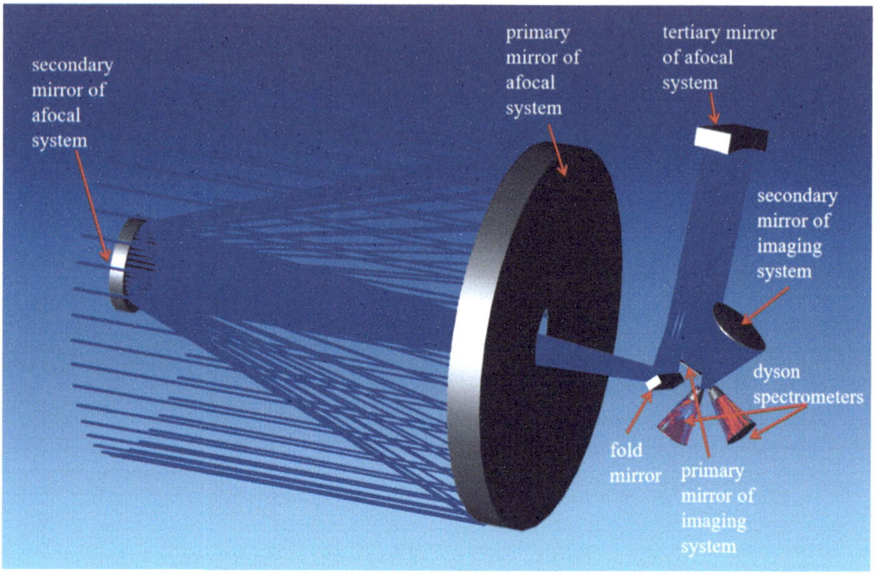

Fig. 16. Overall structure 1 of optical system

The above imaging spectral system utilizes a small F-number optical system combined with satellite platform scanning to achieve the technical indicators of 1m ground spatial resolution, 5 nm spectral resolution, and 10 km cutting width, while ensuring that the average SNR is not less than 100. The spectral line bending and color distortion of the system are within 5% of a single pixel, and the spectral imaging quality is good, making it easy for subsequent data processing.

3.4 Analysis and Discussion

By using a small F-number optical system and combining it with satellite scanning technology, imaging spectral remote sensing with a spatial resolution of 1m was achieved. High spatial resolution and hyperspectral resolution are achieved simultaneously, and the average SNR is not less than 100. Under the existing processing and testing conditions as well as the launch capacity of rockets, this plan has engineering feasibility.

However, the aperture of the optical system in this scheme is still relatively large, resulting in high manufacturing costs, and the flyback scheme requires high accuracy in controlling the angular velocity of the satellite platform.

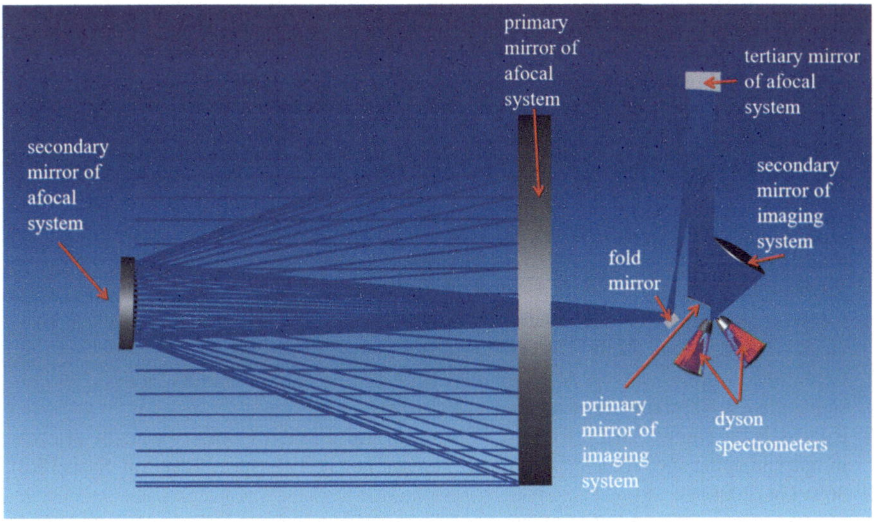

Fig. 17. Overall structure 2 of optical system

Further improvements can be made in the following aspects: adopting detectors with higher quantum efficiency can further reduce the optical system aperture of this scheme. Grating splitting will distribute a portion of the signal energy to other diffraction levels, and the diffraction efficiency will lose a portion of the signal energy. Subsequently, prism splitting can be used to solve the problem of nonlinear dispersion. Adopting a new working mode to increase the residence time of the target scenery signal, it should be noted that while increasing the integration time, the noise is also increasing. Due to the linear independence of the noise, the increase is slower than the signal, and the SNR is still increasing. Detectors with large pixel sizes can be selected based on design parameters. The larger the detector size, the more conducive it is to improving SNR. Adopting new devices, such as the Single Photon Avalanche Diode (SPAD), to achieve weak signal detection at the physical limit of single photon level.

In addition, a new working mode is adopted to increase the dwell time of the target scenery signal. The author has conceived a more innovative technical solution that can further reduce the aperture of the optical system, while ensuring high spatial resolution, high spectral resolution, and high signal-to-noise ratio, which has engineering feasibility.

4 Conclusion

The satellite scanning scheme proposed in this article can achieve imaging spectral detection at a ground level of 1m spatial resolution, while achieving high spatial and spectral resolution while ensuring SNR above 100. It can be applied to high-resolution ground remote sensing scenarios such as urban environmental remote sensing, soil resource monitoring, and coastal zone environmental monitoring. The imaging spectroscopy technology proposed in this paper has higher spatial resolution and spectral resolution compared to publicly reported in orbit or under research imaging spectrometers.

References

1. Li, S.Y., Liu, Z.W., Liu, K., et al.: Advances in application of space hyperspectral remote sensing(invited). Infrared Laser Eng. **48**(3), 9–23 (2019)
2. Mouroulis, P., Green, R.O., Chrien, T.G.: Design of pushbroom imaging spectrometers for op-timum recovery of spectroscopic and spatial information. Appl. Opt. **39**(13), 2210–2220 (2000)
3. Nieke, J., Schlapfer, D., Dell'Endice, F., et al.: Uniformity of imaging spectrometry data products. IEEE Trans. Geosci. Remote Sens. **46**(10), 3326–3336 (2008)
4. Michael, J.B., Jeff, J.S., Mike, A.C., et al.: The PROBA/CHRIS mission: a low-cost smallsat for hyperspectral multiangle observation of the earth surface and atmosphere. IEEE Trans. Geosci. Remote Sens. **42**(7), 1512–1520 (2004)
5. Eesa. OrbView-4 Imaging Minisatellite[EB/OL]. https://www.eoportal.org/satellite-missions/orbview-4#references. Accessed 4 May 2023
6. Ronald, B.L., Thomas, W.C., Richard, M.N., et al.: Advanced responsive tactically-effective military imaging spectrometer(ARTEMIS) design. Int. J. High Speed Electron. Syst. **18**(2), 369–374 (2008)
7. Gao, M., Zhang, S.C., Li, S.Y.: Tiangong-1 hyperspectral remote sensing and application. J. Remote Sens. **18**(Sup): 2–10 (2018)
8. Natale, V.G., Kafri, A., Tidhar, G.A., et al.: SHALOM-Space-borne hyperspectral applicative land and ocean mission. In: 2013 5th Workshop on Hyperspectral Image and Signal Processing: Evolution in Remote Sensing (WHISPERS), June 26–28, 2013. IEEE, 2013: 1–4,Gainesville, FL, USA (2013)
9. Marie, J.L.F., Sylvain, M., Steven, H.: HYPXIM-an advanced hyperspectral satellite for environment, security and defence. In: The XXII Congress of the International Society for Photogrammetry and Remote Sensing, August 25-September 1, 2012, Melbourne, Australia (2012)
10. Zhu, J.C.: Optical system of long-slit imaging spectrometer in geostationary orbit and low earth orbit. Suzhou: Soochow University, 26–37(2020)
11. He, S.K., Wang, G., Bai, Y.L., et al.: Research progress of high-reflection coatings on mirrors of large-aperture space cameras. Laser Technol. **46**(1), 143–148 (2022)
12. Pco.dimax cs[EB/OL]. https://www.pco.de/fileadmin/user_upload/pco-product_sheets/BR_PCODIMAXCS_V105.pdf. Accessed 4 May 2023
13. Wang Y Q. Research on imaging spectrometer motion compensation implementation technology. Changchun Changchun Inst. Opt. Fine Mech. Phys. Chin. Acad. Sci. 29–33 (2013)
14. Li, Q.: Design of airborne thermal infrared dual-band optics with large-aperture and wide-field. Suzhou, Soochow University, 17–29 (2017)

RS-YOLO: A Remote Sensing Ship Target Detection Algorithm Based on Feature Selection Alignment

Hongyu Lin(✉), Xingcheng Zhao, Xinyu Zhang, and Xingjuan Mi

Beijing Institute of Space Mechanics and Electricity, Beijing, China
bisme5085@163.com

Abstract. With the continuous development of remote sensing technology, ship target detection technology based on visible remote sensing images has gradually become a research hotspot. To address challenges related to ship targets in remote sensing images with varying rotational angles and complex backgrounds, this study introduces the RS-YOLO (Rotate Ship-YOLO) algorithm, which is an enhancement of the YOLOv5 (You Only Look Once version 5) algorithm. The paper presents a multi-scale target detection approach utilizing the BiFaPN (Bi-directional Feature-aligned Pyramid Network) that integrates a feature selection module and a feature alignment module. Additionally, the study enhances the regression loss of YOLOv5 and incorporates the regression box WBF (Weighted Boxes Fusion) strategy to enhance the accuracy of the RS-YOLO ship target detector. Experimental results demonstrate that RS-YOLO achieves a 4.05% improvement in mAP0.5 compared to YOLOv5, showcasing its potential for application in remote sensing ship target detection.

Keywords: Remote sensing image · Feature alignment · Weighted fusion · Ship target detection

1 Introduction

Due to the rapid development of science and technology, optical remote sensing images contain richer remote sensing information, which provides a data basis for the field of remote sensing ship target detection [1]. Ship target detection technology in remote sensing scenarios has many application prospects in civil and military fields. In the civil field [2], the ship target detection technology monitors the sea traffic situation and provides real-time traffic conditions for ships to avoid marine traffic accidents. In the military field [3–5], the ship target detection technology can monitor the deployment of enemy ships in the sea area, providing critical information for the operational command. Therefore, it is of great practical significance to research ship target detection technology under a remote sensing scenario.

In recent years, with the rapid development of satellite remote sensing and its related technologies [6], the resolution of satellite remote sensing images is constantly improving, which makes the target image information under the remote sensing image more

H. P. Urbach et al. (Eds.): ISSOIA 2023, SPP 191, pp. 217–226, 2024.
https://doi.org/10.1007/978-981-97-6718-2_20

prosperous and more perfect. However, at the same time, it also brings the problems of complex background information, inconsistent size of the target, and arbitrary direction of the target. In this paper, based on the massive remote sensing image resources accumulated in the field of remote sensing, deep learning technology is applied to the field of remote sensing image ship detection, which improves the accuracy of remote sensing image ship detection and the main research work of this paper is as follows:

(1) To address the issue of uneven scale distribution of ship targets in remote sensing images, we introduce a multi-scale target detection method called BiFaPN. This method is based on feature selection alignment, which integrates a feature selection module and feature alignment module within a bidirectional feature pyramid network;
(2) In addressing the issue of arbitrary ship direction and complex backgrounds in remote sensing images, we propose RS-YOLO. This approach utilizes KFIoU [6] (Kalman Filtering IoU) Loss to replace CIoU [7] (Complete IoU) Loss for border regression, enhances YOLOv5 through a combination of regression box WBF strategy and multi-scale object detection method based on feature selection alignment.

2 Methods

2.1 BiFaPN

BiFaPN converts the feature response of the target object into an attention heatmap through an attention mechanism and passes it on to subsequent detection branches. Then, deformable convolutions are used to align the convolutional features with the attention heatmap adaptively. The overall structure of BiFaPN is shown in Fig. 1.

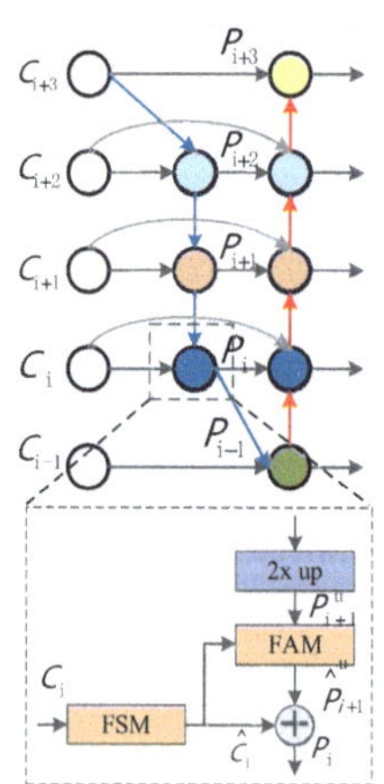

Fig. 1. Overall structure of the BiFaPN.

The structure of the feature alignment module FAM (Feature Alignment Module) is shown in Fig. 2, which applies deformable convolution to the feature map P^u_{i+1} and uses the feature map $\hat{C}_i$ and the feature map P^u_{i+1} as a reference to obtain the offset Δ_i.

Then, based on the calculated offsets, the deformable convolution can adjust the sample positions of the sample convolution; The features can be aligned according to the spatial distance between $\hat{C}_i$ and P_{i+1}^u to better adapt to the shape and size changes of the target.

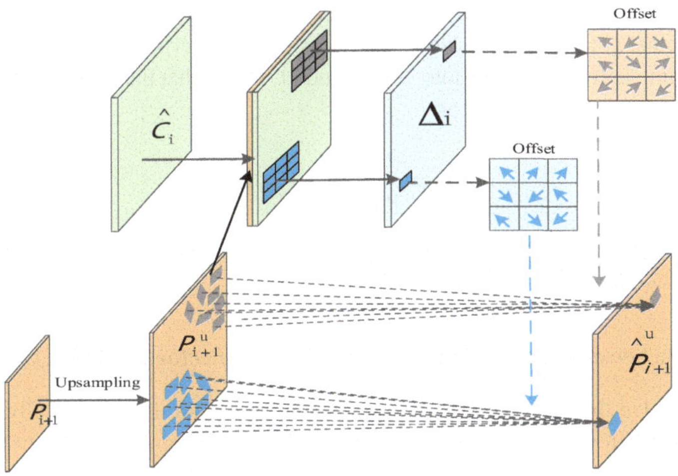

Fig. 2. Schematic diagram of the FAM.

The network structure of the FSM (Feature Selection Module) is shown in Fig. 3, which first transforms the feature maps obtained from global pooling into one-dimensional features for encoding.

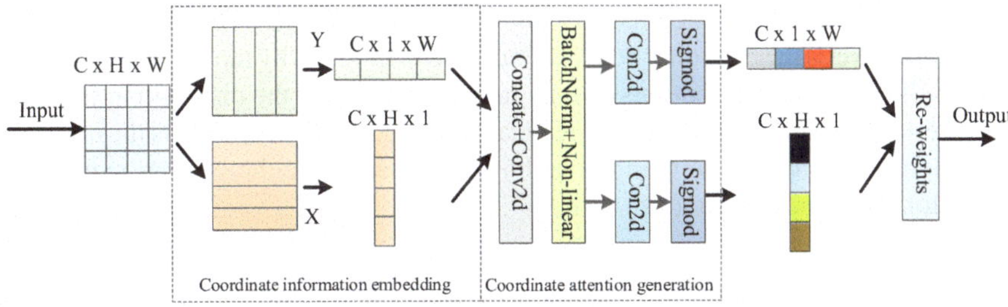

Fig. 3. Schematic diagram of the FSM.

In this case, the process of the feature importance selection layer can be formulated as follows:

$$u = f_m(z) \tag{1}$$

where $z = \left[z_c^h, z_c^w\right]$ decomposes global pooling, transforms into the value obtained from a one-dimensional feature encoding operation, and the feature importance selection layer $f_m(\cdot)$ is a 1×1 convolutional layer, followed by a sigmoid activation function. The formula for global pooling is shown below:

$$z_d = \frac{1}{H_i \times W_i} \sum_{H=1}^{H_i} \sum_{W=1}^{W_i} c_d(h, w) \tag{2}$$

The specific process of global pooling decomposition is as follows: giving the input X, each channel is first encoded along the horizontal and vertical coordinate directions using a pooling kernel of dimensions $(H, 1)$ and $(1, W)$ so that the output of the channel with height h can be formulated as:

$$Z_c^h(h) = \frac{1}{w} \sum_{0 \leq i < w} x_c(h, i) \tag{3}$$

Finally, it provides the features selected from the FSM to the FAM as a reference for learning alignment offsets.

2.2 Remote Sensing Image Rotating Ship Target Detector RS-YOLO

In order to solve the problem that objects in remote sensing images usually have any direction, a rotating ship target detector RS-YOLO is designed based on YOLOv5s Baseline, which is suitable for remote sensing scenarios. Its network structure is shown in Fig. 4. The network first uses the long edge definition method to convert a horizontal box into a rotating box in any direction. Then, the BiFaPN is used to replace the FPN (Feature Pyramid Network) and PANet (Path Aggregation Network) in the feature fusion section so that the feature maps output by the Neck section have more affluent and more accurate features. Finally, this paper uses the KFIoU to address the non-differentiable problem of the rotating bounding box IoU after adding angle information Loss replaces the CIoU Loss, achieves regression of borders, and solves the problem of complex backgrounds in remote sensing image environments by using the WBF instead of NMS.

KFIoU converts the bounding box to a Gaussian distribution by moving the centers of two high-dimensional distributions so that they overlap. Then, the distribution function of the overlapped region is obtained by Kalman filtering. Finally, it converts the obtained distribution function into a rotating bounding box to calculate the overlap region and IoU. KFIoU is defined as Eq. (4):

$$KFIoU = \frac{v_{B_3}(\sum)}{v_{B_1}(\sum_1) + v_{B_2}(\sum_2) - v_{B_3}(\sum_3)} \tag{4}$$

where B_1, B_2, and B_3 are the three different bounding boxes shown in Fig. 4, $v_B(\sum)$ is the area of their corresponding rotated rectangular boxes calculated by Gaussian distribution. The KFIoU loss is shown in Eq. (5):

$$Loss_{KFIoU} = 1 - KFIoU \tag{5}$$

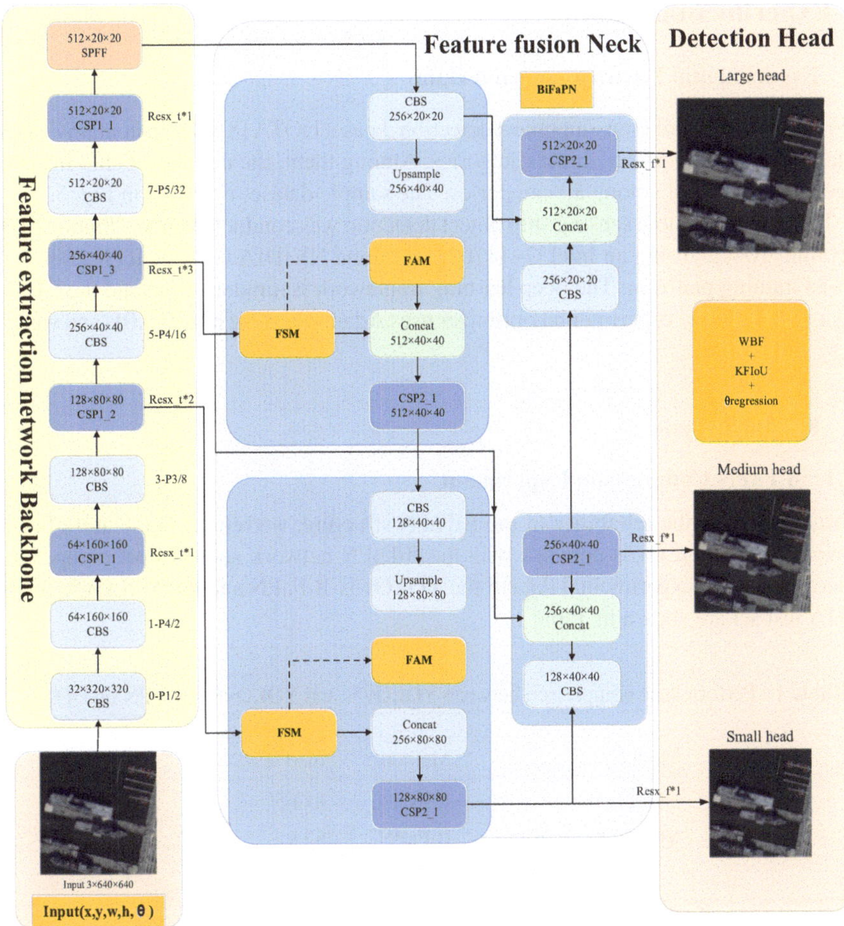

Fig. 4. RS-YOLO network structure diagram.

Fig. 5. Sample DOTA data set

3 Experimental Results

3.1 Experimental Environment and Data

This paper uses remote sensing object detection dataset DOTA [8], which contains 18828 instances and 15 common object categories. Among them, the number of ship instance targets ranks second among all categories. The sample dataset is shown in Fig. 5.

This paper's remote sensing ship object detection was conducted on an Ubuntu 20.04 operating system, using an Intel i7–11700 8-core and NVIDIA Geforce RTX 2080Ti to build a training piatform. The deep learning framework is mmdetection, and the CUDA version is 11.1. The ecperimental optimizer uses Adam, learning rate = 0.01, and weight decsy = 0.0005.

3.2 Result Analysis

3.2.1 BiFaPN Comparison Experiment

In order to verify the reliability of the BiFaPN structure, we replace the feature fusion network of the YOLOv5s network with the BiFaPN network and conduct comparison experiments. The experimental data of the YOLOv5s-BiFaPN and the YOLOv5s on the DOTA test set are shown in Table 1.

Table 1. Performance comparison between YOLOv5s and YOLOv5s-BiFaPN on DOTA.

Model	Method	mAP0.5/%
YOLOv5s	PANet	81.65
YOLOv5s-BiFaPN	BiFaPN	82.97

From Fig. 6 (a) and Fig. 6 (b), it can be observed that YOLOv5s and YOLOv5s-BiFaPN perform well in the detection of medium-sized targets, and YOLOv5s-BiFaPN still has a good performance on some larger ship targets, and the BiFaPN structure has a better detection performance for objects with different scale distributions compared to the existing multiscale strategies.

3.2.2 Ablation Experiments of Rotating Ship Target Detection Algorithm

In Table 2, YOLOv5s_3 indicates that the rotating frame as well as the rotating detection algorithm are added on top of YOLOv5s, YOLOv5s_4 indicates that the BiFaPN structure is used for feature fusion on top of YOLOv5s_3; YOLOv5s_5 indicates that the WBF method is used for predictive edge screening on top of YOLOv5s_3; RS-YOLO indicates that both strategies are used based on YOLOv5s_3.

As can be seen in Table 2, the rotating border detection accuracy of YOLOv5s_3 network is 85.67%, while the rotating border detection accuracy of YOLOv5s_4 network results in an improvement to 87.29%, and the rotating border detection mAP0.5 of YOLOv5s_5 network is 86.06%. However, the introduction of the WBF predictive

(a)YOLOv5-BiFaPN test results (b) YOLOv5 test results

Fig. 6. YOLOv5s and YOLOv5s-BiFaPN visualized detection results on the DOTA dataset.

Table 2. Rotating ship target detection algorithm ablation experiment results.

Model	HBB/OBB	Rotated Head	BiFaPN	WBF	mAP0.5/%
YOLOV5s-Baseline	HBB	no	yes	yes	83.51
YOLOV5s_3	OBB	yes	no	no	85.67
YOLOV5s_4	OBB	yes	yes	no	87.29
YOLOV5s_5	OBB	yes	no	yes	86.06
RS-YOLO	OBB	yes	yes	yes	87.56

bounding box filtering method has limited improvement in rotated border detection accuracy compared to horizontal border detection because rotated border detection adopts the OBB (Oriented Bounding Box) annotation method, which can more accurately match the shape of the object, avoiding the leakage problem caused by NMS, and the IoU between the neighboring detection frames is approximately equal to zero so that there will not be a case of mistaken deletion due to the excessively large IoU. Therefore, the improvement of algorithm accuracy by WBF needs to be improved in this annotation method. The result of the rotated border detection accuracy of the RS-YOLO network is 87.56%, which is higher than other network models. The above experimental results prove that the BiFaPN feature fusion structure and WBF are still effective in the rotating bezel ship detection task and also show that RS-YOLO has excellent detection performance in the remote sensing ship detection task.

Figure 7. Demonstrates the detection effect of YOLOv5s-Baseline (up) and RS-YOLO (down) networks under some complex backgrounds, and it can be seen that RS-YOLO has higher accuracy and better model generalization.

3.2.3 Horizontal Comparison Experiments

The detailed comparison of experimental data of the single-scale OBB test result comparison of different remote sensing ship detection algorithms on the DOTA ship detection test set is shown in Table 3.

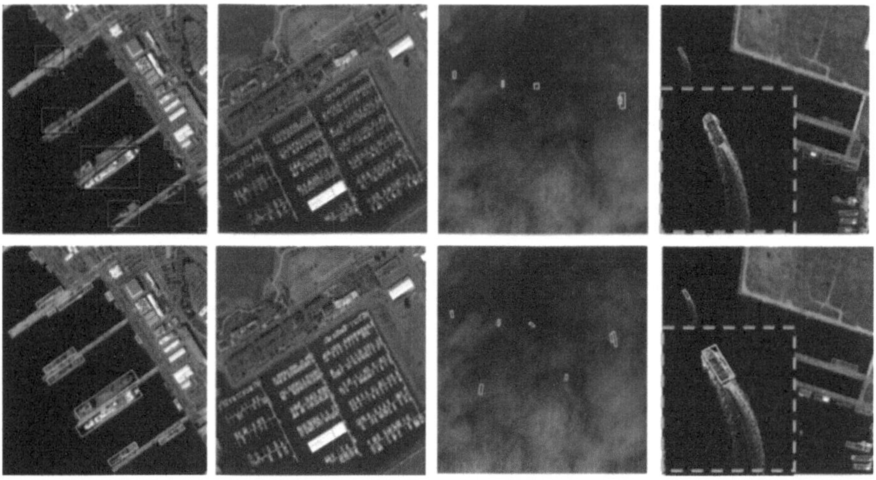

Fig. 7. Ship target detection results under background complexity.

Table 3. Experimental results of remote sensing ship detection algorithms.

Module	HBB/OBB	Backbone	Method	mAP0.5/%
two-stage:				
Mask R-CNN [9]	OBB	ResNet50	FPN	85.57
Faster R-CNN-O [10]	OBB	ResNet50	FPN	77.11
SCRDet ++ [11]	OBB	ResNet50	FPN	86.05
Oriented R-CNN [12]	OBB	ResNet50	FPN	87.52
one-stage:				
RetinaNet-O [13]	OBB	ResNet50	FPN	79.11
R3Det [14]	OBB	ResNet50	FPN	78.21
PolarDet [15]	OBB	ResNet50	FPN	85.84
S^2ANet [16]	OBB	ResNet50	FPN	87.25
RS-YOLO	OBB	CSPDarkNet	BiFaPN	**87.56**

As can be seen from the table, RS-YOLO is the better-performing algorithm among the current mainstream ship detection algorithms in the visible remote sensing ship detection task.

4 Conclusion

This paper introduces the RS-YOLO, a remote sensing ship target detector designed to tackle challenges such as significant changes in ship scale, arbitrary ship direction, and complex background environments in ship images. The proposed method utilizes a bidirectional feature pyramid network to implement a multi-scale object detection approach through feature selection alignment, which combines a feature selection module and feature alignment module. Additionally, the introduction of KFIoU, combined with the Regression Box WBF strategy, enhances YOLOv5. The resulting RS-YOLO is tailored for remote sensing scenarios, leading to improved accuracy in ship target detection even in complex remote sensing backgrounds.

References

1. Song, Z., Sui, H., Li, Y.: A survey on ship detection technology in high-resolution optical remote sensing images. Geomatics Inf. Sci. Wuhan Univ. **46**(11), 1703–1715 (2021)
2. Dubey, S.R., Singh, S.K., Chaudhuri, B.B.: Activation functions in deep learning: a comprehensive survey and benchmark. Neurocomputing **503**, 92–108 (2022)
3. Zhou, Y., Yang, X., Zhang, G., et al.: Mmrotate: a rotated object detection benchmark using pytorch. In: Proceedings of the 30th ACM International Conference on Multimedia, pp. 7331–7334 (2022)
4. Liu, R.W., Yuan, W., Chen, X., et al.: An enhanced CNN-enabled learning method for promoting ship detection in maritime surveillance system. Ocean Eng. **235**, 109435 (2021)
5. Bo, L.I., Xiaoyang, X.I.E., Xingxing, W.E.I., et al.: Ship detection and classification from optical remote sensing images: a survey. Chin. J. Aeronaut. **34**(3), 145–163 (2021)
6. Yang, X., Zhou, Y., Zhang, G., et al.: The KFIoU loss for rotated object detection. arXiv preprint arXiv:2201.12558 (2022)
7. Zheng, Z., Wang, P., Ren, D., et al.: Enhancing geometric factors in model learning and inference for object detection and instance segmentation. IEEE Trans. Cybern. **52**(8), 8574–8586 (2021)
8. Xia, G.S., Bai, X., Ding, J., et al.: DOTA: A large-scale dataset for object detection in aerial images. In: Proceedings of the IEEE Conference on Computer Vision and Pattern Recognition, pp. 3974–3983 (2018)
9. He, K., Gkioxari, G., Dollár, P., et al.: Mask r-cnn. In: Proceedings of the IEEE International Conference on Computer Vision, pp. 2961–2969 (2017)
10. Girshick, R.: Fast r-cnn. In: Proceedings of the IEEE International Conference on Computer Vision, pp. 1440–1448 (2015)
11. Yang, X., Yan, J., Liao, W., et al.: Scrdet++: detecting small, cluttered and rotated objects via instance-level feature denoising and rotation loss smoothing. IEEE Trans. Pattern Anal. Mach. Intell. **45**(2), 2384–2399 (2022)
12. Xie, X., Cheng, G., Wang, J., et al.: Oriented R-CNN for object detection. In: Proceedings of the IEEE/CVF International Conference on Computer Vision, pp. 3520–3529 (2021)
13. Del Prete, R., Graziano, M.D., Renga, A.: RetinaNet: a deep learning architecture to achieve a robust wake detector in SAR images. In: 2021 IEEE 6th International Forum on Research and Technology for Society and Industry (RTSI), pp. 171–176. IEEE (2021)
14. Yang, X., Yan, J., Feng, Z., et al.: R3det: refined single-stage detector with feature refinement for rotating object. In: Proceedings of the AAAI Conference on Artificial Intelligence, vol. 35, no. 4, pp. 3163–3171 (2021)

15. Zhao, P., Qu, Z., Bu, Y., et al.: Polardet: a fast, more precise detector for rotated target in aerial images. Int. J. Remote Sens. **42**(15), 5831–5861 (2021)
16. Han, J., et al.: Align deep features for oriented object detection. IEEE Trans. Geosci. Remote Sens. **60,** pp. 1–11 (2020)

Infrared Small Moving Target Detection Based on Combination of 3D-Conv and 2D-Conv

Cui Shuwei(✉) and Wu Wenbo

Beijing Institute of Space Mechanics and Electricity, Beijing 100090, China
cuishuwei123456@foxmail.com

Abstract. Infrared dim and small targets usually have the characteristics of small target size, few pixels, and low SNR, which are easily submerged by background and noise. By analyzing the characteristics of infrared small targets, we notice that the motion characteristics of infrared targets are different from noise. In addition, 3D Conv in the spatiotemporal domain can effectively extract the motion features of infrared small targets. Based on the characteristics of infrared small target images, this paper proposes an infrared small target detection algorithm that combines 3D Conv and YOLOv5. In experiments, we achieved significant improvements in false alarm rate and detection rate when the number of parameters only increased by 0.3% compared to yolov5s neural network model. In the presence of noise interference, our proposed model can reduce the number of false alarms by at least 40% and achieve a precision rate of over 96%, which proves that the model has good robustness. The experiment proved that our proposed algorithm combining 3D conv and 2D conv can effectively detect Infrared dim small targets, with good detection rate and low false alarm rate.

Keywords: neural network · dim and small target detection · 3D convolutional

1 Introduction

Recently, infrared target detection has been currently widely used in many fields [1], such as fire monitoring and ship tracking. However, infrared dim target images usually have the following characteristics [2]. The imaging distance is far, and the image is very dim, easily submerged in background and noise. Moreover, the distribution of small targets tends to be sparse, and there is a serious imbalance between the target and background distributions. Besides, targets typically possess fewer effective pixels, often manifesting as mere dots, which is difficult to distinguish from noise. Due to the characteristics of infrared target, infrared target detection is usually difficult and is a hot topic in the field of target detection.

In recent years, with the widespread application of neural networks, many methods for detecting infrared small targets have been proposed, broadly categorized into two groups: single frame object detection and multi frame object detection [3].Single frame object detection has the characteristics of fast detection speed and low computational power requirements, and is currently the mainstream object detection method. Reference [4] proposed an RPN (Region Proposal Network) structure, which improved the

H. P. Urbach et al. (Eds.): ISSOIA 2023, SPP 191, pp. 227–235, 2024.
https://doi.org/10.1007/978-981-97-6718-2_21

quality of candidate boxes while reducing the number of candidate boxes. Reference [5] proposed a method of combining shallow and deep network features to address the low detection rate of small targets. In response to the complex steps and slow training speed of the candidate box series algorithm, the YOLO algorithm [6] was proposed. This algorithm eliminates the step of generating candidate boxes, saves computational time, and uses methods such as feature fusion to improve the detection ability of small targets. However, due to the characteristics of infrared imaging, relying solely on a single frame image for detection is difficult under low signal-to-noise ratio conditions, often resulting in low detection accuracy and high false alarm rate [7].

In response to the problems of single frame object detection algorithms, some scholars have proposed multi frame object detection methods. Multi frame object detection has the characteristics of good anti-interference effect and detection effect. Reference [8] proposed an optical flow guided neural network, which obtains motion information of infrared targets by calculating optical flow. Reference [9] extracts moving target information through inter frame feature matching. Reference [10] proposed the use of 3D convolutional structures for infrared moving object detection. However, multi frame object detection algorithms often require combining information from several frames before and after to achieve detection of the current frame, resulting in a certain lag in the target output. At the same time, due to the large amount of information that needs to be considered, multi frame algorithms usually have a relatively large computational workload and generally have poor real-time performance.

This paper focuses on the characteristics of single frame object detection and multi frame object detection, using a combination of 3D residual structure and YOLOv5 model to achieve infrared small and dim target detection. Improve the accuracy of target detection and reduce false alarm rate with almost unchanged parameter quantities.

2 Network Structure

Through our analysis of the characteristics of infrared small targets, we observed distinct differences in motion features between infrared targets and noise. In addition, the motion features of infrared small targets can be well extracted and noise can be weakened through 3D convolution in the spatiotemporal domain [11]. So, we propose an infrared small target detection algorithm that combines 3D convolution and YOLOv5. Within this algorithm, a 3D convolutional residual module is employed to extract spatiotemporal features of infrared moving targets. Subsequently, these features are fed into the YOLOv5 neural network for object detection.

2.1 3D Residual Structure

Resnet [12] is currently widely employed in the field of target detection, proving instrumental in addressing issues such as gradient vanishing, gradient explosion, and overfitting. Traditional residual networks are mostly constructed based on 2D convolutions. Its structure is shown in Fig. 1.

The calculation formula is shown in Eq. 1.

$$y = \sigma(F(x, W) + x) \tag{1}$$

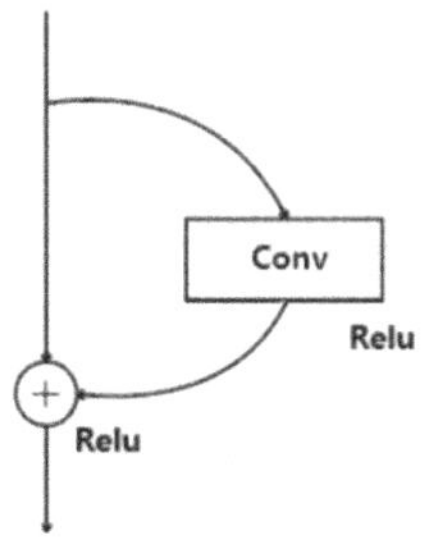

Fig. 1. Schematic diagram of residual structure.

where y means the output of the residual block, σ means the activation function, F means residual function, x means input data, W means all weights within the residual block.

To attain spatiotemporal feature information extraction, we substitute the 2D convolution with 3D convolution, based on the 2D residual structure.

Due to the slow speed of pure 3D convolution operations, we use a P3D (pseudo-3d residual networks) structure [13] to convert a 3D convolution layer with a $3 \times 3 \times 3$ size kernel into two separate layers: one with a $1 \times 3 \times 3$ size kernel and the other with a $3 \times 1 \times 1$ size kernel. This method can reduce the number of model parameters, improve computational speed, and achieve the same effect as pure 3D convolution. The calculation formula is shown in Eq. 2.

$$x_{t+1} = x_t + T(S(x_t)) \tag{2}$$

where, $S(x)$ is a 3D convolution with a convolution kernel of $1 \times 3 \times 3$, $T(x)$ is a 3D convolution with a convolution kernel of $3 \times 1 \times 1$.

Moreover, to ensure dimensional consistency when adding residuals, it's essential to include a convolutional layer with a kernel size of $1 \times 1 \times 1 \times 1$ both before and after the 3D convolutional layer to enhance or reduce the dimensions. The final 3D residual structure is shown in Fig. 2.

After calculating each convolutional layer, the ReLU activation function needs to be used for processing, as shown in Eq. 3.

$$f(x) = \max(0, x) \tag{3}$$

Due to the large number of parameters in deep 3D convolutional networks, significant computational overhead is incurred, and due to the small size of the target, only shallow networks can achieve good feature extraction results. Therefore, we opt for a two-layer 3D conv residual structure for spatiotemporal feature extraction, and replace the first convolutional layer of the YOLOv5s neural network model with the module we designed.

2.2 Connection Between 3D Conv and 2D Conv

YOLOv5 is a classic single stage object detection network [14], which has higher real-time performance and detection speed compared to two-stage object detection algorithms. This network mainly consists of an image input layer, a backbone layer, a feature

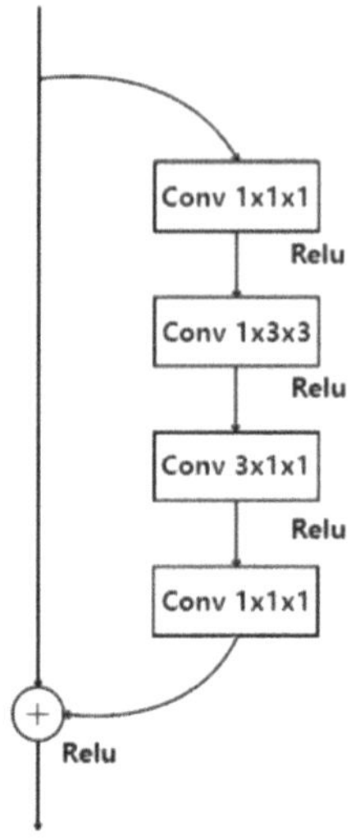

Fig. 2. Schematic diagram of 3D residual structure.

extraction layer (neck), and a target detection layer (head). The image input layer of the network uses data augmentation technology to improve the robustness of the model. The backbone layer uses modules such as Focus and SCP (Spatial Channel-wise Propagation) to increase the adequacy of feature extraction. Furthermore, the feature extraction layer adopts structures such as FPN (Feature Pyramid Network) and multi-scale fusion to enhance the detection accuracy, particularly for small targets.

The YOLOv5 model comes in different variants such as YOLOv5s, YOLOv5m, YOLOv5l, etc., which are categorized based on the number of parameters and model size. Typically, an increase in the number of model parameters corresponds to enhanced detection accuracy. However, this also demands higher computational power and time for both training and testing phases. To enhance real-time performance and maintain a lightweight model, this paper adopts YOLOv5s for object detection, connecting it with the 3D residual structure.

2.3 Connection Between 3D Conv and 2D Conv

After the 3D convolution module, the dimension size can be expressed as $n \times T \times H \times W$. Where, n means the number of channels, T means the length of the image sequence, H, W means the size of the image. The required size for a 2D convolutional layer is $n \times H \times W$. To solve the problem of inconsistent tensor dimensions required for 3D and 2D convolutions, we use a convolution layer with a kernel size equal to the number of input image frames to reduce its temporal dimension to 1. The size of the convolutional kernel is $T \times 1 \times 1$, Integrate time-domain features and connect them with the YOLOv5s two-dimensional convolutional [12] neural network to complete the subsequent target detection task. The final process is shown in Fig. 3.

The final network structure is shown in Table 1.

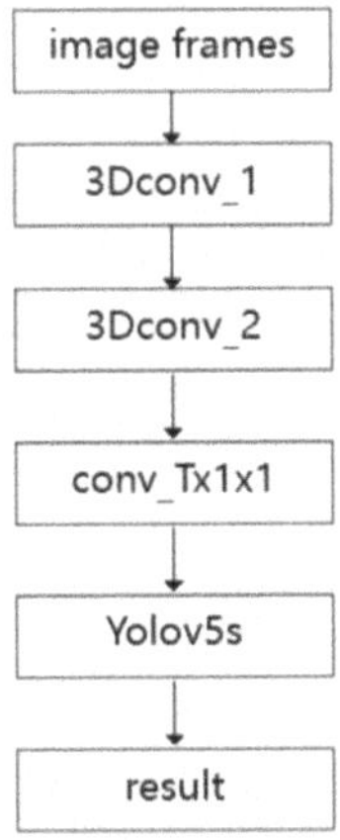

Fig. 3. Algorithm flow chart.

Table 1. Number of neural network parameter.

Layer	Parameter
Conv3d	26336
Backbone	3511040
SPPF	656896
Neck	2834688
Detect	16182
Total	7045142

3 Experimental Process and Results

3.1 Experimental Environment and Datasets

The experimental environment for this paper uses a RTX3070 graphics card with 8GB of memory, and the software environment is Python 3.7, Pytorch 1.10, and Cuda 11.2. The experiment was conducted using the "Infrared Time Sensitive Target Detection and Tracking Dataset for Air Ground Applications" [15] as the dataset, with an image size of 256×256 size infrared image sequence. Because the distribution of dim targets is sparse, a smaller input image size helps the network better capture local features and mitigate the risk of model overfitting. It also reduces processing latency and enhances system responsiveness. Qualitative and quantitative analysis were conducted on the performance of the proposed method.

In the experiment, each image frame is augmented with its two adjacent frames to form a spatiotemporal detection image sequence to train and test the neural network. Besides, the dataset is partitioned into training and testing sets in a ratio of 3:1.

3.2 Experiment Method

In order to evaluate the feasibility of this method from multiple aspects, this paper considers several metrics including precision rate, false alarm rate, mAP50 (mean Average Precision at IOU 0.5), parameter quantity, model size, and FLOPS (Floating Point Operations Per Second) as parameters to compare the proposed model with the YOLOv5 model. Moreover, in order to compare the anti-interference performance of the model, a certain amount of random noise and Gaussian noise were added to the original dataset.

Precision rate often referred to as precision or positive predictive value, represents the probability of a predicted positive sample being correct among all predicted positive samples. The calculation formula is shown in Eq. 4.

The false alarm rate represents the proportion of negative samples incorrectly predicted as positive samples to all actual negative samples, reflecting the network's ability to correctly predict positive samples. The calculation formula is shown in Eq. 5.

$$precision = \frac{TP}{TP + FP} \tag{4}$$

$$falsealarm = \frac{FP}{FP + TN} \tag{5}$$

where, *TP* means predicting positive classes as positive class numbers, *FP* means predicting negative classes as positive class numbers, *TN* means predicting negative classes as negative class numbers.

3.3 Experiment Results

Firstly, the original dataset is used for training and detection, and compared with the yolov5s, yolov5m, and yolov5l network models. The results are shown in Table 2. The analysis results show that the model proposed in this paper can improve precision rate by 3% and reduce the number of false positives by 50% compared to the YOLOv5s model. Moreover, the parameter quantity of the model proposed in this paper only increases by 0.3% compared to the YOLOv5s model. FLOPS only increases by 4%, meeting the lightweight and real-time requirements of the detection model. At the same time, the performance of the model can surpass that of the YOLOv5l model. This discrepancy arises because the noise signal bears resemblance to the target signal in the spatial domain. However, in the time domain, the noise signal deviates significantly from the target signal, as shown in Fig. 4. Partial detection results illustrated in Fig. 5, which shows that the model proposed in this paper can reduce false alarms caused by noise and improve the detection accuracy of the model.

To test the anti-interference performance of the model, Gaussian noise signals with a mean of 0, variance of 0.01 and 0.025, and random noise signals with a noise density of 0.001 and 0.01 were added to the image data, and then trained and compared with yolov5s. The results are shown in Tables 3 and 4. The analysis results indicate that even when the noise intensity is high, the proposed model maintains a high level of detection accuracy. In contrast, the performance of the YOLOv5 model decrease significantly under similar conditions. This proves that the model proposed in this paper can better adapt to detection tasks under high noise and low SNR conditions.

Table 2. Comparison of detection results and performance

model	precision rate	false alarm	mAP_0.5	parameter	Model size/MB	FLOPS
Ours	0.970	0.025	0.966	7045142	14.17	16.6G
YOLOv5s	0.940	0.050	0.936	7022326	14.11	15.9G
YOLOv5m	0.946	0.046	0.945	20871318	41.24	48.2G
YOLOv5l	0.962	0.033	0.962	46138294	90.69	108.2G

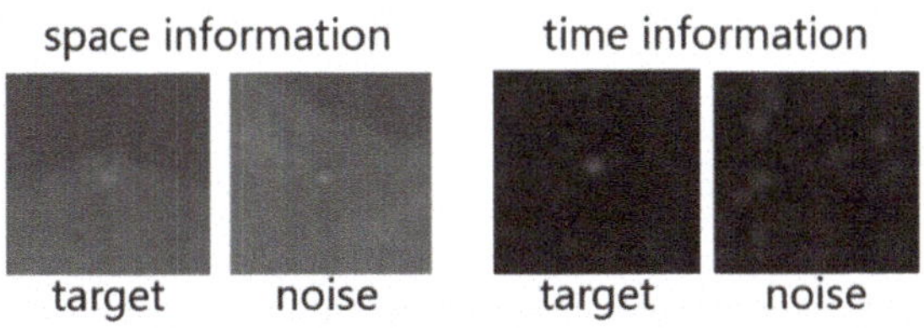

Fig. 4. Comparison of spatial and temporal characteristics between targets and noise.

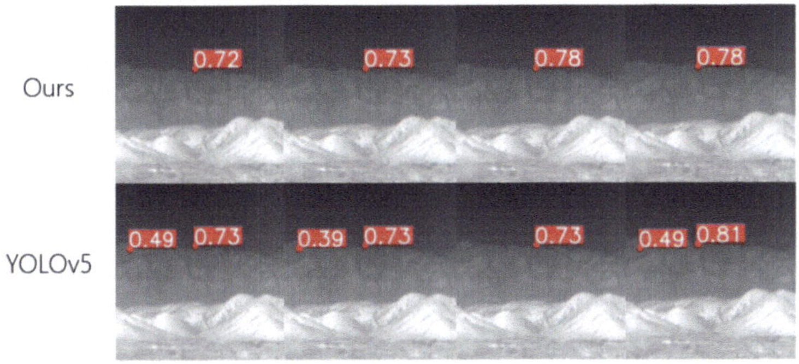

Fig. 5. Partial detection results of ours and YOLOv5 models.

Table 3. Detection results with gaussian noise.

Noise var	Precision rate (our method)	Precision rate (yolov5)
0	0.970	0.940
0.01	0.965	0.925
0.025	0.942	0.878

By observing the above experimental results, spatiotemporal 3D convolution effectively can enhance detection success rates, diminish false alarm rates, and augment anti-interference capabilities. Moreover, the model parameter quantity only increases

Table 4. Detection results with random noise.

Noise num	Precision rate (our method)	Precision rate (yolov5)
0	0.970	0.940
0.001	0.966	0.933
0.01	0.948	0.915

by 0.3% compared to yolov5s, FLOPS increases by 4%, and the average detection speed reaches 28.9ms per frame, meeting the requirements of real-time and fast performance.

4 Conclusion

This paper proposes an infrared small target detection algorithm that combines 3D Conv and 2D Conv. In this algorithm, we design a three-dimensional convolutional residual module to extract the spatiotemporal features of infrared moving targets. Then, the obtained features are input into the CNN neural network for object detection. And a dataset named "A dataset for infrared time sensitive target detection and tracking for air ground application" is selected as training and testing data, and compared with traditional CNN neural network algorithms. In the experiment, compared with the Yolov5s neural network model, we achieved significant improvements in false alarms and detection rates while only increasing the number of parameters by 0.3%. In the presence of noise interference, our proposed model can reduce the number of false alarms by at least 40% and achieve an accuracy of over 96%, which proves that the model has good robustness. Experiments have shown that our proposed algorithm combining 3D-conv and 2D-conv can effectively detect infrared small targets, with good detection rate and low false alarm rate.

References

1. Huang, Y., Xu, J., Bai, S., et al.: Satellite infrared remote sensing technology and its application in disaster prevention and relief. Spacecraft Recov. Remote Sens. **41**(05), 118–126 (2020)
2. Eysa, R., Hamdulla, A.: Issues on infrared dim small target detection .and tracking. In: 2019 International Conference on Smart Grid and Electrical Automation (ICSGEA), pp. 452–456 (2019)
3. Lv, P.Y., Sun, S.L., Lin, C.Q., et al.: Space moving target detection and tracking method in complex background. Infrared Phys. Technol. **91** (2018)
4. Ren, S., He, K., Girshick, R., et al.: Faster R-CNN: towards real-time object detection with region proposal networks. IEEE Trans. Pattern Anal. Mach. Intell. **39**(6), 1137–1149 (2017)
5. Kong, T., Yao, A., Chen, Y., et al.: Hyper net: towards accurate region proposal generation and joint object detection. IEEE (2016).https://doi.org/10.1109/CVPR.2016.98
6. Zhang, R., Wen, C.: SOD-YOLO: a small target defect detection algorithm for wind turbine blades based on improved YOLOv5. Adv. Theory Simul. **7**, 5 (2022)
7. Shao, X., Zhang, C., Wel, Y., et al.: Improved YOLOv3 algorithm for remote sensing image road automatic intersection identification. Spacecraft Recov. Remote Sens. **43**(05), 123–132 (2022)

8. Zhu, W., Tan, Y.: A moving infrared small target detection method based on optical flow-guided neural networks. In: 2023 4th International Conference on Computer Vision, Image and Deep Learning (CVIDL), Zhuhai, China, pp. 531–535 (2023)
9. Han, L., Tan, C., Liu, Y., Song, R.: Research on the on-orbit real-time space target detection algorithm. Spacecraft Recov. Remote Sens. **42**(06), 122–131 (2021)
10. Liu, X., Li, X., Li, L., Su, X., Chen, F.: Dim and small target detection in multi-frame sequence using Bi-Conv-LSTM and 3D-Conv structure. IEEE Access **9**, 135845–135855 (2021)
11. Wen, C., Liu, S., Yao, X., et al.: A novel spatiotemporal convolutional long short-term neural network for air pollution prediction. Sci. Total. Environ. **654**, 1091–1099 (2019)
12. He, K., Zhang, X., Ren, S, et al.: Deep residual learning for image recognition. In: Proceedings of the IEEE Conference on Computer Vision and Pattern Recognition, pp. 770–778 (2016)
13. Qiu, Z., Yao, T., Mei, T.: Learning spatio-temporal representation with pseudo-3d residual networks. In: Proceedings of the IEEE International Conference on Computer Vision, pp. 5533–5541 (2017)
14. Jocher, G., Chaurasia, A., Stoken, A, et al.: ultralytics/yolov5: v6. 2-yolov5 classification models, apple m1, reproducibility, clearml and deci. ai integrations. Zenodo (2022)
15. Hui, B., Song, Z., Fan, H., et al.: A dataset for infrared image dim-small aircraft target detection and tracking under ground/air background. Sci. Data Bank **5**, 12 (2019)

Detector-Based Absolute Radiation Traceability Method in Radiation Calibration

Zheng Xinglin(✉), Guo Yongxiang, Guo Chengliang, Dong Huiwen, Liu Yiliang, and Tang Shijian

Beijing Institute of Space Mechanics and Electricity, Beijing 100094, People's Republic of China
13135341@qq.com

Abstract. Optical radiation is the basic carrier of optical remote sensing information. The geometry and spectral resolution capabilities of optical remote sensors on various satellite platforms are directly related to their ability to accurately measure optical radiation. At present, in the application of space optical remote sensing detection, users are not satisfied with judging the presence or absence of targets, but also solve more elaborate problems. Researchers have been committed to the research of high-precision radiation calibration and standard transmission technology, so as to realize the accurate inversion of the quantitative characteristic parameters of ground targets and environments from optical radiation data, and establish corresponding identification and measurement numerical models.

The radiation standard transfer chain generally includes two categories, one is based on blackbody radiation sources, and the other is based on cry-radiometers and photo-detectors. In this paper, the basic methods of the above two radiation standard transfer chains are described and compared and analyzed.

Keywords: blackbody radiation source · cry radiometer · primary standard · trap detector · traceability link

1 Radiation Standard Transfer Chain

Satellite remote sensor radiation calibration generally includes three categories: prelaunch laboratory calibration, on-board calibration and site calibration [1, 2]. Each calibration test requires high-precision primary standards and working standards as a guarantee, so the establishment of high-precision radiation standards is the basis for all these applications. In the process of measurement traceability, the radiation standard transfer chain generally includes two categories, one is based on the blackbody radiation source, and the other is based on the cryometer and photodetector. Among them, the establishment of traditional radiation standards and radiation standard transfer chains is based on blackbody radiation theory [3–5], and Fig. 1 shows the radiation standard transfer chain based on radiation sources:

Since the spectral radiation of a blackbody radiation source changes with temperature and wavelength, and its variation law conforms to Planck's law of radiation, the greatest uncertainty of this method comes from the measurement of temperature. In the

H. P. Urbach et al. (Eds.): ISSOIA 2023, SPP 191, pp. 236–242, 2024.
https://doi.org/10.1007/978-981-97-6718-2_22

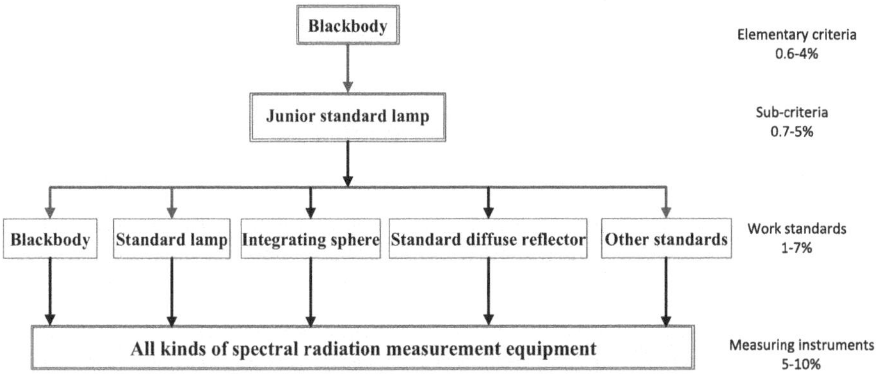

Fig. 1. Radiation standard transfer chain based on radiation sources

actual operation process, because the temperature is difficult to control and the temperature distribution uniformity is not good, the measurement uncertainty is large, which is difficult to meet the actual needs of high-precision radiation measurement.

The uncertainty of measuring radiation power in the range of 250 nm–50 μm is about 1×10^{-4}, and its accuracy is much higher than and blackbody radiation source, so the development of low-temperature absolute radiometer technology has opened up new ideas and space for effectively reducing the uncertainty of secondary work standards. However, low-temperature absolute radiometer has the characteristics of large volume, difficult operation, expensive cost, etc. [6–8], and is not convenient for use in daily work and experiments, so we must conduct research on radiation standard transmission technology on the basis of low-temperature absolute radiometer, and transmit high-precision standard devices and secondary working standard devices.

Silicon photodiode-based trap detector due to the specific small volume, convenient operation and low price and other characteristics and become an excellent choice for standard transmission detector, as shown in Fig. 2, the right side is the uncertainty of all levels of standards, it can be seen that the establishment and transmission of this new type of radiation standard will reduce the uncertainty of all level of standards by 1–2 order of magnitude.

2 Working Principle of Cryoradiometer

In the standard transfer chain, cryoradiometers are the principle of cryoradiometer and absolute radiometer is photoelectric substitution method, but due to the application of new technical means such as low temperature, superconductivity and vacuum, the accuracy of cryoradiometer relative to normal temperature absolute radiometer has been greatly improved.

2.1 Development of Cryoradiometer

Absolute radiometer is the earliest detector used for optical radiation measurement reference, its detection surface is a layer of black substance with a high absorption ratio,

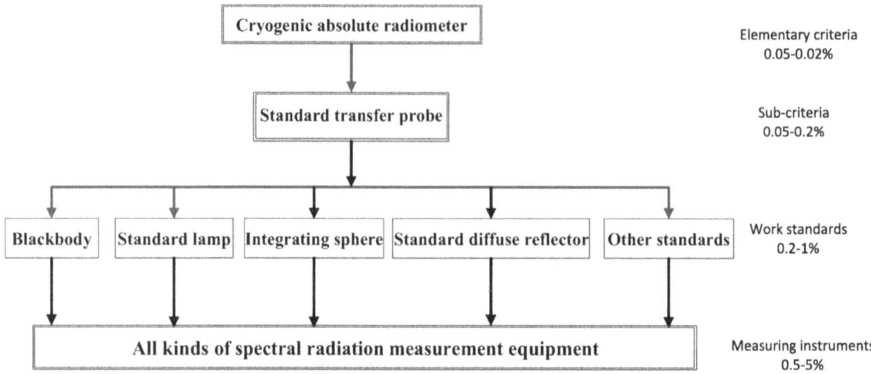

Fig. 2. High-precision radiation standard transfer chain based on cryogenic absolute radiometer and photodector

receiving light irradiation will produce temperature rise, its working principle is to use the equivalence of optical radiation heating and electric heating to measure optical radiation power. The room temperature absolute radiometer has undergone many continuous improvements, and the uncertainty level of measuring optical radiation power has been close to 0.1%, but because of the working in the normal temperature environment, the absolute radiometer is affected by many factors such as materials and the environment, and the measurement uncertainty is difficult to further decrease. The working principle of cryoradiometer is the same as that of absolute radiometer, but because it works in a liquid nitrogen environment, it breaks through the limitations of environment and materials, and its measurement uncertainty is reduced by about an order of magnitude compared with normal temperature absolute radiometer.

2.2 Working Principle of Cryoradiometer

The block diagram of the system structure of the cryoradiometer is show in Fig. 3:

Cryoradiometers measure laser power using the principle of electrical substitution. First, the radiated power F incident into the blackened absorption surface A of the electrical alternative radiometer(ESR) is absorbed; This radiation heats chamber to a temperature b to a temperature T determined by the heat sink d and the thermal conductive ring c, the heat capacity, and the thermal balance of F and the environment, which is measured by the temperature sensor e. After the radiation is balanced, the radiation F is turned off, and then the f is energized and heated to bring the temperature in the cavity to equilibrium, and the temperature of b is equal to T, and the heating power flowing through f is accurately measured, which is the radiant power F [9, 10].

The operating principle formula of the cryoradiometer is shown in Eq. 1, the left side of the equation reflects the effect of light heating, and the right side of the equation reflects the effect of electric heating.

$$F \cdot (ASTN) = V_h \cdot \frac{V_i}{R_i} \tag{1}$$

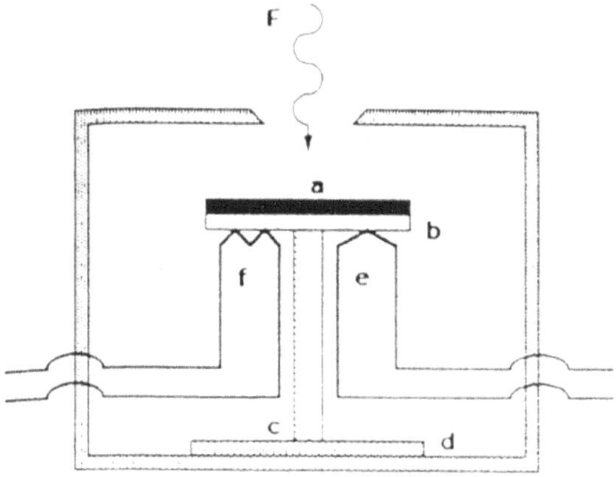

Fig. 3. Block diagram of the cryoradiometer system

In Eq. 1, F is the measured radiant flux(radiant power), A is the absorption coefficient of the cavity, T is the transmittance of the windows, S is the stray light correction coefficient, and N is the photoelectric heating inequality coefficient. V is the voltage across the cavity electric heater, V and R are the terminal voltage and resistance values of the standard resistors in the electric heating circuit(V/R, the current flowing through the electric heater).

2.3 Application of Cryoradiometer

As a radiation reference, the measurement uncertainty of the cryogenic absolute radiometer itself can be guaranteed and reproduced by analysis and debugging, long-term calibration and international comparison. The method of establishing a secondary reference with the detector as the reference is to measure the power of a frequency-stable laser or a highly stable monochromatic light source with a low-temperature absolute radiometer, and ten use the light source as a radiation standard to calibrate the detector and transmit the reference to the detector. The establishment of the working reference is to use the calibrated detector or blackbody secondary reference to transmit the next level of the working reference, which can be standard lamps, detectors, blackbodies and other light sources, and the calibrated amount includes photometric radiation or detector sensitivity, uniformity, etc. In this way, a high-precision radiation standard transfer chain is established, because the low-temperature absolute radiometer reference is more than two orders of magnitude higher than the traditional reference accuracy, which greatly improves the accuracy of the entire photometric radiation measurement.

In addition to physical quantities with a dimension of 1 spectral transmittance and reflectance, important units of optical metrology can be linked to cryoradiometers and thus improve the measurement accuracy of the corresponding quantities. These quantities include laser power and energy, detector absolute responsivity, blackbody radiation temperature, spectral radiation brightness and irradiance, luminous flux, etc. The relationship of cryoradiometers to other major optical quantities and basic optical radiation measurement methods is shown in Fig. 4 [11].

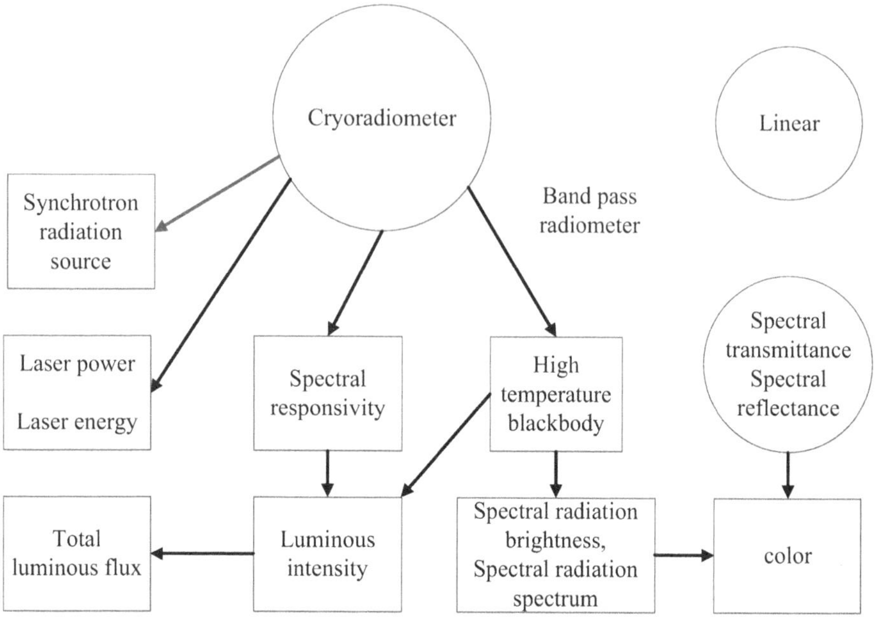

Fig. 4. Block diagram of the relationship between the main magnitudes and cryoradiometers in the field of optical metrology

3 High-Precision Transmission Standard Based on Trap Detector

The high-precision transmission standard based on the trap detector is commonly known as the secondary standard, and the requirements for the transmission standard are high response rate, wide response band, large linear range and small temperature coefficient. In order to reduce the influence of reflection loss on accuracy, the reflectivity of the photosensitive surface should be as small as possible. In the visible ~ near-infrared band, silicon photodiode meets the above conditions, and its manufacturing process is mature, stable and cheap, so the trap detector based on silicon photodiode has become the best choice for standard transmission detectors.

Silicon trap detector is composed of several silicon photodiodes in a certain spatial orientation [12], after the reflection and absorption of multiple surfaces, the light injected into the device is basically and almost completely absorbed, the reflected light only accounts for a few hundred thousandths of the incident light, to a completely negligible extent, so this device is also called a fully absorbed light trap detectors was first proposed by Zalewski and Duda [13], which can be roughly divided into two categories: transmission and reflection.

Figure 5 above is the structure of a four-piece reflective trap detector, and Fig. 6 above is the structure of a three-piece reflective trap detector, both of which are currently the most widely used, and their disadvantage is that the reflected power loss is not easy to measure. The more common transmission trap detectors are mainly six transmission and four transmission types, as shown in Fig. 7 and Fig. 8, respectively. The six-piece transmission trap detector has the advantages of easy measurement of escape

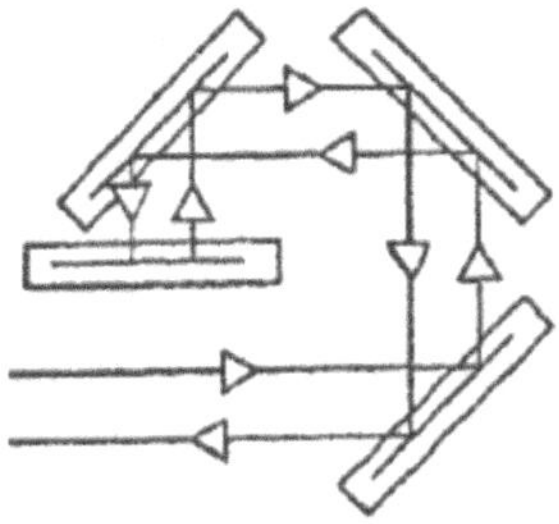

Fig. 5. Four-piece reflective trap detector

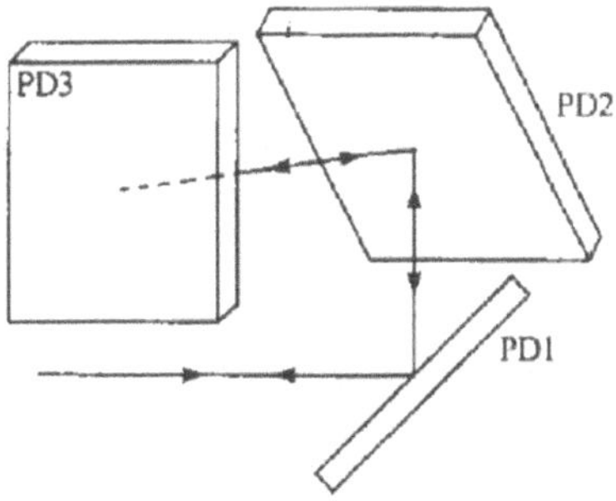

Fig. 6. Three--piece reflective trap detector

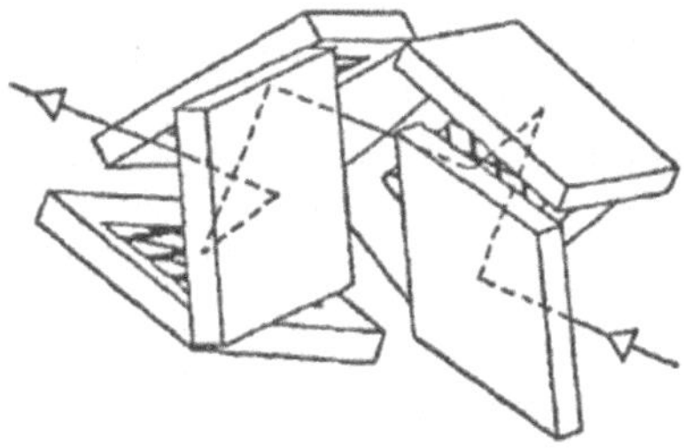

Fig. 7. Six-piece transmission trap detector

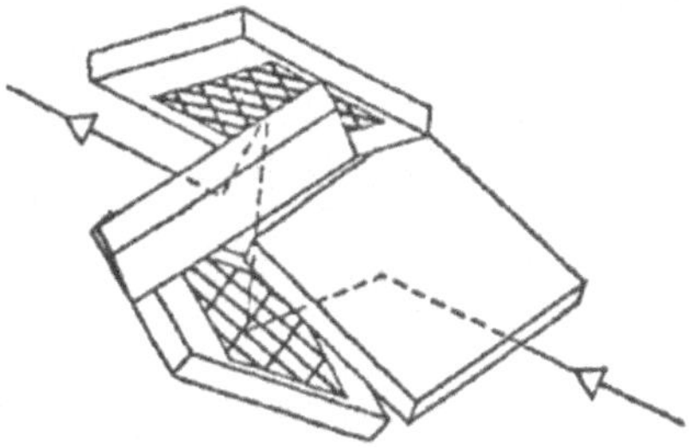

Fig. 8. Four-piece transmission trap detector

power and non-sensitivity to polarization, but its structure is complex and the number of photodiodes use is large; The four-piece transmission trap detector is highly sensitive to the polarization state of the incident light.

The structural design of laboratory standard transfer detectors can be selected taking into account various factors [14].

Standard detectors base on trap detectors are becoming increasingly important in modern photoradiation metrology and photometry. It greatly shortens the transfer chain length between the primary radiation standard and the working standard and the user standard, so that the uncertainty of the final radiation standard passed to the user is greatly reduced.

4 Concluding Remarks

The detector-based radiation standard transfer system has the unparalleled superiority of the traditional standard transmission chain based on radiation sources, so cryoradiometer, as the most accurate optical radiation measurement method recognized in the world today, has been used as the focus of research in many countries around the world. The Chinese Academy of Metrology, Anhui Institute of Optics and Fine Mechanics, Chinese Academy of Sciences and Xi'an Institute of Applied Optics have introduced cryometers from abroad and established corresponding radiation standards and standard transmission chains, and in recent years, Changcun Insitute of Optical, Fin Mechanics and Physics, Chinese Academy of Sciences and other units have also carried out corresponding research work.

References

1. Gong, H., Tian, G.L., Tao, Y., et al.: Vicarious radiometric calibration and validation of CBERS02B CCD data. J. Remote Sens. **14**(1), 1–6 (2010)
2. Xu, W., Zhang, C., Chen, Z., et al.: On-orbit radiometric calibration of optical remote sensing satellite using an array of reflected point sources. Spacecraft Recov. Remote Sens. **43**(1), 98–107 (2022). (in Chinese)
3. Man, Y., Chen, S.: Research progress of on-board absolute radiometric calibration for optical remote sensing satellite. Chin. Space Sci. Technol. **42**(6), 12–22 (2022)
4. Huang, D.: British National physical laboratory (NPL) standard blackbody radiation source. Modern Metrol. Test **3**(6), 59–61 (1995)
5. Wang, T., Yuan, Z., Duan, Y., et al.: Development of BF-50B blackbody radiation source standard device. Modern Sci. Instrum. **5**(4), 23–25 (2003)
6. Martin, J.E., et al.: A cryogenic radiometer for absolute radiometric measurements. Metrologia **21**, 147–155 (1985)
7. Foukal, P.V., et al.: Cryogenic absolute radiometers as laboratory irradiance standards, remote sensing detectors, and pyroheliometers. Appl. Opt. **29**, 988–993 (1990)
8. Jung, T.M., Carter, A.C., Woods, S.I., Kaplan, S.G.: Kaplan. Calibration and deployment of a new NIST transfer radiometer for broadband and spectral calibration of space chambers (MDXR). In: Proceedings of SPIE 8015, 80150C-1 to 80150C-11 (2011)
9. Quan, Z., Gao, Y., Wang, H.: Absolute spectral response rate test of detector based on cryradiometer. Aerospace Meas. Technol. **29**(3), 70–72 (2009)
10. Li, Y., Zhao, Z., et al.: Spectral calibration of large dynamic range dual-channel camera. Spacecraft Recov. Remote Sens. **38**(5), 44–49 (2017). (in Chinese)
11. Quinn, T.J.: Primary methods of measurement an primary standards. Metrologia, 34–63 (1997)
12. Shi, X.S., Chen, K.F., Wang, H.F., et al.: Experimental studies on Si trap detector absolute spectral responsivity calibration. Opt. Optoelectron. Technol. **7**(5), 63–65 (2009)
13. Zalewski, E.F., Hoyt, C.R.: Silicon photodiode device with 100% external quantum efficiency. Appl. Opl. **22**, 2867–2873 (1983)
14. Liu, H.: Application of detector as transmission standard in photometric measurement. Modern Metrol. Test **4**, 11–13 (1999)

Research on Coherent Detection Technology of Four Quadrant Photodetectors Based on Monte Carlo Method

Ziqi Zhang[1,2], Shoufeng Tong[1(✉)], Xiaonan Yu[1], Peng Lin[1], and Baiqiu Zhao[1,2]

[1] Institute of Space Optoelectronics Technology, Changchun University of Science and Technology, 7089 Weixing Road, Changchun 130022, China
tsf1998@sina.com

[2] School of Photo-Electronic Engineering, Changchun University of Science and Technology, 7089 Weixing Road, Changchun 130022, China

Abstract. Space gravitational wave detection reflects the optical path changes between satellites by sensing small changes in phase information, thereby inverting gravitational wave signals. The gravitational wave detection system places high demands on the four quadrant photodetector as a photoelectric conversion device. The position detection accuracy of the four quadrant detector will greatly affect the perception of phase and optical path information. In order to improve the position detection accuracy of the detector for receiving light spots, this article first builds a mathematical model for spot position detection based on the principle of four-quadrant detector position detection and the Gaussian distribution characteristics of light spots in space laser communication. Secondly, based on the Monte Carlo method, a four quadrant detector coherent detection simulation system is constructed by characterizing the spot power information with random points. Analyze the impact of spot radius, system signal-to-noise ratio, and spot centroid position on position detection accuracy. The simulation results show that increasing the system signal-to-noise ratio, appropriately reducing the spot radius, and placing the spot centroid position close to the detector origin can effectively improve the detector position detection accuracy. And the simulation results are consistent with the theoretical results of the mathematical model, which has certain guiding significance for engineering applications.

Keywords: gravitational wave detection · four quadrant detector · monte carlo method · position detection accuracy

1 Introduction

The concept of gravitational waves was first proposed by Einstein [1]. Mainly including random gravitational waves, pulse gravitational waves, and periodic gravitational wave signals. Among them, periodic gravitational wave signals mainly come from black holes and massive dense binary stars, which have the advantages of stable frequency and long duration, which is more conducive to signal extraction. Therefore, they have become the main target source for people to detect gravitational waves.

H. P. Urbach et al. (Eds.): ISSOIA 2023, SPP 191, pp. 243–252, 2024.
https://doi.org/10.1007/978-981-97-6718-2_23

Foreign scholars have conducted extensive research and achieved fruitful results in the field of gravitational wave detection. In 2016, American scientists successfully detected the gravitational wave signal generated by the merger of two black holes in the universe for the first time through the Ground Laser Interferometer Gravity Wave Observatory (LIGO) [2]. The Space Antenna Laser Interferometer Program (LISA), jointly developed by NASA and ESA, is currently one of the most mature space gravity wave detection programs. LISA has a detection sensitivity of 10 pm/$\sqrt{}$Hz within the frequency range of 0.1 mHz–100 mHz, and the measurement accuracy of the testing quality angle is higher than 20 nrad/$\sqrt{}$Hz. The allocated noise budget is 2 $\pi\mu$rad/$\sqrt{}$Hz, [3–5] laying the foundation for the space gravitational wave detection plan.

Domestic scholars started their research on gravitational wave detection relatively late, but now they have also achieved breakthrough results. The Chinese Academy of Sciences formulated the "Tai Ji Plan" project, proposed a three-step development strategy of "single star, double star and three-star", and successfully launched in 2019. The "Tai Ji Plan" plans to launch the "Tai Ji 2" binary satellite to orbit the sun in 2024, with an arm length of more than 500000 km and a ranging accuracy of 10 pm/$\sqrt{}$Hz; In 2030, the "Tai Ji 3" three-star spacecraft was launched into a preset orbit with an arm length of 3 million kilometers, and the ranging accuracy was improved to 8 pm/$\sqrt{}$Hz, breaking through key technologies such as time delay interference [6].

In this paper, based on the background of gravitational wave detection, a coherent detection simulation system of four-quadrant detector based on Monte Carlo simulation method is built, and the position and shape information of light spot in the detector's photosensitive surface is simulated by random point model. Combined with the optical front heterodyne interference and the four-quadrant detector position detection principle, the influence of spot radius, system signal-to-noise ratio, spot centroid position and heterodyne efficiency on the detector position detection accuracy was simulated, and the theoretical results were verified by the simulation results. It provides theoretical basis and simulation results verification for the subsequent desktop experiment system of laser heterodyne interference.

2 Coherent Detection Principle of Four Quadrant Detectors

The block diagram of the coherent detection system for the four quadrant detector is shown in the following figure: the signal light from the transmitting end is coherently mixed with the local oscillator signal generated by the receiving end at the optical beam splitter, and the generated intermediate frequency signal propagates to the photosensitive surface of the four quadrant detector and undergoes photoelectric conversion. After small signal amplification and analog-to-digital signal conversion, the phase locking of each quadrant photocurrent signal is carried out through a digital Costa phase-locked loop. After phase locking, position detection and calculation are carried out through the output signals of each optocurrent signal Q channel (Fig. 1).

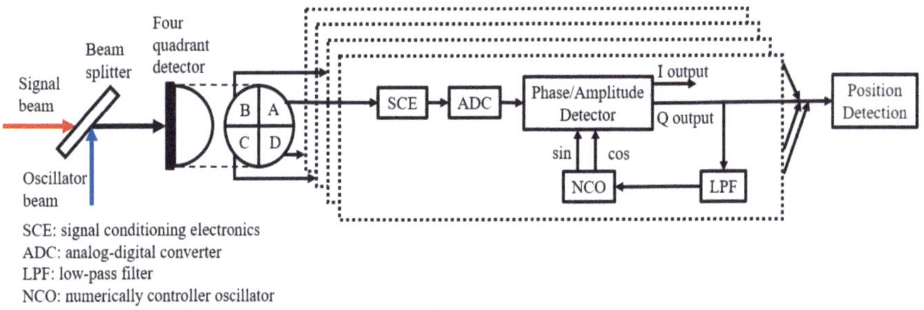

Fig. 1. Block Diagram of Four quadrant Detector Coherent Detection System

2.1 Four Quadrant Detector Position Detection Principle

The four quadrant detector arranges four identical photodiodes in a Cartesian coordinate system, and can be made of materials such as Si and InGaAs suitable for different wavelengths of detectors. For the light spot incident on the photosensitive surface of the detector, the photodiodes in each quadrant of the four quadrant detector perform photoelectric conversion on the receiving power, generating a photoelectric current. The detector uses the magnitude of the photoelectric current generated in each quadrant to substitute it into the relative position calculation formula of the centroid of the light spot to obtain the relative position (x_r,y_r) of the light spot in the photosensitive surface of the detector [7].

$$\begin{aligned} x_r &= \frac{(P_A + P_D) - (P_B + P_C)}{P_A + P_B + P_C + P_D} = \frac{(I_A + I_D) - (I_B + I_C)}{I_A + I_B + I_C + I_D} \\ y_r &= \frac{(P_A + P_B) - (P_C + P_D)}{P_A + P_B + P_C + P_D} = \frac{(I_A + I_B) - (I_C + I_D)}{I_A + I_B + I_C + I_D} \end{aligned} \tag{1}$$

Among them, P_A,P_B,P_C,$P_D$represents the received optical power of the four quadrants and I_A,I_B,I_C,I_D represents the photocurrent generated by the four quadrants[8].

The relative position (x_r,y_r) of the centroid of the light spot needs to be converted from relative position to absolute position based on the distribution characteristics of the light spot model. In free space laser communication transmission systems, the distribution of spot energy can be approximated as a Gaussian distribution[9]. Therefore, combining the distribution characteristics of Gaussian light spots, the relative position (x_r,y_r) of the light spot centroid is converted into an absolute position (x_0,y_0):

$$\begin{aligned} x_0 &= \frac{\sqrt{\pi}\omega}{2\sqrt{2}} x_r \\ y_0 &= \frac{\sqrt{\pi}\omega}{2\sqrt{2}} y_r \end{aligned} \tag{2}$$

Using the standard deviation of the absolute position (x_0,y_0) of the centroid of the spot as the measurement standard for position detection accuracy, combined with the Gaussian spot distribution model and mathematical transformation, the expression form of spot position detection accuracy is constructed:

$$\begin{aligned}\sigma_{x_0} &\approx \frac{\sqrt{\pi}\omega}{2\sqrt{2}}\cdot\frac{1}{\sqrt{R_{SN}}}\cdot\sqrt{1+erf^2\left(\frac{\sqrt{2}x_0}{\omega}\right)}\\ \sigma_{y_0} &\approx \frac{\sqrt{\pi}\omega}{2\sqrt{2}}\cdot\frac{1}{\sqrt{R_{SN}}}\cdot\sqrt{1+erf^2\left(\frac{\sqrt{2}y_0}{\omega}\right)}\end{aligned} \tag{3}$$

Based on this, a four quadrant detector spot centroid position detection accuracy model is established. Among them, $\boldsymbol{\omega}$ represents the waist radius of the light spot, R_{SN} represents the system signal-to-noise ratio, and (x_0,y_0) represents the absolute value of the centroid position of the light spot. The above three parameters are the main factors affecting the accuracy of position detection.

2.2 Coherent Detection Principle

The four quadrant detector coherent detection system coherently mixes the weak signal light from the transmitting end with the locally generated strong local oscillator light from the receiving end. On the one hand, coherent mixing can amplify the power of the intermediate frequency signal and have a certain mixing gain, amplifying the received weak signal light above the noise layer. On the other hand, high-frequency received signals (THz) can be converted into intermediate frequency signals (MHz), which can be well applied in subsequent electronic signal processing. The light fields of signal light and local oscillator light are:

$$\begin{aligned}E_S(\boldsymbol{r},t) &= |E_S(\boldsymbol{r})|\cdot\exp(i\psi_S(\boldsymbol{r}))\cdot\exp(i[\omega_S t+\varphi_S])\\ E_{LO}(\boldsymbol{r},t) &= |E_{LO}(\boldsymbol{r})|\cdot\exp(i\psi_{LO}(\boldsymbol{r}))\cdot\exp(i[\omega_{LO} t+\varphi_{LO}])\end{aligned} \tag{4}$$

Among them, $|E_S(r)|$,$|E_{LO}(r)|$ represents the intensity of the two laser beams, $\psi_S(r)$, $\psi_{LO}(r)$ represents the wavefront incidence angle, ω_S, ω_{LO} represents the angular frequency, φ_S, φ_{LO}represents the optical phase, and t represents time.

After mathematical calculations such as reflection and transmission of the splitter prism and photoelectric conversion of the detector, the expressions for the received light power and generated photocurrent of the detector are obtained as follows:

$$\begin{aligned}P(t) &= \frac{P_S}{2}+\frac{P_{LO}}{2}+\sqrt{P_S P_{LO}\gamma}\cdot\cos(\omega_{het}t+\phi)\\ I(t) &= \xi\frac{q_e\lambda}{hc}P(t)=\eta\cdot\left[\frac{P_S}{2}+\frac{P_{LO}}{2}+\sqrt{P_S P_{LO}\gamma}\cdot\cos(\omega_{het}t+\phi)\right]\end{aligned} \tag{5}$$

Among them, P_S represents signal optical power, P_{LO} represents local optical power, γ represents heterodyne efficiency, ω_{het} represents intermediate frequency signal angular frequency, ϕ represents phase difference information, ξ represents photodiode efficiency, q_e represents electron charge, λ represents wavelength, h represents Planck constant, c represents speed of light, and η represents detector response.

2.3 Heterodyne Efficiency Simulation

Heterodyne efficiency is used to characterize the efficiency of coherent mixing between optical front-end signal light and local oscillator light. According to formula (5), heterodyne efficiency also has a certain impact on the amplitude of the intermediate frequency signal generated by coherent mixing. In the mathematical model, it can be seen as the overlapping area of two circular light spots formed by signal light and local oscillator light. Build a mathematical model in the simulation system and simulate the heterodyne efficiency parameters by calculating the overlapping area between two circles. However, the results calculated using mathematical area can only characterize the heterodyne efficiency under a uniform spot model. In actual space laser communication and detection systems, the spot mostly presents a Gaussian distribution. Therefore, it is necessary to combine the Gaussian distribution characteristics of the light spot on the basis of the mathematical area calculation results, in order to obtain the heterodyne efficiency of the light spot that is suitable for practical application scenarios. The simulation results of heterodyne efficiency are shown in the following Fig. 2.

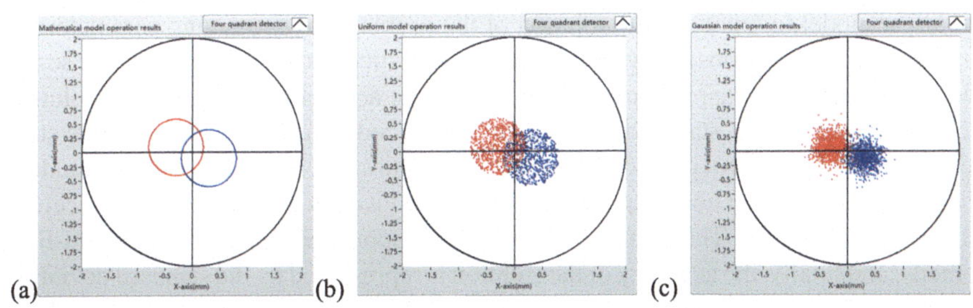

Fig. 2. Heterodyne efficiency operation results. (a) Mathematical model (b) Uniform model (c) Gaussian model

In the simulation system, when the radii of the signal light and the local oscillator light are both 0.5 mm, and the centroid positions of the two light spots on the photosensitive surface of the four quadrant detector are (±0.3,0), (±0.2,0), and (±0.1,0), the heterodyne efficiency results between the two light spots simulated using the above three methods are shown in the table below (Table 1).

Table 1. Heterodyne efficiency operation results

Spot radius	Spot centroid	Mathematical model	Uniform model	Gaussian model
0.5mm	(±0.3,0)	0.2847	0.2734	0.3687
	(±0.2,0)	0.5046	0.5061	0.7638
	(±0.1,0)	0.7471	0.7518	0.9382

From the simulation results in the table above, it can be seen that under the uniform distribution model of the light spot, the heterodyne efficiency simulation results are approximately equal to the calculation results of the mathematical model, proving the correctness of using the Monte Carlo method in the simulation system. Under the Gaussian distribution model of the light spot, the simulation results of heterodyne efficiency are higher than those of the uniform distribution model of the light spot. This is due to the spectral density characteristics of the light spot energy under different distribution models. The simulation results provide certain guidance for the subsequent heterodyne interference experimental system.

2.4 Monte Carlo Simulation Analysis Method

The Monte Carlo method, also known as statistical experiment method, is a basic method for describing various random phenomena in the system simulation process. This method is suitable for solving problems that are difficult to solve using analytical methods [10]. The basic idea of Monte Carlo method for solving problems is: a) Establish a random process or probability model, and use the parameters or mathematical features of the process or model to represent the solution of the problem; b) Estimating parameters or numerical features by observing or conducting statistical experiments on processes or models, and obtaining approximate solutions; c) The accuracy of the problem solution is characterized by the standard error of the estimated value. The simulation system uses Monte Carlo random points to represent Gaussian spot energy, constructs a random model based on photon counting and statistics, and visually characterizes the received light power of the spot. The centroid position of the spot is estimated through statistical methods, and finally, the accuracy of the solution is represented by the standard deviation of the estimated value, which is the detection accuracy of the centroid position of the spot.

The Monte Carlo random point simulation results adopted by the system are shown in the following figure. Using a two-dimensional Gaussian distribution random number to

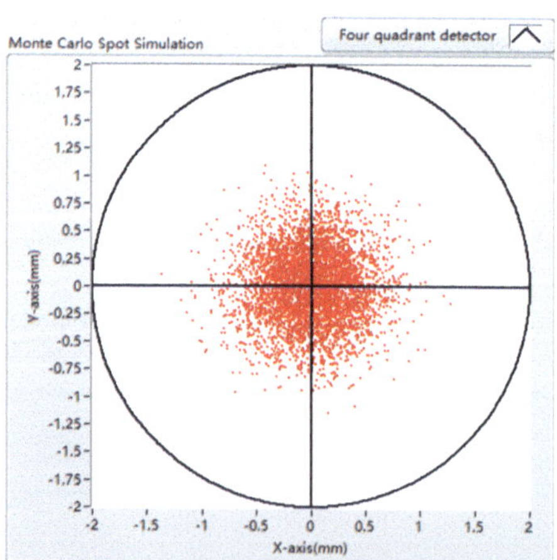

Fig. 3. Monte Carlo Random Point Simulation Results of Four Quadrant Detectors

simulate and simulate Gaussian light spots, the number of random points is the number of photons, establishing a connection between photon energy and the received light power in the actual system (Fig. 3).

3 Analysis of Simulation Results

3.1 The Influence of System Signal-To-Noise Ratio on Position Detection Accuracy

Analyze the impact of system signal-to-noise ratio on the accuracy of spot position detection. Set the radius of the detector photosensitive surface to 1.5 mm, the radius of the spot to 0.5 mm, and the true position of the spot centroid is located at the origin of the detector photosensitive surface. In non coherent and coherent architectures, gradually increase the system signal-to-noise ratio and observe the variation of position detection accuracy with the system signal-to-noise ratio. Among them, the noise factor in the coherent detection process of the four quadrant detector mainly comes from the shot noise of the local oscillator laser. In addition, it also includes thermal noise of optoelectronic devices, dark current noise, laser relative intensity noise in laser components, ADC quantization error, etc (Fig. 4).

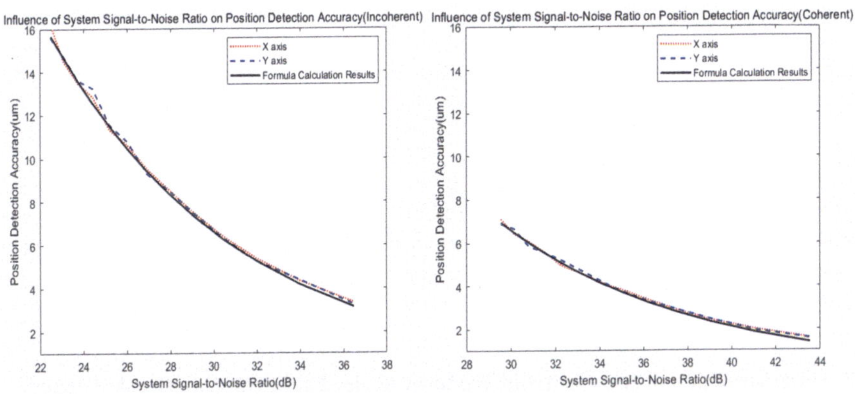

Fig. 4. Curve of the influence of detector noise on position detection accuracy.(a) Incoherent; (b) Coherent

Among them, the black curve represents the calculation results of the mathematical model for position detection accuracy, while the red and blue curves represent the simulation results of the X and Y axes, respectively. Due to the mixing gain brought by local oscillator light, the signal-to-noise ratio of the system in coherent architecture will be greater than that in non coherent architecture. From the simulation results, it can be seen that the position detection accuracy improves with the increase of the system signal-to-noise ratio, and the simulation results are in good agreement with the results of the mathematical model, with a high level of confidence.

3.2 The Influence of Spot Radius on Position Detection Accuracy

Analyze the impact of spot radius on the accuracy of spot position detection. Set the radius of the detector photosensitive surface to 1.5 mm, the system signal-to-noise ratio to 37.5 dB, and the true position of the spot centroid is located at the origin of the detector photosensitive surface. In non coherent and coherent architectures, the radius of the controlled spot gradually increases within the range of 0.2 mm−0.8 mm, with an increase step of 0.05 mm, and the variation of position detection accuracy with the radius of the spot is observed (Fig. 5).

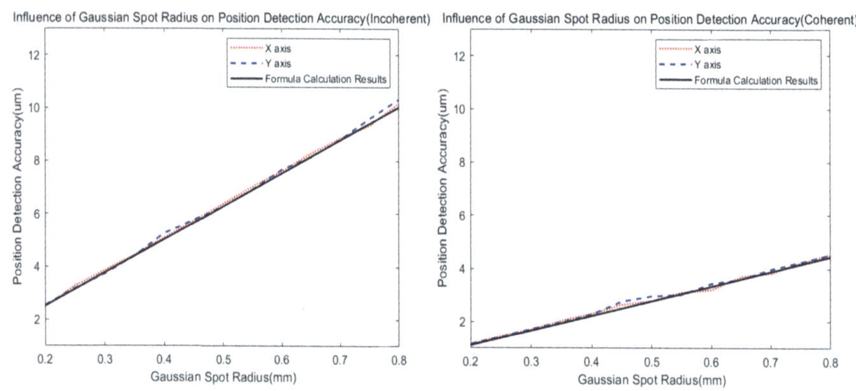

Fig. 5. Curve of influence of Gaussian spot radius on position detection accuracy.(a) Incoherent; (b) Coherent

From the simulation results, it can be seen that the position detection accuracy improves with the decrease of the spot radius, and the simulation results are in good agreement with the results of the mathematical model, with a high level of confidence. However, due to factors such as detector dead zone, the spot radius should not be too small.

3.3 The Influence of the Centroid Position of the Light Spot on the Accuracy of Position Detection

Analyze the impact of the centroid position of the light spot on the detection accuracy of the light spot position. Set the radius of the detector's photosensitive surface to 1.5 mm, the system signal-to-noise ratio to 37.5 dB, and the spot radius to 0.5 mm. In non coherent and coherent architectures, the true values of the centroid positions of the x-axis and y-axis light spots gradually increase within the range of −0.15 mm to 0.15 mm, with a step size of 0.01 mm. Observe the variation of position detection accuracy with the radius of the light spot (Fig. 6).

From the simulation results, it can be seen that the closer the centroid position of the light spot is to the origin of the detector, the higher the accuracy of position detection, and the simulation results are in good agreement with the results of the mathematical model, with a high degree of confidence. Considering factors such as the dynamic range

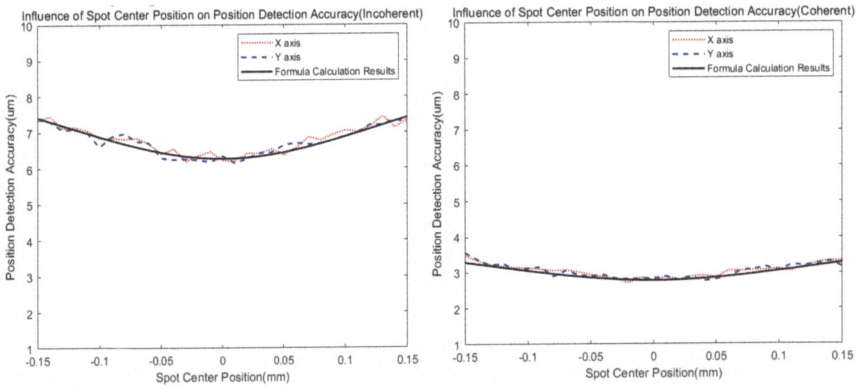

Fig. 6. Curve of influence of spot center position on position detection accuracy.(a) Incoherent; (b) Coherent

of the light spot on the detector, it is necessary to locate the light spot as close as possible to the origin of the detector.

4 Conclusion

Build a four quadrant detector coherent detection simulation system based on Monte Carlo method. The simulation results show that improving the system signal-to-noise ratio, appropriately reducing the spot radius, and keeping the spot centroid position close to the detector origin can effectively improve the position detection accuracy, and the simulation results are consistent with the mathematical model calculation results. Provide certain guiding significance for coherent detection experiments and engineering applications of four quadrant detectors.

Most of the traditional four-quadrant detector position detection methods and literatures only reflect the influence of key factors on the position detection accuracy through theoretical analysis or desktop experiment phenomenon, and do not build a simulation system for the four-quadrant detector position detection. The coherent detection simulation system of the four-quadrant detector built in this paper can simulate the position and shape of the light spot in the detector's photosensitive surface by Monte Carlo random point simulation method, which can be corresponding to the experimental phenomena of the desktop. Based on the relevant theoretical derivation, the accuracy of position detection is calculated and the influencing factors are analyzed and simulated. The establishment of the simulation system has a certain guiding effect on the actual desktop experiment phenomena and results, and has laid a certain foundation for the precise phase measurement technology of gravitational wave detection.

References

1. Liang, Y.R.: High Precision Phase Measurement for Heterodyne Laser Interferometer, pp. 1–2. Huazhong University of Science and Technology, Wu Han (2013)

2. Abbott, B.P., Abbott, R., Adhikari, R., et al.: LIGO: the laser interferometer gravitational-wave observatory. Prog. Phys. **72**(7), 076901 (2009)
3. Hough, J., Robertson, D., Ward, H., et al.: LISA -the interferomter. Space Res. **32**(7), 1247–1250 (2003)
4. Danzmann, K., Rudiger, A.: LISA technology-concept, status, prospects. Class. Quantum Gravity **20**(10), S1–S9 (2003)
5. Universidad, D.G.: Laser Ranging and Data Communication for the Laser Interferometer Space Antenna, 28–30 (2012)
6. Luo, Z.R., Zhang, M., Jin, G., et al.: Introduction of Chinese space-borne gravitational wavedetection program"Taiji"and""Taij-1"satelite mission. J. Deep Space Explor. **7**(1), 3–10 (2020)
7. Li, X.P.: Research on wide range and high precision spot position detection system based on four ouadrant detector. Chang Chun Changchun Inst. Opt. (Fine Mech. Phys. Chin. Acad. Sci.) 7–9 (2022)
8. Shang, T., Li, X., Liu, Z.J., et al.: Novel four-quadrant design and the method for obtaining spot parameters. Infrared Laser Eng. **41**(04), 1034–1040 (2012)
9. Qin, L.C., He, W.: Laser spot center location algorithm based on four-quadrant detector. Appl. Laser **41**(03), 561–568 (2021)
10. Hu, H.P., Zhang, J., Yang, L.: Application of monte Carlo method in communication system simulation. Digit. Commun. World **05**, 43–45 (2010)

Spaceborne Ultraviolet Sky Survey Telescope with Large Field of View

Nian Wei(✉)

Beijing Institute of Space Mechanics and Electricity, Beijing 100094, China
nianwei1@126.com

Abstract. Time domain astronomy is a field that contains important opportunities for astronomical discovery, involving the detection and physical mechanism research of various transient sources in the universe. These transient sources include flares, supernovae, gamma ray bursts, black hole tidal collapse events, fast radio bursts, and gravitational wave electromagnetic counterparts, etc. The ultraviolet band plays an important role in the study of astronomical transient events, while the space ultraviolet survey telescope is an important instrument in time domain astronomy. This article introduces the current status of time-domain sky surveys in the ultraviolet band both domestically and internationally, analyzes the scientific objectives of imaging observations in the mid ultraviolet spectral range, and design a spaceborne ultraviolet sky survey telescope with a large field of view. According to the performance requirements, an enhanced ultraviolet detector was used to design the optical and imaging system of a large field of view remote sensing camera. The instrument has an effective aperture of 20 mm and a field of view angle of $15° \times 15°$, with a smaller volume and lighter weight, meeting the requirements for carrying small satellite platforms. The simulation results show that the system has excellent image quality in the operating spectral range of 270-300 nm.

Keywords: spaceborne · ultraviolet sky survey · large field of view · remote sensing

1 Introduction

Astronomy is a fundamental science driven by observations. The rapidly astronomical detection technology (such as large target detectors) and data processing technology (such as GPU acceleration) have also expanded from the traditional optical band to the full electromagnetic band, which has led to the rapid development of the electromagnetic sky with time changes and has grown into a new and important frontier field of international discovery opportunities, involving the detection and physical mechanism research of various temporary source celestial bodies in the universe, such as flares, supernovae (SN), gamma ray bursts (GRBs), black hole tidal collapse events (TDEs), fast radio bursts (FRBs), and gravitational wave electromagnetic counterparts [1].

The ultraviolet band has special significance in astronomy, with a band from 10 nm to 400 nm, as shown in Fig. 1 [2]. It contains rich information about the physical

H. P. Urbach et al. (Eds.): ISSOIA 2023, SPP 191, pp. 253–263, 2024.
https://doi.org/10.1007/978-981-97-6718-2_24

properties of gases, stars and galaxies. This band is crucial for describing the discovery of young and hot stars, as well as strong non heat sources. In addition, the ultraviolet spectral range has a very dark sky background, making it easy to achieve good signal-to-noise ratio. At present, the time-domain sky survey observation projects mainly focus on optical and high-energy bands, and the international large-scale ultraviolet band by the Earth's atmosphere, observations in the ultraviolet band mainly involve high-altitude exploration(near ultraviolet exploration by high-altitude balloon equipment) and deep space exploration (space ultraviolet satellites). In the 1980s and the beginning of this century, several space ultraviolet telescopes were built and operated internationally, such as the International UltravioletExplorer (IUE) and the Galaxy Evolution Explorer (Galex) [4]. However, their field of view was not large and they were not used for time-domain sky surveys, and they have been out of operation for a long time. The only devices currently in operation that can carry out space ultraviolet observation are the Hubble Space Telescope's ultraviolet detector and the Swift satellite's optical ultraviolet telescope [4]. However, these two ultraviolet telescopes have a very small field of view and can only perform follow-up observations, which cannot be used for sky observation, and the ultraviolet band is only 2500 angstroms. Therefore, it is very necessary to establish a wide field of view ultraviolet band observation platform as soon as possible [5]. The important astrophysical issues such as late stellar evolution and explosion, dense celestial accretion and merging processes, interstellar medium properties, and cosmic dark energy properties.

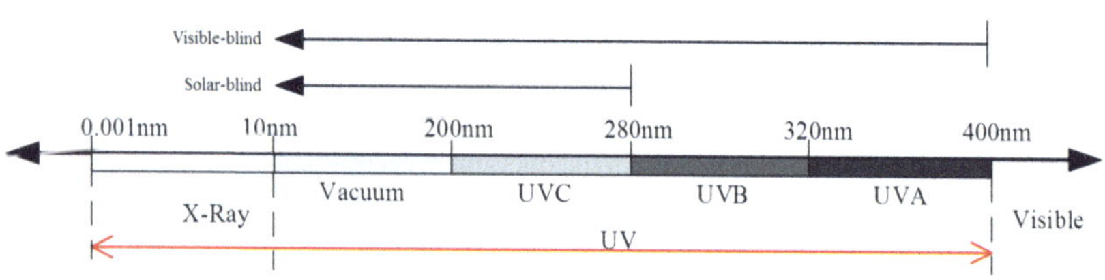

Fig. 1. Schematic diagram of ultraviolet band range.

This article develops a large field of view ultraviolet imaging remote sensor that can perform transient imaging observations in the mid ultraviolet spectral range. On the premise of meeting the requirements of carrying space and heavy distance, according to the performance indicators, the optical and imaging system of a large field of view remote sensing camera was designed using an image intensifier type ultraviolet detector component. It can observe a large range of sky areas according to the instructions of the ground command system and cooperate with the entire satellite pointing machine, meeting the requirements of a large field of view ultraviolet sky survey. The main achievable task objectives include [6]:

(1) Conduct astronomical observations on different celestial targets.
(2) Detection of shock wave bursts in the late stage of stellar evolution.
(3) Detection of Superbright Supernova, Interacting Supernova, and Strange Temporal Sources.
(4) Detection of tidal disintegration of black holes.

2 Overall Plan

2.1 System Composition

The large field ultraviolet imaging camera mainly consists of a main body and a signal processor, as shown in the following figure. It can be used for detecting ultraviolet radiation from small celestial bodies and processing the obtained data on the ground. On this basis, it can also obtain ultraviolet radiation intensity data of the Earth's atmospheric background observed in space (Fig. 2).

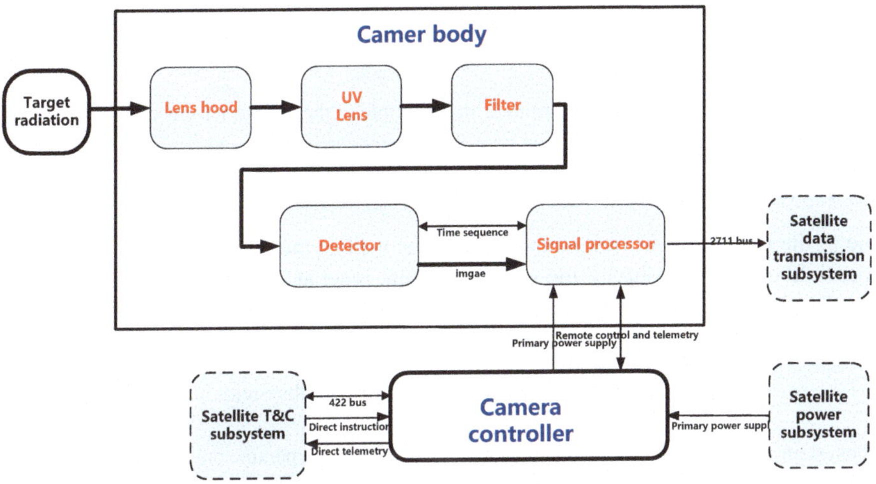

Fig. 2. Camera composition and working principle diagram.

The camera body is composed of a light shield, lens, bandpass filter, detector, etc. Among them, the camera body and signal processor are designed as one, installed on the opposite floor of the satellite platform through structural interfaces. The signal processor is connected to the power supply subsystem and measurement and control subsystem of the satellite platform through electrical interfaces. The data collected by the signal processor is sent to the satellite data transmission subsystem through the 2711 bus.

2.2 Detector Selection

Due to the extremely weak ultraviolet radiation signal of the target itself and limited by the loss of the ultraviolet light path, the signal reaching the detector is very weak, thus requiring high sensitivity of the detector. The system adopts an image intensifier type ultraviolet detector, as shown in the following figure. The ultraviolet light signal collected by the UV camera lens is converted into an electrical signal through the photocathode of the image intensifier, amplified by the image intensifier, output to the rear fluorescent screen, and transmitted to the CCD image sensor for imaging through a fiber optic cone [7] (Fig. 3).

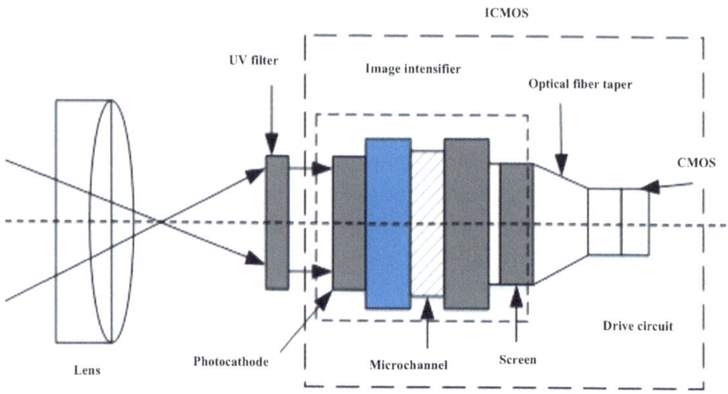

Fig. 3. Schematic diagram of the principle of the ultraviolet sensor.

2.3 Parameter Requirement

Due to the limitations in weight and volume, as well as meeting the requirements of a large field of view sky survey, the system aperture is set at 20 mm and the field of view angel is set at 15° × 15°, the parameters of each subsystem are as follows (Table 1).

Table 1. Technical indicators of each subsystem.

Subsystem	title	parameters
Optical system	Aperture	20 mm
	Focal length	55 mm
	FOV	15° × 15°
Filter	Center wavelength	285 nm
	Passband	270–300 nm
Detector	Image size	Φ18 mm
	Pixel size	5.5 μm × 5.5 μm
	Pixel numbers	2048 × 2048

3 Optical System Design

3.1 Optical System Design and Results

Due to the large field of view angel of the system, it is difficult to implement coaxial reflection systems, and it is also very difficult to implement off axis reflection systems [8]. Additionally, there are high requirements for processing and assembly, resulting in complex systems and low reliability. 15° × 15° field of view is not very large for small aperture transmission systems, and there are many transparent ultraviolet materials. The working spectral bandwidth is only 30 mm, and color difference is easy to correct [9]. The tolerance is loose, and the risk is low, making it easy to achieve. Therefore, the optical system adopts a transmission type [10].

The size of the UV optical system is Φ29 mm × 85 mm, with a total of 7lenses in the system, using fused quartz and calcium fluoride (Fig. 4).

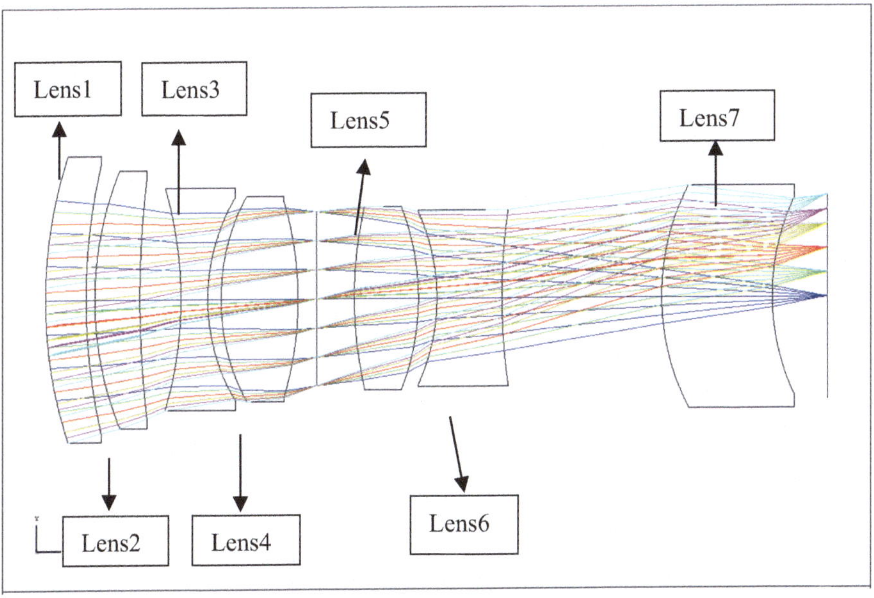

Fig. 4. UV optical system diagram.

The indicators achieved by the design are:

(1) Wavelength: 270 nm–300 nm
(2) Aperture: 20 mm
(3) Fov: 15° × 15°
(4) Image size: 19.9 mm
(5) Focal length: 54.8874 mm
(6) Lens size: Φ29 mm × 85 mm

3.2 Image Quality Evaluation

The MTF curve of optical system is shown in the following figure, with an average MTF value of 0.3590@25lp (Fig. 5).

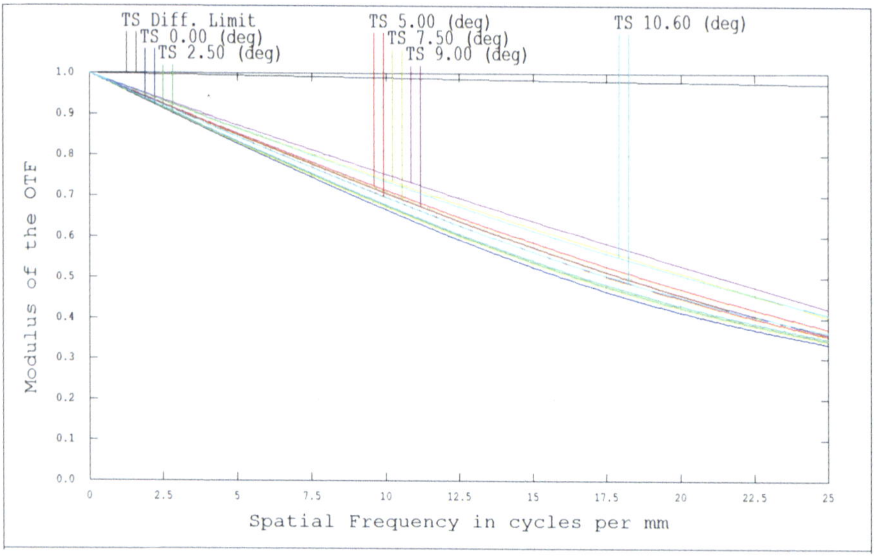

Fig. 5. Optical system MTF.

The optical system spot diagram is shown in the following figure, with a maximum RMS value of 31.57 μm for each field of view, meeting the criteria requirements (Fig. 6).

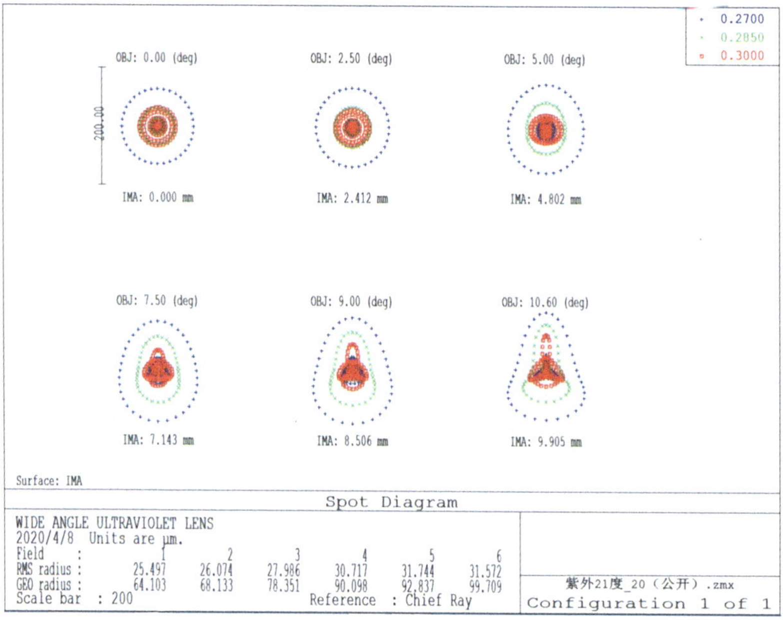

Fig. 6. UV system spot diagram.

The relative illumination of the optical system is shown in the following figure, which is superior to the design requirements (Fig. 7).

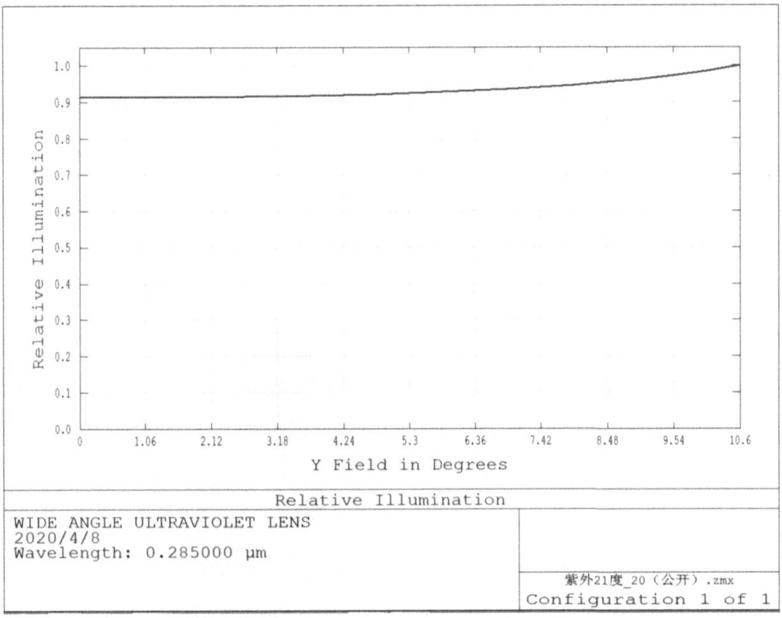

Fig. 7. Relative illumination of UV optical system.

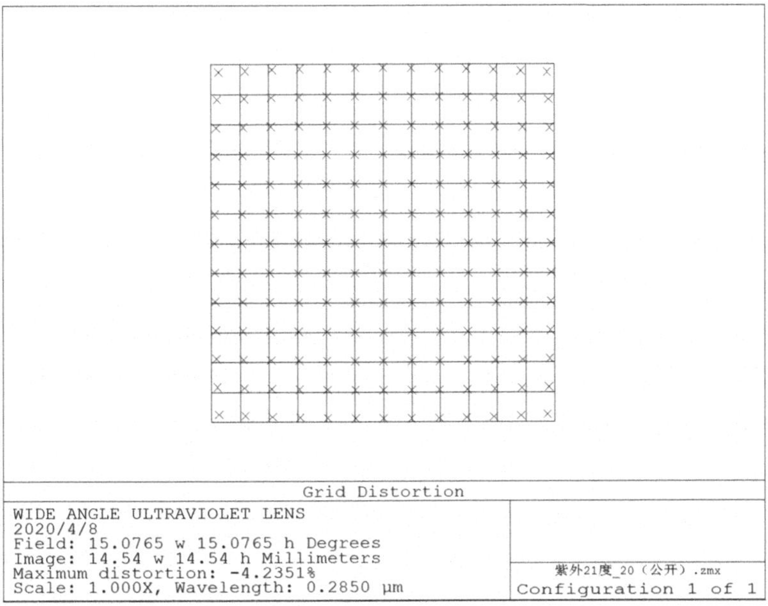

Fig. 8. UV optical system distortion.

The optical system distortion is shown in the figure below, and the maximum distortion of the system is 4.2351%, which meets the indicator requirements (Fig. 8).

4 Structure Design

The overall structure and physical image of the camera are shown in the following figure. Considering the requirements of mechanical, thermal, and radiation environments, as well as weight, aluminum alloy is selected as the structural material [11]. The total weight of the camera is less than 2 kg. The outer envelope size of the camera (optical axis direction) is 200 mm × 90 mm × 90 mm, including light shield and thermal control components (Fig. 9).

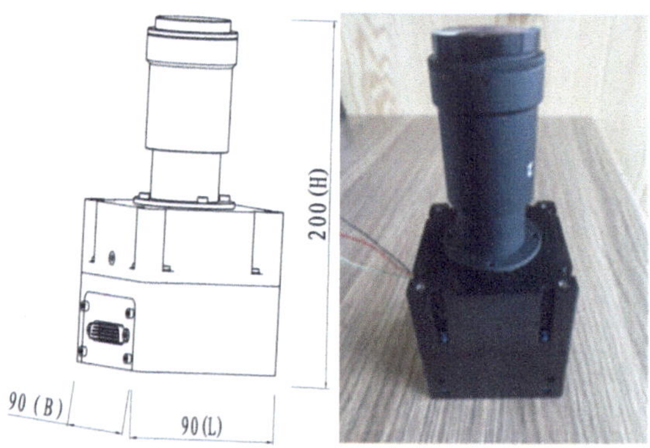

Fig. 9. Camera envelope size and physical image.

5 Electronics Design

The electronic design consists of three circuit boards, namely focal plane circuit board, signal processing circuit board, and interface circuit board. The boards are connected by connectors [12] (Fig. 10).

The circuit schematic diagram is shown in the above figure, with one external interface connector: including power input, remote control command and telemetry, and image data output interface.

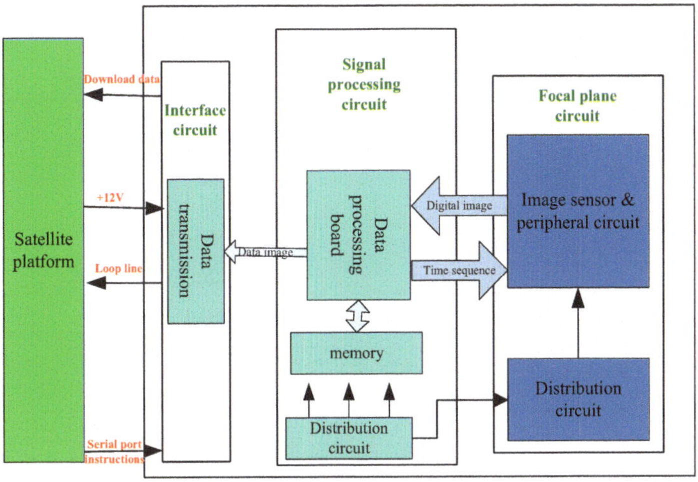

Fig. 10. The circuit schematic diagram.

6 Analysis of Detection Capability

The signal-to-noise ratio is the ratio of the signal mean to the noise standard deviation [13]. According to theory of point target detection, the signal-to-noise ratio of an image can generally be detected when it is greater than 3. The formula for calculating the signal-to-noise ratio can be obtained from the formula for the number of photons in the image intensifier per unit time, the formula for the number of signal electrons obtained from the detector, and the formula for the noise value output from the detector [14].

$$SNR = \sqrt{\frac{I\frac{\pi D^2}{4R^2}\tau_1\tau_2\frac{\lambda}{hc}Q_{pc}t\eta\tau_3 t_{\text{int}}Q_{cmos}\left(\delta^2-\delta\right)}{6\delta^2-2}} \tag{1}$$

Among them, I is the target radiation intensity, R is the distance, D is the aperture, τ1is the lens transmittance, τ2 is the filter transmittance, Q_{pc} is the efficiency of the photocathode in the image intensifier, t is the effective integration time of the image intensifier, η is the detection efficiency of MCP, τ3 is the transmittance of the optical fiber, t_{int} is the integration time of the detector, Q_{cmos} is the quantum efficiency of the detector and δ is the second multiplication coefficient.

The SNR is 6 [15]. Based on the optical system and the selected detector indicators, the detection ability of the UV camera can be calculated as shown in the table below, which shows the detection ability of the remote sensor at different integration times. When the integration time is 4800 ms, the detectable magnitude is 14.6 (Table 2).

Table 2. Camera detection capability under different integration times.

Integral time (ms)	Magnitude	Illuminance (W/m^2)
50	9.6	2.4e-12
100	10.3	1.22e-12
300	11.5	4e-13
500	12.1	2.5e-13
1000	12.9	1.2e-13
2000	13.6	6e-14
4000	14.3	3e-14
4800	14.6	2.7e-14

7 Conclusion

The ultraviolet (UV) band plays an important role in the study of astronomical transient events [16]. However, the ultraviolet band in space is still largely unknown. We design a spaceborne ultraviolet telescope, which has a 15° × 15° field of view and can be used to study transient events in the medium wave ultraviolet (NUV) spectrum. The optical mechanical design considers the main limitations such as volume, weight constraints, launch load vibration, compactness, and cost-effectiveness, meeting the configurations of various satellite platforms such as cubic satellites and small satellites.

References

1. Liu, Z., Yuan, W., Sun, H., et al.: Massive black holes and tidal disruption events at the center of galaxies. Scientia Sinica Physica, Mechanica & Astronomica **48**(3), 22–34 (2018)
2. Meinel, A.B., Shannon, R.R., Whipple, F.L., et al.: A large multiple mirror telescope (MMT) project. Opt. Eng. **11**(2), 110233 (1972)
3. Yang, M., Zheng, Y., Ni, Y., et al.: Progress and prospect in planetary exploration of solar system. Chinese Space Science and Technol. **43**(5), 112 (2023)
4. Ruan, N., Su, Y.: Summarization of overseas space ultraviolet instrument development. Spacecraft Recovery & Remote Sensing **29**(3), 71–78 (2008)
5. Cai, Z., Zhang, C., Fan, F.: History and prospect of astronomical telescopes: Introducing Tsinghua multiplexed survey telescope(MUST). Experimental Technol. Manage. **38**(5), 1–9 (2021)
6. Wang, F., Zheng, X.-Z.: The evolution of star formation in early-type galaxies. Acta Astronomica Sinica **52**(2), 105–114 (2011)
7. Luo, L., Tang, L., Zuo, W.: Research progress in ultraviolet enhanced image sensors. Infrared Technol. **43**(11), 1023–1033 (2021)
8. Zheng, G., Jiao, J., Yu, Y., et al.: Design of solar-blind ultraviolet warning optical system with large FOV and large relative aperture. Infrared and Laser Eng. **50**(2), 159164 (2021)
9. Chen, L., Hu, Z., Qiu, K., et al.: Development of scientific ultraviolet grating for spectrum survey. Chinese J. Lasers **49**(6), 0611002–1–0611002–12 (2022)

10. Gao, X.: Research on Optical Design of Space Deep Ultraviolet Camera System. Master's thesis of Changchun University of Science and Technology (2021)
11. Yang, L., Nian, W., Dong, L., et al.: Research on solar-blind ultraviolet warning system with large field of view. Spacecraft Recovery & Remote Sensing **43**(3), 88–95 (2022)
12. Sachkov, M., Sichevsky, S., Shustov, B., et al.: Field camera unit of the WSO-UV mission. SpaceTelescopes and Instrumentation 2020: Ultraviolet to Gamma Ray (2020)
13. Fluder, G.: Design of a hybrid refractive-diffractive telescope for observations in UV. Exp. Astron. **11**, 1–10 (2020)
14. Wang, X., Fang, X., Nie, J.: Military ultraviolet technology. Infrared and Laser Eng. **42**(S01), 58–61 (2013)
15. Yi, T., Xuli, Z., Jing, W., et al.: Spectrum detection technology of martian ozone. Spacecraft Recovery & Remote Sensing **33**(6), 1–9 (2012)
16. Yan, D., Guo, J.: Research and development of hydrodynamic escape from exoplanetary atmosphere. Progress in Astronomy **37**(4), 384–407 (2019)

Hyperspectral Compression Based on Tucker Decomposition

Binbin Chen(✉), Weiping Kong, Lin Mi, Guang Yang, and Xinfang Yuan

Aerospace Information Research Institute, Chinese Academy of Sciences, Beijing 100094, China
chenbb@aircas.ac.cn

Abstract. Tucker decomposition stands out as an effective method for hyperspectral data compression, with the core tensor playing a pivotal role in determining its quality . Consequently, the selection of an appropriate core tensor is a focal point in the realm of data compression. This study conducted three experiments, examining spatial size, waveband number, and comprehensive analyses to discern the optimal core tensor. Our findings reveal a significant relationship between the size of the core tensor decomposed by Tucker and the data compression rate. Notably, the spatial size of the core tensor was found to be inversely proportional to the compression rate. Specifically, a core tensor size constituting 2/5 of the data's spatial dimensions and a waveband number comprising 3/10 of the data were identified as optimal. This configuration demonstrated the ability to effectively preserve both spatial and spectral information while maintaining a commendable compression rate. In conclusion, selecting a core tensor size of 2/5 of the data's spatial size and 3/10 of the data's waveband number emerged as a favorable choice, striking a balance between information retention and compression efficiency in the context of hyperspectral data compression using Tucker decomposition.

Keywords: Hyperspectral · Tucker Decomposition · Compression · Remote Sensing

1 Introduction

As one of the emerging directions of remote sensing science, Hyperspectral remote sensing technology has the advantages of high spectral resolution, high feature dimension, and image-spectrum integration [1]; which makes it be widely used in various fields in recent years, such as land monitoring, agricultural and forestry survey, disaster warning, ecological environment evaluation, military defense, etc. With the development of science and technology, hyperspectral imagers are developing towards higher spatial and spectral resolution, higher detection sensitivity, wider spectral coverage. However, hyperspectral remote sensing data is becoming larger and larger, making the storage and transmission of data more challenging [2–5].

In recent years, a large number of researchers have studied hyperspectral data compression and developed a series of hyperspectral data compression methods. The main methods include wavelet transform, principal component analysis (PCA), JPEG2000,

H. P. Urbach et al. (Eds.): ISSOIA 2023, SPP 191, pp. 264–276, 2024.
https://doi.org/10.1007/978-981-97-6718-2_25

Karhunen-loeve transform (KLT) [6, 7], three-dimensional tar based coding with classification for embedding (3D-TCE) [8], three-dimensional set partitioned embedded block (3D-SPECK) [9], and the combination of PCA andJPEG2000 [10], etc.

Tensor is a generalization of vector and matrix, which means that vector is a first-order tensor, matrix is a second-order tensor, three-dimensional data is a third-order tensor. Because hyperspectral image is three-dimensional data, it can be represented by a three-dimensional tensor. Based on this fact, tensor representation has become a hotspot in hyperspectral image processing. Tucker decomposition is one of the most widely used methods for tensor representation, which can decompose high-dimensional tensors into a core tensor and a dictionary of each dimension. Besides, it can separate the information of each dimension in each tensor, and the kernel tensor represents the correlation between a pair of the information of each dimension, which can retain the hyperspectral spatial and spectral information at the same time. Due to the above advantages, Tucker decomposition has been widely used in the field of hyperspectral compression [11–13]. The existing researches on Tucker decomposition mainly focus on compression performance, but there have been fewer studies on how to choose the core tensor of Tucker decomposition.

The objectives of this study were: i) to analyze the core tensor of Tucker compression from the spatial and spectral dimensions based on the trends of running time, compression ratio, peak signal-to-noise ratio, and mean spectral angle; ii) To compare the quality of compressed image from different spatial and spectral dimension combinations; iii) To propose an optimal scheme for selecting the core tensor based on the comprehensive analysis of trends in the core tensor and the quality of the compressed image.

2 Materials and Methods

2.1 Tensor Definitions

Tensor $X \in R^{I_1 \times I_2 \times \cdots I_N}$ is a multi-dimensional array, where N is the order of tensor X, I_n $(1 \le n \le N)$ represents the magnitude of n-order of the tensor [14]. Hyperspectral data can be considered as a third-order tensor, i.e. $X \in R^{I_1 \times I_2 \times I_3}$, in which I_1, I_2 and I_3 represent the height, width and waveband of hyperspectral data, respectively. The inner product of the two tensors with the same size is the sum of the products of their entries, i.e. $< X, Y >= \sum_{i_1=1}^{I_1} \sum_{i_2=1}^{I_2} \cdots \sum_{i_N=1}^{I_N} x_{i_1 i_2 \cdots i_N} y_{i_1 i_2 \cdots i_N}$. The distance between tensors with the same size is defined as $D(X, Y) = \|X - Y\|_F$.

2.2 Tucker Decomposition Compression Algorithm

The third -order tensor Tucker decomposition is described as a decomposition of a given third-order tensor $X \in R^{I_1 \times I_2 \times I_3}$ into an unknown core tensor $G \in R^{J_1 \times J_2 \times J_3}$ multiplied by a set of 3 unknown component matrices [15]. The equation and the decomposition diagram (Fig. 1) are as follows.

$X = G \times_1 A \times_2 B \times_3 C$, where $A \in R^{I_1 \times J_1}$, $B \in R^{I_2 \times J_2}$,$C \in R^{I_3 \times J_3}$.

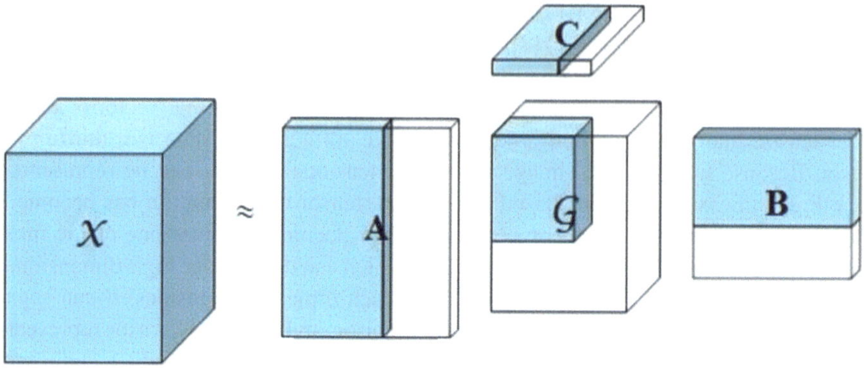

Fig. 1. Tucker decomposition of a third-order tensor.

2.3 Compression Evaluation Index

Compression Ratio (CR). Hyperspectral compression capability is generally expressed by CR. The CR refers to the ratio of the amount of data in the original image to the amount of data in the compressed image. The larger CR value, the more compressed data are reduced from the original image, and the stronger the compression capability is.

$$CR = \frac{Input\_Size}{Output\_Size} \tag{1}$$

where *Output_Size* represents the size of the compressed image and *Input_Size* represents the size of the original image.

Peak Signal to Noise Ratio (PSNR). PSNR is used to measure the image recovery of decompression. The formula is expressed as:

$$\begin{aligned} PSNR &= 10\log_{10}(\frac{MAX^2}{MSE}) \\ MSE &= \frac{1}{m \times n \times l}\sum_{i=1}^{m}\sum_{j=1}^{n}\sum_{k=1}^{l}\|r(i,j,k) - \hat{r}(i,j,k)\|^2 \end{aligned} \tag{2}$$

where MSE is the mean square error between the two images. MAX is the maximum value of image gray. r is the original image, $\hat{r}$ is the reconstructed image after compression and decompression. The larger the PSNR, the higher the similarity between the compressed and decompressed image and the original image, indicating that the recovery effect will be better.

Mean Spectral Angle (MSA). MSA is the average of the spectral angle (SA) of all points after compression and decompression at the original data, which can represent the spectral information of hyperspectral images. The formula is expressed as:

$$MSA = \frac{1}{m \times n} \sum_{i=1}^{m} \sum_{j=1}^{n} SA(\mathrm{i,j})$$
$$SA = \cos^{-1} \frac{\sum_{k=1}^{l} t_i r_i}{\sqrt{\sum_{k=1}^{l} t_i^2} \sqrt{\sum_{k=1}^{l} r_i^2}} \tag{3}$$

where t_i represent the spectral curve of the original data, r_i and l the spectral curve and the number of spectral segments of the compressed and decompressed data, respectively. The smaller the MSA, the smaller the spectral difference before and after compression.

3 Results and Analysis

Two sets of actual flight data obtained from the airborne hyperspectral imager were selected (Fig. 2). Specifically, Fig. 2(a) is a large scene including complex surface features, with a spatial dimension of 500x500, a number of wavebands of 218, and an original data size of 103 MB, which is referred to as 500–500-218. Figure 2 (b) is a small scene containing relative simple surface features, with a spatial dimension of 400x400, a number of wavebands of 218, and an original data size of 66.5 MB, which is referred to as 400–400-218. The test computer configuration we used is Intel (R) Xeon (R) W-2255 CPU @ 3.7 GHz, 64 G RAM.

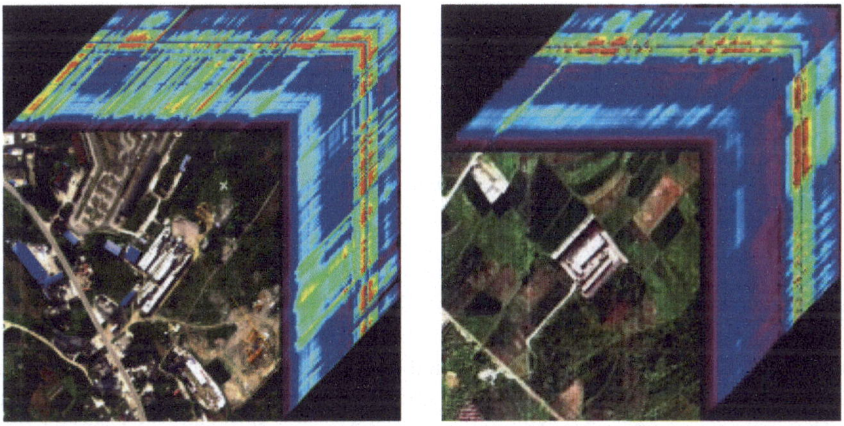

(a) (b)

Fig. 2. Flight data with (a) large scene (500–500-218) and (b) small scene (400–400-218).

3.1 Spatial Direction

Tucker compression was performed on the two scenes from the spatial direction. The initial space size of the core tensor was 1/10 of the original image space size, i. e. 50x50 and 40x40 respectively, and the band number was 45. The space size was doubled every time until 9 times. In order to reduce the randomness of the data, each space size was calculated 10 times respectively, and its average value was used for experimental analyses.

The compression algorithm has been iterated many times to make the data meet the compression requirements, the result was shown in Fig. 3. We can see that the compression time of both images reaching the maximum when the space size was 4 times of the initial size, indicating that the initial size 4 times is the smallest core tensor spatial dimension. Figure 4 shows the PNSR rised sharply with the increase of space size before 4 times, but rised gently after 4 times. Figure 5 shows the MSA decreased sharply with the increase of space size before 4 times, and then decreased gently. Figure 5 shows the compression ratio core MSA is similar. The experiment shows that when the spatial size of the core tensor is 2/5 of the data that to be compressed, a larger compression rate and a large amount of original data information can be obtained at the same time, which will be the better spatial size of the core tensor (Fig. 6).

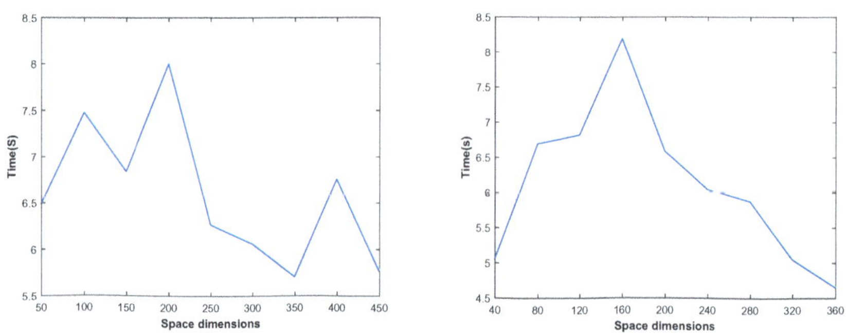

(a) 500-500-218 (b) 400-400-218

Fig. 3. Running time of core tensors with different space sizes.

3.2 Band Direction

Tucker compression was also performed on the two scene images from band direction. The initial space size of the core tensor was 1/10 of the number of image bands, i.e. about 21. The space size was fixed to 2/5 of the original data, i.e. 200*200 and 160*160. The band size was doubled each time until it was 9 times. In order to reduce the randomness of the data, 10 operations were performed for each band, and the average value was used for experimental analyses.

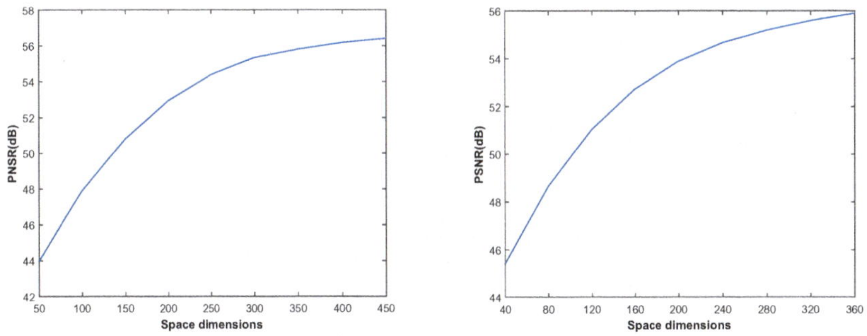

(a) 500-500-218 (b) 400-400-218

Fig. 4. PNSR of core tensors with different space sizes.

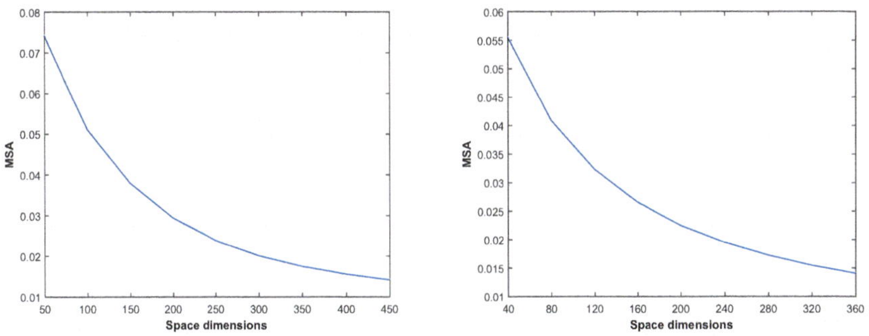

(a) 500-500-218 (b) 400-400-218

Fig. 5. MSA of core tensors with different space sizes.

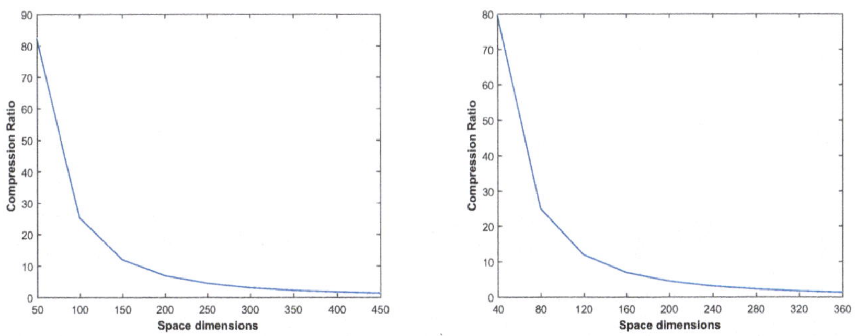

(a) 500-500-218 (b) 400-400-218

Fig. 6. Compression ratio of core tensors with different space sizes.

Figure 7 shows the compression time of both scenes increased linearly with the increasing of band number. Figure 8 shows the PNSR increases sharply before the band number was 3/10 of the data to be compressed, and then increased gently. Figure 9 shows the MSA decreased linearly with the increasing of band number. Figure 10 shows the compression rate decreased sharply before the band number is 3/10 of the data to be compressed, and then decreased gently. All the figures presented above, when the band number of the kernel function is 3/10 of the data that to be compressed, it can keep a large amount of original data information while ensuring the compression ratio, so it is a better band number of the core tensor.

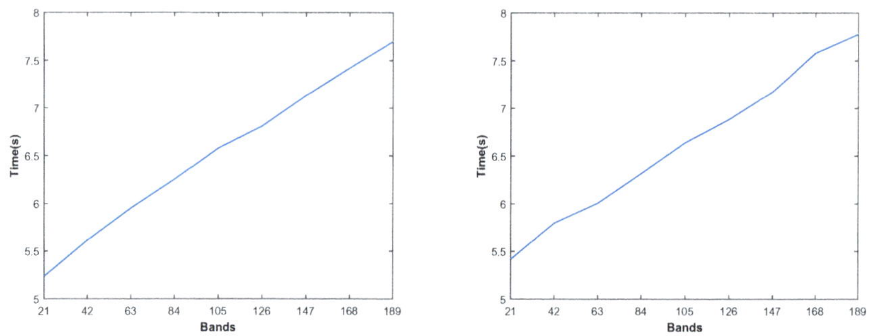

(a) 500-500-218 (b) 400-400-218

Fig. 7. Running time of core tensor with different bands.

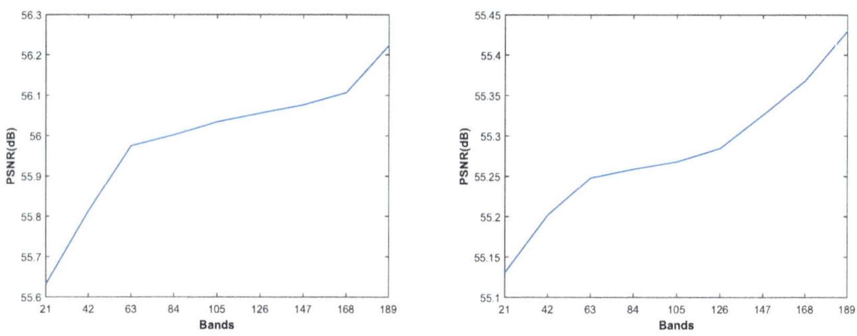

(a) 500-500-218 (b) 400-400-218

Fig. 8. PNSR of core tensor with different bands.

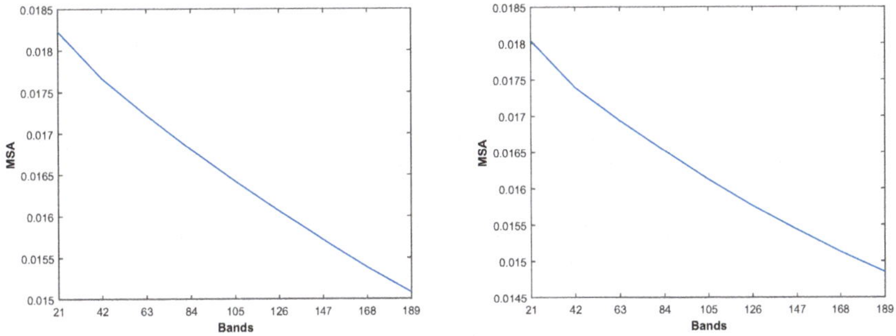

(a) 500-500-218 (b) 400-400-218

Fig. 9. MSA of core tensor with different bands.

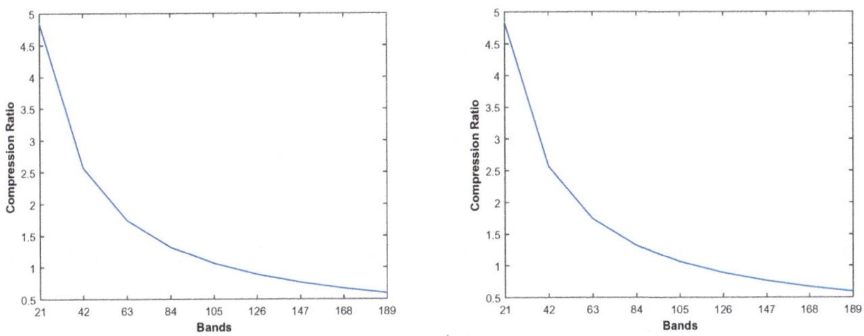

(a) 500-500-218 (b) 400-400-218

Fig. 10. Compression ratio of core tensor with different bands.

3.3 Comprehensive Analysis

Based on the results of the above tests, the kernel functions with spatial resolution of 0.2, 0.4, 0.5, 0.8 times and band number of 0.2, 0.3, 0.5 and 0.8 times of the original image are respectively selected for data compression tests, and then 0.2, 0.4, 0.5, 0.8 times of the original image are used to describe the results. The compression results of the two scene images are shown in the Fig. 11.

Figure 11(a) shows the data compression time of both of the two scene images increased sharply from 0.2 to 0.4 times of the space compression time, then the data of 500 * 500 * 218 slowly increased, but the data of 400 * 400 * 218 slowly decreased, which may be due to the simple data content of the small scene. Figure 11(b) shows the PNSR of two scene images rised sharply from 0.2 to 0.4, and then rised gently, indicating that the quality of 0.4 times of image compression was relatively high. Figure 11(c) shows the trend of MSA was just opposite to PNSR, but also implying that o.4 times of image compression can obtain the best result.. The trend of the compression ratio was similar to the MSA (Fig. 11(d)). Due to the consistency of the compression ratio of the two scenes,

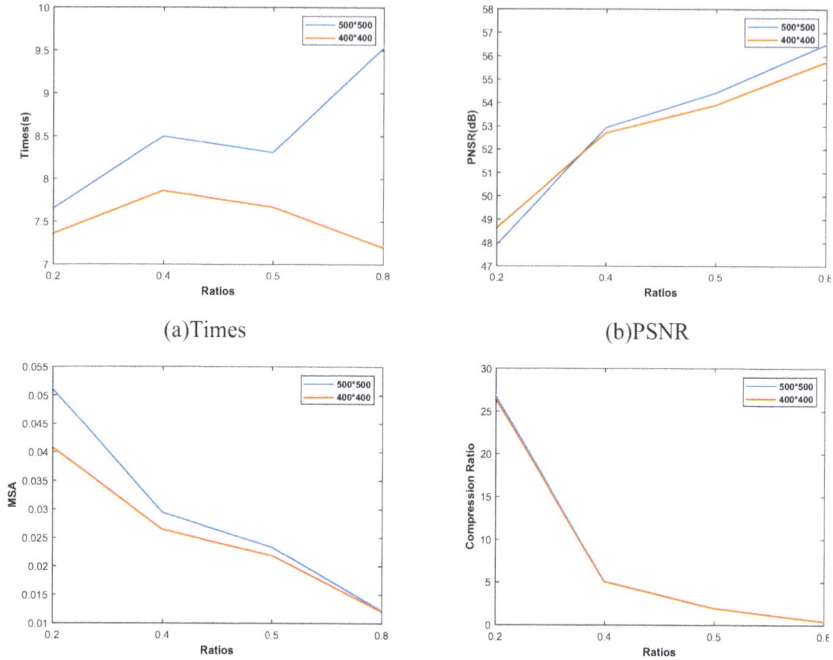

(a)Times (b)PSNR

(c)MSA (d)Compression Ratio

Fig. 11. Comprehensive analysis.

it can be seen that the compression ratio of the image is not related to the complexity of the image, but is closely related to the size of the core tensor. At the same time, when the core tensor is 0.8, the effect of data compression can no longer be achieved.

Figures. 12 and 13 show the data of 500–500-218 and 400–400-218 in the 10th channel after the kernel function is 0.2, 0.4, 0.5 and 0.8 times decompressed. The comparison shows that the kernel function has a better recovery effect when the kernel function is 0.4 times the data size. To sum up, when the space size of the kernel function size is 0.4 times of the size of the treatment compressed data and the number of bands is 0.3 times, a large compression ratio and effective retention of data information can be achieved at the same time.

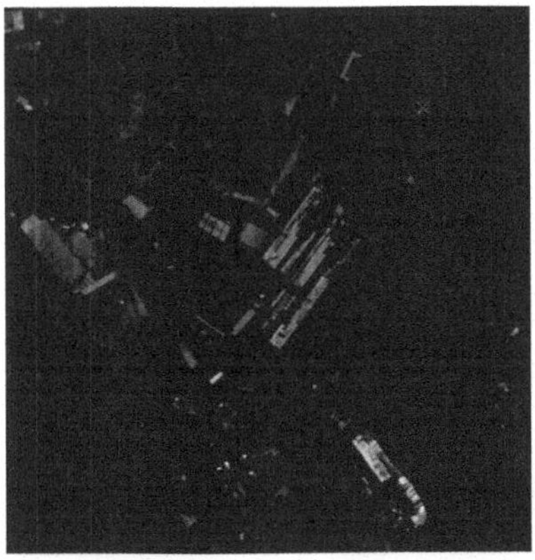

(a) Original image

(b) 0.2 times

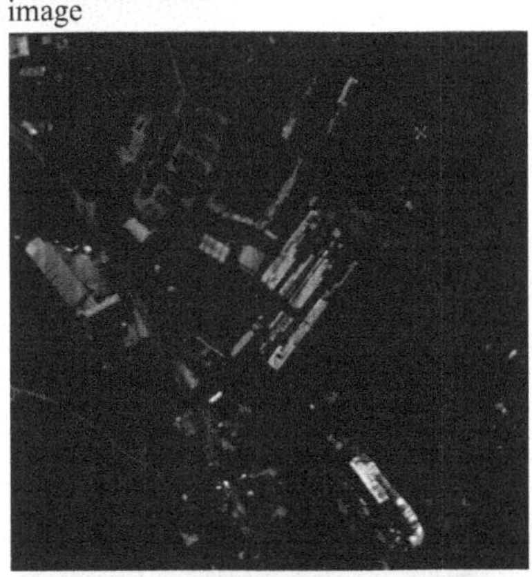

(c) 0.4 times

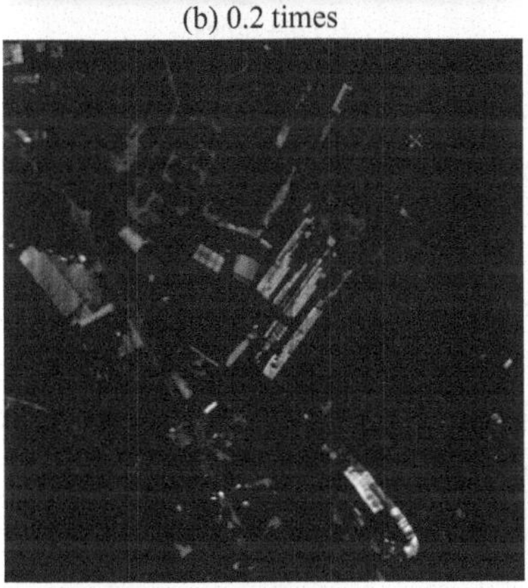

(d) 0.5 times

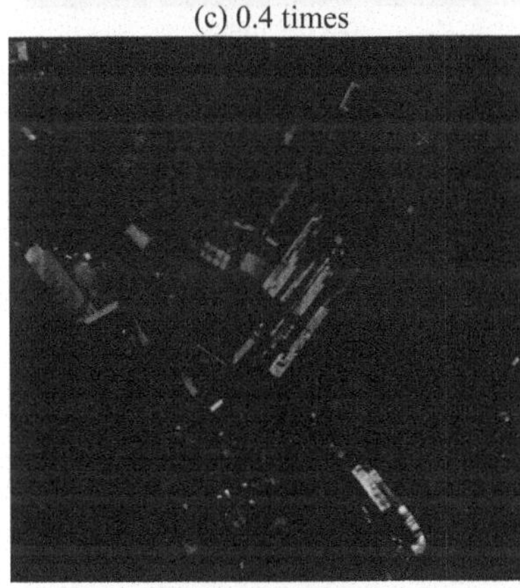

(e) 0.8 times

Fig. 12. Image of the tenth channel after decompression of complex ground objects.

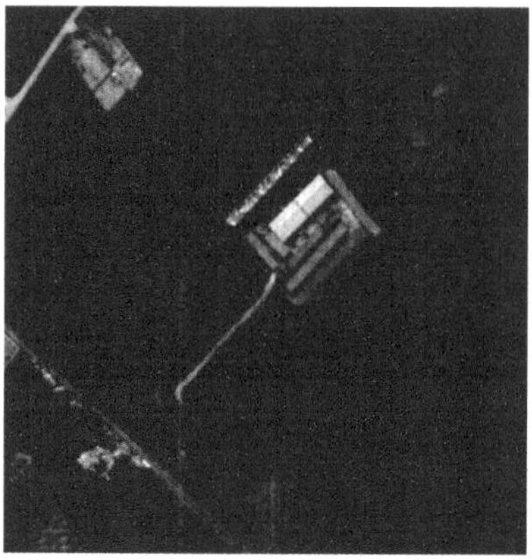

(a) Original image

(b) 0.2 times

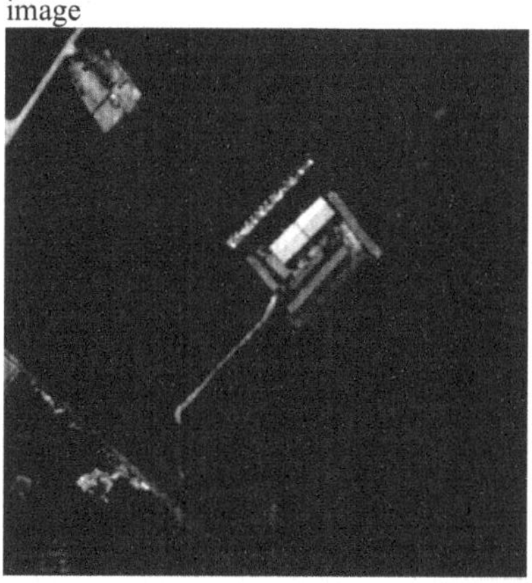

(c) 0.4 times

(d) 0.5 times

(e) 0.8 times

Fig. 13. Image of the tenth channel after decompression of simple ground objects.

4 Conclusions

Hyperspectral imagers are developing in the direction of high resolution, high sensitivity, high color depth, wide spectrum and wide coverage, which means that the size of hyperspectral data is getting larger and larger. Effective data compression methods can achieve high compression ratio, while retaining the original information of hyperspectral data and removing the redundancy. As an effective data compression method, Tucker decomposition has been widely used in the field of hyperspectral compression. Our results indicate that the size of the core tensor of Tucker decomposition is inversely proportional to the data compression rate. When the space size of the core tensor is 2/5 of the original data, and the number of bands is 3/10 of the data to be compressed, it is a better core tensor size, which can effectively retain the maximum spatial and spectral information of the data to be compressed, and maintain a higher compression rate. The selection of adaptive core tensor will improve the automation of the whole algorithm and further enhance the application depth of hyperspectral data.

Funding. This study is supported by the Strategic Priority Research Program of the Chinese Academy Sciences (Grant No: XDA26010304).

References

1. Tong, Q.X., Hang, B.Z., Zheng, L F.: Hyperspectral Remote Sensing: the Principle, Technology and Application. Higher Education Press, Beijing (2006)
2. Goetz, A.F.H., Vane, G., Solomon, J.E., Rock, B.N.: Imaging spectrometry for earth remote sensing. Science **228**, 1147–1153 (1985)
3. Green, R.O., et al.: Imaging spectroscopy and the airborne visible/infrared imaging spectrometer AVIRIS. Remote Sens. Environ. **65**, 227–248 (1998)
4. Murchie, S., et al.: CRISM (compact Reconnaissance Imaging Spectrometer for Mars) on MRO (Mars Reconnaissance Orbiter. In: Proceedings SPIE 5660, pp. 66–77 (2004)
5. Mouroulis, P., et al.: Portable Remote Imaging Spectrometer coastal ocean sensor: design, characteristics, and first flight results. Appl. Opt. **53**(7), 1363–1380 (2014)
6. Penna, B., Tillo, T., Magli, E., Olmo, G.: Transform coding techniques for lossy hyperspectral data compression. IEEE Trans. Geosci. Remote Sens. **45**, 1408–1421 (2007)
7. Blanes, I., Serra-Sagristà, J.: Cost and scalability improvements to the Karhunen-Loève transform for remote-sensing image coding. IEEE Trans. Geosci. Remote Sens. **48**, 2854–2863 (2010)
8. Zhang, J., Fowler, J.E., Liu, G.: Lossy-to-lossless compression ofhyperspectral imagery using three-dimensional TCE and an integer KLT. IEEE Geosci. Remote Sens. Lett. **5**, 814–818 (2008)
9. Tang, X., Pearlman, W.A.: Three-dimensional wavelet-based compression of hyperspectral images. In: Hyperspectral Data Compression, Chap. 10, Springer (2005)
10. Du, Q., Fowler, J.E.: Hyperspectral image compression using JPEG2000 and principal component analysis. IEEE Geosci. Remote Sens. Lett. **4**(2), 201–205 (2007)
11. Feng, L, Sun, H., Sun, Q., et al.: Compressive sensing via nonlocal low-rank tensor regularization. Neuroomputing **216**, 45–60 (2016)
12. Yang, S., Wang, M., Li, P., et al.: Compressive hyperspectral imaging via sparse tensor and nonlinear compressed sensing. IEEE Trans. Geosci. Remote Sens. **53**(11), 5943–5957 (2015)

13. Zhang, L., Zhang, L., Tao, D., Huang, X., Bo, D.: Compression of hyperspectral remote sensing images by tensor approach. Neurocomputing **147**, 358–363 (2015)
14. Kolda, T., Bader, B.: Tensor decompositions and applications. SIAM Rev. **51**, 455–500 (2009)
15. Cichocki, A., Zdunek, R., Amari, S.-I.: Nonnegative matrix and tensor factorization [Lecture Notes]. In: IEEE Signal Processing Magazine **25**(1), 142–145 (2008)

An On-Satellite Real Time Fusion Technology for HDR Images Acquired by Large Array CMOS Sensors

Yan Wang(✉), Linan Yuan, and Shuli Dong

Beijing Institute of Space Mechanics and Electricity, Beijing 100094, China
77993762@qq.com

Abstract. The high dynamic-range (HDR) imaging technology has become an important topic in remote sensing imagery in recent years as it can provide more meticulous information. However, real time HDR fusion on satellites is still an engineering knot, mainly for two reasons: first, the weakened timeliness of remote sensing images caused by the earthward image transmission as complex fusion algorithms are restricted on satellites with only limited hardware processing platforms available thereon; second, the possible smearing for moving objects caused by multiple exposures of on-satellite imaging may reduce the definition and usability of images acquired. The paper studies a HDR real time fusion system for a type of large area array CMOS sensors. With the system, a real time HDR fusion is achieved for two images of different brightness with high gain and low gain, respectively, output by CMOS image sensors through one exposure. The system can not only use the two images exposed at the same time point to avoid smearing caused by asynchronous imaging, but also abstain from complex fusion algorithms like image registration, greatly reducing the hardware requirements for on-satellite real time fusion of HDR images, and improving the equipment employment feasibility on satellites.

First, the paper elaborates the fusion algorithm based on the working principle of CMOS sensors . Then, as general on-chip memories are unable to handle the large area array CMOS sensor's massive pixels and data processing, the design in the paper combines off-chip large-capacity flash memories and DDR memories to meet the requirement of massive storage and achieve a rapid real time fusion. Finally, an imaging verification test is carried out. The results show that the designed HDR real time fusion system can meet the on-satellite use demands of large area array CMOS sensors.

Keywords: CMOS · HDR · real time fusion

1 Overview

High-dynamic-range (HDR) images, compared to ordinary ones, can provide a greater dynamic range and more meticulous images. Finally, HDR images are synthesized based on low dynamic range (LDR) images of different exposure time and the best details of LDR images at different exposure, and can better reflect the visual effect in a real environment.

H. P. Urbach et al. (Eds.): ISSOIA 2023, SPP 191, pp. 277–286, 2024.
https://doi.org/10.1007/978-981-97-6718-2_26

Traditional HDR images are typically obtained via fusing two or more LDR images of different exposures which are got through exposure time modulation. For example, with a single camera, multiple LDR images of different exposures for the same scene are obtained through multiple exposures, and then the HDR images can be obtained through the fusion algorithm. Traditional HDR images are typically acquired through multiple imaging and the follow-up fusion. For remote sensing cameras, due to the platform stability and other reasons, the results of multiple imaging cannot be exactly the same, and image registration is required before fusion. There are two difficulties in realizing on-satellite real-time fusion of HDR images: first, the weakened timeliness of remote sensing images caused by the earthward image transmission as complex fusion algorithms are restricted on satellites with only limited hardware processing platforms available thereon; second, the possible image motion for stationary objects and smearing for moving objects caused by multiple exposures of on-satellite imaging may reduce the definition and usability of images acquired. In view of the said difficulties, this paper proposes a system that realizes real-time fusion of HDR images based on the two images output by a large area array CMOS sensor through one exposure [1].

2 HDR-Image Fusion Algorithm

2.1 Introduction to the Large Area Array CMOS Sensor

In this paper, a large area array CMOS sensor with a pixel count greater than 10 k × 10 k is studied. It can output two images - a high gain image and a low gain image through one exposure. The data flow is shown in Fig. 1:

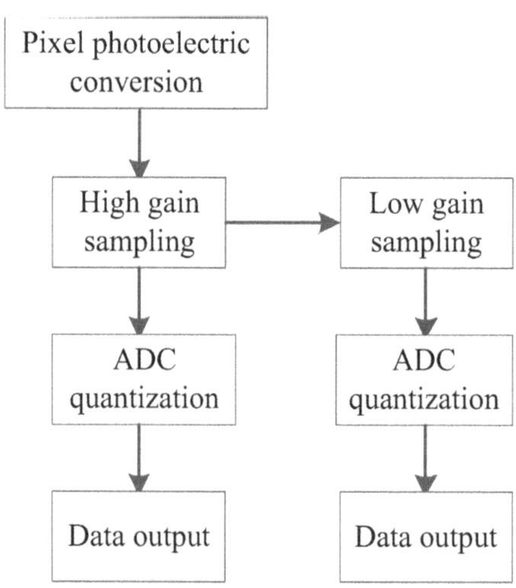

Fig. 1. Data flow of the CMOS sensor.

Figure 1 shows that the CMOS sensor can output two images of different brightness with high gain and low gain in one exposure, and achieve real-time HDR fusion with the two images. The advantages of this technology are the application of two images exposed at the same moment can avoid both the problems of image motion and smearing caused by asynchronous imaging, and the complex fusion algorithms such as registration, greatly reducing hardware requirements for on-satellite real-time fusion of HDR images, thus improving the feasibility.

2.2 CMOS Sensor HDR-Image Fusion Algorithm Based on the Laplacian Pyramid

The HDR-image fusion algorithm based on the Laplacian pyramid makes up four processes - image details enhancement, establishment of the Laplacian pyramid of images, fusion operator selection and image contrast modulation. Image details enhancement refers to the gray enhancement operation performed on high gain and low gain source images. The fusion of details enhancement-processed images will create higher image quality. The establishment and reconstruction of the Laplacian pyramid is carried out based on a theoretical approach. The selection of fusion operators directly affect the process of image fusion, while the contrast modulation is for a better display of the images preliminarily fused and reconstructed by the Laplacian pyramid. Compared with the classical Gaussian-Laplacian pyramid fusion, this algorithm has three differences: first, image details enhancement is added; second, the fusion weight is not created based on parameters such as contrast, saturation and exposure; and, third, the image contrast is modulated after fusion [2].

(1) Image details enhancement
Low gain images have a low gray value in general, and are dark even with a relatively longer exposure time, making us impossible to effectively identify details of bright and dark ground objects. High gain images present a good display effect to dark objects while an overexposure to bright objects in low gain images, In summary, whether it is a high- or low-gain image, the details of bright or dark objects are hidden in "pure black" or "pure white" pixels. If the high- and low-gain original images are directly used for fusion, the experimental results show that the display effect of HDR images is often poor. Therefore, the high- and low- gain original images are initially enhanced to fully mine the details of bright and dark objects and secure a high accuracy for the image fusion. On the basis of analyzing the characteristics of high- and low- gain images, normalization and nonlinear stretching are used to enhance the gray of the images.

(2) Establishing the Laplacian pyramid of images [3–5]
The Laplacian pyramid sequences are established for the high- and low-gain images, and the high gain images are used as an example below. The first layer of the pyramid is the original high gain image. Then, based on the Gaussian filter operator, it is low-pass filtered, then down-sampled to shrink to 1/2 of the original size, creating the second layer image of the pyramid. Low-pass filter convolution and 1/2 down-sampling are performed sequentially to obtain the third, the fourth to the Nth layer, from which the Gaussian pyramid image sequence is constructed. The Gaussian

filter operator is set as follows:

$$pyr = \begin{pmatrix} 1 & 4 & 6 & 4 & 1 \\ 4 & 16 & 24 & 16 & 4 \\ 6 & 24 & 36 & 24 & 6 \\ 4 & 16 & 24 & 16 & 4 \\ 1 & 4 & 6 & 4 & 1 \end{pmatrix} \cdot \frac{1}{256} \tag{1}$$

It can be seen from the establishment process of the Gaussian pyramid that the image pixel size decreases in order from the first layer to the Nth layer of the pyramid, respectively $m \times n$, $\frac{m}{2} \times \frac{n}{2}$, $\frac{m}{2^2} \times \frac{n}{2^2}$, ……$\frac{m}{2^{N-1}} \times \frac{n}{2^{N-1}}$. To establish the Laplacian pyramid, we first need to perform low-pass filtering and up-sampling on the second layer of the Gaussian pyramid, and make the sampling result be twice of the original size. Then, subtracting the images after up-sampling from the first layer of the Gaussian pyramid to obtain the first layer, with the size of m x n, of the Laplacian pyramid. Next, the third layer of the Gaussian pyramid is subjected to low-pass filtering and up-sampling, and subtracted from the second layer of the Gaussian pyramid to obtain the second layer of the Laplacian pyramid, with the size of $\frac{m}{2} \times \frac{n}{2}$.. And so on, the size of the third layer to the (N-1)th layer are obtained. And finally, the size of the Nth layer of the Gaussian pyramid is directly assigned the value of the Nth layer of the Laplacian pyramid. In this way, the Laplacian pyramid is established for low-sensitivity images [6].

(3) Fusion operator selection [7]
The initial (N-1) layers and the Nth layer of the Laplacian pyramid are processed separately. Two fusion operators are used here: the gradient operator for the initial N-1 layers, and the filter operator for the Nth layer with a filter template size of 5*5. A gradient operator is defined as:

$$\nabla f(x, y) = (\partial f / \partial x)^2 + (\partial f / \partial y)^2 \tag{2}$$

When computing digital images, we use difference to replace differential to obtain an approximate gradient value expression:

$$\nabla f(x, y) = [f(x, y) - f(x + 1, y)]^2 + [f(x, y) - f(x, y + 1)]^2 \tag{3}$$

Finally, the 1 to N layers of the Laplacian pyramid are added together to reconstruct the pyramid. Through the reconstruction, a preliminary fusion result based on high- and low-sensitivity images can be obtained.

(4) Image contrast modulation [8–11]
In order to make the fused images display more detailed information of various ground objects, the contrast modulation to the images obtained from the pyramid reconstruction is carried out based on the exposure time and image acquisition time. Theoretically, exposure time has a great influence on the quality of remote sensing images. Generally speaking, the overall brightness of images with a short exposure time is low, and the information is not fully mined through the pyramid fusion or similar means, resulting in high difficulty in visual interpretation to dark objects.

However, remote sensing imaging with a long exposure time is likely to lose information related to bright objects due to the high brightness, leading to a slightly worse display effect after fusion. In addition, the moment of image acquisition also affects the overall gray value of the image, and images acquired on daytime or at night often have deviations even they are processed with the same fusion method. In order to improve the image fusion results, it is necessary to modulate the contrast of the fused images after the reconstruction with the Laplacian pyramid. And the formula for the modulation is as follows:

$$y(x) = \begin{cases} f(x), x < 0.2 \\ g(x), x \geq 0.2 \end{cases} \quad (4)$$

In the formula, y is the gray value of the image after contrast modulation, and x is the normalized value of the image gray before contrast modulation. The discriminant threshold 0.2 is determined based on the gray values of the original images with high and low sensitivity. Images with a gray value below 0.2 are acquired on daytime, while those above 0.2 are at night. The gamma correction algorithm is used to modulate image contrast to obtain a better visual image. Gamma correction, also known as power transformation, is essentially a function introduced to overcome the nonlinearity of the image acquisition equipment's input and the image display equipment's output. If the pixel gray value is $x\prime\epsilon[0, x_{max}\prime]$, while x_{max} refers to the maximum pixel gray value, the procedure for contrast modulation using the gamma correction algorithm will be as follows:

1) Normalize the gray values of images to obtain the normalized pixel gray value $x\varepsilon$ [0,1];
2) Fit the gamma function $y(x) = ax^{\gamma}$, with γ as the gamma factor, the value of which determines the gray mapping method from the input image to the output image, to enhance the low gray or the high gray.
3) Conduct gray conversion to x the pixel gray value processed with the gamma correction to obtain the converted gray scale value.

3 Design of HDR-Image Real-Time Fusion Circuit System

This system realizes a real-time fusion of HDR images with the fusion algorithm based on the FPGA hardware platform. Due to the large size of images, the FPGA platform cannot meet the storage space requirement of fusion coefficients. And Flash memories are used to store these coefficients.

Flash memories are non-volatile even in case of power-offs, which ensures the full storage of the correction coefficients. But the low data readout speed of Flash memories cannot satisfy the system's real-time correction requirement. Therefore, DDR memories featuring fast readout and programming speeds are used as a buffer of correction coefficients to meet the speed requirement of real-time correction. DDR memory is a device developed on SDRAM storage technology and has been widely used in the computer industry.

DDR memory is characterized by a double data rate structure for high-speed operation, a maximum data transmission clock rate up to 666MHz, and a huge storage capacity.

Therefore, for high-speed and large-capacity data processing systems, the application of DDR memories is suitable in terms of both cost and performance. Figure 2 is the principle block diagram of the fusion circuit system:

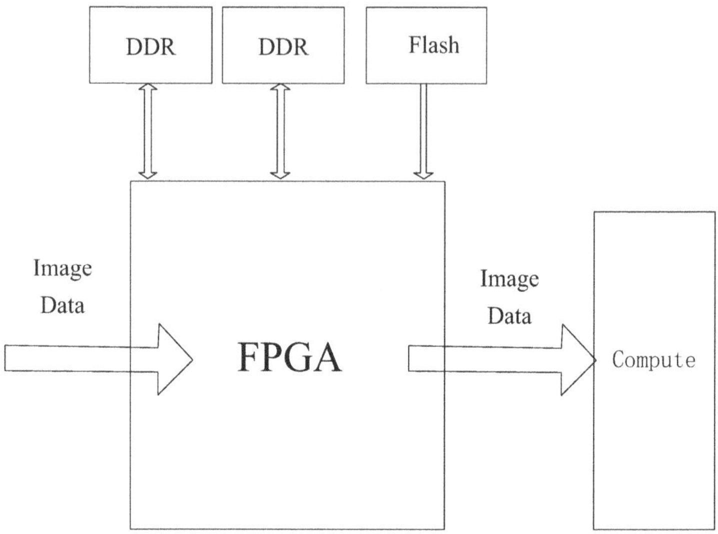

Fig. 2. The real-time fusion circuit of HDR images.

The fusion coefficient has been programmed into Flash memories before the system's startup. When the system is powered up, the fusion coefficients will first be read out from the Flash memories and programmed into a DDR memory to complete the preparation before the real-time fusion. The image signals transmitted from the CMOS sensor are stored in another DDR memory through the FPGA platform. The fusion module reads the pixel data and fusion coefficients from these two DDR memories respectively, completes the fusion algorithm inside the module, and outputs the fused pixel data for display. The entire fusion process is conducted under the same clock domain in the FPGA platform, which can ensure the smooth progress of the fusion algorithm pipeline.

4 Test Results

We used the system to conduct an outdoor scene imaging test, and tested HDR- image fusion effects on daytime and at night. Figures 3, 4 and 5 give the daytime test results, covering pre- and post-fusion images.

Figures 6, 7 and 8 present the nighttime test results, covering pre- and post-fusion images.

Fig. 3. Daytime high gain image.

Fig. 4. Daytime low gain image.

Fig. 5. Image after real-time fusion of daytime HDR images.

Fig. 6. Nighttime high gain image.

Fig. 7. Nighttime low gain image.

Fig. 8. Image after real-time fusion of nighttime HDR images.

Since the dynamic range of the image sensor is much smaller than that of natural light, as can be seen from Figs. 3, 4, 6 and 7, the images cannot cover all the brightness levels in the scene at the same time, making the objects overexposed or unidentifiable due to darkness. After real-time fusion of HDR images, Fig. 5 and Fig. 8 can provide more detailed information, and reflect the objects in the scene more clearly. So, the technology will be quite helpful to improve the application of remote sensing cameras.

5 Conclusion

With the system, a real time HDR fusion is achieved for the two images of different brightness with high gain and low gain, respectively, output by CMOS sensors through one exposure. The system not only reduces the complexity of the fusion algorithm, but also avoids the image motion for stationary objects and smearing for moving objects caused by multiple exposures, which greatly promotes the engineering progress of real-time fusion of HDR images on satellites.

References

1. Cui, Y.F., Liu, J.Y., He, H.Y., et al.: Influence of micro vibration on radiation quality of high resolution TDICCD images. Chinese Space Science Technol. **38**(6), 28–35 (2018)
2. Yang, B.W.G., et al.: New algorithm of infrared image enhancement based on histogram nonlinear extension. Infrared and Laser Eng. **32**(1), 1–4 (2003)
3. Yun, S.K., et al.: Image enhancement using a fusion framework of histogram equalization and laplacian pyramid. IEEE Transactions on Consumer Electronics **56**(4), 2763–2771. https://doi.org/10.1109/TCE.2010.5681167 (2010)
4. Ravisankar, P., et al.: Acoustic image enhancement using Gaussian and Laplacian pyramid -a multiresolution based technique. Multimedia Tools and Appl. **77**(5), 5547–5561. https://doi.org/10.1007/s11042-017-4466-7 (2018)
5. Li, X.Q.Z., et al.: A sum-modified-Laplacian and sparse representation based multimodal medical image fusion in Laplacian pyramid domain. Medical & Biological Engineering &Comput. **57**(10), 2265–2275. https://doi.org/10.1007/s11517-019-02023-9 (2019)
6. Kaur, H.R.: Image fusion on digital images using laplacian pyramid with DWT. In: 2015 Third International Conference on Image Information Processing (ICIIP), 21–24 December 2015,Waknaghat, India. IEEE, pp. 393–398 (2016). https://doi.org/10.1109/ICIIP.2015.7414803
7. Zhong, H.Y., et al.: Photon counting image enhancement algorithm based on improved regional energy fusion rules. J. Data Acquisition and Process. **34**(7), 111–121 (2019)
8. Yuan, L., et al.: A high dynamic range image fusion method based on dual gain image. Int. J. Image Data Fusion **14**(1), 15–37 (2023)
9. Chang, M., Chen, G.H.: Image compression and reconstruction based on Laplacian pyramid. Packaging Engineer **41**(15), 239–244 (2020)
10. Redondo, P.: Correction of trapdoor effect following nasal pyramid reconstruction with a transposition flap – Science Direct. Actas Dermo-Sifiliográficas (English Edition) **110**(1), 32–36 (2019). https://doi.org/10.1016/j.adengl.2018.10.020
11. Tan, W., Yang, B.X., He, H.Y.: A new definition evaluation method for remote sensing images of satellite. Chinese Space Science Technol. **36**(2), 50–57 (2016)
12. Guo, Y., et al.: Design and implementation of staring camera video circuit on GF-4 satellite. Spacecraft Recovery & Remote Sensing **37**(4), 49–58 (2018)
13. Wang, X.L.C., et al.: Application and development of multi-source information fusion in space situational awareness. Spacecraft Recovery & Remote Sensing **42**(1), 11–20 (2021)

Radiation Consistency Correction Method of Optical High Resolution Dual Camera Image

Hongqiang Wang(✉) and Li Liu

Space Star Technology CO., LTD., Beijing 100095, China
gzwhq@qq.com, 284499899@qq.com

Abstract. With the improvement of the resolution of the optical camera on the satellite, the method of carrying two cameras on the satellite is often used to improve the width and realize the wide coverage. The ground processing can get the dual camera image. The dual camera image avoids the geometric and radiation reprocessing process of the two cameras, and has great advantages in the quantitative application of remote sensing data. However, because the photoelectric system of the two cameras on the satellite is independent and the temperature control is different, compared with the single camera products, the image processing of the two cameras has difficulties in geometric splicing and radiation consistency correction. In this paper, aiming at the problem of dual camera image radiometric consistency correction, a method combining single camera radiometric correction model and dual camera radiometric correction model is proposed to correct dual camera image radiometric consistency. On the one hand, the radiometric calibration model of single camera is obtained by using side-slither calibration of multi scene. On the other hand, the radiation consistency correction model of the right camera is established based on the left camera. The dual camera image data of PRSS-1 remote sensing satellite are used for comparison and analysis before and after the use of the modified model. Finally, a large number of on orbit data tests show that there is no color difference between the dual camera images after correction, and the correction accuracy is better than 3% in the range of one scene. The accurate radiation consistency correction model of the dual camera images in any orbit can be obtained. The method in this paper has good applicability.

Keywords: Dual Camera Image · Relative Radiation Calibration · Side-slither · PRSS-1 Satellite

1 Introduction

With the improvement of the resolution of the satellite optical camera, the method of carrying dual cameras on the satellite is often used to improve the width and achieve wide coverage. For example, the GF-2 satellite launched on August 19, 2014 is equipped with two high-resolution 1m panchromatic and 4 m multispectral cameras, and the combined width of the dual cameras can reach 45 km [1]. And Pakistan Remote Sensing Satellite One (PRSS-1), which was successfully launched on July 9, 2018, is equipped

H. P. Urbach et al. (Eds.): ISSOIA 2023, SPP 191, pp. 287–305, 2024.
https://doi.org/10.1007/978-981-97-6718-2_27

with two high-resolution 1m panchromatic and 3m multispectral cameras, with a combined width of up to 60km [2]. Because the photoelectric systems of the two cameras on the satellite are independent, and the temperature control is different, compared with the single camera image, the dual camera image processing is difficult in geometric stitching and radiation consistency correction. Ground processing systems often provide single camera image products [1], while dual camera image products can only be provided after further geometric and radiation processing such as splicing, uniform light and uniform color. In this paper, a method combining single camera side-slither calibration and dual camera radiometric correction model is proposed to perform dual camera image radiometric consistency correction, which can directly provide dual camera image products, avoid users from carrying out complex processing of geometry and radiation of two cameras during application, and has great advantages in remote sensing data quantitative application.

In terms of calculation of radiation correction model of single camera, statistics method is generally used on orbit, and histogram matching method is more used, which can solve the problem of CCD response nonlinearity, but a large number of different ground objects need to be photographed on orbit [3]. In recent years, many satellites have a side-slither radiometric calibration mode, where the satellite axis is rotated 90 degree in yaw relative to normal acquisitions. The method of acquiring calibration data by using the side-slither calibration radiometric calibration mode on the satellite can theoretically be used as a calibration ground object compared with the way of shooting the ground calibration field with the traditional normal imaging mode. Moreover, the attitude agility of the satellite itself is adopted, and no additional on-board calibration devices are needed. More and more of them are applied to the relative radiometric calibration of cameras, and good results are obtained [4–7]. In terms of dual camera radiation consistency correction model, Han Jie proposed to eliminate the radiation inconsistency in the overlapping area of adjacent camera images by using the cross absolute radiation calibration coefficient [8, 9]. Affected by the absolute radiation calibration coefficient model, the radiation inconsistency cannot be completely eliminated. This paper focuses on the research and verification of dual camera radiometric consistency correction model acquisition and processing effect verification.

2 Radiation Consistency Correction Method for Dual Cameras

2.1 Dual Camera Radiation Consistency Correction Process

In this paper, the combination of single camera side-slither calibration and dual camera image radiation consistency correction model per track automatic correction method can achieve high-precision dual camera image radiation consistency correction. It can be divided into two major processes: side-slither correction and dual camera model correction. The correction process is shown in Fig. 1.

First, the relative radiation calibration (RRC) of the left and right cameras can be obtained through the side-slither calibration. For the left and right camera image data acquired per track, the relative radiation correction is performed according to the relative radiation correction model of the left and right cameras. Then, based on the histogram of the overlapping area of the left and right cameras, the relative radiation correction

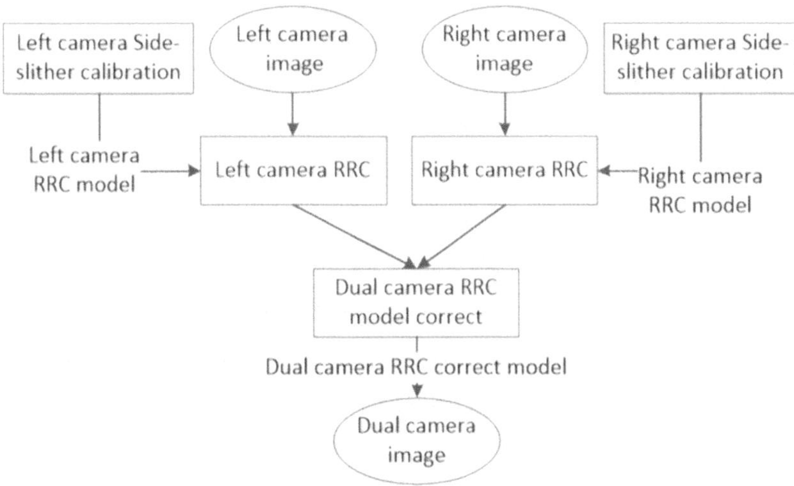

Fig. 1. Radiation consistency correction process of dual cameras

model of the right camera is established with the left camera as the benchmark. And the correction model is used to correct the image data of the right camera, and finally the dual camera radiometric correction image is obtained.

2.2 Side-Slither Calibration Principle and Influence Factor Analysis

Compared with the traditional normal ground imaging mode, the side-slither radiometric calibration theoretically means that each pixel on the detection line will image the same ground object in turn, and any ground object can be used as a calibration ground object. And it adopts the attitude agility characteristics of its own satellite, without additional on-board calibration devices. Because of these two outstanding advantages, Ikonos-2, Quickbird-2, RapidEye, Landsat-8 and other foreign satellites all use side-slither for relative radiometric calibration, which has gradually become a substitute for the traditional ground calibration field relative radiometric calibration [4–7] (Figs. 2, 3).

The relative radiometric calibration on orbit generally adopts statistical method, and the histogram matching method is often used, which can solve the problem of CCD response nonlinearity. The use of side-slither radiometric calibration can reduce the shooting of a large number of traditional on orbit calibration fields or ground objects, and improve the calibration efficiency. In this paper, the histogram matching method is used to obtain the relative radiometric calibration coefficient for side slither radiometric calibration. For any pixel i, according to the cumulative probability density function $p_i'(k)$, and the cumulative probability density function P of the expected histogram $P(n)$ and $P(n+1)$compare, find the appropriate DN value, and generate a histogram lookup table F(i, k). F(i, k) refers to the lookup table where the grayscale value of the ith pixel is k.

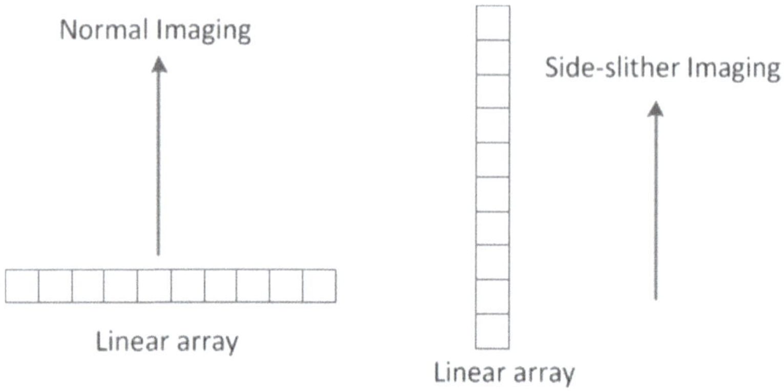

Fig. 2. Schematic diagram of normal imaging and side-slither calibration imaging

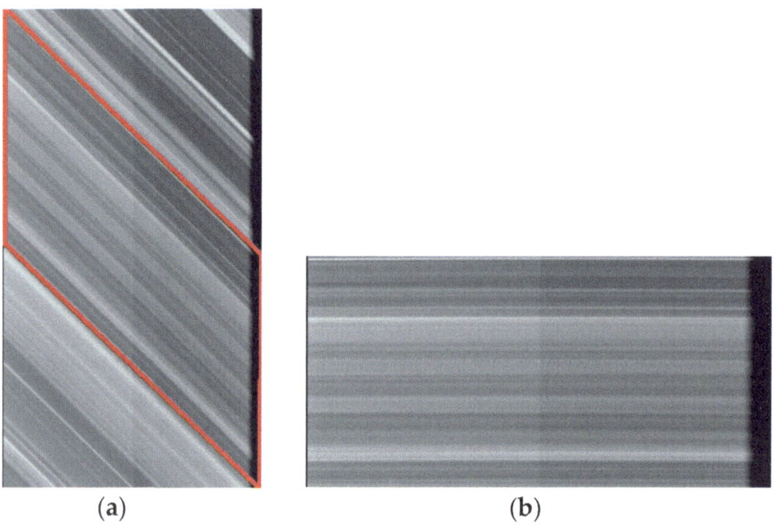

Fig. 3. Before and after data sorting of side-slither calibration.(a) Side-slither calibration raw data; (b) Collated calibration data.

While $P(n) \leq p_i{\prime}(k) < P(n+1)$

$$F(i,k) = \begin{cases} n, p_i{\prime}(k) - P(n) \leq P(n+1) - p_i{\prime}(k) \\ n+1, p_i{\prime}(k) - P(n) > P(n+1) - p_i{\prime}(k) \end{cases} \quad (1)$$

Considering that there are few samples at the low end and high end of the pixel response, linear interpolation is used to calculate the coefficients at the low end and high end of the pixel response during histogram matching. Assume that the minimum and maximum values of DN in the expected histogram are k_{min}, k_{max}, and F(i, k) are expressed as follows, and the value range of DN gray value is 0–1023, 1023 is the maximum DN of the CCD.

$$F(i,k) = \begin{cases} \frac{F(i,k_{min})\times k}{k_{min}}, k = 0,1,2,\ldots,k_{min}-1 \\ F(i,k_{max}) + \frac{(1023-F(i,k_{max}))\times(k-k_{max})}{1023-k_{max}}, k = k_{max}+1,\ldots,1023 \end{cases} \quad (2)$$

side-slither radiometric calibration method is mainly affected by earth rotation, non-uniformity of optical system illuminance, CCD probe element splicing method, satellite attitude control accuracy, etc. The influence of earth rotation can be corrected by offset angle compensation. Due to the small field angle of the camera, the influence of the nonuniformity of the illumination of the optical system is small. n terms of satellite attitude, the pointing deviation and stability of pitch angle and roll angle will cause the deviation of photographed objects, and the pointing deviation and stability of yaw angle will affect the consistency of all pixels on the CCD linear array to observe the same object.. The pitch angle has the greatest influence [10]. Assuming that the pitch angle stability is 0.001°/s, the velocity of the lower point of the low orbit satellite is about 7.5 km/s, the TDICCD pixel is 32,000 pixels, and the resolution is 1 m, then the time difference between the first pixel and the last pixel passing the same ground point is 1.6 s, the maximum error of the pitch angle is 0.0016°, and the maximum error caused by the pitch angle is 37 pixels. Therefore, when using the side-slither radiometric calibration method, a variety of uniform scenes should be selected as the calibration field for the ground scene as far as possible.

2.3 Dual Camera Radiometric Correction Model

Although side slither can reduce the sample size of on orbit statistical ground objects, due to the influence of attitude, it can not ensure that all pixel observation ground objects are completely consistent, which is more obvious for dual cameras with large width. The splicing area of left and right cameras often has obvious splicing lines. We use the dual camera radiometric correction model for correction.

Since the overlapping area of two cameras takes pictures of the same ground object, and the shooting time is very short, it can be assumed that the pixel input of the overlapping area of two cameras is consistent, and the right camera is corrected based on the left camera. F_r (k_r) It is the histogram lookup table of the right camera. The formula for histogram correction is as follows. Among them, $delta_r$ (k_r) represents the relative radiation correction of the right camera,the formula for histogram correction is as follows, k_r represents the pixel gray value of the right camera, with a value range of 01023.

$$F_r(k_r) = k_r + delta_r(k_r) \quad (3)$$

Calculate the cumulative probability density function p of the left and right camera overlap $p_l'(k_l)$ and $p_r'(k_r)$, representing the cumulative probability density of histograms of two overlapping regions, k_l 、 k_r represents the gray value of the left and right camera pixels, with a value range of 0–1023.

According to the principle that the pixel input and output in the overlapping area of two cameras should be consistent, find the DN value of the left and right cameras corresponding to the same cumulative probability density of the overlapping area of the left and right cameras, and the difference between the two is $\mathrm{delta_r(k)}$. Since the cumulative probability density cannot be completely equal, the histogram matching method is used to find the closest value of the cumulative probability density function.

Calculation model of the minimum and maximum range of DN in the overlap area of the right camera:

While: $p\prime(k_l) \leq p\prime(k_r) < p\prime(k_l + 1)$

$$\mathrm{delta_r(k_r)} = \begin{cases} \mathrm{k_l - k_r,\ p'_r(k_r) - p'_l(k_l) \leq p'_l(k_l + 1) - p'_r(k_r)} \\ \mathrm{k_l + 1 - k_r,\ p'_r(k_r) - p'_l(k_l) > p'_l(k_l + 1) - p'_r(k_r)} \end{cases} \quad (4)$$

From the formula (4), the formula (3) can be expressed as follow:

While: $p\prime(k_l) \leq p\prime(k_r) < p\prime(k_l + 1)$

$$\mathrm{F_r(k_r)} = \begin{cases} \mathrm{k_l,\ p'_r(k_r) - p'_l(k_l) \leq p'_l(k_l + 1) - p'_r(k_r)} \\ \mathrm{k_l + 1,\ p'_r(k_r) - p'_l(k_l) > p'_l(k_l + 1) - p'_r(k_r)} \end{cases} \quad (5)$$

The overlapping area needs to be strictly consistent, which needs to be carried out after sensor calibration. Sensor correction is often aimed at single scene image processing. The number of pixels in the overlapping area is small, and the samples in the overlapping area of single scene image are not enough to cover the full response range of pixels, such as 0–1023. To obtain accurate $\mathrm{delta_r(k_r)}$ model requires multi track data for calculation in order to obtain a corrected model with a complete response range. For the lack of samples at the high and low ends, the correction amount shall be calculated by linear interpolation. The linear correction model can be use follows, and the linear coefficient a and b can be obtained by the least square method.

$$delta_r(k_r) = a \times k_r + b \quad (6)$$

Namely:

$$F_r(k_r) = (1 + a) \times k_r + b \quad (7)$$

Or expressed as:

$$F_r(k_r) = a\prime \times k_r + b \quad (8)$$

$$a\prime = 1 + a$$

3 Correction Results and Evaluation Analysis

The image data of PRSS-1 dual camera were used for testing and verification analysis. PRSS-1 remote sensing satellite is the first remote sensing satellite exported by China to Pakistan. It was successfully launched on July 9, 2018. PRSS-1 remote sensing satellite is equipped with two panchromatic multispectral high-resolution cameras with 1 m/3 m spatial resolution. The width of a single camera is about 31km, and the width of double cameras after splicing is more than 60km. It has the characteristics of high resolution and large width. The remote sensing data obtained by satellite can be widely used in Pakistan's land and resources survey, environmental protection, disaster monitoring and management, agricultural survey and urban construction and other business fields, can play a positive role in Pakistan's development of national economy, improving people's lives, promoting social progress, etc., and will also provide spatial remote sensing information services for the construction of the China Pakistan Economic Corridor and the "the Belt and Road".

3.1 Radiation Consistency Correction Model of Dual Cameras

The relative radiation correction model of the left and right cameras is obtained by using the side-slither data of PRSS-1 satellite, and the image data of the left and right cameras are corrected respectively. The relative radiation correction model of the right camera is obtained by using the 2.3 section method for the overlap area. Table 1 is the list of side-slither calibration data in 2018. It lists a gain and integration series of multispectral and panchromatic. Figure 4 shows the relative radiation correction coefficient of the central pixel (1536th pixel) of eight CCD cameras in band 1 blue band.

Table 1. Side-slither Calibration Data List.

S/N	Band	Imaging date	Track numbe	Gain	Integral series
1	PAN\MSS-BAND1\2\3\4	20180725	244	6\6\10\6\8	3\4\3\4\3
2	PAN\MSS-BAND1\2\3\4	20180730	319	6\6\10\6\8	3\4\3\4\3
3	PAN\MSS-BAND1\2\3\4	20180806	422	6\6\10\6\8	3\4\3\4\3
4	PAN\MSS-BAND1\2\3\4	20180902	820	6\6\10\6\8	3\4\3\4\3

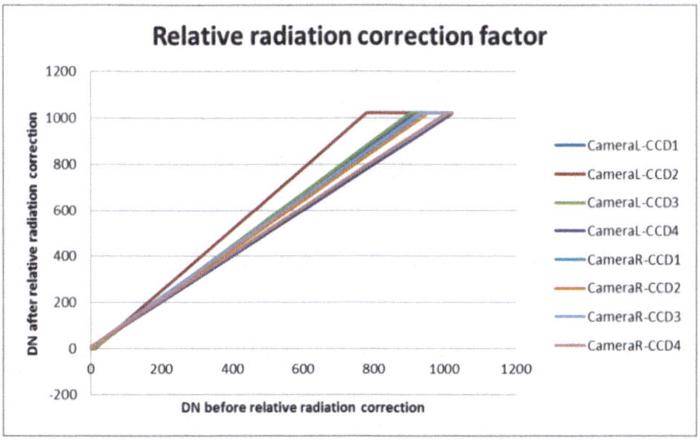

Fig.4. Relative Radiation Correction Coefficient.

Figure 5 shows the original data of the multispectral blue spectrum segment of the dual cameras in Pakistan with orbit 2133 taken by PRSS-1 satellite on November

30, 2018. Figure 6 shows the overlapping area of the scene data with a red box. The overlapping area is 616 pixels.

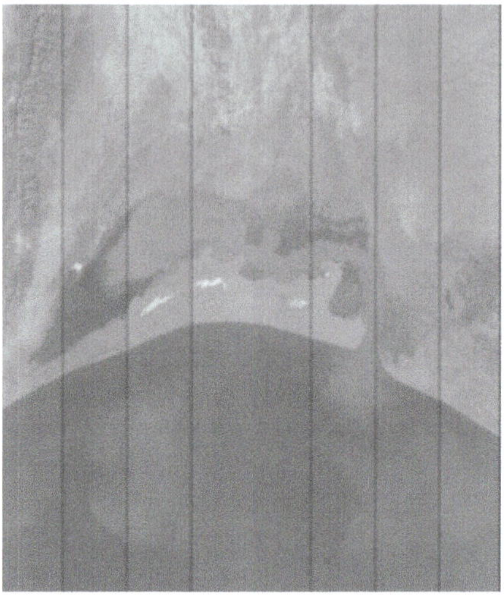

Fig. 5. Raw Data of PRSS-1 Satellite (Scene 6 of orbit 2133)

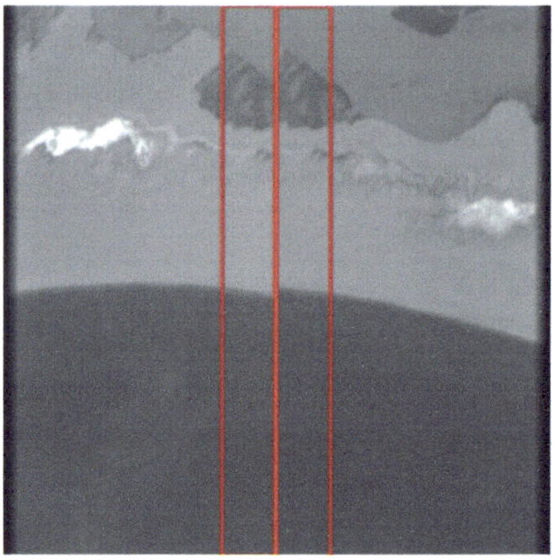

Fig. 6. Schematic Diagram of Double Camera Overlap Area (Scene 6 of 2133 Track).

The relative radiation correction model of the left and right cameras is obtained by using the side-slither data of PRSS-1 satellite, and the image data of the left and right cameras are corrected respectively. The relative radiation correction model of the right camera is obtained by using the 2.3 section method for the overlap area. Figure 7 shows the overlapping area histogram of the left and right cameras and the calculated $delta_r$. The multi spectral bands are blue, green, red, near red and panchromatic.

It can be seen from the double camera overlap area correction model in Fig. 7 that the single scene correction model cannot cover the full range of pixel response. Figure 8 uses the whole track data and linear interpolation method to obtain the correction model of the full range of pixel response. In the same way, the 1589, 2188 and other two tracks are processed. Taking the multispectral blue band as an example, the dual camera correction model is obtained as follows.

It can be seen from the double camera overlap correction model in Fig. 9 and Table 2 that the trend of the double camera correction model obtained from the multi track data is consistent, but not identical. Within the response range of 0–1023 pixels, the deltar range is 7.01−29.41, 12.82−36.59, 13.54−21.65 respectively, and the deltar difference range is 6.53−7.75.

3.2 Radiation Consistency Correction Results of Dual Cameras

The dual-camera radiation consistency correction model obtained above is used for correction, and the results of 2133 and 2281 are taken as examples to illustrate.

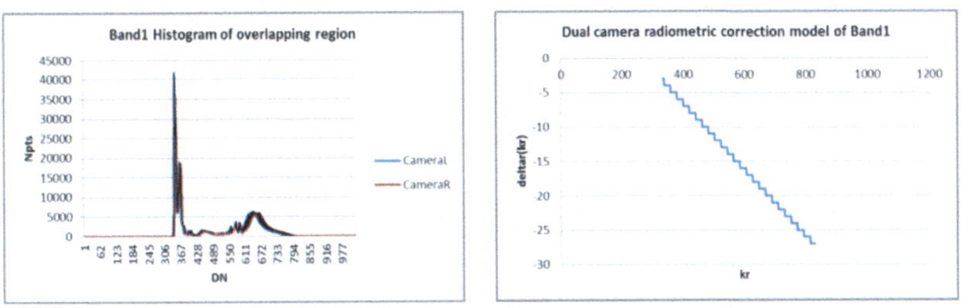

(a) Blue band

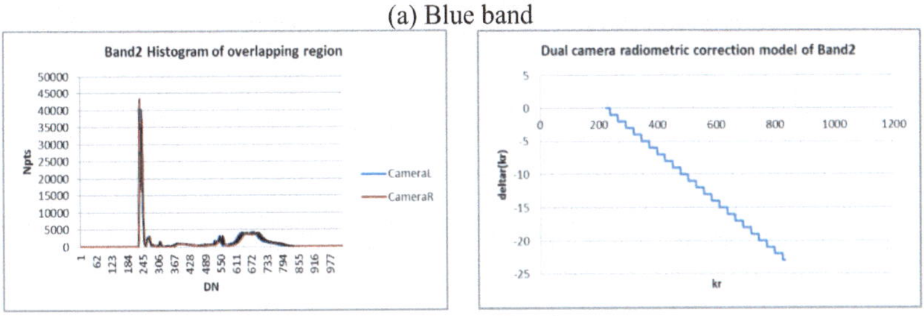

(c) Green band

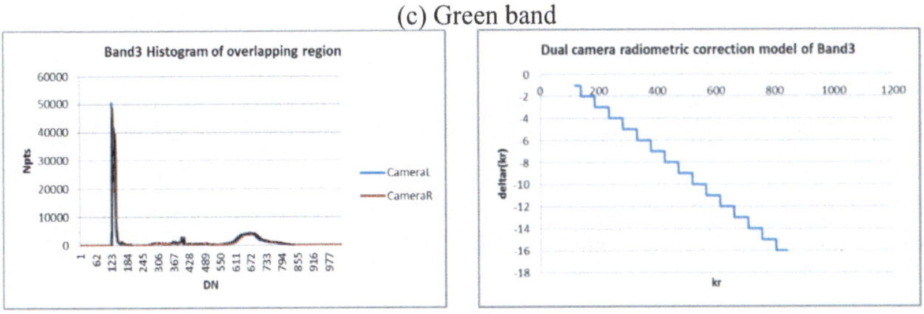

(c) Red band

Fig. 7. Histogram of the overlapping area of two cameras and the modified model (scene 6 of track 2133). (a),(b),(c),(d),(e) are the histogram of overlapping region and dual camera radiometric correction model of blue, red, near red and Panchromatic band respectively; (f) Dual-camera's $delta_r$ of all bands.

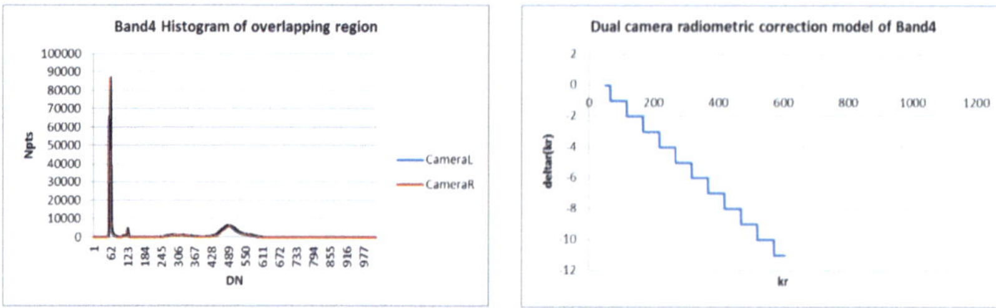

(d) Near red band

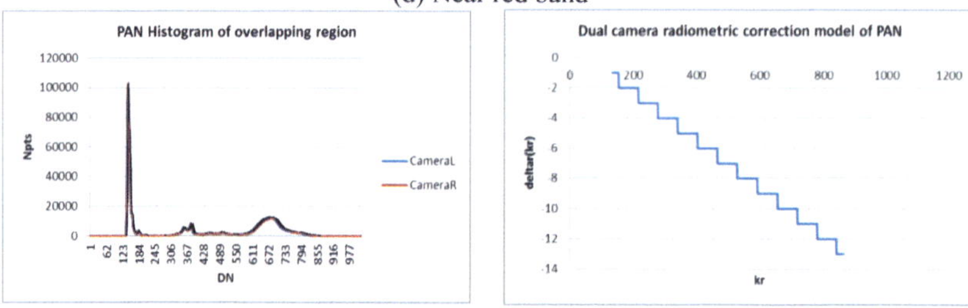

(e) Panchromatic band

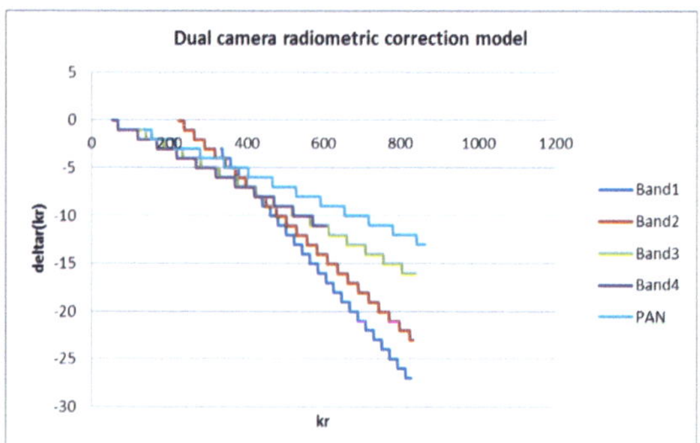

(f) Dual-camera's delta$_r$ of all bands

Fig. 7. (*continued*)

It can be seen from Fig. 10 and Fig. 11 that after the radiation consistency correction of the dual cameras in this paper, the dual camera image has no color difference visually, and the black stripes in the CCD mosaic area of the original image and the nonuniformity between the two cameras have been well corrected. Figure 12 and Fig. 13 are the results of mosaic area processing before and after radiation consistency correction of dual-camera images of 2133 and 2281, respectively.

3.3 Evaluation Method for Radiation Consistency Correction of Dual Cameras

The evaluation method of dual-camera radiation consistency correction can refer to the common radiation consistency evaluation method. The streaking metrics is often used internationally for evaluation [4, 6], which is also called the adjacent pixel method.

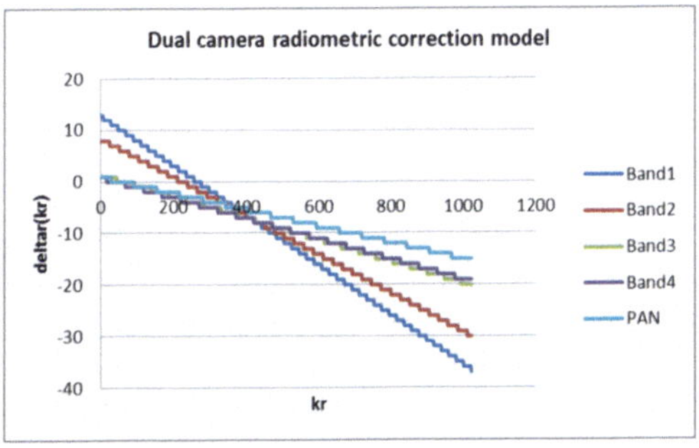

Fig. 8. Double camera overlap correction model (2133 rails).

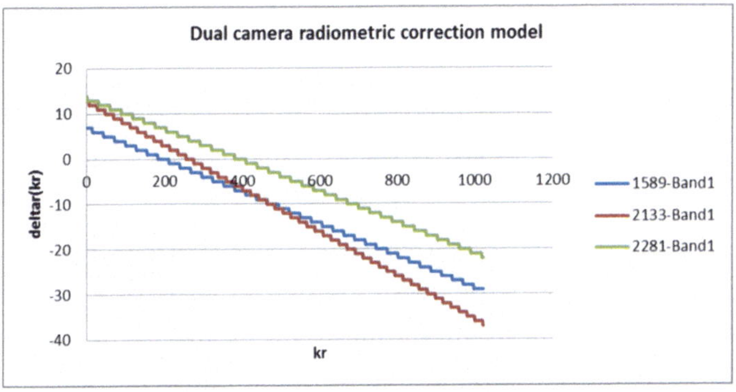

Fig. 9. Double camera overlap correction model (blue band, 3-track).

Table 2. Double camera overlap correction model (blue band, 3-track).

S/N	Band	imaging date	track number	a	a′	b
1	MSS-BAND1	20181024	1589	-0.0356	0.9644	7.010
2	MSS-BAND1	20181130	2133	-0.0483	0.9517	12.820
3	MSS-BAND1	20181210	2281	-0.0344	0.9656	13.537

In China, the average row standard deviation method, the average standard deviation method and the generalized noise method are also used for evaluation [11].

(1) Adjacent pixel method, namely streaking metrics: calculate the mean value of each column of pixels, and calculate the difference between the mean value of each column of pixels and the adjacent column of pixels to obtain the streaking metrics of each pixel. Take the mean value of the streaking metrics of all pixels to get the streaking

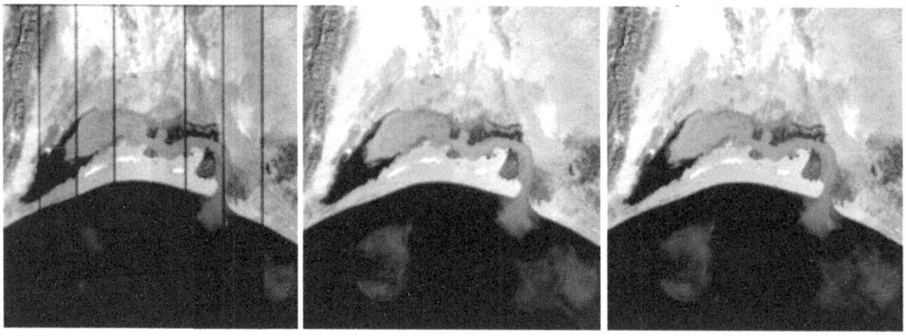

(a) (b) (c)

Fig. 10. Before and after radiation consistency correction of dual-camera images (scene 6 of track 2133). (a) Original image; (b) relative radiometric correction result;(c) dual-camera radiometric correction result.

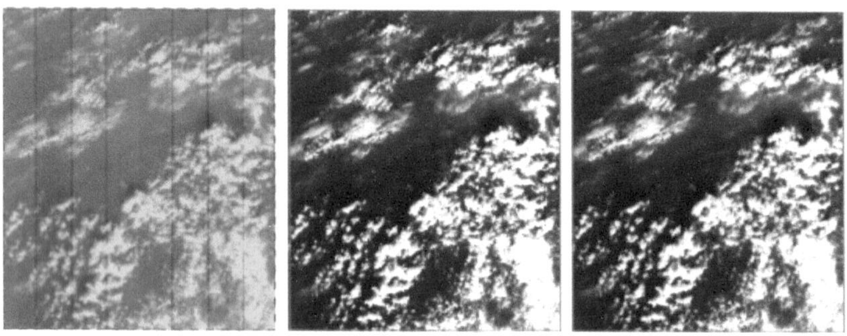

(a) (b) (c)

Fig. 11. Before and after radiation consistency correction of dual-camera images (scene 6 of track 2281). (a) Original image; (b) relative radiometric correction result;(c) dual-camera radiometric correction result.

metrics of the whole image.

$$\text{Streaking}_{\text{Det}_i} = \frac{\left|\overline{DN}_i - \frac{\overline{DN}_{i-1}+\overline{DN}_{i+1}}{2}\right|}{\frac{\overline{DN}_{i-1}+\overline{DN}_{i+1}}{2}} \times 100\% \quad (9)$$

where, $\text{Streaking}_{\text{Det}_i}$ is the streaking metrics of the i-th pixel, and $\overline{\text{DN}}_\text{i}$ is the column mean value of the i-th pixel.

(2) Average row standard deviation method (RA): calculate the mean and standard deviation of each column of pixels, and then calculate the mean value of the whole image to obtain the relative radiometric calibration accuracy.

$$\text{RA} = \frac{\sqrt{\sum_{i=1}^{M}\left(\overline{DN}_i - \overline{DN}\right)^2/M}}{\overline{DN}} \times 100\% \quad (10)$$

where, RA is the relative radiometric calibration accuracy, (DN)-_i is the mean value of the i-th pixel, and (DN)- is the mean value of the entire image.

(a) (b) (c)

Fig.12. Before and after radiation consistency correction of dual-camera images (2133 splicing area). (a) Original image; (b) relative radiometric correction result; (c) dual-camera radiometric correction result.

(3) Average standard deviation method (R): first calculate the mean and standard deviation of each column of pixels, and then divide by the mean value of the column to obtain the relative radiometric calibration accuracy of the column of pixels. The accuracy of the whole image is obtained by taking the average of the accuracy of all

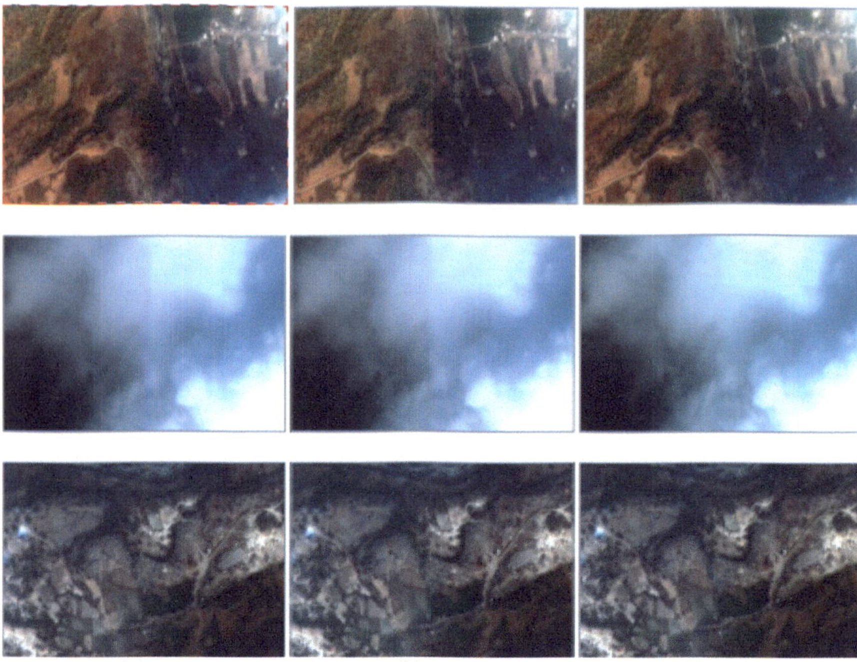

(a) (b) (c)

Fig.13. Before and after radiation consistency correction of dual-camera images (2281 splicing area). (a) Original image; (b) relative radiometric correction result; (c) dual-camera radiometric correction result.

pixels.

$$R_i = \frac{\sqrt{\sum_{j=1}^{N} (DN(j, i) - \overline{DN}_i)^2 / N}}{\overline{DN}_i} \times 100\% \tag{11}$$

where, R_i is the relative radiometric calibration accuracy of the i-th pixel,DN(j,i) represents the DN value of the i-th column in the j-th row of the image, and (DN)-_i is the mean value of the i-th pixel.

(4) Generalized noise method: taking the average gray value of all pixels of the image as a reference, calculate the sum of squares after the difference of the gray value of each pixel point, and then use the total number of pixels to standardize, which represents the average degree of the gray change of the image.

$$\varepsilon = \frac{1}{\overline{DN}} \sqrt{\frac{\sum_{i=1}^{M} \sum_{j=1}^{N} [DN(j, i) - \overline{DN}]^2}{\mathrm{M} \times \mathrm{N}}} \tag{12}$$

where, ε is the radiation consistency accuracy of dual cameras; M and N are the number of columns and rows of the image; DN(i, j) denotes the DN value of row j and column i of the image, and $\overline{\mathrm{DN}}$ represents the image mean value of the selected area.

3.4 Evaluation Results of Dual-Camera Radiation Consistency Correction

Six groups of typical natural homogeneous ground objects with low, medium and high brightness areas were selected to evaluate the accuracy of dual-camera radiometric consistency correction. The natural ground objects include uniform rivers, oceans, crops and beaches. The evaluation samples are shown in Fig. 14. The six groups of evaluation samples are the original blue-band images, relative radiometric correction results and dual-camera radiometric correction results. See Table 3, Table 4 and Table 5 for the evaluation results of the radiation consistency of the splicing area of the original image, the relative radiometric correction result and the dual-camera radiometric correction result, respectively. Table 6 and Table 7 show the improvement of the dual-camera radiometric correction accuracy and the proportion of improvement.

According to "Table 3 Evaluation Results of Radiation Consistency of the Original Image Mosaic Area", the relative radiometric calibration accuracy of the original image's two-camera mosaic area is 3.61% for the adjacent pixel method, 3.64% for the average row standard deviation method, 3.58% for the average standard deviation method, and 0.96% for the generalized noise method.

According to "Table 4 Evaluation Results of Radiation Consistency in the Splicing Area of Relative Radiation Correction Results", it can be concluded that the relative radiation calibration accuracy of the dual-camera splicing area of the relative radiation correction results is 1.26% for the adjacent pixel method, 1.36% for the average row standard deviation method, 1.24% for the average standard deviation method, and 0.34% for the generalized noise method, all better than 2%.

According to "Table 5 Evaluation results of radiation consistency of the splicing area of the dual-camera radiometric correction results", the relative radiometric calibration accuracy of the dual-camera splicing area of the dual-camera radiometric correction results is 0.24% for the adjacent pixel method, 0.54% for the average row standard deviation method, 0.21% for the average standard deviation method, and 0.14% for the generalized noise method, all of which are better than 1%.

According to "Table 6 Accuracy improvement of dual-camera radiometric correction" and "Table 7 Accuracy improvement ratio of dual-camera radiometric correction", it can be concluded that the accuracy of dual-camera radiometric correction results has been greatly improved compared with the relative radiometric correction results, which are respectively: adjacent pixel method is 1.14%, average row standard deviation method is 0.92%, average standard deviation method is 1.16%, generalized noise method is 0.18%, and the evaluation accuracy of the four methods has been improved by 0.18%-1.14%. The improvement ratio is 85.08% for the adjacent pixel method, 63.79% for the average row standard deviation method, 87.45% for the average standard deviation method, and 61.21% for the generalized noise method. The improvement ratio of the evaluation accuracy of the four methods is 61.21%-85.08%.

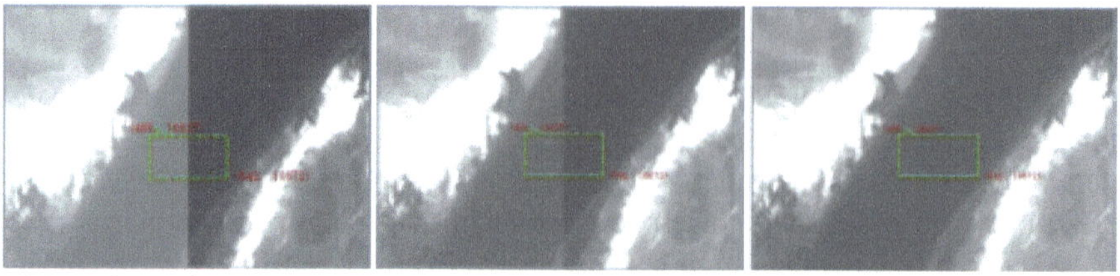

(a) Sample 1: River

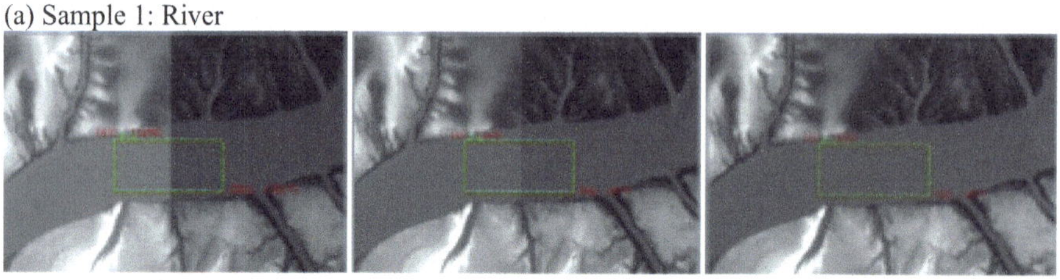

(b) Sample 2: River

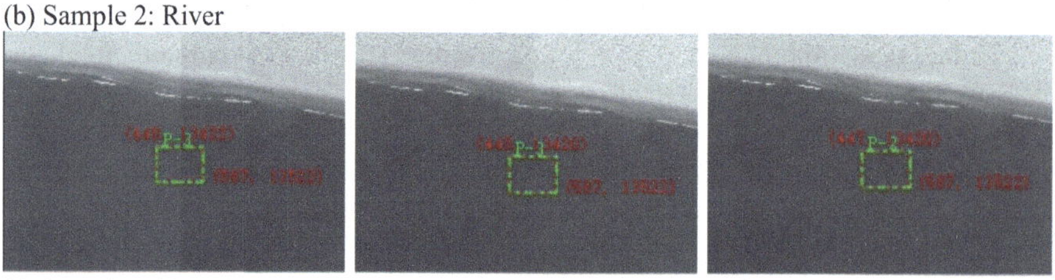

(c) Sample 3: Ocean

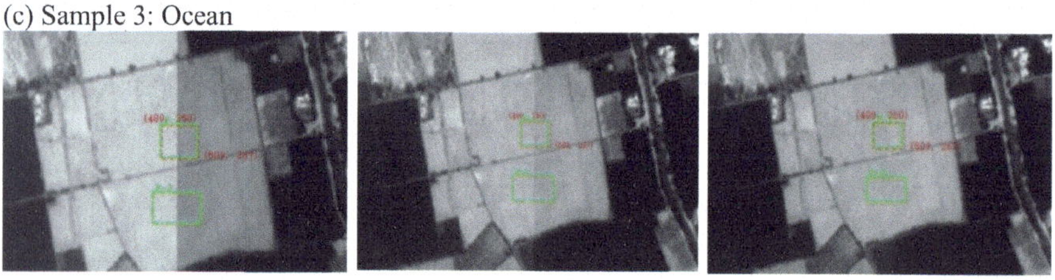

(d) Sample 4: Cropland

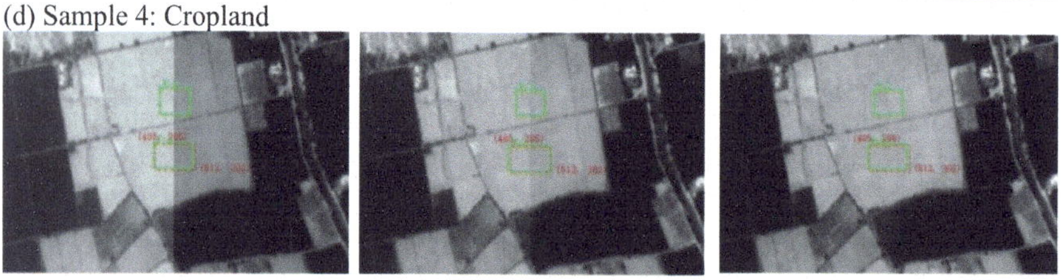

(e) Sample 5: Cropland

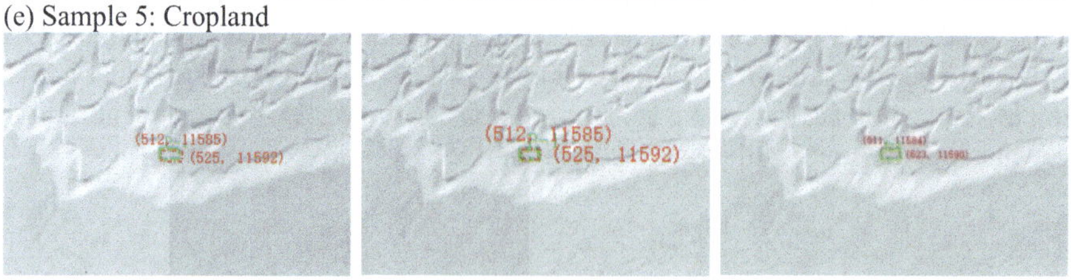

(f) Sample 6: Beach

Fig.14. Sample image of dual-camera radiation consistency evaluation. From left to right: original image, relative radiometric correction result, dual-camera radiometric correction result.

Table 3. Radiation consistency evaluation results of original image mosaic area.

No	Figure Type	Brightness Range	Streaking	RA	R	ε
Sample 1	River	Low	4.46%	4.48%	4.45%	0.92%
Sample 2	River	Low	4.53%	4.54%	4.53%	0.74%
Sample 3	Ocean	Low	2.32%	2.41%	2.32%	0.39%
Sample 4	Cropland	Medium and high	3.86%	3.89%	3.74%	1.81%
Sample 5	Cropland	Medium and high	4.29%	4.32%	4.26%	1.60%
Sample 6	Beach	Medium and high	2.17%	2.20%	2.16%	0.32%
	average		3.61%	3.64%	3.58%	0.96%

* Note: In Table 3, low brightness refers to radiance less than or equal to100 $W/m^2/sr/m$, and medium and high brightness refers to radiance greater than100 $W/m^2/sr/m$.

Table 4. Radiation consistency evaluation results of the splicing area of the relative radiation correction results.

No	Figure Type	Brightness Range	Streaking	RA	R	ε
Sample 1	River	Low	1.55%	1.60%	1.54%	0.34%
Sample 2	River	Low	1.63%	1.67%	1.63%	0.28%
Sample 3	Ocean	Low	0.97%	1.17%	0.96%	0.19%
Sample 4	Cropland	Medium and high	0.85%	0.99%	0.78%	0.57%
Sample 5	Cropland	Medium and high	1.26%	1.37%	1.23%	0.46%
Sample 6	Beach	Medium and high	1.28%	1.34%	1.28%	0.21%
	average		1.26%	1.36%	1.24%	0.34%

Table 5. Evaluation results of radiation consistency in the splicing area of dual-camera radiation correction results.

No	Figure Type	Brightness Range	Streaking	RA	R	ε
Sample 1	River	Low	0.16%	0.42%	0.13%	0.14%
Sample 2	River	Low	0.12%	0.39%	0.10%	0.10%
Sample 3	Ocean	Low	0.21%	0.67%	0.18%	0.08%
Sample 4	Cropland	Medium and high	0.58%	0.79%	0.55%	0.22%
Sample 5	Cropland	Medium and high	0.27%	0.57%	0.21%	0.21%
Sample 6	Beach	Medium and high	0.12%	0.39%	0.10%	0.06%
	average		0.24%	0.54%	0.21%	0.14%

satellite constellation launched by Digital Aerospace Corporation in 2017 to monitor plant growth [10]; Japan's Axel Space launched the Gray Crane micro-nano satellite constellation in 2018, which consists of five micro-nano satellites with a single mass of 80 kg, which can be used for agriculture, forestry, terrain and disaster monitoring. Similarly, two other systems were reported, the Vivid-i micro-nano satellite constellation of the British Earth Imaging Company, and the NIUSAT micro-nano satellite for ecological disaster monitoring and control developed by Islam University of Narul in India.

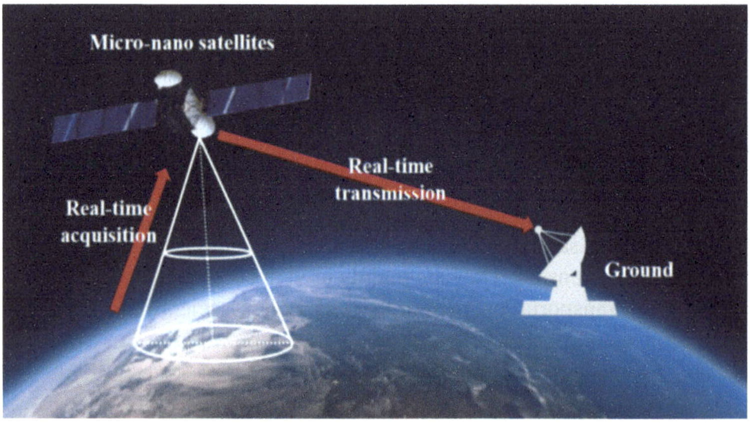

Fig. 3. Schematic diagram of laser communication monitoring of micro-nano remote sensing satellites

The research and development of micro-nano satellite technology started late in China, and the number of satellites in orbit is much lower. At present, China's relatively mature micro-nano satellite constellation systems used in the monitoring of the earth's ecological environment include "Zhuhai-1" and "Jilin-1". In 2017, the first satellite of the "Zhuhai-1" remote sensing micro-nano satellite constellation was successfully launched, which consists of 34 capsules of 55 kg micro-nano satellites and is equipped with optoelectronic devices such as lidar [11], which are mainly used in environmental monitoring fields such as agriculture, forestry and oceans [12]; The first generation of satellites of the "Jilin-1" micro-nano satellite constellation was launched in 2015, and its first phase of construction was completed with 60+ satellites in orbit. In June 2023, the star-to-ground laser communication rate reached 10 Gbps, and the constellation will be networked to form an all-weather and full-spectrum data acquisition capability to provide services in resources monitoring, forestry, environmental protection and disaster monitoring [13]; In August 2023, with the help of the space station project, China realized the launch of micro-nano satellites from the space station.

Laser communication technology based on high-power semiconductor lasers has played an important role in the detection of the ecological environment by micro-nano-satellites, and has provided abundant observation data for forests, oceans, topography and ecological disaster. Micro-nano satellite is small in size, and has strict requirements on the weight and power consumption of communication loads, etc. Compared with traditional microwave communication, laser communication has the advantages of large bandwidth, anti-electromagnetic interference and anti-interception, which are orders of magnitude higher in frequency and rate. Its terminal light source adopts high-power semiconductor lasers with the characteristics of small size, simple structure, high efficiency and so on, with fast modulation response and high repetition frequency, which is more in line with the application requirements of micro-nano satellite. High-power semiconductor lasers are truely the "eye of the falcon" for ecological environ-ment detection by micro-nano satellites.

3.2 Marine Ecological Environment Monitoring

Under the double pressure of human activities and climate change, the overall situation of the marine ecological environment is not optimistic. Lidar technology based on high-power LD has demonstrated its real-time and continuity in marine environmental monitoring, and can effectively serve the monitoring and protection of marine ecosystems, with advantages that are incomparable to traditional monitoring means.

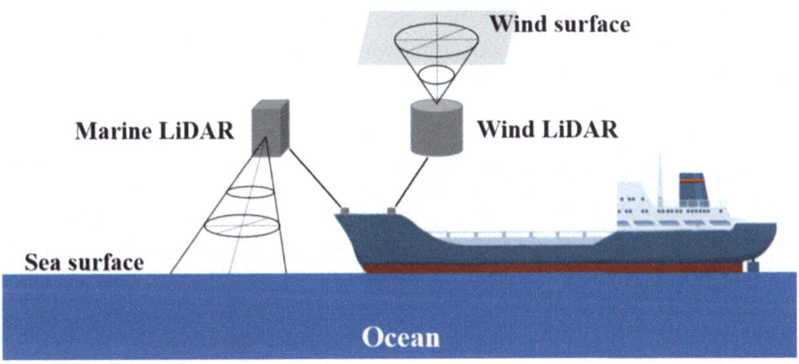

Fig. 4. Working diagram of semiconductor laser wind measurement radar and ocean monitoring lidar

At present, marine LiDAR is commonly used to monitor the marine water environment. Its principle is that a narrow-pulse high-power solid-state laser, LD-pumped, is directed to the ocean, and the reflection, scattering and induced luminescence occur, and the optical characteristic information is obtained by receiving the reflected light signal to analyze the physical or chemical parameters of the marine environment. Coherent Doppler Lidar is mainly aimed for the monitoring of the sea-surface wind field, which is realized by the monitoring of the sea-surface wind field by the LD-pumped solid-state or fiber laser, and generating echo signals from the atmospheric aerosol particles, and using the Doppler shift phenomenon to obtain information on atmospheric motion. The working modes of two kinds of LiDAR are shown in Fig. 4. The development of LIDAR technology and semiconductor laser technology is inseparable, at present, marine LIDAR commonly adopts high-power Nd^{3+}: YAG solid lasers pumped by 808nm semiconductor laser. High-power solid-state and fiber lasers pumped by 8xx–9xx nm semiconductor lasers are gradually applied to the field of coherent Doppler LIDAR monitoring of the sea surface wind field.

Australia is the first country using lidar to detect the marine environment. In 1986, Billard observed the stratification of seawater through the WRELADS II system [14]. Over the past two decades, the National Oceanic and Atmospheric Administration (NOAA) has used the airborne lidar FLOE based on high-power LD-pumped Nd: YAG solid-state lasers to continuously observe and study the ocean, collecting a large amount of marine monitoring data on marine fish stocks, water characteristics, internal waves, and plankton layers [15]; At the beginning of the 21st century, the U.S. Naval Aviation Systems Command developed the 532 nm Sea TROLL marine lidar system to study the thin layer of marine phytoplankton. The research on marine wind lidar technology has been carried out in the early 21st century. In 2004, the National Oceanic and Atmospheric Administration (NOAA) conducted a comparative test of a 2 μm laser wind radar with a variety of detection equipments [16]. In 2012, Japan's Mitsubishi Corporation used 3D laser wind radar to conduct long-term sea observation experiments to study its performance in the marine environment.

China's marine lidar research began at the end of the 20th century. Ocean University of China has developed a marine lidar BLOL based on Nd: YAG solid-state laser, and obtained the vertical corridor of the water body in the Bohai Sea of China [17]. In 2017, Zhejiang University developed a dual-wavelength ocean lidar system, which can indirectly analyze ecological conditions such as marine phytoplankton density, seawater eutrophication and red tide [18]. In 2017, the airborne lidar Mapper 5000 based on a 532 nm Nd: YAG solid-state laser developed by the Shanghai Institute of Optics and Mechanics, Chinese Academy of Sciences, has successfully detected the subsurface phytoplankton layer and sea-water depth in the South China Sea [19]. China's Doppler wind lidar technology has made some progress in the 21st century. In 2014, Wu Songhua's research group at Ocean University of China designed a 1550nm shipborne coherent Doppler wind radar to measure the wind field of the ocean-atmosphere boundary layer in the Yellow Sea [20]. In 2016, the airborne Doppler lidar developed by Ocean University of China realized the first unmanned helicopter-borne sea observation test in China, and carried out investigation on atmospheric wind field inversion and data analysis of the observed sea area [21]. In 2019, Yang Zehou and others from Beijing Institute of Technology carried out offshore wind field measurement and analysis of the 1550 nm shipborne light wind radar [22].

Compared with the traditional means of marine monitoring, LiDAR technology based on high-power semiconductor lasers have advantages such as non-contact telemetry, high spatial and temporal resolution, long monitoring distance, capable of all-weather marine environmental monitoring in a timely way to grasp the marine environment changes. And it can be flexibly mounted on the platforms of ships, airplanes, satellites and other platforms, providing an important basis for marine ecological environment monitoring, being of great significance in eco-nomic construction, social benefits, national defense, environmental protection and other aspects. Therefore, high-power semiconductor laser is considered as the "heart of remote sensing" for marine ecological environment detection.

3.3 Heavy Metal Contamination Detection

Currently, the world is facing serious heavy metal pollution of water environment, soil, nuclear waste and food, which jeopardizes the ecological environment and human life. Laser-induced breakdown spectroscopy (LIBS) based on high-power semiconductor lasers is a new method in the field of spectroscopic analysis, which can realize rapid, contact-free, real-time online monitoring of heavy metal contamination elements.

LIBS is currently a more advanced technique for heavy metal analysis. Figure 5 shows the schematic diagram of LIBS for heavy metal detection. The light source of this technology generally requires high-energy lasers, and at present, high-power LDs are commonly used as the pump source for 1064 nm or 532 nm Nd: YAG solid-state lasers, with pulse energies of about 10–1000 mJ, which are widely used in LIBS detection due to its high energy density of the laser pulses, high reliability, and low maintenance costs, and can be converted to different wavelengths through frequency conversion nonlinear crystals [23]. 1064 nm wavelength laser has a stronger breakdown spectrum at the same energy, there-fore, LIBS system usually adopts 1064 nm lasers [24], but this laser wavelength has serious attenuation in the water, so for heavy metals monitoring in the water, the most favorable laser wavelength is 532 nm [25].

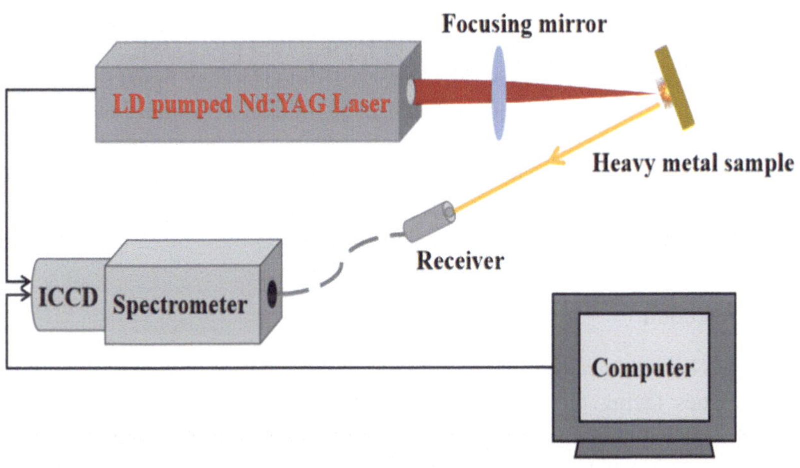

Fig. 5. Schematic diagram of LIBS heavy metal elements detection

With the gradual maturity of laser and spectrometer technology, LIBS technology has developed rapidly in environmental pollution monitoring and material classification and pretreatment at home and abroad. In 1984, Davida et al. took the lead in applying LIBS to the detection of metal element concentration in static liquids, which proved the feasibility of LIBS for the detection of different elements in water [26]. In 1994, Alexander et al. first published their LIBS detection results for heavy metal contamination including As, Cd, Cr, Hg, Pb and Zn in soil [27]. In 2016, Popov, a researcher at the Russian Academy of Sciences, used the LIBS system to detect Sr element in seawater [28]. In 2018, Rehan etc. used the LIBS system to detect toxic heavy metals such as Cu, Cr, and Ni in sea salt

[29]; In 2019, Sperança etc. used LIBS technology to detect Ti content in sunscreens, demonstrating that LIBS technology has great potential in detecting metallic elements in sunscreens [30].

The rapid development of LIBS technology in China began after the first international LIBS conference held by Ocean University of China in 2011 [31]. In 2013, Wan Xiong et al. from the Chinese Academy of Sciences, used the LIBS system to detect Pb and Ba elements in fish in heavy metal pollution areas, proving the feasibility of LIBS technology in the detection of marine biological metal elements [32]. In 2015, Zhong Shilei et al. from Qingdao University conducted a series of studies on Mn, Zn, Cu, Pb, Fe, Mg and Na in liquid samples using LIBS combined with ultrasonic atomization [33]. In the same year, Gong Gangmeng et al. from Jiangxi Agricultural University used LIBS to quantitatively analyze the heavy metal Cr element in milk powder, and the study showed that the high-resolution spectrometer can significantly improve the detection accuracy of Cr element [34]. In 2017, the Anhui Institute of Optics and Mechanics of the Chinese Academy of Sciences successfully developed a portable LIBS online detection system for heavy metal elements in soil [35]. From 2018, China began to study LIBS detection of nuclear pollution, the main research units engaged in this aspect are groups from Xi'an Jiaotong University and Dalian University of Technology [36]; In 2019, Ye Song et al. from Guilin University of Electronic Science and Technology used LIBS to detect trace amounts of Pb in drinking water [37]; In 2020, Rong Kai et al. from Xi'an Jiaotong University used LIBS for the measurement of Hg in mixed gases [38].

Compared with other analytical techniques, LIBS technology based on high power semiconductor lasers has the unique advantages of simultaneous multi-element detection, simple structure, fast detection speed, high sensitivity, independent of sample morphology, etc. It is capable of accurately and precisely online detecting heavy metal contaminating elements without contact. High-power semiconductor lasers are the "sensitive fruit flies" for metal detection in ecological environment.

3.4 Gas Detection

Currently, the detection of various gases in the ecological environment is one of the important issues facing the world. Tunable Diode Laser Absorption Spectroscopy (TDLAS) based on high-power LD has become one of the most promising technologies for gas detection, capable of qualitative and quantitative online real-time analysis.

TDLAS has the advantages of high detection performance, strong anti-interference ability, non-contact detection and low cost. The principle is to use a tunable LD as the light source with narrow linewidth to detect the absorption lines of specific gas molecules. Figure 6 shows the basic framework of the TDLAS system. TDLAS technology based on high-power semiconductor lasers began to be applied to the field of environmental gas detection in the 1990s. In the twenty-first century, there are more and more high-power semiconductor lasers applied to TDLAS technology, such as Distributed Feedback Lasers (DFBs), Vertical Cavity Surface Emitting Lasers (VCSELs), and Quantum Cascade Lasers (QCLs), etc., and the advent of these lasers greatly improved the detection limit of the gas concentration of TDLAS technology.

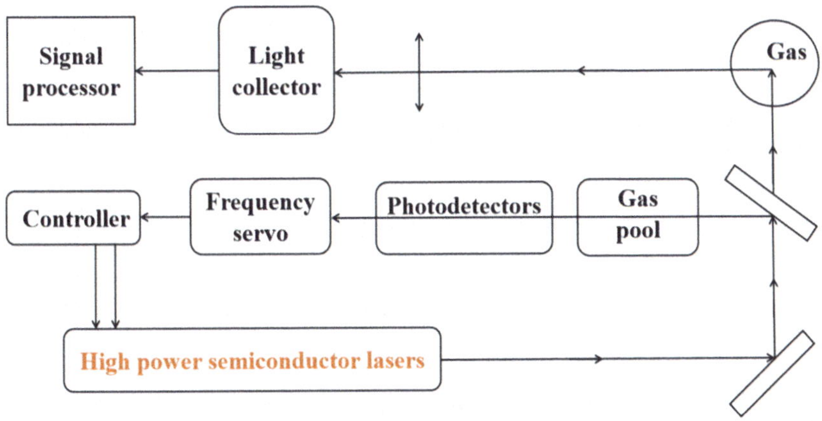

Fig. 6. Basic framework diagram of TDLAS system

In 1998, Monitor successfully developed the world's first TDLAS gas detector based on 1550 nm InGaAsP high-power LD, which can be used to detect ammonia concentration [39]; In 2009, Engel et al. used TDLAS technology to detect δ^{13} C(CO_2) in herbivore respiration to analyze the use of different plant re-sources [40]; In 2013, Frish et al. developed a miniaturized b-TDLAS sensor that can be used to monitor methane emissions from landfill plants [41]; In 2017, Ghorbani et al. used a gas sensor based on TDLAS to conduct real-time monitoring experiments on CO concentrations before and after smoking [42]; In 2019, Diemel et al. from the Technical University of Darmstadt, Germany, developed an in-situ TDLAS sensor for distinguishing the exhaust air temperature and molar fraction in the engine, which can measure six gas molecules at the same time, with a wavelength covering 1.4–5.2 μm [43]; In 2021, Woods Hole Oceanographic Institution and Massachusetts Institute of Technology developed a TDLAS sensor based on a 5 μm tunable diode laser to detect the concentration of carbon dioxide dissolved in the ocean [44].

TDLAS technology was introduced into China in the late 90s of the last century, and many universities and research institutes in China have caught up to the leading level in this field. In 2005, F. Dong et al. from the Anhui Institute of Op-tics and Mechanics of the Chinese Academy of Sciences used TDLAS technology based on 1.65 μm 5 mW tunable LD to detect the concentration of methane gas in the atmosphere [45], and in 2013, Zhang Jiawei et al. from Northeast Forestry University built a CO gas detection system based on 1.58 μm DFB LD for monitoring early forest fires, which is conducive to forest fire prevention in China [46]; In 2019, Shi Jintao of the University of Electronic Science and Technology of China developed a multi-component high-precision harmful gas detection prototype, using a single-mode DFB LD of 2–8 mW, which can simultaneously measure multiple gases [47]; In 2021, Dai Tongxin and others from the Changchun Institute of Optics and Mechanics of the Chinese Academy of Sciences dveloped a gas detection system based on InGaAs semiconductor lasers to indirectly control the greenhouse effect by accurately detecting the concentration of CO_2 gas in the atmosphere [48]; In the same year, the Ocean University of China reported a portable TDLAS system to study the Earth's carbon biological cycle by detecting dissolved CO_2 gas from the seawater surface [49].

The tunable semiconductor laser has high power and small line width com-pared with the traditional detection light sources, and the frequency and wave-length can be tuned to avoid other interferences when detecting gases. TDLAS technology with tunable semiconductor lasers as the light source has made re-markable progress in the field of ecological environment. Compared with other absorption spectroscopy technologies, it has high detection sensitivity, suitable for detection of a wide range of gases, fast response time and real-time online monitoring. High-power semiconductor lasers are the "Kiwi bird" of ecological environment gas detection.

3.5 Geological Monitoring

Geologic disasters often lead to huge destructive forces, which brings a great threat to the social and economic development and even to the survival of human beings. Three-dimensional laser scanning (TLS) based on high-power semiconductor lasers has the advantages of rapid response, high efficiency and reliability in geologic monitoring, and has special application value.

TLS is a remote sensing monitoring tool that emerged in the 1990s and works on the principle of laser ranging technology based on high-power LD. Figure 7 shows a schematic diagram of the geological monitoring by this technology. The system has many carriers, like satellight-borne, airborne, ship-borne, and ground-based, among which the ground-based carrier has outstanding features and wide application fields, and is the main application mode. In the field of geological monitoring, due to the limitation of terrain conditions, the most representative 3D laser scanning system is the terrestrial three-dimensional laser scanner and three-dimensional laser radar. For the present, for more and more cases, the technology has been favorably applied for landslide monitoring, soil erosion, geomorphological deformation and other geological monitoring.

The 3D laser scanner is the core component of the 3D LIDAR system. The measurement work of 3D laser scanner is mainly accomplished by ranging, angular measurement, scanning and image acquisition and other parts working in a coordinated way. In addition, the laser ranging unit is the core component of the 3D laser scanner, and its ranging principle can be divided into four ways: triangulation, pulsed, phased and pulsed-phased, of which the pulsed ranging laser scanning system is mainly used in geological monitoring applications. The system usually adopts four types of laser sources: liquid, gas, solid-state and LD. With the rapid development of semiconductor technology, GaAs LD are generally used.

The development of 3D laser scanning technology has always been attached great importance at home and abroad, and with the gradual maturity of this technology, the application of geological monitoring and disaster prediction has gradually emerged. In 2003, Stephen A. White and Yong Wang of East Carolina University used 3D laser scanning technology to monitor the erosion of the coastline of North Carolina, USA, and obtained preliminary data on coastal erosion, which proved the feasibility of this technology in soil erosion monitoring [50]; In 2006, Dong Xiujun et al. from Chengdu University of Technology applied ground-based TLS to the geological monitoring of high and steep slopes without vegetation cover, and put forward the application value

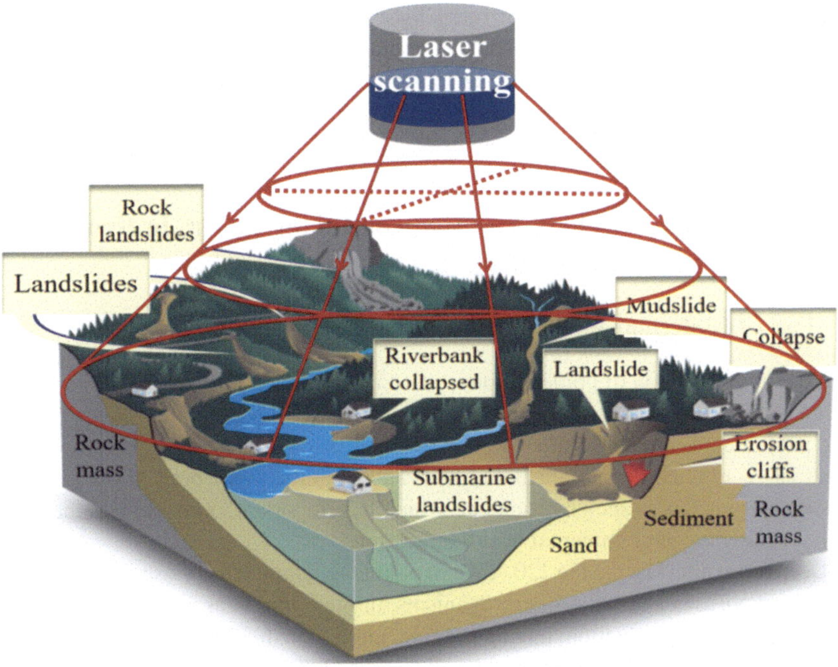

Fig. 7. Schematic diagram of geological monitoring

and prospect of this technology in geological landslide monitoring [51]. In 2009, Ding Liangang et al. from Beijing Normal University reported the use of a 3D laser scanner (maxi-mum scanning distance 350 m, speed 5000 pts/s) to monitor the erosion change state in the grass lattice sand barrier, which promoted the application of this technology in environmental desertification control [52]. In 2006–2010, R. Kenner et al. used ground-based laser scanning technology to observe the melting of a glacier on a permafrost rock wall in the Alps, and successfully monitored the millimeter rate of glacier ablation [53]; In 2012, the China Center for Aerogeophysical Prospecting and Remote Sensing of Land and Resources reported an airborne lidar system (3000 m) being used to carry out dynamic monitoring and application experiments on geological landslides in the reservoir area of the Three Gorges Project of the Yangtze River, which provided strong technical support for the monitoring of geological disasters such as landslides [54]. In 2015, Kromer et al. conducted long-term geological monitoring of rockfall disasters in the northern mountains of Canada based on TLS technology [55]. In 2017, Hu Lei et al. obtained, in a quick way, the topographic data of sudden geological disasters through a 3D laser scanner (1200 m, 59 pts/s). In 2019, Jian Ruimin used a 3D laser scanner to monitor the deformation of tunnels, comparing it with traditional monitoring technology, testified the application prospect of 3D laser scanning technology in geological deformation monitoring [56].

Three-dimensional laser scanning technology based on high-power semiconductor lasers is both reliable and highly valuable. Compared with the traditional measurement technology, it has the advantages of being lightweight, fully auto-mated, non-contact, high precision, high speed and all-weather measurement, capable of all-round monitoring of complex geological environments, enhancing prevention and mitigation of disasters and protection of the ecological environment. High-power semiconductor laser is the "throat bag of bats" in geological monitoring.

4 Application of High-Power Semiconductor Lasers in Ecological and Environmental Protection

4.1 Laser Disinfection and Sterilization

With the normalization of epidemic prevention and control, the disinfection and sterilization of air, water, public places and other environmental health fields have gradually been paid attention to. Laser disinfection and sterilization technology based on high-power LD is efficient, economical and safe, and can effectively prevent the spread of bacteria and viruses in the environment, which has important application value in the field of ecological environmental health protection.

Deep ultraviolet laser in the 220–280 nm wavelength range can be used for irradiation of bacteria, viruses and other microorganisms, through the breakage of their chemical bonds to prevent reproduction, having a significant role in disinfection and sterilization. Near-infrared light and heat source with wavelength of 700–1000 nm can be used to sterilize and disinfect bacteria through the photo-thermal effect and deterioration of bacterial proteins. Figure 8 shows a schematic diagram of laser sterilization. Currently used for sterilization laser light source is mainly the LD-pumped deep ultraviolet all-solid-state laser, III-nitride deep ultraviolet semiconductor laser and near-infrared semiconductor laser, of which the deep ultraviolet all-solid-state laser, III-nitride deep ultraviolet semiconductor laser can work on the surface of objects and the viruses and bacteria in the air for rapid elimination, with little or no limitation. Near-infrared semiconductor laser is mainly applied in medical fields such as laser therapy and rapid assisted sterilization of medical devices.

At present, the R&D and application of laser devices for disinfection and sterilization have become a research hotspot both at home and abroad. In 2006, Su Yanli et al. from Shandong Normal University used semiconductor lasers to pump Nd: YVO_4 crystals to develop a 213 nm acousto-optic Q-switched all-solid-state deep ultraviolet laser with an LD pumping power of 10.3 W [57], and in 2014, Zhu et al. used a 115.1 W high-power semiconductor laser to side-pump Nd: YAG/Cr^{4+}: YAG/YAG composite crystals to obtain a 266 nm deep ultraviolet laser output with an average power of 1.27 W [58]; In 2016, Shahid Beheshti Medical University in Iran used an 810 nm high-power semiconductor laser to conduct auxiliary experiments in root canal retreatment, and the results showed that this technology has effective disinfection and antiseptic effect [59]. In 2017, Cheng et al. used a 980 nm semiconductor single-mode fiber-coupled laser to emit a 5–10 W high-power infrared laser to sterilize the surface of metal medical devices [60]; In 2019, Zhang et al. of Japan developed a 271.8 nm deep ultraviolet semiconductor laser with disinfection and sterilization effect, with a pulse duration of 50 ns and a repetition rate of 2 kHz [61]; In 2022, Zhao Zhibin et al. of Hainan Normal

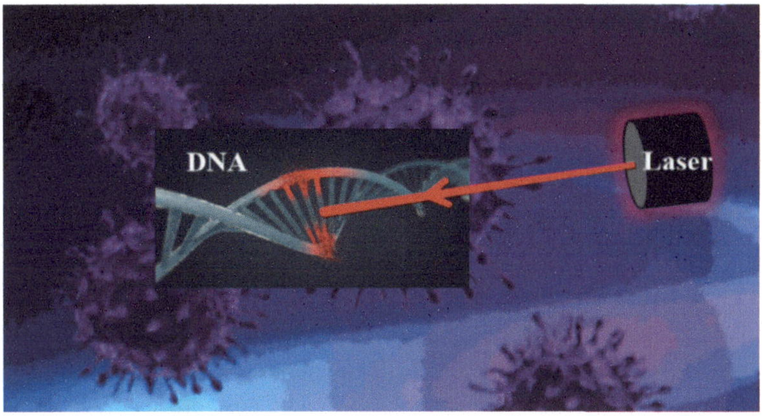

Fig. 8. Schematic diagram of laser sterilization

University used a self-developed all-solid-state 228 nm far-ultraviolet pulsed solid-state laser to conduct low-light irradiation experiments on Escherichia coli and Bacillus, and the bacterial inactivation rate was as high as 100%, and the results proved that the 228 nm far-ultraviolet pulsed laser has strong antibacterial properties. In Zhao's experiments, the solid-state laser they used adopts a high-power diode laser to pump an Nd^{3+} laser crystal to achieve laser output [62].

The development of laser sterilization technology cannot be separated from the advancement of semiconductor lasers. Compared with the traditional disinfection and sterilization method, the laser disinfection and sterilization technology based on high-power semiconductor lasers can be carried out in a lower-temperature environment, avoiding the damage that may be caused to items by sterilization in a high temperature environment. Secondly, it does not require the assistance of chemical substances, avoiding the potential risk of chemical residues. In addition, the laser disinfection and sterilization technology has the characteristics of high efficiency, rapidity and reliability, and the sterilization process can be completed in a short period of time, effectively preventing the spread of bacteria and viruses in the ecological environment, and has a wide range of application prospects in environmental protection and medicine. High-power semiconductor lasers are the "holy light" for sterilization of viruses in the ecological environment.

4.2 Laser Weeding

Traditional weeding methods mainly rely on manual weeding and chemical herbicides, obviously, manual weeding is laborious, time-consuming and labor-intensive, and cannot meet the actual demand, while the use of chemical herbicides may cause pollution to soil and water sources. Laser weeding technology based on high-power semiconductor lasers is a new type of safe, efficient and labor-saving contactless physical weeding method, which is of great value in the field of environmental protection.

Figure 9 shows the schematic diagram of laser weeding. Through advanced image recognition and machine learning algorithms, the technology is able to identify weeds with a high degree of accuracy, capable of focusing high-power laser beams on the roots of weeds, destroying the cellular structure of weeds through high-energy laser beams, avoiding damage to crops and realizing the effect of weeding. Laser weeding technology has a very wide range of applications and can cope with various types of weeds without producing chemical residues harmful to the environment. The laser light source used in previous laser weed control technologies was the CO_2 laser and fiber lasers. However, CO_2 lasers and fiber lasers are large and expensive, so CO_2 and fiber lasers are not advantageous in terms of portability and cost. Blue light band laser has the characteristics of short wavelength, small diffraction, etc. Therefore, at present, the widely used high-power LD-pumped all-solid-state blue laser and blue light LD have unique characteristics in terms of size, weight, cost, high efficiency and reliability, which make them more suitable for use in the field of laser industry compared with the CO_2 and fiber lasers. Their uniqueness in terms of size, weight, cost, efficiency, and reliability makes them easier to integrate into application systems than CO and fiber lasers. At the same time, the use of relatively low power LD-pumped all-solid-state blue lasers and blue diode lasers offers fewer operational hazards and a higher safety margin than higher power and more complex CO_2 and fiber lasers, considering that the laser beam is converted to heat when it acts a surface, and high-energy lasers can even ignite the application site and potentially cause thermal injuries to people.

Fig. 9. Schematic diagram of laser weeding

Laser weeding technology based on high-power semiconductor lasers has been applied at home and abroad in recent years. In 2020, Russian researchers developed a laser weeding device for removing weeds around tomato plants, using a LD-pumped solid-state blue laser with an output power of 5 W [63]; In 2022, Mu Danlei of Kunming University of Science and Technology proposed a weeding scheme based on blue laser and designed a laser weeding robot system to test its weeding effect, and the weeding light source used in the system is a blue high-power LD with an output power of 10–18 W [64]; In the same year, Mwitta Canicius et al. from the University of Georgia in the United States used blue LDs in six different powers (1.2 W, 1.35 W, 4.2 W, 4.5 W, 5.1 W, 6.1 W) to verify the effectiveness of laser weeding [65].

Compared with traditional weed control methods, laser weeding technology based on high-power LDs has unique advantages such as high efficiency, precise positioning, low seedling injury rate and low environmental impact. It reduces the use of chemical herbicides while improving crop quality and yield, effectively reduces the risk of pollution to the ecological environment and human society, and playing increasing important role in environmental protection. High-power semiconductor lasers are the "invisible scissors" of ecological environment and vegetation protection.

4.3 Laser Bird Repellent

With the improvement of the ecological environment at home and abroad, human awareness of the protection of birds has increased, laser bird repellent technology based on high-power semiconductor lasers has been widely used in the field of ecological prevention and control, aviation, power grids, agriculture and bird safety and other areas of ecological protection. This is a kind of environmentally friendly bird repellent technology that is completely harmless to birds and the ecological environment.

It has been found by scientific research that the vision of birds has four or five primary colors, and they can see in the spectral range of 300–680 nm, and different birds' vision has some differences in sensitivity to the spectral wavelengths. But most birds are most sensitive to the 532 nm wavelength of the green laser. The technique generally uses a 532 nm green laser as the light source in the laser bird repellent device. A schematic diagram of laser bird repellent is shown in Fig. 10. The principle of the technology is that in the case of birds near an area, as birds are highly sensitive to the green spectrum, using an expanded, collimated 532 nm green laser beam, and continuously sweeping through the surrounding area, like a huge green stick, then by the stimulation of bright green laser beams, and out of the survival instinct, the bird flew away from the protected area. The laser for the laser bird repeller complies with the national IEC60825–1 Editional 1.2 standard, and is not harmful to human eyes.

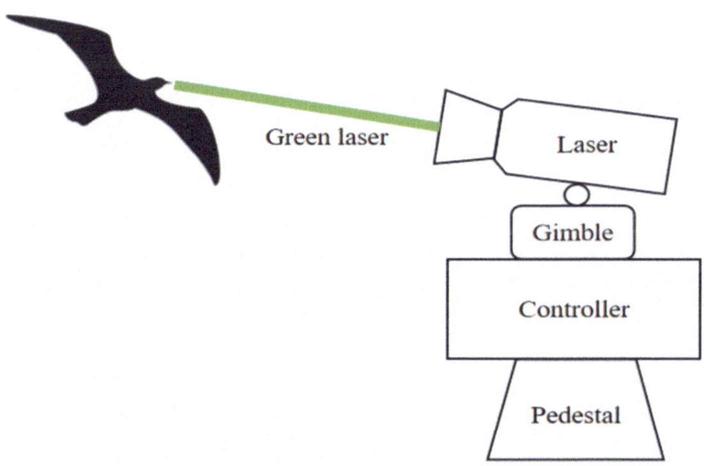

Fig. 10. Schematic diagram of laser bird repellent

At present, the 532 nm green laser used in the laser bird repellent device is generally produced by the 808 nm high-power LD pumping solid-state laser. From the perspective of economy and practicability, this green laser has the best bird repellent effect, not only cheaper, the technology is more mature, with good beam quality, compared with other lasers. The commercialization of semiconductor lasers in the 90s of the 20th century led to the gradual development of laser bird repellent technology. At present, laser bird repellent technology based on high-power LD-pumped solid-state lasers at home and abroad has gradually been commercialized. In 2017, Brown et al. from the University of Rhode Island in the United States designed a 532 nm green solid-state laser using an 808 nm LD-pumped Nd:YVO_4 crystal, which can drive away pest birds that harm sweet corn and reduce the degree of crop loss in the field [66]; Suzhou Aisilan Optoelectronics Co., Ltd., a domestic enterprise, has developed a variety of intelligent auto-matic laser bird repellers based on handheld or tower type, the light sources are all 520–532 nm green LDs with output power in the range of 0.5–3 W, which are suitable for indoor and outdoor environments such as air-ports, power grids, orchards, etc.; The KY-JGQNQ PTZ laser bird repellent developed by Beijing Shuimu Kaiyuan Technology Co., Ltd. is such an equipment designed and produced for airport bird prevention, which disperses birds through a specific wavelength laser, so as to achieve the purpose of bird repellent without harm. The equipment uses a 532 nm green laser light source, 1000 mW power, which has a long working distance and can work continuously around the clock, effectively ensuring the operation safety of the airport environment.

Compared with other methods, the laser bird repellent technology based on high-power semiconductor lasers can achieve the purpose of bird repellent by stimulating the birds' vision, with wide coverage and lasting effect, which can effectively scare away birds to look for new, safe habitat. This ecological control method is conducive to reducing bird harm, with good environmental effects. High-power semiconductor laser is the "invisible shield" of ecological environment protection.

4.4 Laser Removal

In recent years, with the wide application of laser technology in various industries, the trend of miniaturization and cheapening of semiconductor lasers is obvious. Laser removal technology based on high-power semiconductor lasers can be applied to remove pollutants that can be found everywhere in daily life and plays a very important role in environmental protection.

Figure 11 shows the various application scenarios of laser removal technology. Its principle is to shine a high-energy laser beam on the target object where the laser thermal evaporation effect occurs, making the target temperature rise to melting, evaporation. In existing research and applications, the core of laser removal technology is the high-power laser. The light sources commonly used in laser removal technology are mainly CO_2 lasers, fiber lasers, and semiconductor lasers. CO_2 laser with wavelength 10.6 μm, works better for non-metallic materials generally, though, CO_2 laser without water-cooling can not work continuously and its beam quality is not good, and can be used only for close-range clearing. For fiber lasers, the atmospheric transmittance in the 1.06 μm band is greater than 90%, and it works most efficiently in laser clearance applications. Currently, erbium-doped fiber laser pump source is mainly 980 nm or

1480 nm LDs, ytterbium-doped fiber laser pump source is mainly 915 nm or 976 nm LDs. Compared with CO_2 lasers, fiber lasers have the advantages of miniaturization, convenient for field operations, high efficiency, especially the LD technology is getting increasingly mature, high-power LD pumped fiber lasers have gradually begun large scale applications. Therefore, semiconductor lasers and LD-pumped fiber lasers have gradually replaced CO_2 lasers for laser removal applications.

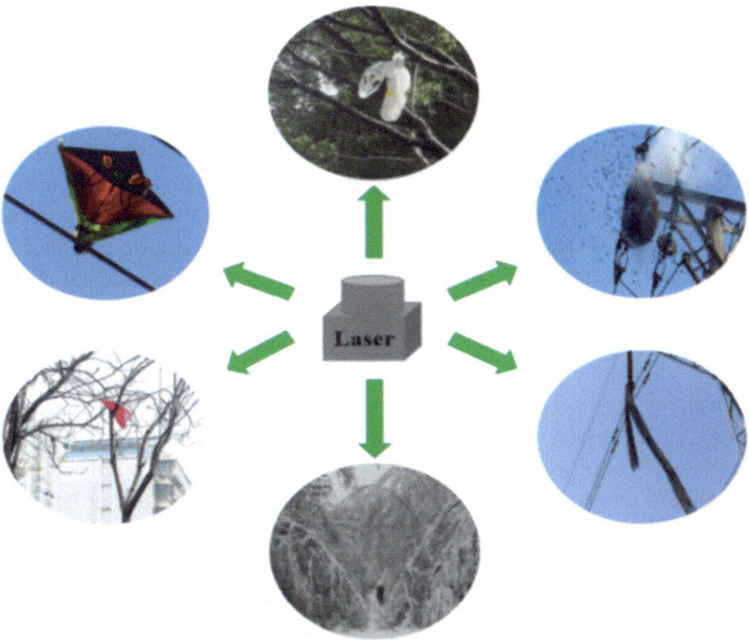

Fig. 11. Application scenarios of laser removal technology

Tsinghua University and China Southern Power Grid Research Institute has co-operated to develop a semiconductor laser wrecker de-icing instrument with a total output power of 300 W, the laser device in the equipment has a total of five sub-modules, each sub-module uses 980 nm wavelength LDs with a maximum output power of 60 W, and the equipment can effectively remove the frozen ice on the road surface and the icing on the power equipment in bad weather [67]; In 2017, the Shenzhen Power Supply Bureau of China Southern Power Grid reported, for the first time at home and abroad, that a laser wrecker removal device based on the UAV platform was successfully developed, which used end-face pumping technology to pump the doped Yb fiber with a high-power semiconductor laser array, and obtained a high-efficiency, 100 W high-output power fiber laser source, the final laser output power of the device is 80 W, which can re-move a variety of substances that pollute the environment such as fishing lines, kite lines, and nylons [68].

Compared with other types of lasers, semiconductor lasers have remarkable directionality, instantaneous high energy and long transmission distance. Com-pared with other methods such as manual and robotic removal, the laser removal technology based on high-power semiconductor lasers has the advantages of non-contact, high speed and high accuracy, which can significantly reduce labor cost and save material resources, enhance the safety of the work, and efficiently clean up obstacles or pollutants in the ecological environment, which is conducive to the protection and improvement of the ecological environment. High-power semiconductor lasers are the "Vimalakirti - God of Purification" of ecological environment protection.

5 Summary and Outlook

Semiconductor lasers are very versatile optoelectronic devices that can be assembled and combined with other systems and equipments better than other types of lasers due to their compactness and direct electric injection pumping. This paper focuses on the unique advantages of high-power LDs, such as small size, easy tuning, high power output and energy conversion efficiency, and briefly describes the domestic and international research on its use as ideal light sources in ecological and environmental monitoring and protection fields such as micro-nano satellite ecological remote sensing, marine environmental monitoring, heavy metal pollution detection, gas detection, geological monitoring, laser disinfection, laser sterilization, laser weed control, laser bird repellent, laser removal, and so on. The current development status in respective fields have been summarized and discussed. It can be seen that the applications of high-power semiconductor lasers cover almost every where from space to the ocean, from macro to micro. From the point of view of the development level, China's semiconductor laser technology, in the early stage of development, kept in synchronization with those of advanced countries. Although the process was hampered by lagging industrial development, with the rapid development of progress of the industry in the new century, has finally caught up with the world leading level. In general, with the continuous development of high-power LDs, the technology exploration of ecological environment monitoring and protection is in full swing the world over with the fierce market competition.

With the continuous increase of the demand for laser light sources in various fields, the pursuit of high-power semiconductor lasers is never-ending. For example, the future development will be in the direction of single wavelength or a narrow line width, together with pioneering new process and new type of combining technology to make a single wavelength laser output, with order of magnitude increase in the spectral characteristics. The efforts can also put to strive for further optimization of beam quality and brightness by optical design and cavity coating technology to improve the intensity of the excitation beam on the surface of the target, so as to achieve faster and more efficient monitoring and protection. In addition, we can focus on research and development of new materials, new structure of the semiconductor lasers, from the ultraviolet to the far infrared to achieve the full wavelength laser output, in that case, much more molecules can be fully covered for convenient detection, with "one device, multiple inspection" of the commercial value. In the future, the device performance parameters, internal structure and appearance device can also be personalized according to application requirements for customized, modular configuration. In addition, the future of high-power semiconductor lasers can not only be applied in the field of ecological environment monitoring and protection, but also to develop other ap-plication scenarios, active in the infrared countermeasures, laser guidance, laser communications, laser medical, laser cladding, and other industrial production, national defense and military, aerospace, medical and scientific research fields. High-power semiconductor laser will be our new tool to create a better world.

References

1. Elahi, P., AkÇaalan, Ö., Ertek, C., Eken, K., Ilday, F.Ö., Kalaycoğlu, H.: High-power Yb-based all-fiber laser delivering 300 fs pulses for high-speed ablation-cooled material removal. Opt. Lett. **43**(3), 535–538 (2018)
2. Hall, R.N., Fenner, G.E., Kingsley, J.D., Soltys, T.J., Carlson, R.O.: Coherent light emission from GaAs junctions. Phys. Rev. Lett. **9**(9), 366–368 (1962)
3. Nathan, M.I., William, P.D., Burn, G., Lasher, G.: Stimulated emission of radiation from GaAs p-n junctions. Appl. Phys. Lett. **1**(3), 62–64 (1962)
4. Quist, T.M., et al.: Semiconductor maser of GaAs. Appl. Phys. Lett. **1**(4), 91–92 (1962)
5. Li, H., Reinhardt, F., Chyr, I., et al.: High-efficiency, high-power diode laser chips, bars, and stacks. In: Proceedings of Lasers and Applications in Science and Engineering, **6876**, pp. 68706G. San Jose, California (2008)
6. Liu, G., Lehkonen, S., Xu, Z., et al.: High power 808 nm to 1060 nm CW and QCW laser diode bars. SPIE LASE, **10900**, pp. 10900B. San Francisco, California (2019)
7. Skidmore, J., Peters, M., Rossin, V., Guo, J., Xiao, Y., Cheng, J., et al.: Advances in high-power 9XX nm laser diodes for pumping fiber lasers. SPIE LASE, **9733**, pp. 97330B. San Francisco, California (2016)
8. Xia, F., et al.: Spaceborne miniaturized laser communication terminal's current situation and development trend. Space Electronic Technol. **17**(03), 73–80 (2020)
9. d'Humières, B., Esmiller, B., Gouy, Y., Steck, E., Quintana, C., et al.: The C3PO project:a laser communication system concept for small satellites. SPIE LASE, **10096**, p. 1009611. San Francisco, California (2017)
10. Richards, G.: Satellite digest 554. Space Flight **3**, 38–42 (2019)
11. Yun, X.: First star of "Zhuhai-1" Remote Sensing Micro-Nano-Satellite constellation. Satellite Application (06), 78 (2017)
12. Li, X., Fan, H., Pan, S., Jiang, X., Wu, J.: Overview of Zhuhai-1 hyperspectral satellite data and application. Satellite Appl. **08**, 12–18 (2019)
13. Sun, W., et al.: Development status and literature analysis of China's earth observation remote sensing satellites. J. Remote Sensing **24**(05), 479–510 (2020)
14. Billard, B.: Remote sensing of scattering coefficient for airborne laser hydrography. Appl. Opt. **25**(13), 2099–2108 (1986)
15. Churnside, J.H., Marchbanks, R.D.: Subsurface plankton layers in the Arctic Ocean. Geophys. Res. Lett. **42**(12), 4896–4902 (2015)
16. Wolfe, D.E., Brewer, W.A., Tucker, S.C., et al.: Shipboard multisensor merged wind profiles from the New England air quality study 2004. J. Geophysical Research Atmospheres **112**(D10), 7344(2007)
17. Wu, D.: Ocean Lidar (BLOL) for Measuring Chlorophyla Concentration, Diffuse Attenuation Coefficient and Water Leaving Radiance. ActaOptica Sinica, 1001 (1998)
18. Li, K., He, Y., Hou, C., et al.: Detection of chlorophyll profiles from coastal to oceanic water by dual-wavelength ocean lidar. Chin. J. Lasers **48**(20), 156–166 (2021)
19. He, Y., Hu, S., Chen, W., et al.: Research progress of domestic airborne dual-frequency LiDAR detection technology. Progress in Laser and Optoelectronics **55**(08), 7–17 (2018)

20. Zhai, X., Wu, S., Liu, B., Song, X., Yin, J.: Shipborne wind measurement and motion-induced error correctionby coherent doppler lidar over yellow sea in 2014. Atmospheric Measurement Techniques **11**(03), 1313–1331 (2018)
21. Wang, Q., Wu, S., Zhang, H., et al.: Observation and data processing of offshore wind field based on UAV-borne doppler lidar. Infrared Millim. Waves **40**(04), 516–529 (2021)
22. Zhou, J., Yang, Z., Song, S., et al.: Shipborne wind lidar for regional 3D wind field measurement. Infrared and Laser Eng. **49**(S2), 49–55 (2020)
23. Huang, Y.: Research on the Method of Improving the Detection Accuracy of Heavy Metals in Soil Based on LIBS. Changchun University of Technology, Changchun (2023)
24. Bi, Z.: Research on Laser Spectrum Detection System for Marine Oil Pollution and Metal Elements. Harbin Institute of Technology, Harbin (2020)
25. Li, Y., Wang, Z., Wu, J., et al.: Effects of laser wavelength on detection of metal elements in water solution by laser induced breakdown spectroscopy. Spectroscopy and Spectral Analysis **32**(03), 582–585 (2012)
26. Cremers, D.A., Radziemski, L.J., Loree, T.R.: Spectrochemical analysis of liquids using the laser spark. Appl. Spectrosc. **38**(5), 721–729 (1984)
27. Alexander, D.R., Poulain, D.E., Ahmad, M.U., Kubik, R.D., Cespedes, E.R.: Environmental monitoring of soil contaminated with heavy metals using laser-induced breakdown spectroscopy. In: IGARSS'94–1994 IEEE International Geoscience and Remote Sensing Symposium **2**, pp. 767–769, Pasadena, CA, USA (1994)
28. Popov, A.M., Drozdova, A.N., Zaytsev, S.M., et al.: Rapid, direct determination of strontium in natural waters by laser-induced breakdown spectroscopy. J. Anal. At. Spectrom. **31**(5), 1123–1130 (2016)
29. Rehan, I., et al.: Determination of toxic and essential metals in rock and sea salts using pulsed nanosecond laser-induced breakdown spectroscopy. Appl. Opt. **57**(2), 295–301 (2018)
30. Speranca, M.A., Andrade, D.F., Castro, J.P., Pereira-Filho, E.R.: Univariate and multivariate calibration strategies in combination with laser-induced breakdown spectroscopy (LIBS) to determine Ti on sunscreen: a different sample preparation procedure. Opt. Laser Technol. **109**, 648–653 (2019)
31. Wang, Q.: Highly Sensitive Analysis of Heavy Metals in Water by Discharge-Assisted Laser-Induced Breakdown Spectroscopy. Jilin University, Jilin (2023)
32. Wan, X., Wang, J., Ye, J., et al.: Analysis of distribution and contents of heavy metal pollution in fish body with laser-induced breaddown spectroscopy. Spectroscopy and Spectral Analysis **33**(01), 206–209 (2013)
33. Zhong, S., Zheng, R., Lu, Y., Cheng, K., Xiu, J.: Ultrasonic nebulizer assisted LIBS: a promising metal elements detection method for aqueous sample analysis. Plasma Sci. Technol. **17**(11), 979 (2015)
34. Gong, G., Yang, J.: Quantitative analysis of Cr in milk powder by laser induced breakdown spectroscopy. J. Anhui Agri. Sci. **43**(22), 219–222 (2015)
35. Meng, D., et al.: Application of a mobile laser-induced breakdown spectroscopy system to detect heavy metal elements in soil. Appl. Opt. **56**(18), 5204–5210 (2017)
36. Wu, J., Yu, H., Qiu, Y., et al.: Plasma characteristics and element analysis of steels from a nuclear power plant based on fiber-optic laser-induced breakdown spectroscopy. J. Phys. D Appl. Phys. **52**(1), 014006 (2019)
37. Ye, S., Tao, Y., Dong, D.: Rapid measurement of trace lead in water by laser-induced breakdown spectroscopy coupled with super-hydrophobic nano-substrate. J. Lasers **40**(02), 35–39 (2019)
38. Rong, K., Wang, Z., Hu, R., et al.: Experimental study on mercury content in flue gas of coal-fired units based on laser-induced breakdown spectroscopy. Plasma Sci. Technol. **22**(07), 88–95 (2020)

39. He, X., Pan, Y.: Research on the development status of TDLAS technology and its application. J. Ezhou University **24**(01), 102–104+109 (2017)
40. Engel, S., Lease, H.M., McDowell, N.G., Corbett, A.H., Wolf, B.O.: The use of tunable diode laser absorption spectroscopy for rapid measurements of the $\delta^{13}C$ of animal breath for physiological and ecological studies. Rapid Commun. Mass Spectrom. **23**(9), 1281–1286 (2009)
41. Frish, M.B., Wainner, R.T., Laderer, M.C., et al.: Low-cost lightweight airborne laser-based sensors for pipeline leak detection and reporting. SPIE Defense, Security, and Sensing, **8726**, p. 87260, Baltimore, Maryland, United States (2013)
42. Ghorbani, R., Schmidt, F.M.: ICL-based TDLAS sensor for real-time breath gas analysis of carbon monoxide isotopes. Opt. Express **25**(11), 12743–12752 (2017)
43. Diemel, O., Honza, R., Ding, C., Böhm, B., Wagner, S.: In situ sensor for cycle-resolved measurement of temperature and mole fractions in IC engine exhaust gases. Proc. Combust. Inst. **37**(2), 1453–1460 (2018)
44. Colson, B., Michel, A.P.M.: Tunable diode laser absorption spectroscopy instrument for flow-through measurement of dissolved carbon dioxide in the ocean. 2021 Conference on Lasers and Electro-Optics, pp. 1–2, San Jose, CA, USA (2021)
45. Dong, F., Kan, R., Liu, W., et al.: Tunable diode laser absorption spectroscopic technology and its applications in air quality monitoring. Chinese J. Quantum Electronics **2005**(03), 315–325 (2005)
46. Zhang, J., Zhang, H., Li, M.: TDLAS-based early-stage forest fire detection system. Forest Eng. **29**(02), 139–142 (2013)
47. Shi, J.: Research on Concentration Detection Technology of High-Precision Toxic and Harmful Gas based on TDLAS Technology. University of Electronic Science and Technology of China, Chengdu (2019)
48. Dai, T., Wang, B., Lian, H., et al.: Development of photoelectric detection circuit for TDLAS CO_2 gas detection system. J. Lasers **42**(04), 41–45 (2021)
49. Zhang, Z., Li, M., Guo, J., Du, B., Zheng, R.: A portable tunable diode laser absorption spectroscopy system for dissolved CO_2 detection using a high-efficiency headspace equilibrator. Sensors **21**(5), 1723 (2021)
50. White, S.A., Wang, Y.: Utilizing DEMs derived from LIDAR data to analyze morphologic change in the North Carolina coastline. Remote Sens. Environ. **85**(1), 39–47 (2003)
51. Dong, X., Huang, R.: Application of 3D laser scanning technology to geologicsurvey of high and steep slope. Chin. J. Rock Mech. Eng. **S2**, 3629–3635 (2006)
52. Ding, L., Yan, P., Du, J., et al.: Monitoring the state of erosion and deposition in straw checkerboard barriers based on 3D laser scanning technique. Science of Surveying and Mapping **34**(02), 90–92 (2009)
53. Kenner, R., Phillips, M., Danioth, C., Denier, C., Thee, P., et al.: Investigation of rock and ice loss in a recently deglaciated mountain rock wall using terrestrial laser scanning: Gemsstock, Swiss Alps. Cold Reg. Sci. Technol. **67**(3), 157–164 (2011)
54. Liu, S., Guo, D., Chen, W., Zheng, X., Wang, S.: The application of airborne lidar technology in landslide investigation and monitoring of three gorges reservoir area. China Geology **39**(02), 507–517 (2012)
55. Kromer, R.A., Hutchinson, D.J., Lato, M.J., Gauthier, D., Edwards, T.: Identifying rock slope failure precursors using LiDAR for transportation corridor hazard management. Eng. Geol. **195**(2015), 93–103 (2015)
56. Jian, R.: Research on deformation monitoring method based on 3D laser scanning technology. Western Resources **2019**(03), 197–198 (2019)
57. Su, Y., He, J., Jiang, Q., et al.: Efficient 213 nm radiation fifth harmonic generation of a laser diode pumped Nd: YVO_4 laser. Chin. J. Lasers **2006**(12), 1590–1592 (2006)

58. Zhu, S., Chen, Z., Chen, Z., Jiang, W., Wang, S., et al.: A LD side-pumped deep ultraviolet laser at 266 nm by using a Nd: YAG/Cr^{4+}: YAG/YAG composite crystal. Opt. Laser Technol. **63**(2014), 24–28 (2014)
59. Mohammad, A., lMostafa, G., Saranaz, A.M., Sadat, T.F., Maryam, B.: Comparison of the antibacterial effect of 810 nm diode laser and photodynamic therapy in reducing the microbial flora of root canal in endodontic retreatment in patients with periradicular lesions. J. Lasers in Medical Sci. **7**(2), 99104 (2016)
60. Cheng, H., Zeng, L., Huang, Z., Chen, Y., Chen, T.: Sterilizing effect of 980nm continuous laser on metal surface of medical equipment. J. Third Military Medical Univ. **39**(09), 878–883 (2017)
61. Zhang, Z., Kushimoto, M., Sakai, T., et al.: A 271.8 nm deep-ultraviolet laser diode for room temperature operation. Applied Physics Express **12**(12), 124003 (2019)
62. Zhao, Z., Cheng, C., Jin, Y., et al.: Inactivation effect of all-solid-state 228 nm far-UVC pulsed laser. Chin. J. Lasers **49**(15), 212–215 (2022)
63. Ildar, R., Christian, A.: A concept of a compact and inexpensive device for controlling weeds with laser beams. Agronomy **10**(10), 1616 (2020)
64. Mu, D.: Research on the Actuator and Recognition Algorithm of Laser Weeding Robot. Kunming University of Science and Technology, Kunming (2022)
65. Canicius, M., Glen, R.C., Eric, P.: Evaluation of diode laser treatments to manage weeds in row crops. Agronomy **12**(11), 2681 (2022)
66. Nelson, B.R., David, B., Robotic laser scarecrows: a tool for controlling bird damage in sweet corn. Crop Protection **146**, 105652 (2021)
67. Zhao, Y., Zhang, G., Luo, B., et al.: The technology of high power semiconductor laser deicing. China Southern Power Grid Technol. **5**(05), 60–64 (2011)
68. Yu, P., Tian, J.: Integrated design of UAV airborne laser obstacle removal. Environmental Technol. **38**(05), 195–200+206 (2020)

A Color Calibration Method for Multispectral Camera of Long Focal Length and Large Aperture

Yong-Xiang Guo(✉), Xing-Lin Zheng, Bing-Jie Liu, Mu-Qing Wang, Jun-Yu Ke, Jing-Yi Wang, Yun-Fei Li, and Mai-Ling Xing

Beijing Institute of Space Mechanics and Electricity, Beijing, China
gyxllr605@163.com

Abstract. With the development of technology in remote sensing, a lot of multispectral cameras with large aperture and long focal length are widely applied. The traditional color calibration method is no longer applicable for this kind of cameras. A new method of color calibration is introduced. The lighting system of this method is composed of integrating sphere, transmission color standards and collimator. Incident light of different colors can be provided by the system, and the chromaticity coordinates can be measured. There are many adjustable parameters for cameras, for example, gain, integral stage and integral time. The digital number (DN) values of different wave bands can be changed by adjusting these parameters. In this paper, the mathematical model is established by relating the RGB values after color synthesis with equivalent radiance incepted by each wave band of the camera. The difficulty caused by many adjustable parameters which can change the DN values of the camera can be avoided in this way. In this method, the chromaticity coordinates of incident light from the source system must be tested accurately . This method can be used in the space field.

Keywords: Remote Sensing · Multispectral Camera · Color Calibration · Chromaticity Coordinates · Equivalent Radiance

1 Introduction

With the development of technology in remote sensing, a lot of multispectral cameras with large aperture and long focal length are widely applied. The images of them can be obtained in five wave bands including blue, green, red, near infrared, and panchromatic. True color remote sensing images are synthesized with blue, green and red wave bands. Compared with gray-scale images, more information can be got from true color images. Because of the influence of light source, remote sensing cameras and so on, true colors for ground objects can't be reappeared by acquired color remote sensing images. Reliability and validity of the information expressed by data will be affected seriously.

The traditional method of camera color calibration is based on the CIE standard illuminations D65 and reflecting standard color-plate [1, 2]. The reflecting standard color-plate is imaged in the focus plane of digital cameras. The RGB values of all color

H. P. Urbach et al. (Eds.): ISSOIA 2023, SPP 191, pp. 329–336, 2024.
https://doi.org/10.1007/978-981-97-6718-2_29

blocks in the images of cameras can be obtained. They could be built a relationship with standard colorimetric values of the standard color-plate. Then the images of the cameras can be color-corrected in real time. This method is also applicable for partial short focal length cameras in remote sensing.

In space field, the imaging method and focal length of multispectral cameras have their specialty. First, measurement distance is very long for some large focal length cameras. This kind of measurement can't be finished in the laboratory. Second, for some large aperture and wide field of view cameras, the standard color-plate required for testing will be much bigger. The cost of large color-plate is very high, and processing difficulty of them is very high too. So the traditional color calibration method is not applicable for this kind of remote sensing camera. There are many adjustable parameters for multispectral cameras, for example, gain, integral stage and integral time [3]. The digital number (DN) values of different wave bands can be changed by adjusting these parameters. It brings much difficulty to the color calibration. Based on this, a new method of color calibration for multispectral cameras is presented in this paper.

2 Basic Theory of Colorimetry

According to colorimetry mechanism, the object color mainly depends on spectral power distribution of optical radiation into the eyes of the observer. The equation for calculating tri-stimulus values X, Y, Z is as follow [4].

$$\begin{aligned} X &= K\sum_{\lambda}\phi(\lambda)\bar{x}(\lambda)\cdot\Delta\lambda \\ Y &= K\sum_{\lambda}\phi(\lambda)\bar{y}(\lambda)\cdot\Delta\lambda \\ Z &= K\sum_{\lambda}\phi(\lambda)\bar{z}(\lambda)\cdot\Delta\lambda \end{aligned} \tag{1}$$

where: X, Y, Z—Tri-stimulus values;

K—Normalization factor;
$\phi(\lambda)$— Spectral power distribution of optical radiation into the eyes of the observer;
$\bar{x}(\lambda), \bar{y}(\lambda), \bar{z}(\lambda)$— Spectral tri-stimulus values of a CIE standard colorimetric observer;
$\Delta\lambda$— Wavelength interval.

The colors of filter glasses are one kind of transmission object color. So the colors of them depend on relative spectral power distribution $S(\lambda)$ of the optical source and transmission $\tau(\lambda)$ of filter. The equation is $\phi(\lambda) = S(\lambda)\cdot\tau(\lambda)$. So Eq. (1) can be changed as follow [5, 6].

$$\begin{aligned} X &= K\sum_{\lambda}S(\lambda)\tau(\lambda)\bar{x}(\lambda)\cdot\Delta\lambda \\ Y &= K\sum_{\lambda}S(\lambda)\tau(\lambda)\bar{y}(\lambda)\cdot\Delta\lambda \\ Z &= K\sum_{\lambda}S(\lambda)\tau(\lambda)\bar{z}(\lambda)\cdot\Delta\lambda \end{aligned} \tag{2}$$

According to Eq. (2), tri-stimulus values can be calculated. Then chromaticity coordinates (CIE-xyz) can be calculated by using Eq. (3).

$$\begin{aligned} x &= X/(X+Y+Z) \\ y &= Y/(X+Y+Z) \\ z &= Z/(X+Y+Z) \end{aligned} \tag{3}$$

where: x, y, z are chromaticity coordinates of CIE-xyz coordinate system.

3 Color Calibration Scheme

The traditional method of camera color calibration is based on the CIE standard illuminations D65 and reflecting standard color-plate. It's not applicable for remote sensing multispectral cameras of large aperture, wide field and long focal length. Based on this case, a new method of color calibration is presented in this paper.

There are many adjustable parameters for this kind of cameras, for example, gain, integral stage and integral time. The digital number (DN) values of different wave bands can be changed by adjusting these parameters. In this paper, the mathematical model for data processing is established by relating the RGB values after color synthesis with equivalent radiance incepted by each wave band of the camera. The formula is as follow.

$$\begin{pmatrix} R \\ G \\ B \end{pmatrix} = \begin{pmatrix} a_1 & a_2 & a_3 \\ b_1 & b_2 & b_3 \\ c_1 & c_2 & c_3 \end{pmatrix} \cdot \begin{pmatrix} L_{B1} \\ L_{B2} \\ L_{B3} \end{pmatrix} \tag{4}$$

where: B1, B2 and B3 correspond to blue, green and red bands, respectively. 3×3 coefficient matrix in Eq. (4) is desired, and there are nine unknowns. L_{B1}, L_{B2} and L_{B3} are the equivalent radiance values incepted by different wave bands when one color block is imaged by the camera. R, G and B are the RGB values after color synthesis. Schematic diagram of optical path is as shown in Fig. 1.

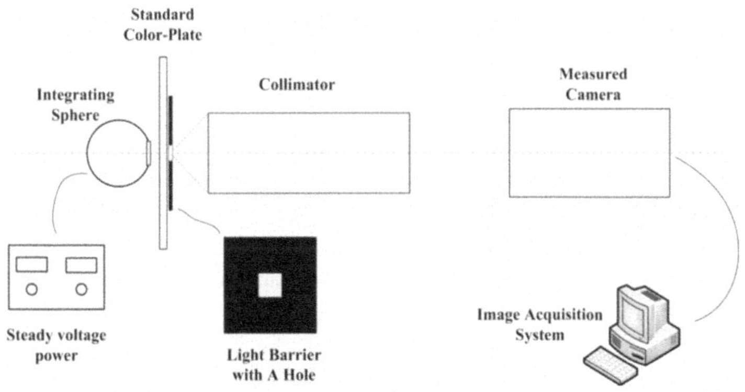

Fig. 1. Schematic diagram of optical path

The light source system of this method is composed of integrating sphere, transmission standard color-plate and collimator. The color-plate should be placed in the focal plane of collimator. The choice of focal length and optical aperture of the collimator is related to optical parameters of the measured multispectral cameras. Incident light of different colors can be provided by the source system. By this way, color calibration in the laboratory can be completed.

The plan in this paper includes the following steps.

(1) Set up the devices, including integrating sphere, standard color-plate, light barrier with a hole and collimator. Then put the color-plate into the focal plane of the collimator.
(2) Turn on the light source. After some time, place each color block into the light path by adjusting the horizontal and vertical position of the color-plate, and measure spectral radiance distribution and chromaticity coordinates of the light source system by spectral radiometer.
(3) According to Fig. 1, put the measured camera in place. Make sure that one color block can be imaged to CCD devices of B1–B3 wave bands, and acquire the images.
(4) By adjusting horizontal and vertical position of the color-plate, acquire the images of every color block.

The chromaticity coordinates (CIE-xyz) of emergent light from the source system must be tested accurately. They can be converted into tri-stimulus values of incident light. Then the coordinates of RGB color space are calculated with tri-stimulus values of output light [7, 8]. Input equivalent radiance is calculated with spectral radiance distribution data from Step 2 and relative spectral response curve of different spectral bands of camera [9]. Then color calibration coefficient matrix can be obtained by solving the Eq. (4) from the mathematical model combined with coordinates of RGB color space from source system and input equivalent radiance.

Calculate the DN values using the images data from Step 3 and Step 4. Combined with absolute calibration equation of corresponding camera parameter combination state in radiometric calibration and 3×3 coefficient matrix from color calibration, the camera images after color synthesis can be obtained.

4 Use of Coefficient Matrix for Color Calibration

Color synthesis using images acquired by camera t includes the following steps.

(1) For example, a certain scene was imaged by B1–B3 wave bands of multispectral camera. The images should be corrected by using results of relative radiometric calibration.
(2) According to gain, integral stage and integral time of the camera, the corresponding absolute calibration equations of B1–B3 wave bands can be obtained from results of absolute radiometric calibration [10–12].

$$DN_{B1} = A_{B1} \times L_{B1} + B_{B1} \tag{5}$$

$$DN_{B2} = A_{B2} \times L_{B2} + B_{B2} \tag{6}$$

$$DN_{B3} = A_{B3} \times L_{B3} + B_{B3} \tag{7}$$

where: DN_{Bi} is the output DN value of each wave band. L_{Bi} is input equivalent radiance of each wave band, and its unit is W/m^2/sr/μm. A_{Bi} and B_{Bi} are slope and intercept of absolute calibration equation of each spectrum.

(3) Eq. (5) – Eq. (7) can be changed as follow.

$$L_{B1} = \frac{DN_{B1} - B_{B1}}{A_{B1}} \tag{8}$$

$$L_{B1} = \frac{DN_{B1} - B_{B1}}{A_{B1}} \tag{9}$$

$$L_{B1} = \frac{DN_{B1} - B_{B1}}{A_{B1}} \tag{10}$$

(4) Combined with Eq. (8) – Eq. (10), Eq. (11) can be obtained.

$$\begin{pmatrix} R \\ G \\ B \end{pmatrix} = \begin{pmatrix} a_1\ a_2\ a_3 \\ b_1\ b_2\ b_3 \\ c_1\ c_2\ c_3 \end{pmatrix} \cdot \begin{pmatrix} \frac{DN_{B1}-B_{B1}}{A_{B1}} \\ \frac{DN_{B2}-B_{B2}}{A_{B2}} \\ \frac{DN_{B3}-B_{B3}}{A_{B3}} \end{pmatrix} \tag{11}$$

(5) This paper assumes that the number of pixels of array CCD is N.

$$\begin{pmatrix} R \\ G \\ B \end{pmatrix}_m = \begin{pmatrix} a_1\ a_2\ a_3 \\ b_1\ b_2\ b_3 \\ c_1\ c_2\ c_3 \end{pmatrix} \cdot \begin{pmatrix} \frac{(DN_{B1})_m-B_{B1}}{A_{B1}} \\ \frac{(DN_{B2})_m-B_{B2}}{A_{B2}} \\ \frac{(DN_{B3})_m-B_{B3}}{A_{B3}} \end{pmatrix}, m = 1, 2, 3, \ldots, N \tag{12}$$

5 Verification Experiment

The verification experiment of color correction method was carried out by using a planar array multispectral camera. Different spectral segments [13] of the multi-spectral camera were realized by independent cameras. The camera parameters are as follow.

1) Focal length: 37 mm
2) Field of view: 2ω=64°
3) Spectral range:

B1(blue): 0.45 μm–0.52 μm
B2(green): 0.52 μm–0.60 μm
B3(red): 0.61 μm–0.69 μm

The transmission standard color plate is placed in the focal plane position of the reflection collimator, and the D65 light source is used to illuminate the color plate. The collimating beam of different colors can be realized by adjusting the horizontal and vertical position of color plate. In this experiment, the transmission standard color plate is TE226, with a total of 45 color blocks. The spectral radiance and color coordinates of collimated beam of each color were measured by PR735 spectroradiometer. Three

Table 1. Tristimulus values and conversion results of sRGB color space

No. of color blocks	XYZ		RGB	
S03	X	0.101131	R	0.064421
	Y	0.112068	G	0.119820
	Z	0.182613	B	0.175783
S11	X	0.362842	R	0.373643
	Y	0.459431	G	0.518236
	Z	0.192537	B	0.129962
S17	X	0.288230	R	0.549200
	Y	0.148838	G	0.012894
	Z	0.313139	B	0.316650

Table 2. Equivalent radiance data of camera corresponding to different color blocks

Spectrum	S03	S11	S17
B3(Red)	0.081039	0.252944	0.401608
B2(Green)	0.111570	0.428778	0.118620
B1(Blue)	0.189039	0.339499	0.189203

color blocks of S03, S11 and S17 are selected to carry out the following analysis and calculation work. The tristimulus values (XYZ) were calculated according to the spectral radiance curve measured by PR735 and normalized. The sRGB color space conversion results of different color blocks in the standard color plate were calculated using Formula (2), as shown in Table 1.

The white balance of the camera can be carried out by using standard white light. The coefficient of B1 is 1.1820013. That of B2 is 0.9095566, and B3 is 0.9482804.

The DN values of various color blocks from B1–B3 spectrum of camera were collected. The equivalent radiance data were calculated after white balance processing, average and normalization processing. The results were as follow (Table 2).

The result of color correction coefficient matrix is as follow.

$$\begin{pmatrix} a_1 & a_2 & a_3 \\ b_1 & b_2 & b_3 \\ c_1 & c_2 & c_3 \end{pmatrix} = \begin{pmatrix} 1.503239 & 0.422462 & -0.552977 \\ -0.366472 & 1.499115 & -0.093827 \\ 0.461398 & -1.030117 & 1.340050 \end{pmatrix} \tag{13}$$

The aerial images taken during a flight test of the planar array multispectral camera were used for verifying the color correction method. The images of a certain scene in three spectral segments were processed according to the results of radiation calibration in the laboratory. The image of each spectral segment after processing was shown as follow (Fig. 2).

Fig. 2. The image of B1–B3 band after flat field

First of all, the image of each spectrum is matched [14, 15]. And then, according to "4. Use of coefficient matrix for color calibration", color correction processing was carried out combining with the absolute radiometric calibration results in laboratory. The final synthetic color image is as shown in Fig. 3.

Fig. 3. Final synthetic color image

6 Conclusions

Firstly, based on the theory of colorimetry, a new method of color calibration is presented in this paper according to the characteristics of multispectral cameras in remote sensing. Incident light of different colors can be provided by that source system based on integrating sphere, transmission standard color-plate and collimator. By this way, color calibration in the laboratory can be completed. It will be applicable for most of multispectral cameras in remote sensing. The color calibration test of a multi-spectral camera shows that this method has a good correction effect. According to the actual requirements, increasing the order of the coefficient matrix for color calibration is valid for improving the color corrected effects.

References

1. Haeghen, Y.V., Naeyaert, J.M.A.D., Lemahieu, I., Philips, W.: An imaging system with calibrated color image acquisition for use in dermatology. IEEE Trans. Med. Imaging **19**(7), 722–730 (2000)

2. Funt, B., Bastani, P.: Intensity independent RGB-XYZ colour camera calibration. Proceedings AIC, pp. 128–131 (2012)
3. Qiao, L., Xiayun, X., Mi, W., et al.: Precision analysis of satellite multi-spectral image registration based on geometric positioning consistency. Spacecraft Recovery & Remote Sensing **36**(6), 64–71 (2015)
4. Jing, Q., Jiao, S., Yu, B., et al.: Chromaticity. Science Press, Beijing, pp. 324–337 (1979)
5. CIE. Colorimetry. Second Edition, Vienna. Austria, Publication CIE NO. 15.2 (1986)
6. Bai, Y., Yuanjin, C., Mengjiao, Z., et al.: Method and demonstration of true-color imaging by EMCCD. Infrared Technol. **39**(4), 329–340 (2017)
7. Wyszecki, G.: Development of new CIE standard sources for colorimetry. Die Farbe **19**(1/6), 43–76 (1970)
8. Wyszecki, G., Stiles, W.S.: Color Science. Wiley, New York (1982)
9. Yuhuan, Z., Feng, Y., Hongwei, L., et al.: Measurement technology for relative spectral responsivity of the ultraviolet ICCD. Spectroscopy and Spectral Analysis **29**(5), 1371–1374 (2009)
10. Liu, Y., Hao, Z.: Research on the radiometric calibration of TDI CCD remote camera. SPIE, 6033: 60330K-1~11 (2005)
11. Yaxia, L.: Research on the laboratory radiometric calibration of the TDI CCD camera. Opto-Electronic Eng. **34**(5), 71–74 (2007)
12. Yiyun, M., Shiping, C.: Research progress of on-board absolute radiometric calibration for optical remote sensing satellite. Chinese Space Science and Technol. **42**(6), 12–22 (2022)
13. Zhenfen, Z., Yuanjin, C., Mengjiao, Z., et al.: Color correction of true-color night vision image by extracting green plants with differential depth of field. Infrared Technol. **42**(9), 886–892 (2020)
14. Qinwei, W., Zhang, L., Yuntian, J., et al.: An optical-to-SAR remote sensing image matching method for wide area scene. Spacecraft Recovery & Remote Sensing **43**(5), 109–122 (2022)
15. Nian, H., Sun, L., Yu, Y.: Remote sensing image matching and mosaic based on harris-laplace combined with scale-invariant algorithm. Spacecraft Recovery & Remote Sensing, **37**(6), 95–101 (2016)

Analysis and Research on TT&C Coverage of Typical Constellations Based on Land-Based TT&C System

Yiping Zhao(✉), Zhijun Liu, Junxiao Ma, and Feng Li

China Academy of Space Technology, Beijing, China
andyzhao@126.com

Abstract. Aiming at the coverage of typical constellations based on land-based TT&C system, this paper analyzes the visibility of constellations from three dimensions: single station, single satellite, TT&C network and TT&C network. Combined with the constellation configuration of a typical "1100 km orbit constellation" and the feasibility observation station, the analysis method of TT&C coverage is introduced with the visible time, the number of visible links and the maximum return visit interval as the demand indicators, and constructive suggestions are put forward on five design indicators such as satellite TT&C arc segment, real-time control of satellite number, single station equipment configuration, maximum emergency response time and telemetry return rate required for TT&C design in subsequent constellation design.

Keywords: Constellation · TT&C coverage analysis · Visible time · Visible link number · Maximum return visit interval

1 Overview

Research on global satellite communication constellations began in the 1980s, characterized by fast information transmission, low information loss, high communication service quality, and low cost and portability of ground terminal devices. [1] Currently, the long-term in-orbit management of various types of communication satellites in China relies on ground-based control stations, limiting the long-term, effective, and real-time control of communication satellites beyond the scope of ground control stations. The traditional satellite TT&C coverage analysis is based on the land-based TT&C system, and all of them are aimed at the TT&C coverage analysis of a single satellite. The TT&C coverage analysis of a single satellite only needs to calculate the TT&C arc to the maximum extent and give the visible time according to the TT&C station resources available to users. [2] However, the orbit of constellation satellites is generally low, and the visible arc of TT&C is relatively short for single station and single satellite, which puts forward new requirements for coverage analysis of satellite constellation TT&C based on land-based TT&C system. [3] The analysis of constellation TT&C coverage based on land-based TT&C system needs to provide five design inputs for the design of real-time control of

H. P. Urbach et al. (Eds.): ISSOIA 2023, SPP 191, pp. 337–351, 2024.
https://doi.org/10.1007/978-981-97-6718-2_30

satellite number, single station equipment configuration, emergency maximum response time and telemetry return rate, except for the TT&C arc of a single satellite. However, there are few studies on the methods of constellation TT&C coverage analysis.

In this paper, the visibility analysis of constellation by TT&C network, the visibility analysis of constellation by single station and the visibility analysis of TT&C network by single satellite are mainly analyzed, and the design ideas for five design inputs needed for TT&C of constellation are put forward by investigating three technical indexes: visibility time, number of visible links and maximum return visit interval.

2 Visibility Analysis Model of Satellite TT&C

The TT&C ground station equipment can only observe satellites above the ground level. When the elevation angle is less than5°, the observation effect of TT&C ground station is not good, because the smaller the elevation angle, the greater the attenuation of electromagnetic waves, the usable signals received by the ground station will be attenuated and the noise signals will be enhanced. Because of the limitation of the antenna itself, the TT&C ground station can not reach a relatively low elevation angle, so in practical application, only the area with elevation angle greater than 5° is observed. Visibility analysis is an essential condition for monitoring and controlling satellites observed by ground stations. The observation satellite is shown in the Fig. 1. Let the center of the earth be O, radius of the earth be Re, the height of the satellite be H, the ground monitoring point be M, and the satellite i be S_i.

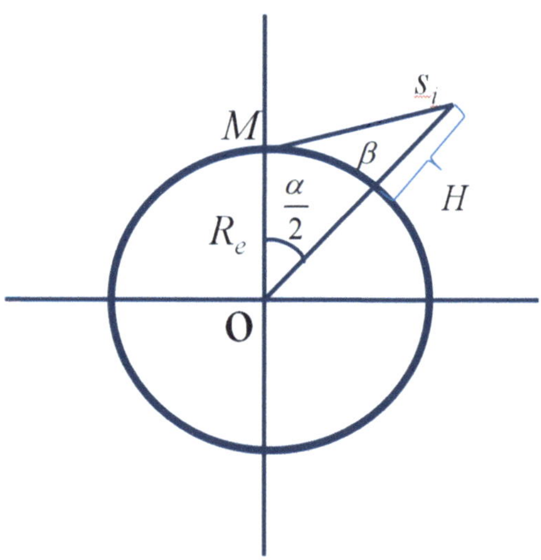

Fig. 1. Schematic diagram of observation satellites

At present, the satellite orbit of constellation is a circular orbit, that is, the eccentricity $e = 0$, and the perigee angle of the corresponding satellite orbit becomes an uncertain value, so it may be set $\omega = 0$.

According to the theory of dynamics, the relations of six orbital elements (semi-major axis a, eccentricity e, orbital inclination i, perigee radiation angle ω, ascending intersection right ascension Ω and true perigee f) are as follows:

Satellite operating period:

$$T = 2\pi\sqrt{\frac{a^3}{GM}} \tag{1}$$

GM is the gravitational constant of the earth;

Average angular velocity of satellite motion:

$$n = \frac{2\pi}{T} = \sqrt{\frac{GM}{a^3}} \tag{2}$$

Assuming that the satellite passes through a certain point at time T0, and this point is taken as the starting point, the angle between the position of the satellite at any time t and this point with respect to the center of the earth is:

$$M(t) = n(t - T0) \tag{3}$$

The perigee angle of the satellite at any time t:

$$E(t) = M(t) + e \sin E(t) \tag{4}$$

Then the true perigee angle of the satellite at any time t:

$$f(t) = 2\arctan(\sqrt{\frac{1+e}{1-e}} \cdot \tan(\frac{E(t)}{2})) \tag{5}$$

The coordinates of the satellite in the rectangular coordinate system of the satellite orbit plane are:

$$\begin{pmatrix} x \\ y \\ z \end{pmatrix} = \begin{pmatrix} acos(f(t)) \\ asin(f(t)) \\ 0 \end{pmatrix} \tag{6}$$

From orbit coordinate system to station center coordinate system (Fig. 2),

The coordinates of the available satellite in the station coordinate system are:

$$\begin{aligned}\begin{pmatrix} x \\ y \\ z \end{pmatrix} &= \begin{pmatrix} 0 \\ 0 \\ -(R_e + h) \end{pmatrix} + R_y(\frac{\pi}{2} - \varphi)R_z(\lambda)R(ERA)R_z(-\Omega)R_x(-i)\begin{pmatrix} a\cos(f(t)) \\ a\sin(f(t)) \\ 0 \end{pmatrix} \\ &= \begin{pmatrix} 0 \\ 0 \\ -(R_e + h) \end{pmatrix} + R_y(\frac{\pi}{2} - \varphi)R_z(\lambda + ERA - \Omega)R_x(-i)\begin{pmatrix} a\cos(f(t)) \\ a\sin(f(t)) \\ 0 \end{pmatrix}\end{aligned} \tag{7}$$

Where, R_x, R_y and R_z respectively represent the transformation matrix when rotating along the x-axis, y-axis and z-axis, Re is the average radius of the earth, and the latitude, longitude and altitude of the observation station are, then $(\lambda\varphi h)$.

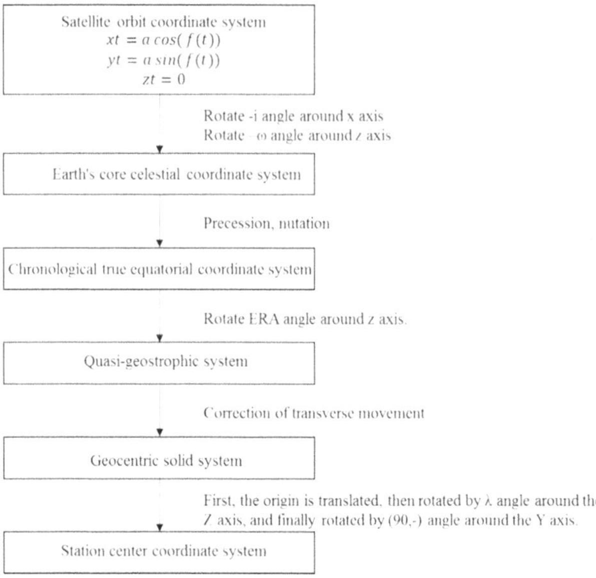

Fig. 2. The transform relationships of coordinates

Azimuth Az and elevation El of the satellite can be calculated in the station center coordinates as follows:

$$Az = \begin{cases} arctan(\frac{x}{y}), y > 0 \\ \pi + arctan(\frac{x}{y}), y < 0 \\ 90^\circ \ or \ 270^\circ, y = 0 \end{cases} \tag{8}$$

$$El = \arctan(\frac{z}{\sqrt{x^2 + y^2}}) \tag{9}$$

2.1 Measurable Time

The observable conditions of satellite are as follows: $\begin{matrix} Az > \beta > 5^\circ \\ El > \beta > 5^\circ \end{matrix}$, Through step-by-step ergodic simulation, the measurable time of satellite can be obtained.

2.2 Visible Link Number

For different observation stations, the actual observable arc segments are different in time domain due to different geographical locations, and multi-stations can realize complementary observation in time domain, thus realizing the calculation of the number of visible links. For the observable arc segment of the whole constellation, it is assumed that there are n satellites to be observed in a certain time window [Tm, Tn], and the set

Si is the visible time of the ground observation station to the satellite Si and the azimuth and elevation angles corresponding to the observation, including:

$$S_i = \left\{(t_1^{S_i}, Az_1^{S_i}, El_1^{S_i}), (t_2^{S_i}, Az_2^{S_i}, El_2^{S_i}) \cdots (t_r^{S_i}, Az_r^{S_i}, El_r^{S_i})\right\} \tag{10}$$

Assuming that the set of observation time, azimuth angle and elevation angle in the optimal observation strategy of the ground observation station is finally obtained, then the visible link value is obtained. $S = \left\{(t_1, Az_1, El_1), (t_2, Az_2, El_2) \cdots (t_r, Az_r, El_r)\right\}$, then $S \subseteq \bigcup_{i=1}^{N} S_i$.

3 Satellite TT&C Design Requirements

Universal satellite measurement and control refers to satellite telemetry, remote control and measurement. The coverage analysis of satellite constellation measurement and control affects the design of satellite constellation, but the design of satellite constellation is constrained not only by satellite constellation measurement and control, but also by control, power supply and distribution and load business. This paper only studies the TT&C coverage analysis of satellite constellations, not the coverage analysis of payload services. Taking the satellite constellation without inter-satellite link as an example, this paper mainly relies on the ground monitoring and control station to monitor and control the satellite constellation, and analyzes the monitoring and control coverage of the satellite with the selected ground station as input, without involving the site selection and optimization of the ground station.

The analysis of constellation TT&C coverage needs to find suitable technical indicators to complete the design input conditions in five aspects: ① TT&C arc ② Real-time control of satellite number ③ Single station equipment configuration ④ Maximum emergency response time ⑤ Telemetry return rate. In this paper, the visibility time, the number of visible links and the maximum return visit interval are used to analyze the visibility of constellation by single station, ground TT&C network by single satellite and constellation by ground TT&C network.

3.1 Measurement and Control Arc Segment

In this paper, the satellite constellation without inter-satellite link is studied, so the measurement and control arc segment is still dominated by single satellite, which is measured by the index of satellite visibility time. The TT&C arc takes the existing TT&C station resources as input to calculate the visible time of a single satellite. When using multiple stations to calculate the total visible time, after the measurement and control time of multiple stations is accumulated, it is necessary to deduct the time when the satellite is simultaneously visible by two or more measurement and control stations.

3.2 Real-Time Control of Satellite Number

The number of real-time control satellites is measured by the number of visible satellites, which can be the number of real-time visible satellites of a single station or the number of real-time visible satellites of the ground monitoring and control network. It can be seen that the index of satellite link number reflects two kinds of measurement and control capabilities: when there is no inter-satellite link in the constellation to be measured, this index represents the maximum number of satellites that can be controlled in real time, and it is also the maximum number of satellites that can be rescued in real time when the constellation fails; When the measured constellation has inter-satellite links, this index affects the number of available channels for real-time routing between Skynet and ground network, that is, the number of available routing nodes. When calculating the number of satellites visible in real time in the ground TT&C network, it is necessary to deduct the fact that satellites are visible by two or more TT&C stations at the same time after accumulating the number of satellites visible at multiple stations.

3.3 Single Station Equipment Configuration

Satellite constellation TT&C coverage analysis needs to provide two inputs for single station equipment configuration, the number of conFig.d antennas and the number of satellite signal processing channels. This paper uses single station to measure the maximum number of visible links of the constellation. Commonly used ground station measurement and control antennas are divided into mechanical scanning antenna and electrical scanning antenna. If the mechanical scanning antenna is used, the maximum number of visible links of a single station to a constellation is the minimum number of antennas that a single station needs to conFig.. If an electric scanning antenna is used, the number of satellites tracked simultaneously by one or more antennas is greater than or equal to the maximum number of visible links. The number of single-station satellite signal processing channels should be equal to the maximum number of visible links. The reliability of antennas and satellite signal processing channels is different, so this paper only gives the minimum number of configurations. According to the requirements of reliability, the number of antennas and processing channels needed to be conFig.d is calculated by backup.

3.4 Maximum Emergency Response Time

This paper studies the satellite constellation without inter-satellite link, which is not real-time controllable in the whole network. When a satellite in the satellite constellation fails, it is necessary to analyze the coverage of measurement and control to give the maximum corresponding emergency time, and each satellite subsystem designs the safety mode based on this. In this paper, the maximum return visit interval is used to measure this. Usually, the maximum invisible time of each satellite for the ground monitoring and control network is calculated first. Then take the largest invisible time of all satellites.

3.5 Telemetry Return Rate

For satellite constellations without inter-satellite links, time-delay telemetry technology is usually used for measurement and control, that is, when the satellite is invisible, the telemetry on the satellite is stored, and when the measurement and control station is visible, the stored telemetry is downloaded through a high-speed channel. Under normal circumstances, the downlink telemetry rate of satellites is V. In this paper, the visible time of satellites on the same orbital plane is accumulated by the ground TT&C network within a certain time t to obtain the visible time of joint measurement and control of TT&C stations on different orbital planes, and then the average of these times is obtained, and finally the telemetry return rate needs to be greater than or equal to $\frac{T}{t \cdot V}$.

4 Use Case Input Condition

The low-orbit constellation launched by SpaceX's star chain plan has an orbital height of 1150 km. According to the distribution of satellites in orbit provided by the "Scientists Data Alliance", 1100 km is a relatively empty orbital resource, so this orbital height is taken as an example for analysis. It consists of six orbital planes, each with nine satellites. Detailed constellation parameters are shown in Table 1. The schematic diagram of constellation configuration is shown in Fig. 3. [4] As we all know, Beijing is the core of the country. According to Baidu map, the satellite monitoring and control center is located in Xi'an, so these two locations are selected for analysis, but there is still a lack of high latitude coverage at this time. Kiruna Station abroad is an open TT&C station that provides TT&C resources for commercial satellites. Finally, these three stations are selected as typical representatives for analysis. The latitudes and longitudes are listed in Table 2.

Table 1. Constellation parameter table

	Configuration	True anomaly	RAAN	Perigee orbit height	Apogee orbit height	Inclination
Values	6 × 9	20°	32.1°	1100 km	1100 km	85°

For the analysis of in-orbit emergency mode control of the mobile constellation, taking the main star field range as ±90°, the ground station antenna's field of view without solar wing obstruction is calculated based on ±80°. This includes the three stations in Beijing, and the ground station's visible range is calculated based on ±80°. Theoretically, when the ground station antenna is inclined by 5°, ground noise enters, and actual control station antennas are limited to either limited control or full control, with limited control having a smaller range of motion. The simulation interval is 20 s, simulated for 24 h, and the satellite parameters, as shown in Table 1, indicate a satellite orbit time of 107 min.

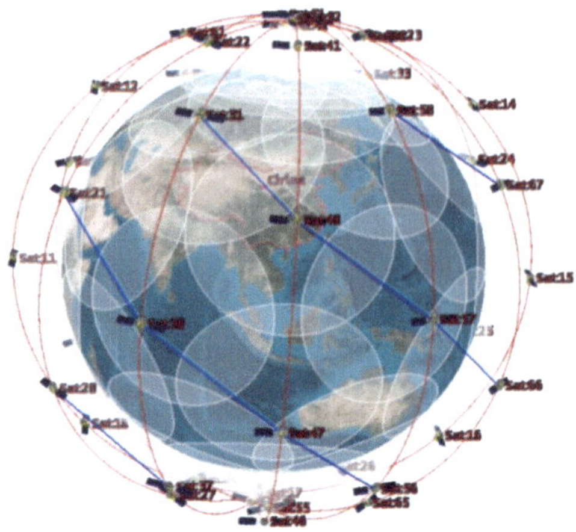

Fig. 3. Mobile satellite constellation configuration

Table 2. Control station latitudes and longitudes

Station location	Xi'an	Beijing	Kiruna
Longitude (°E)	108.93	116.0	20.27
Latitude (°N)	34.26	40.0	67.85

Note: The provided latitudes and longitudes are for the centers of Xi'an, Beijing, and Kiruna, and may differ from the actual station positions. However, this does not affect the description of the measurement and control coverage analysis method

5 Simulation Analysis

5.1 Single Station Visibility Analysis of the Constellation

The analysis of single station visibility of the constellation focuses on two aspects: the measurable control time for the constellation by a single station and the visible link count for an individual station. The analysis of the measurable control time for the constellation by a single station provides recommendations for the selection and construction of control stations, while the analysis of the visible link count for an individual station provides suggestions for the scale of control station construction, such as the number of antennas needed [5].

Single Station Measurable Control Time for the Constellation

As shown in Table 3 and Fig. 4, adding Kiruna greatly increases the control arc, ensuring that each orbit has a segment of time available for control, with a maximum of no more than two orbits. Kiruna has two advantages: in terms of indicators, as shown in Table 3, Kiruna's average visibility time is 7073s, compared to Beijing's average visibility time of 3003s and Xi'an's average visibility time of only 2663s. On the other hand, Fig. 4 clearly shows significant differences in visible arc segments for individual stations. Within 24 h, Kiruna is visible approximately 11 times, while Beijing and Xi'an only have four times each. Kiruna's control station significantly shortens the maximum access time. If Kiruna is excluded, relying solely on Beijing or Xi'an would result in up to 11 h of non-visibility each day [6].

Table 3. Statistical analysis of accumulated visible time for satellite control

	Orbit 1	Orbit 2	Orbit 3	Orbit 4	Orbit 5	Orbit 6	Average
Xi'an (s)	2651	2778	2462	2741	2759	2592	2663.83
Kiruna (s)	7333	7307	7426	6909	6738	6730	7073.83
Beijing (s)	2901	3431	3109	2933	2804	2843	3003.5

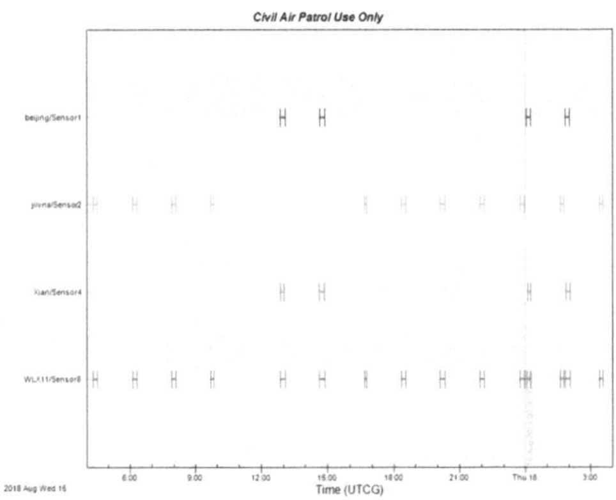

Fig. 4. Illustration of control arc segment for orbit 1

Single Station Visible Link Count for the Constellation

Kiruna Ground Station Visible Link Count

Table 4 provides the statistical information on the real-time visible link count for Kiruna control station over 24 h. This section focuses on the real-time visible link count for a single station, primarily used to analyze the number of antennas required for an individual station. The percentage of cases where the link count is less than or equal to 6 is 96.09%. As commonly used mechanical scanning control antennas have a narrow beam, tracking only one antenna at a time, at least 6 antennas are needed, up to 8. With an electronically scanned antenna, it can simultaneously track at least 6 satellites, with a capability of up to 8 satellites. Figure 5 illustrates the real-time variation of link count with the inclusion of time factors.

Beijing Ground Station Visible Link Count

Table 5 provides the statistical information on the real-time visible link count for the Beijing control station over 24 h. In the case of mechanical scanning antennas, the percentage of link counts less than or equal to 3 is 99.44%. Therefore, at least 3 antenna sets are needed, up to 4. With electronically scanned antennas, it can simultaneously track at least 3 satellites, with a capability of up to 4 satellites. Figure 6 illustrates the real-time variation of link count with the inclusion of time factors.

Table 4. Statistical analysis of real-time visible link count for Kiruna control station

Received Link Count	1	2	3	4	5	6	7	8	Total
Frequency	14	71	1056	1263	1023	725	105	64	4321
Percentage (%)	0.32	1.64	24.44	29.23	23.68	16.78	2.43	1.48	100

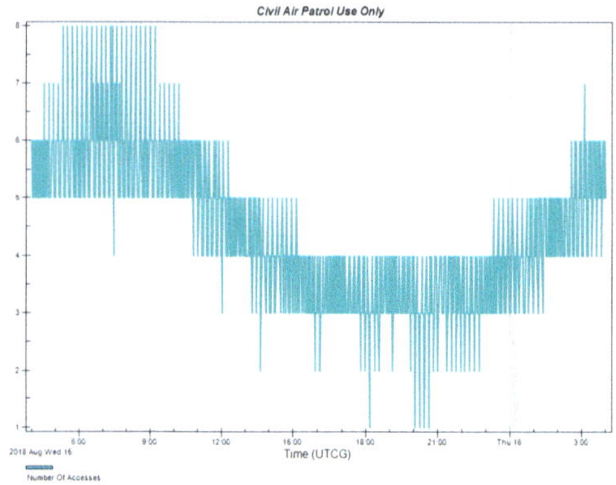

Fig. 5. Real-time visible link count for Kiruna control station with time variation

Table 5. Statistical analysis of real-time visible link count for Beijing control station

Link count	1	2	3	4	Other	Total
Frequency	1346	2504	447	24	0	4321
Percentage (%)	31.15	57.95	10.34	0.56	0	100

Xi'an Ground Station Visible Link Count

Table 6 provides the statistical information on the real-time visible link count for the Xi'an control station over 24 h. The percentage of link counts less than or equal to 3 is 99.70%. Therefore, at least 3 antenna sets are needed, up to 4. With electronically scanned antennas, it can simultaneously track at least 3 satellites, with a capability of up to 4 satellites. Figure 7 illustrates the real-time variation of link count with the inclusion of time factors.

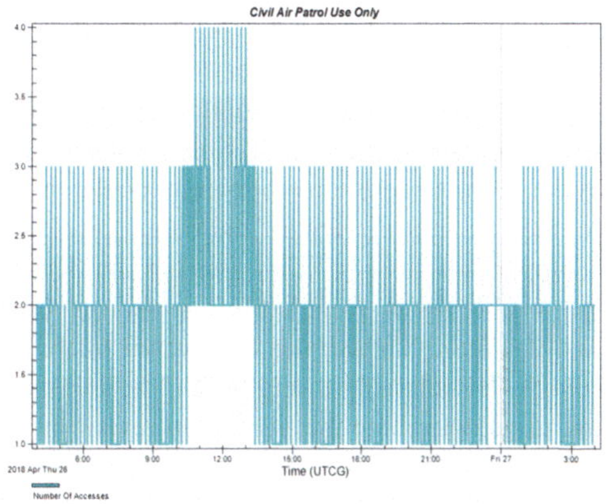

Fig. 6. Real-time visible link count for Beijing control station with time variation

Table 6. Statistical analysis of real-time visible link count for Xi'an control station

Link count	1	2	3	4	Other	
Frequency	1766	2290	252	13	0	4321
Percentage (%)	40.87	53.00	5.83	0.30	0	100

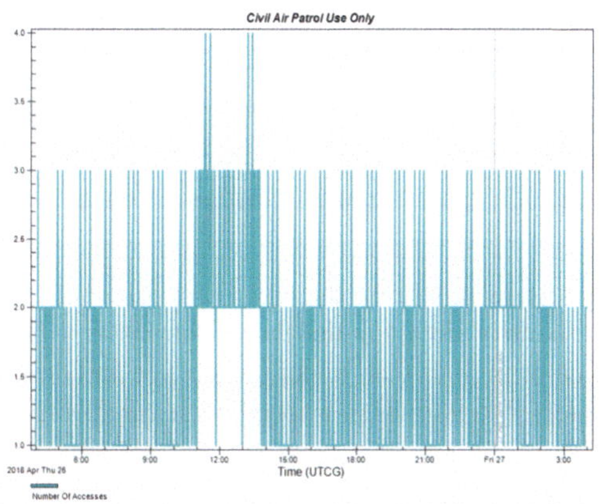

Fig. 7. Real-time visible link count for Xi'an control station with time variation

5.2 Single Satellite Visibility Analysis of the Measurement and Control Network

In this section, the visibility analysis of a single satellite to the ground measurement and control network is conducted to further understand the constellation's strong control coverage capabilities. The analysis focuses on three indicators: control coverage visible cumulative time, control arc segment duration, and maximum revisit interval.

Table 7 provides a 24-h cumulative time analysis for six orbits. It illustrates the accumulated time for single satellites passing over three control stations and the joint control time over these three stations. Due to the overlapping control coverage between Beijing and Xi'an, where satellites are simultaneously visible to both stations, the cumulative time for joint control is not a simple addition of the individual control times for the three stations. The table provides an overall view of the cumulative visible time for single satellite control arcs. Comparisons among Xi'an, Beijing, and Kiruna reveal that higher latitudes result in longer visible times. Detailed information on control arc visibility is presented in Figs. 8 and 9. Figure 8 illustrates the overall visibility of all satellites to the ground control network. By referring to Fig. 9, a clear and intuitive understanding can be gained, allowing us to determine the maximum revisit interval. For the ground network composed of three stations, there is, on average, at least one visible station in each orbit, ensuring coverage in almost every orbit, with occasional instances of two orbits in one pass. Consequently, the maximum revisit interval is less than two satellite orbits, approximately 214 min. Figure 9, using a single satellite in one orbit as an example, provides a detailed depiction of the satellite's visibility over three control stations. It reveals that, on average, Beijing or Xi'an has around four control opportunities in a day. Kiruna, with its higher latitude, significantly enhances the control visibility duration and reduces the maximum revisit interval for satellite control.

Table 7. Statistical analysis of cumulative visible time for satellite control

	Orbit 1	Orbit 2	Orbit 3	Orbit 4	Orbit 5	Orbit 6	Average
Xi'an (s)	2651	2778	2462	2741	2759	2592	2663.83
Kiruna (s)	7333	7307	7426	6909	6738	6730	7073.83
Beijing (s)	2901	3431	3109	2933	2804	2843	3003.5
Joint control (s)	10541	11249	10974	10425	9934	10003	10521

The joint average control time for the three stations is 10521 s, approximately 175.35 min. This indicator implies that, under the storage and downlink mode, a speed increase of approximately 8 to 9 times or a speed decrease during downlink is needed.

5.3 Three-Station Joint Control Constellation

Sections 5.1 and 5.2 analyze the impact of single stations and single satellites on the visibility of the constellation, respectively. This section examines the capabilities of a three-station joint control constellation from a macro perspective, further enhancing the understanding of the constellation's control coverage capabilities. The analysis focuses on three aspects: control coverage area, joint control visible satellite link count, and control link duration.

The control coverage areas of the Beijing, Xi'an, and Kiruna ground stations are shown in Fig. 10. The red color represents the control coverage area of Xi'an station, light blue for Beijing control coverage area, and dark blue for Kiruna control coverage area.

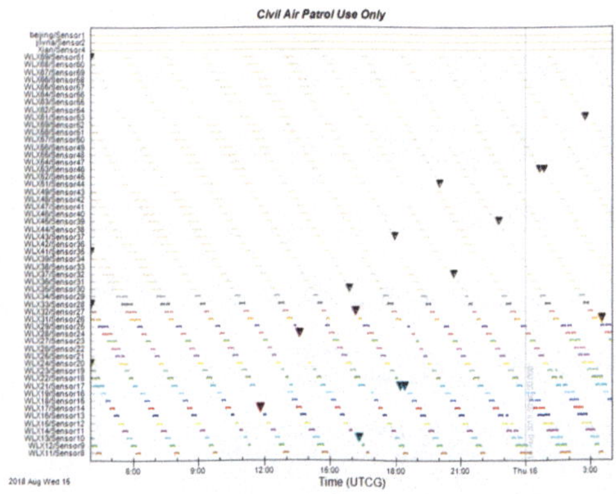

Fig. 8. Constellation control arc segment diagram

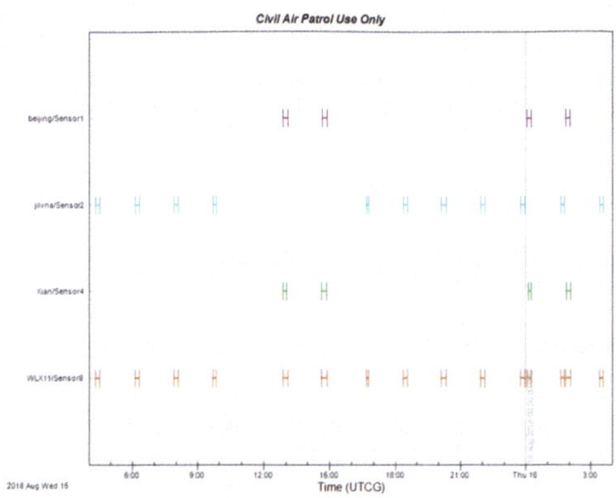

Fig. 9. Control arc segment diagram for orbit 1

The importance of the Kiruna station is evident in the control coverage area, effectively filling the blank control coverage area. However, this Fig. Does not directly reflect all constellation control indicators.

Within 24 h, the joint control visible satellite link count for the three stations is provided in Table 8. The purpose of this statistical link count is used to grasp the number of link nodes between the satellite constellation network and the ground control network on a macro scale. This allows for the analysis of routing or emergency processing capabilities. The visible satellite link count reflects two control capabilities: when there are no inter-satellite links in the controlled constellation, this indicator represents the maximum number of satellites that can be controlled in real-time. In the case of constellation failure, it also represents the maximum number of satellites that can be rescued in real-time. When there are inter-satellite links in the controlled constellation, this indicator affects the number of available channels for real-time routing between the celestial and terrestrial networks, i.e., the number of available routing nodes. Figure 11 shows the relationship between the real-time visible satellite link count and time.

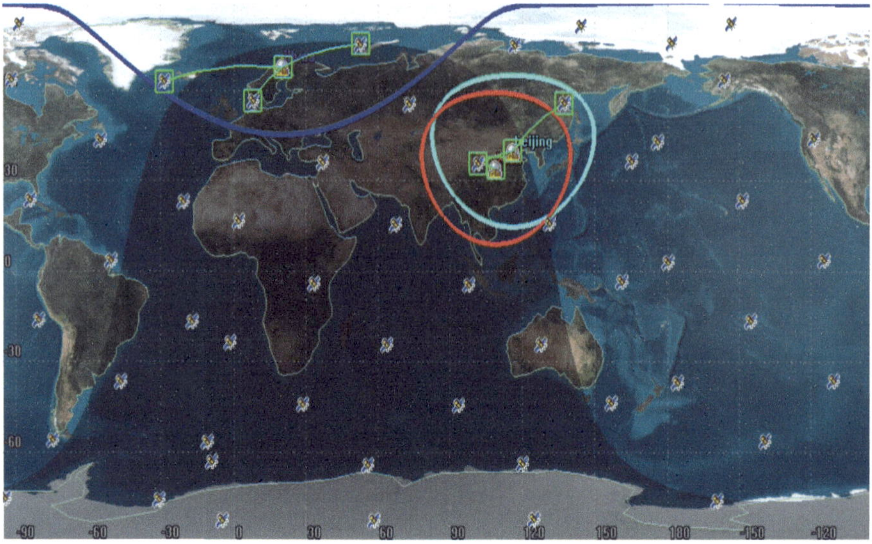

Fig. 10. Simulation of control station visible area

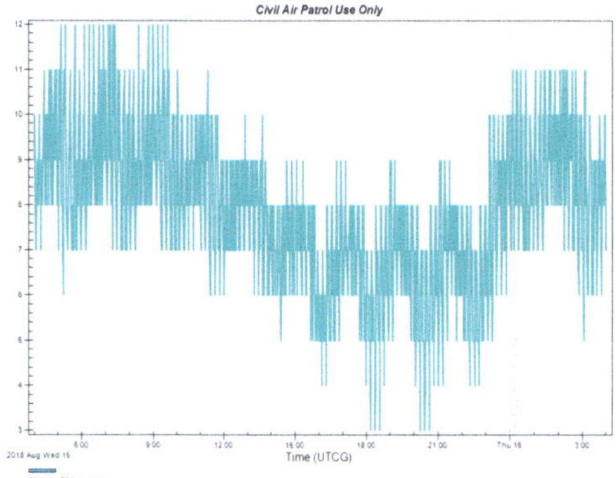

Fig. 11. Real-time visible satellite link count for three-station joint control constellation with time variation

As shown in Tables 5 and 6, due to a simulation error of 20 s, it is not suitable to use a 20 s interval for statistics. A 50 s interval is chosen for statistical analysis. The average duration is 641.2737 s, with a maximum duration of 815.235 s. Durations above 300 s account for approximately 95.95%. In a 24-h simulation, the joint average control time for the three stations is 10521s, approximately 175.35 min (see Table 2). This indicator constrains storage and downlink design, requiring a speed increase of approximately 8 to 9 times or a speed decrease during downlink, particularly in the storage and downlink mode, where telemetry is transmitted only within the country.

Table 8. Statistical analysis of joint control visible satellite link count

Link count	3	4	5	6	7	8	9	10	11	12	Total
Frequency	10	30	295	465	1003	995	831	478	184	30	4321
Frequency (%)	0.23	0.69	6.83	10.76	23.21	23.03	19.23	11.06	4.26	0.69	100

6 Conclusion

Through the coverage analysis of typical constellations by three land base stations in Beijing, Xi'an and Kiruna, it can be found that the ability of land-based TT&C system is improved obviously by high-latitude TT&C stations, and the following conclusions are drawn:

1) For large inclination orbit constellations, high latitude land-based TT&C stations have obvious effects on increasing TT&C radian, shortening the maximum return visit interval and improving the maximum emergency response time index.
2) The maximum number of visible links in high-latitude TT&C base stations is large, so the number of antennas required for single-station equipment configuration is large. Xi'an Station in Beijing needs 3–4 antennas, and Kiruna needs 6–8 antennas. Electric scanning antennas can be used to relieve the antenna processing pressure of TT&C station.
3) In order to meet the needs of satellite telemetry download, the land-based TT&C system puts forward higher requirements for the satellite telemetry download rate, and the satellite telemetry download needs to have 8–9 times the speed-up download ability.

References

1. Yang, T., Li, J.L., Zhou, A.S.: General concept of satellite mobile communication system in China. In: 61st International Astronautical Congress, 27 September – 1 October 2010, Prague, Czech (2010)
2. An, Y., Li W., Wang, W., Yang, X., Wei, P., Wang, X.: A study on the scheduling strategy of ground stations for low Earth orbit satellite constellations. J. Time Freq. **44**(2), 120–131 (2021)
3. Xiao, N., Liang, J., Zhang, J.: Design and planning of China's low Earth orbit satellite constellation. Telecommun. Technol. **50**(12), 14.18 (2010)
4. Yang, X., Li, J.: Walker constellation inter-satellite link analysis. Geod. Geodyn. **32**(2), 143–147 (2012)
5. Li, H., Cheng, C., Huang, L., Chen, S., Li, Y., Qiang, L., Cai, K.: Design of X-band control tasks during the launch and in-orbit phase of Chang'e-5. J. Deep Space Explor. **9**(2), 183–190 (2022). (in Chinese and English)
6. Zhang, Z., Chen, G., Liang, Z., Li, J., Wu, C.: Design of an attitude-adaptive sub-system for astronomical satellite control. Spacecr. Eng. **27**(5), 65–71 (2018)

Peculiarities of SVM-Based Classification of BPG Compressed Three-Channel Images

Vladimir Lukin[1], Fangfang Li[1,2], Jiawen Zhu[2], and Sergii Kryvenko[1(✉)]

[1] National Aerospace University, 17 Chkalova Street, Kharkov 61070, Ukraine
v.lukin@khai.edu, krivenkos@ieee.org

[2] Nanchang Hangkong University, Fenghe South Avenue 696, Nanchang 330063, China

Abstract. Remote sensing systems provide a great amount of useful data for various applications. To transfer these data via communication lines and/or to store them, compression is applied. Lossy compression techniques are used more often since a variable and quite large compression ratio can be attained. Meanwhile, introduced distortions usually result in reduction of probability of correct classification. Then, it is desired to establish connection between compressed image quality and classification characteristics. The latter also depend on a used coder and an applied classifier. In this paper, we study the influence of compression carried out by better portable graphics (BPG) coder applied to three-channel images where a trained support vector machine (SVM) classifier is employed. Several quality metrics describing image quality and parameters that deal with classification accuracy are used. We show that reduction of classification accuracy is almost negligible for small values of parameter Q that controls introduced losses and it starts to quickly increase for Q > 17. Dependences of classification accuracy on Q are individual and depend on image properties. Some dependencies have fluctuating character. The recommendations on Q setting are given.

Keywords: Remote Sensing · Lossy Compression · Image Quality · BPG-coder · SVM-classifier

1 Introduction

Remote sensing (RS) from airborne and spaceborne platforms has demonstrated its usefulness for different applications, including agriculture, ecology, meteorology and so on [1, 2]. Modern RS systems produce an enormous number of images due to increased number of observations with improved spatial resolution. Then, there is a need to store, process, transfer via communication lines, and classify the acquired data [2, 3]. Thus, it often becomes necessary to compress RS images [3–5].

Existing image compression techniques are usually divided into lossless and lossy [6–8]. Lossless compression is known to introduce no distortions into compressed data. However, the techniques of this group are unable to provide quite large values of compression ratio (CR). Besides, CR cannot be varied and controlled. Thus, practical application of lossless compression is limited.

H. P. Urbach et al. (Eds.): ISSOIA 2023, SPP 191, pp. 352–365, 2024.
https://doi.org/10.1007/978-981-97-6718-2_31

On the contrary, lossy compression techniques can produce quite large values of CR and an opportunity of its variation. Meanwhile, distortions (losses) are introduced. They can result in negative effects such as inappropriate visual quality [9] and classification accuracy reduction [10]. Then, it is desired to provide a certain compromise between the produced CR and the provided quality (in wide sense) of compressed RS images [4, 10–14]. By appropriate quality, it is possible to mean such quality that classification accuracy is high enough (does not reduce too much compared to possible maximum).

The influence of lossy compression on classification accuracy of multichannel images has been studied in several publications (see [10, 11, 14–21] and references therein). The problem of comparing the results is that the authors consider different applications including terrain type classification [10, 11, 21], crop mapping [14, 16], forest classification [15], identifying land-cover objects [18], etc. Different techniques of lossy compression have been considered: JPEG is studied in [11]; the Enhanced Compressed Wavelet format is considered in [18]; the authors of [14, 15] employ JPEG and JPEG2000; the 3D version of JPEG2000 is used in [20]; the coders based on discrete cosine transform (DCT) are studied in [10, 21], the High Efficiency Video Coding (HEVC) based approach is applied in [16]. In addition, there are many classifiers that can be used. The Random Forest, Nearest Neighbor, and Gradient Boosting Machine are studied in [18]; a multiple kernel learning approach is proposed in [17]; the maximum likelihood method is considered in [10]; the neural network, support vector machine (SVM), and logistic regression are compared in [21], etc. Finally, multichannel RS data considered in aforementioned papers have different number of components: three-channel optical-band images taken from multispectral data are employed in [10, 21], hyperspectral images are considered in [20]; Landsat-8 and Sentinel-2 data are used in [16] and so on.

Despite of all these differences, several interesting effects have been observed. First, lossy compression with quite small CR does not lead to reduction of classification accuracy [10, 14–16]. Moreover, it is sometimes possible that lossy compression leads to improvement of classification accuracy [10, 11, 14]. Meanwhile, after certain CR, which is individual for a given image and a given coder, the classification accuracy starts to rather quickly decrease with further CR increasing. Second, it is difficult to say what is the best classifier since this depends on several factors such as methodology of learning, number of classes, image complexity, and so on. Third, quality metrics can be calculated and even predicted in advance [22] for compressed images. However, there are no already established strict dependences between image quality metrics and classification accuracy for lossy compressed images [23].

To get them, a lot of additional studies have to be carried out for different coders, classifiers, and RS data types. In this paper, we consider better portable graphics (BPG) encoder [24, 25] (which is one of the best nowadays [16]), SVM classifier (that is quite efficient and simple [21]), and three-channel optical images acquired by Gaofen-2 satellite (that are the simplest case of multichannel RS data). Our goal is to determine the main dependencies of classification accuracy parameters for a set of images of different complexity and analyze them for establishing the main tendencies.

This paper is organized as follow. The problem of the lossy compression controlling is defined, and the solution method is described in Sect. 2. The experiment set is given in

Sect. 3, then the results and discussion are presented in Sect. 4. Finally, the conclusions are given.

2 Used Coder, Classifier, Quality Metrics, and Classification Characteristics

Our experiment was conducted on three-channel remote sensing images acquired by the sensor installed on Gaofen-2 satellite. The BPG coder was employed to compress the obtained data. The reasons for this are the following. The BPG (Better Portable Graphics) is a new and efficient image coder (and format). It outperforms JPEG and JPEG2000 sufficiently as well as is usually better than other modern coders [25, 26]. The compression control parameter is Q, which is an integer within the range of [1… 51], that has a quite strict connection with image quality characterized by peak signal-to-noise ratio (PSNR) and visual quality metrics [26] such as MDSI [27]. A larger Q leads, in general, to a larger CR, a smaller PSNR, and a larger MDSI, i.e. worse quality according to both metrics.

The BPG coder supports RGB, YCgCo, and CMYK color spaces (here we suppose that three-channel RS images can be associated with R, G, and B components). In addition, the BPG coder is able to compress 8–14 bit data (in each component). Finally, the coder has three main formats that deal with color subsampling – 4:4:4, 4:2:2, and 4:2:0 where we used 4:2:2 in our studies as a reasonable compromise.

In this study, the support vector machine classifier was employed to the original and decompressed images.

The peculiarity of support vector machine classification is that the vector is mapped to a higher dimensional space, constructed with the maximum range of hyperplanes in this space. This is the way to construct two parallel hyperplanes on both sides of the separated data hyperplane to maximize the distance between the two parallel hyperplanes. It is assumed that the greater the distance between parallel hyperplanes, the smaller the total error of the support vector machine classifier.

The advantage of the classification method based on support vector machine is that it does not require any assumptions about the distribution of basic data, and has good generalization ability. Due to this, it is very suitable for studying the impact of distortion caused by the BPG lossy compression on RS image classification accuracy.

It has been shown earlier [10] that there is a certain dependence between quality of compressed images characterized by certain metrics and classification accuracy. Recall that lossy compression can achieve considerably better compression rates compared to lossless techniques, but the visual quality of images is inevitably affected. Although it is intuitively clear that compressed image quality and classification accuracy of remote sensing images are considerably correlated, it is still unclear what metrics to choose and calculate to provide appropriate classification accuracy (or its appropriate reduction due to lossy compression). Then, in this study we rely on one standard metric (namely, PSNR) and one visual quality metric (MDSI [27]). Note that MDSI (Mean Deviation Similarity Index) is one of the best quality metrics for characterizing the compressed images [26]. It uses gradients and colors as features to obtain the local quality map, and finally uses deviation pooling to perform the pooling operations on this map. In fact,

MDSI measures the similarity of images (original and compressed in our case), and the smaller the index, the higher the compressed image quality.

3 Experiment Set

The remote sensing images used in our study were cropped from the large-scale land cover dataset of the Gaofen-2 satellite images [28]. We selected a large portion of the large size image (with the total number of 7200 × 6800 pixels), and then cropped nine image fragments of the size 1024 × 1024 pixels. Just these 9 fragments were used as the original images in our experiment (see Fig. 1, the images are further denoted as RSa1 – RSa9). As one can see, these are typical RS images with different content and slightly different complexity.

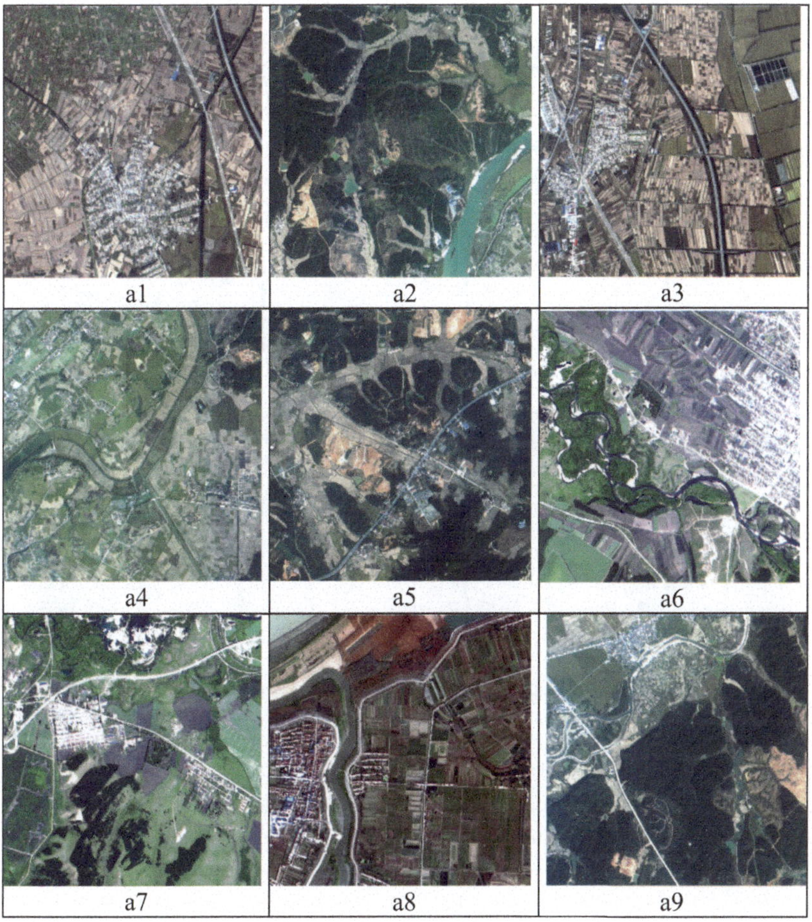

Fig. 1. The nine original remote sensing images

The image lossy compression was divided into two stages. At the first stage, an initial compression, with a step of 5 for Q, was applied and, thus, each image was compressed 11 times to obtain compressed images of different image quality. The compressed image examples of the RSa1 image fragment are shown in Fig. 2. Under each compressed image we give parameters, namely, Q, CR, MDSI, and PSNR (in dB). The compressed image size decreases if Q increases, and the image quality reduces (PSNR decreases, MDSI grows). When Q increases to 26, CR is about 9.5, which is considerably higher than for lossless compression, and distortions due to lossy compression are still difficult to observe.

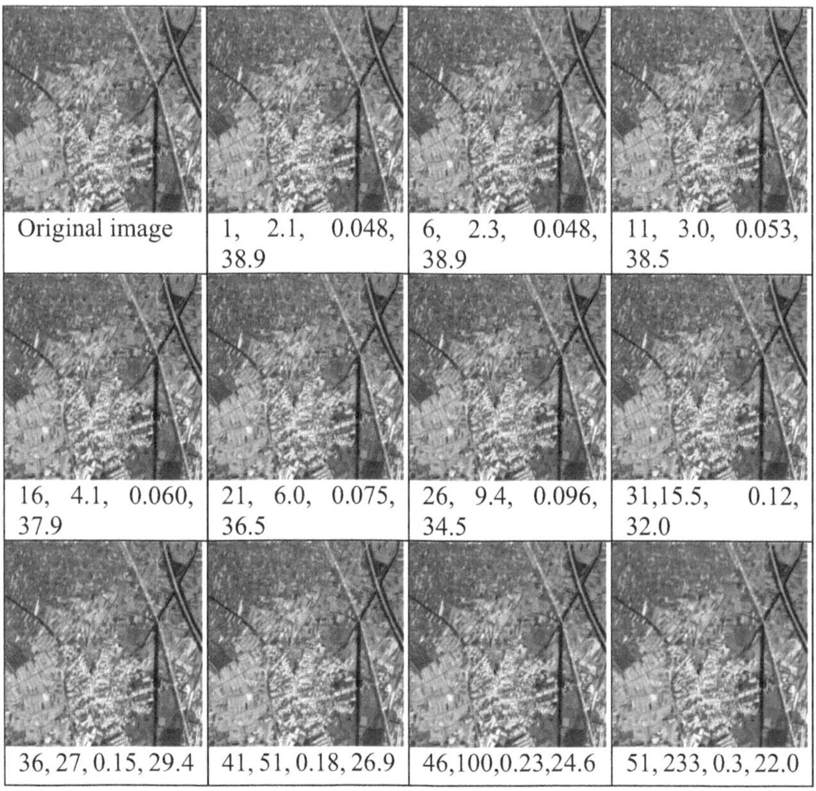

Fig. 2. The image RSa1 and the decompressed images (Q, CR, MDSI, PSNR)

The BPG coder can provide considerably larger CR – for example, it reaches 100 for Q = 46. Meanwhile, in this case, image quality is too low (both metrics evidence in favor of this conclusion). Because of this, we have decided to concentrate on QS $\leq$ 25.

The example in Fig. 2 shows the BPG coder can achieve a quite high CR with maintaining good visual quality. However, there are two questions arisen. First, what is the dependency relationship between the image quality and image classification accuracy? Second, is it feasible to set image compression parameters based on metrics designed with taking into account human visual system?

In order to determine the impact of lossy compression on image classification, we conducted more detailed experiments. We set the compression parameter in the range from 1 to 25 with the step of Q equal to minimal possible value of 1. Nine images were compressed, decompressed, and the decompressed images and their visual quality was evaluated. These images were also classified and analyzed in the sense of total probability of correct classification (PCC) and Kappa coefficients. The areas that can be used for calculating these indicators for the image RSa6 are presented in Fig. 3(a) (the areas used for training are shown in Fig. 3(b) and the areas employed for actual verification are represented in Fig. 3(c)). Based on the actual data on the land use presented in these images, relatively homogeneous fragments of images representing separate classes have been identified by experts. Each of the selected fragments has been marked with a conditional color corresponding to a certain class: Urban (Build-up) - red, Farmland - green, Forest - cyan, Meadow - yellow, Water - blue.

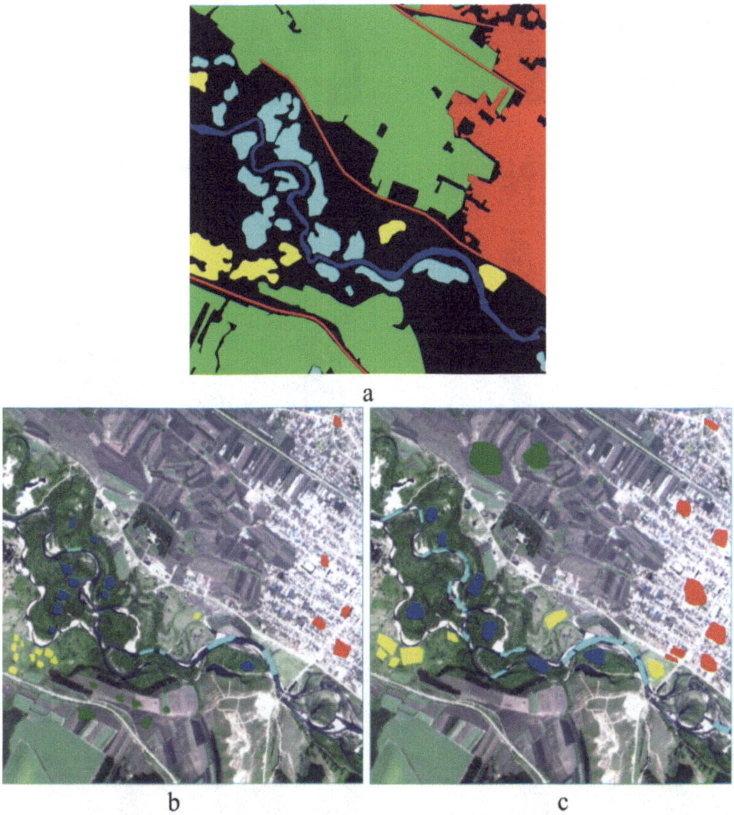

Fig. 3. Fragments used for classifier verification for the image RSa6(a) and the areas used for SVM training

After obtaining the feature sets (three-element vectors of the image values for the corresponding pixels) of the considered classes, the classification has been carried out using the trained SVM. The examples of the obtained classification maps are show in Fig. 4. The decompressed images to be classified correspond to Q = 1, 6, 11, 16, 21, and 30, respectively. The classification results seem quite similar but attentive analysis shows certain differences.

Fig. 4. The classification maps for the image RSa6 compressed with different values of Q

4 Accuracy Analysis

We have calculated the confusion matrices for the classified compressed images. One example for the image fragment RSa6 is presented in Table 1 (Q = 10). As one can see, all the classes are recognized well enough where the classes Forest, Farmland, and Water are recognized in the best manner. The largest probability of misclassifications is observed for the classes Farmland and Meadow, which is not surprising because they have quite similar features.

Table 2 gives data concerning dependencies of probabilities of correct classification for particular classes and in aggregate (overall) on Q. Analysis of these data allows concluding the following:

1) For very small Q (<5), the results are almost identical (not dependent on Q);
2) There are some values of Q (e.g., 5 and 10) for which there are "jumps" of probabilities of correct classification for particular classes but the jumps of the overall PCC are not so essential;
3) For Q < 11, the overall PCC is almost stable, for the larger Q the PCC starts to decrease (in general);
4) If we set an "allowed" reduction of PCC equal to 2%, then the largest Q is equal to 17; for a larger Q, PCC starts to decrease faster and with a larger amplitude jumps (variations).

Recall here that since the default mode of the BPG encoder (the format 4:2:2) has been studied, PSNR for Q = 1 was in the limits from 37 dB to 42 dB (for nine image fragments that were considered). This corresponds to invisible distortions. Meanwhile, for Q = 17, PSNR was in the limits from 36 dB to 40 dB. For Q = 25, the limits were from 34 dB to 36.5 dB, that is a considerable reduction of image quality took place compared to the case of Q = 1.

To control losses, we have also used the visual quality metric MDSI (the larger the worse) that was in the limits from 0.04 to 0.06 for Q = 1 and from 0.082 to 0.094 for Q = 25. This also means that a sufficient increase of MDSI took place and this corresponded to reduction of compressed image visual quality.

Certainly, it is not fair to draw conclusions based only on analysis of data for one image fragment. Because of this, we have also obtained the classification results for other image fragments. They are partly given below.

For all nine test images, we have obtained the dependencies of classification accuracy characterized by probability of correct classification (PCC) and kappa parameter on the parameter Q (Fig. 5). Their analysis shows the following:

1) Whilst PCC for Q = 1 varies in the limits from 91% to 98%, its variation limits are from 86% to 97% for the data compressed with Q = 25.

Table 1. Confusion matrix for the image RSa6 compressed by the BPG coder using Q = 10.

Class	Probability				
	Urban	Farmland	Forest	Meadow	Water
Urban	0.92	0.02	0	0	00.1
Farmland	0.08	0.94	0.02	0.08	0.01
Forest	0	0.02	0.97	0.03	0.04
Meadow	0	0.02	0	0.89	0
Water	0	0	0.01	0	0.94

Table 2. Probability of correct classification for particular classes and in aggregate for the image RSa6 (in %)

a6	Built-up	Farmland	Forest	Meadow	Water	Overall accuracy	Kap-pa
1	92.24	93.49	97.04	89.23	94.19	93.52	90.75
2	92.25	93.48	97.05	89.24	94.21	93.52	90.75
3	92.25	93.48	97.05	89.24	94.21	93.52	90.75
4	92.25	93.47	97.05	89.18	94.19	93.51	90.74
5	92.20	93.42	97.15	89.3	94.31	93.51	90.74
6	92.16	93.42	97.06	89.17	94.28	93.47	90.68
7	92.24	93.33	96.93	89.60	94.33	93.47	90.68
8	92.06	93.39	97.03	88.74	93.77	93.37	90.54
9	92.11	93.03	97.27	89.46	92.49	93.31	90.46
10	91.95	93.12	96.82	88.95	94.15	93.22	90.32
11	91.95	92.99	97.10	87.99	93.71	93.10	90.16
12	91.87	92.89	96.88	88.42	93.91	93.04	90.07
13	91.87	92.87	96.81	87.61	93.71	92.93	89.91
14	91.77	92.62	96.92	88.03	94.01	92.87	89.83
15	91.38	92.76	96.82	85.89	93.65	92.59	89.43
16	91.46	92.5	96.87	86.78	93.04	92.60	89.44
17	91.40	92.13	96.97	86.04	92.55	92.34	89.07
18	91.26	91.96	96.88	85.74	92.26	92.17	88.84
19	90.84	91.37	96.80	86.05	91.87	91.83	88.35
20	90.43	91.72	96.57	83.35	91.85	91.56	87.94
21	90.21	91.31	96.66	82.47	91.29	91.24	87.49

(continued)

Table 2. (*continued*)

a6	Built-up	Farmland	Forest	Meadow	Water	Overall accuracy	Kap-pa
22	90.28	90.63	96.58	82.51	91.57	90.95	87.09
23	89.66	90.70	96.48	81.01	90.99	90.65	86.65
24	89.76	89.91	96.16	82.19	90.38	90.36	86.26
25	89.30	89.45	96.29	77.75	90.16	89.61	85.18

2) Similar reduction of the kappa parameter has been observed; thus, compression by the BPG encoder with Q = 25 often leads to sufficient (not allowed) reduction of PCC and kappa parameter, especially for such images for which PCC for images compressed with Q = 1 was quite low.
3) Meanwhile, the reductions of PCC and Kappa are mostly in the allowed limits for $Q \leq 17$. This approximately corresponds to PSNR about 38 dB and MDSI of about 0.06.

It is also interesting that the dependencies of PCC and Kappa parameter on Q can be both monotonous and fluctuating (the local fluctuations reach a few %).

We have also calculated the Spearman rank order correlation coefficients (SROCCs) between the considered quality metrics (PSNR and MDSI) and the classification accuracy (PCC). It occurred to be rather high (exceeds 0.957) for particular image fragments. Meanwhile, SROCC values for all image fragments considered together are not so "optimistic" – SROCC magnitude values are about 0.5 (they are positive for PCC and PSNR and negative for PCC and MDSI). This means that that there are other factors influencing PCC except of image quality (probably, image complexity).

We have also analyzed the dependencies of probability of correct classification for particular classes on Q. The dependence for the class Water is presented in Fig. 6 and for the class Build-up in Fig. 7.

As one can see here, more dependencies have a fluctuating behavior where amplitudes of fluctuations can reach 7–8% and they are larger than for overall PCC (Fig. 5). We associate these fluctuations with two factors. First, numbers of pixels corresponding to classes in image fragments can be small (in statistical sense). Second, there can be effects in classification accuracy like those with quality metrics for strange images discovered in [29].

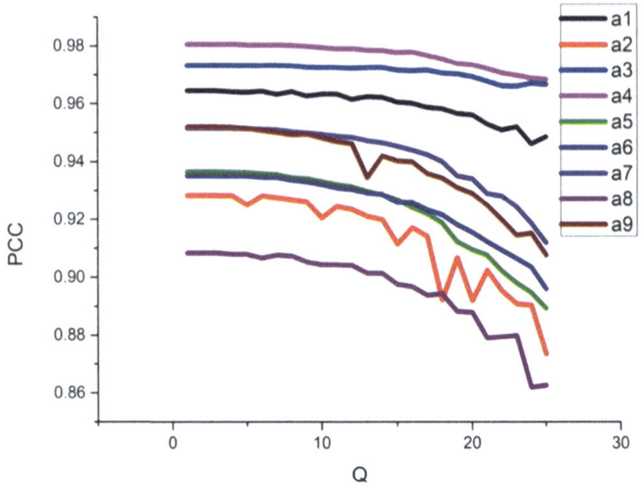

a

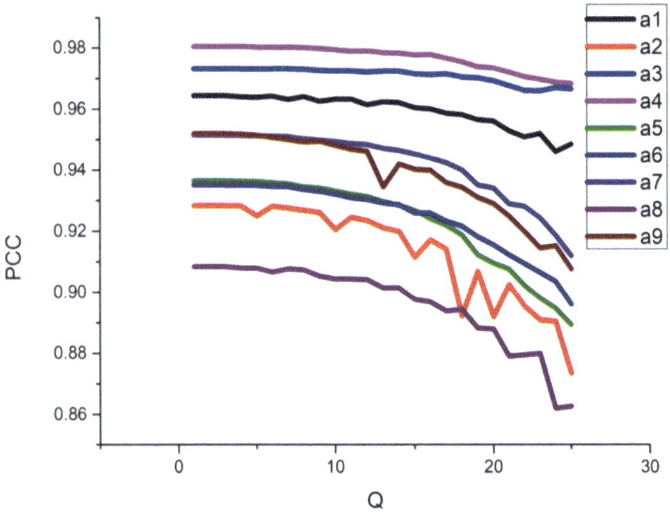

b

Fig. 5. PCC and Kappa for nine fragments as functions of Q

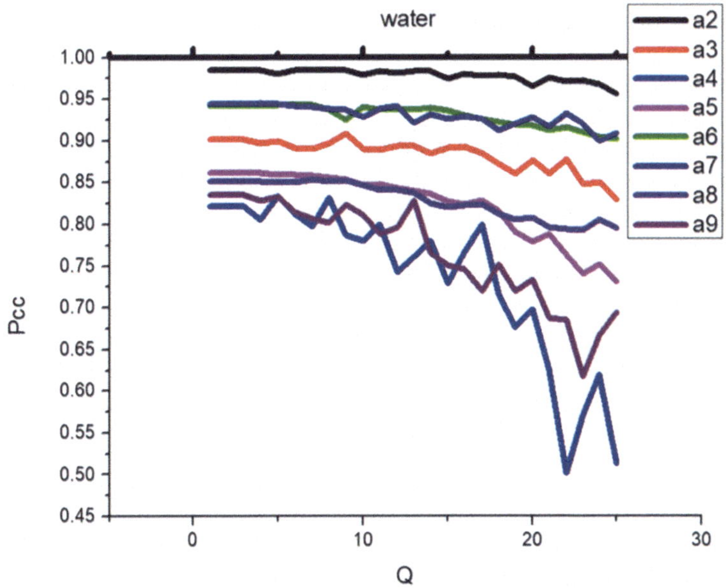

Fig. 6. Dependence of PCC for the class Water on Q for the image fragments

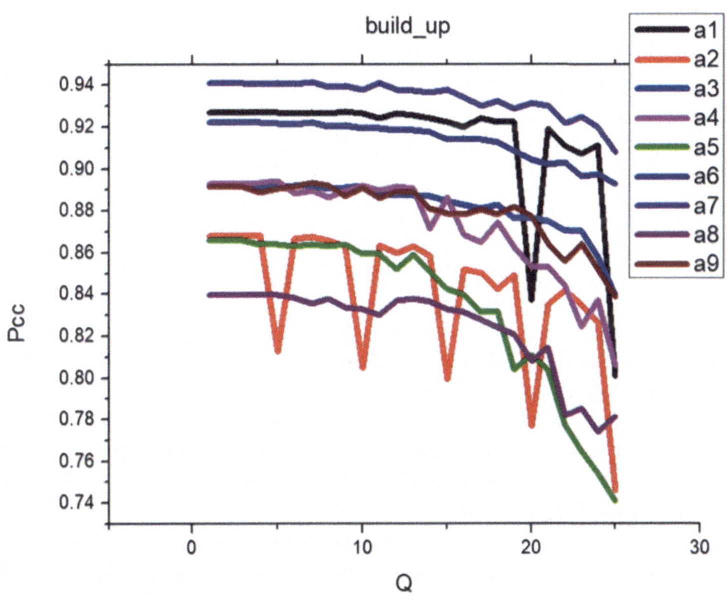

Fig. 7. Dependence of PCC for the class Build-up on Q for the image fragments

5 Conclusions and Future Work

We have considered the use of the SVM classifier in processing of three-channel images compressed by the BPG coder employing the 4:2:2 format where the classifier has been trained for uncompressed data. It has been demonstrated that the compression, in general, leads to reduction of classification accuracy. For $Q < 5$, the reduction is negligible. For $Q \leq 17$, it is in allowed limits. For $Q = 25$, PCC reduction might reach 5%, and this can be treated as a too large payment for RS data size reduction due to lossy compression. It has been also shown that dependencies of the overall PCC and PCC for particular classes on Q might have fluctuations where PCCs for particular classes have more intensive fluctuations. The reasons for this effect have to be additionally studied.

In the future, it is worth considering other formats of the BPG-based lossy compression as well as variants of the SVM training using the already compressed images.

References

1. Li, D., Shao, Z., Zhang, R.: Advances of geo-spatial intelligence at LIESMARS. Geo-Spat. Inf. Sci. **23**(1), 40–51 (2020)
2. Joshi, N., et al.: A review of the application of optical and radar remote sensing data fusion to land use mapping and monitoring. Remote Sens. **8**, 70 (2016)
3. Zhang, B., et al.: One-two-one networks for compression artifacts reduction in remote sensing. J. Photogramm. Remote Sens. **145**, 184–196 (2018)
4. Christophe, E.: Hyperspectral data compression tradeoff. In: Prasad, S., Bruce, L., Chanussot, J. (eds.) Optical Remote Sensing. Augmented Vision and Reality, vol. 3, pp. 9–29. Springer, Berlin, Heidelberg (2011). https://doi.org/10.1007/978-3-642-14212-3_2
5. Penna, B., Tillo, T., Magli, E., Olmo, G.: Transform coding techniques for lossy hyperspectral data compression. IEEE Trans. Geosci. Remote Sens. **45**(5), 1408–1421 (2007)
6. Hussain, A.J., Al-Fayadh, A., Radi, N.: Image compression techniques: a survey in lossless and lossy algorithms. Neurocomputing **300**, 44–69 (2018)
7. Balasubramanian, R., Ramakrishnan, S.S.: Wavelet application in compression of a remote sensed image. In: 2013 the International Conference on Remote Sensing, Environment and Transportation Engineering (RSETE 2013), pp. 659–662 (2013)
8. Syahrul, E.: Lossless and nearly-lossless image compression based on combinatorial transforms. Université de Bourgogne (2011)
9. Blau, Y., Michaeli, T.: Rethinking lossy compression: the rate-distortion-perception tradeoff. In: International Conference on Machine Learning, pp. 675–685. PMLR (2019)
10. Lukin, V., et al.: Lossy compression of multichannel remote sensing images with quality control. Remote Sens. **12**(22), 3840 (2020)
11. Ozah, N., Kolokolova, A.: Compression improves image classification accuracy. In: Meurs, M.J., Rudzicz, F. (eds.) Advances in Artificial Intelligence. Canadian AI 2019. LNCS, vol. 11489, pp. 525–530. Springer, Cham (2019). https://doi.org/10.1007/978-3-030-18305-9_55
12. Lukin, V., Zemliachenko, A., Krivenko, S., Vozel, B., Chehdi, K.: Lossy compression of remote sensing images with controllable distortions. In: Satellite Information Classification Interpretation, pp. 27–43. Intech Open, London (2018)
13. Valsesia, D., Magli, E.: A novel rate control algorithm for onboard predictive coding of multispectral and hyperspectral images. IEEE Trans. Geosci. Remote Sens. **52**(10), 6341–6355 (2014)

14. Zabala, A., Pons, X.: Impact of lossy compression on mapping crop areas from remote sensing. Int. J. Remote Sens. **34**(8), 2796–2813 (2013)
15. Zabala, A., Pons, X.: Effects of lossy compression on remote sensing image classification of forest areas. Int. J. Appl. Earth Obs. Geoinf. **13**(1), 43–51 (2011)
16. Radosavljević, M., et al.: Lossy compression of multispectral satellite images with application to crop thematic mapping: a HEVC comparative study. Remote Sens. **12**(10), 1590 (2020)
17. Chen, Z., Hu, Y., Zhang, Y.: Effects of compression on remote sensing image classification based on fractal analysis. IEEE Trans. Geosci. Remote Sens. **57**, 4577–4590 (2019)
18. Iswanto, B.H., Fauzan, A., Yudha, G.D.: Classification of satellite imagery for identifying land-cover objects using ECW compression images: the case of Makassar city area. J. Phys. Conf. Ser. **2377**, 012017 (2022)
19. Giannopoulos, M., Aidini, A., Pentari, A., Fotiadou, K., Tsakalides, P.: Classification of compressed remote sensing multispectral images via convolutional neural networks. J. Imaging **6**, 24 (2020)
20. Blanes, I., Zabala, A., Moré, G., Pons, X., Serra-Sagristà, J.: Classification of hyperspectral images compressed through 3D-JPEG2000. In: Lovrek, I., Howlett, R.J., Jain, L.C. (eds.) Knowledge-Based Intelligent Information and Engineering Systems, KES 2008. LNCS, vol. 5179, pp. 416–423. Springer, Berlin, Heidelberg (2008). https://doi.org/10.1007/978-3-540-85567-5_52
21. Proskura, G., Vasilyeva, I., Fangfang, L., Lukin, V.: Classification of compressed multichannel images and its improvement. In: 2020 30th International Conference Radioelektronika (RADIOELEKTRONIKA), Bratislava, Slovakia, pp. 1–6. IEEE (2020)
22. Li, F., Kryvenko, S., Lukin, V.: Remote sensing image lossy compression based on JPEG with controlled visual quality. In: Urbach, H.P., Jiang, H. (eds.) Proceedings of the 7th International Symposium of Space Optical Instruments and Applications, ISSOIA 2022. Springer Proceedings in Physics, vol. 295, pp. 8–19. Springer, Singapore (2023). https://doi.org/10.1007/978-981-99-4098-1_2
23. Makarichev, V., Vasilyeva, I., Lukin, V., Vozel, B., Shelestov, A., Kussul, N.: Discrete atomic transform-based lossy compression of three-channel remote sensing images with quality control. Remote Sens. **14**, 125 (2021)
24. BPG Image format. https://bellard.org/bpg/. Accessed 05 May 2022
25. Albalawi, U., Mohanty, S.P., Kougianos, E.: A hardware architecture for better portable graphics (BPG) compression encoder. In: 2015 IEEE International Symposium on Nanoelectronic and Information Systems, Indore, India, pp. 291–296. IEEE (2015)
26. https://www.researchgate.net/publication/359869096_Quality_Control_for_the_BPG_Lossy_Compression_of_Three-Channel_Remote_Sensing_Images
27. Ziaei Nafchi, H., Shahkolaei, A., Hedjam, R., Cheriet, M.: Mean deviation similarity index: efficient and reliable full-reference image quality evaluator. IEEE Access **4**, 5579–5590 (2016)
28. Tong, X.-Y., Lu, Q., Xia, G.-S., Zhang, L.: Large-scale land cover classification in Gaofen-2 satellite imagery. In: IGARSS 2018 - 2018 IEEE International Geoscience and Remote Sensing Symposium, Valencia, pp. 3599–3602. IEEE (2018)
29. Li, F., Lukin, V., Kryvenko, S., Bondzulic, B., Bujakovic, D., Pavlovic, B.: Strange images in remote sensing and their properties. UJRS **10**, 12–18 (2023)

High-Resolution Spectrometers in Exploring Planetary Atmospheres: Current Status and Advancements

Yiqi Li[1,2,3], Xiaobing Sun[2,3](✉), and Xiaobing Zheng[1,2,3]

[1] School of Environmental Science and Optoelectronic Technology, University of Science and Technology of China, Hefei 230026, China

[2] Anhui Institute of Optics and Fine Mechanics, Hefei Institutes of Physical Science, Chinese Academy of Sciences, Hefei 230031, China
xbsun@aiofm.ac.cn

[3] Key Laboratory of Optical Calibration and Characterization, Chinese Academy of Sciences, Hefei 230031, China

Abstract. High-resolution spectrometers have ushered in a new era of planetary exploration, offering high resolving power that enables the detection of narrow absorption lines and subtle spectral features. Recent breakthroughs in component technologies have facilitated the development of compact, lightweight spectrometers tailored for space missions. Noteworthy examples include CIRS on Cassini and TEXES, as well as SPICAV on Venus Express, which have significantly enhanced our understanding of Saturn, Titan, Jupiter and Venus. Despite their remarkable capabilities, these instruments face challenges such as prolonged exposure times and narrow slit widths, which can impact temporal resolution and calibration accuracy. To tackle these obstacles, researchers are actively exploring solutions like micro-electro-mechanical systems (MEMS) and transition edge sensors (TESs). High-resolution spectrometers with expanded spectral coverage are indispensable tools in unraveling the complexities of planetary atmospheres. Ongoing missions, such as the James Webb Space Telescope, will leverage these instruments to scrutinize exoplanet atmospheres and search for potential biosignatures. Continual technological advancements ensure that high-resolution spectrometers remain at the forefront of planetary science, continually unlocking the mysteries of the cosmos.

Keywords: High-resolution spectrometer · spectrograph · planetary atmosphere · technological advancements

1 Introduction

High-resolution spectroscopy has revolutionized our ability to study and characterize planetary atmospheres. By resolving molecular absorption and emission lines into very narrow spectral features, high-resolution spectrometers provide critical insights into atmospheric composition, thermal structure, chemistry, and dynamics. Recent technological advancements have enabled the development of high-resolution spectrometers with exceptional capabilities for space-based and ground-based studies of planetary atmospheres.

H. P. Urbach et al. (Eds.): ISSOIA 2023, SPP 191, pp. 366–381, 2024.
https://doi.org/10.1007/978-981-97-6718-2_32

In the realm of planetary exploration, high-resolution spectroscopy stands as a beacon of scientific progress, offering unprecedented access to the intricate details of celestial atmospheres. The power of this technique lies in its capacity to dissect the spectra of planetary atmospheres, unraveling the unique fingerprints of molecules and processes. The precision of high-resolution spectrometers opens a window into the chemical intricacies, unveiling the composition and interactions shaping these distant worlds.

2 Classification of High-Resolution Spectrometers in Exploring Planetary Atmospheres

High-resolution spectrometers employ diverse techniques to disperse light into its constituent wavelengths and accurately measure the intensity of narrow spectral features. These techniques are grounded in fundamental principles, categorizing high-resolution spectrometers into three main types: Fourier transform spectrometers, grating spectrometers, and prism spectrometers. Each of these techniques offers unique advantages and applications in spectroscopic analysis, contributing significantly to our understanding of planetary atmospheres and their complex compositions.

2.1 Fourier Transform Spectrometers

Fourier transform spectrometers [1] utilize a Michelson interferometer with a moving mirror, these spectrometers generate interference patterns, which, upon Fourier transformation, yield intensity spectra. This technique offers superior spectral resolution, facilitating the measurement of even far-infrared spectra relevant to planetary atmospheric studies, provided suitable IR optics and detectors are employed. Fourier transform spectrometers enable the precise identification and quantification of atmospheric components, such as trace gases, aerosols, and temperature variations. The schematic of a Fourier transform spectrometer is shown in Fig. 1.

In planetary atmospheric research, the high spectral resolution achieved by Fourier transform spectrometers is invaluable for discerning subtle spectral signatures associated with atmospheric constituents and phenomena. By exploiting interference phenomena within the Michelson interferometer, these spectrometers can elucidate the composition and dynamics of atmospheres, contributing to our understanding of fundamental processes governing atmospheric chemistry, dynamics, and climate.

Moreover, Fourier transform spectrometers excel in providing broad spectral coverage and high throughput, thanks to their multiplex advantage. This allows simultaneous measurement of all frequencies, facilitating comprehensive investigations of various spectral features across the electromagnetic spectrum. Consequently, Fourier transform spectrometers significantly contribute to advancing our understanding of planetary atmospheres and their role in shaping planetary environments and climate dynamics.

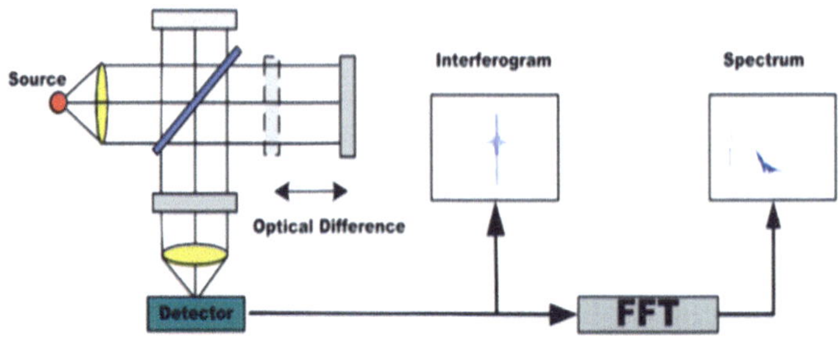

Fig. 1. The schematic of a Fourier transform spectrometer

2.2 Grating Spectrometers

Grating spectrometers [2] provide versatile capabilities for precise analysis of spectral characteristics. By utilizing a diffraction grating as the dispersive component, these spectrometers optimize groove density and geometry to enhance both efficiency and resolution. While the simplest configuration involves a plane grating operating in the first order, enhanced resolution is attainable through the utilization of higher orders and more intricate optical setups like the Czerny-Turner mounting. The schematic of a grating spectrometer is shown in Fig. 2.

In planetary atmospheric research, grating spectrometers serve as indispensable tools for scrutinizing atmospheric compositions and dynamics. The diffraction grating, acting either reflectively or transmissively, possesses parallel grooves etched into its surface. Constructive interference of diffracted light occurs when the optical path difference between grooves aligns with integer multiples of the wavelength, facilitating precise spectral analysis. By varying the incident angle, spectral orders are spatially separated, enabling detailed examination of atmospheric constituents, such as trace gases and particulates.

Gratings are often optimized to maximize efficiency in a desired spectral order, enhancing the sensitivity of atmospheric measurements. Echelle gratings, characterized by coarse rulings and steep blaze angles, offer exceptionally high dispersion and are

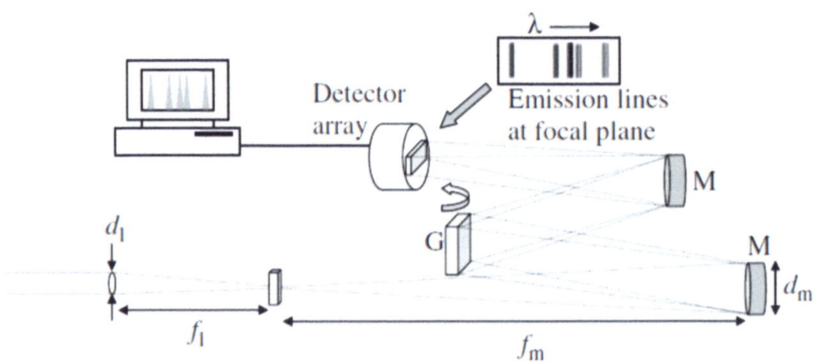

Fig. 2. The schematic of a grating spectrometer

employed in cross-dispersed spectrometers to further augment resolution. Grating spectrometers offer invaluable understanding into the intricate composition and dynamics of planetary atmospheres, aiding in the understanding of atmospheric processes, chemical reactions, and climate dynamics.

2.3 Prism Spectrometers

Prism spectrometers [3] operate on the fundamental principle of dispersion, separating incoming light into its constituent wavelengths, enabling detailed analysis of spectral features. A prism, typically crafted from high-quality optical glass, serves as the dispersing element, refracting light at varying angles depending on wavelength. This dispersion effect facilitates the separation of colors, revealing distinct spectral signatures associated with different atmospheric constituents and phenomena. The schematic of a grating spectrometer is shown in Fig. 3.

In planetary atmospheric exploration, prism spectrometers play a crucial role in characterizing atmospheric compositions and understanding atmospheric processes. By dissecting incoming light into its spectral components, prism spectrometers unveil the presence of various gases, aerosols, and other atmospheric constituents. The distinctive spectral signatures obtained facilitate the identification and quantification of trace gases, aiding in the study of atmospheric chemistry and dynamics. In addition, prism spectrometers can help us understand atmospheric temperature fluctuations, cloud patterns and atmospheric scattering phenomena.

Prism spectrometers offer several advantages in planetary atmospheric research. Their simplicity and reliability make them suitable for remote sensing applications in space missions, providing essential data for understanding the atmospheres of distant planets and moons. Additionally, prism spectrometers can operate across a wide spectral range, from ultraviolet to near-infrared, offering comprehensive coverage of atmospheric features.

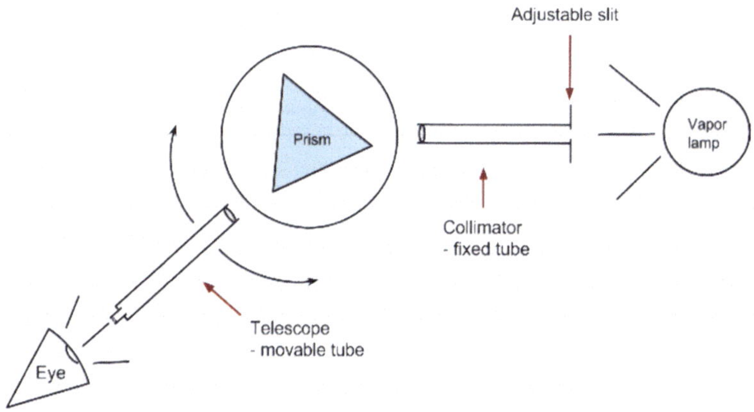

Fig. 3. The schematic of a prism spectrometer

Furthermore, prism spectrometers are often combined with grating spectrometers for enhanced performance. For instance, Zhang et al. [4] proposed a multi-channel hyperspectral imaging system design tailored for simultaneous detection of multiple substances within the ultraviolet range of Venus' atmosphere. This includes the detection of SO_2 and unidentified UV absorber substances within the H_2SO_4 cloud layer. As shown in Fig. 4, the system employs pupil separation prisms and grating multilevel spectra to facilitate multi-channel detection. By splitting the entrance pupil of the telescope using prisms and diffracting spectra from various channels to the same detector through different orders of the grating, the system achieves simultaneous detection of spectra from distinct channels. Featuring a single spectrometer and detector, it not only ensures simultaneous multi-channel detection but also offers advantages such as compact size and ultra-high spectral resolution. According to the system design results, the spectral resolution within the working spectral range of three channels surpasses that of previous or current in-orbit planetary atmospheric detection spectrometers, boasting a resolution better than 0.15 nm.

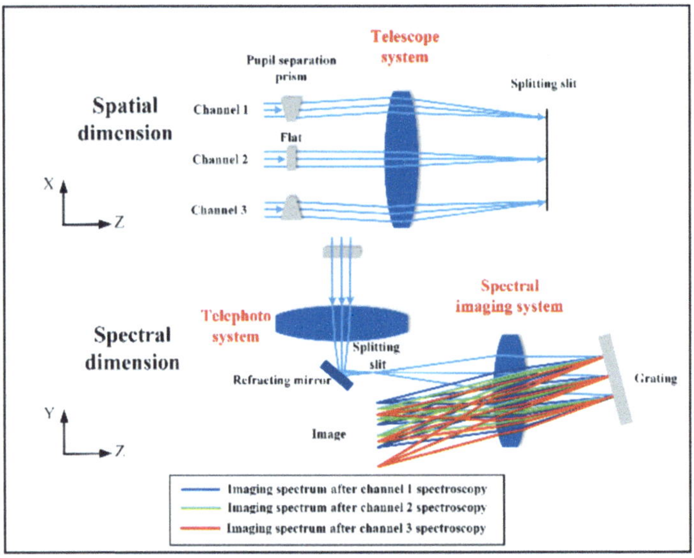

Fig. 4. Multi-channel hyperspectral imaging spectrometer system layout diagram. Note: the blue light represents Channel 1: the distribution schematic of the imaging spectrum on the image plane after spectral dispersion in the 180–220 nm range; the green light represents Channel 2: the distribution schematic of the imaging spectrum on the image plane after spectral dispersion in the 275–305 nm range; and the red light represents Channel 3: the distribution schematic of the imaging spectrum on the image plane after spectral dispersion in the 360–400 nm range.

3 Development of Existing High-Resolution Spectrometers in Exploring Planetary Atmospheres

High-resolution spectrometers have revolutionized our exploration of planetary atmospheres. By capturing detailed data on the composition and dynamics of these atmospheres, these instruments enable scientists to unravel the complexities and nuances of celestial bodies. For instance, high-resolution infrared spectroscopy has been employed in studying planetary atmospheres, as evidenced by research on Venus and various exoplanets. Additionally, the combination of low- and high-resolution spectroscopy has proven instrumental in examining the atmospheres of transiting exoplanets, aiding in tasks like radial velocity measurement and line broadening analysis. Furthermore, ground-based telescopes utilizing high-resolution near-infrared spectroscopy have provided valuable insights into exoplanetary atmospheres, offering a powerful tool for comprehensive studies. The following introduces the significant advancements and key instruments of high-resolution spectrometers in recent years for the exploration of planetary atmospheres.

3.1 Composite Infrared Spectrometer (CIRS) on Cassini

The Cassini spacecraft [5], in orbit around Saturn since July 2004, carries the Composite Infrared Spectrometer (CIRS) designed to analyze thermal emissions from Saturn, its rings, and moons. Functioning as a Fourier Transform Spectrometer (FTS), CIRS is a crucial part of the payload, providing unique measurements and synergies with other instruments. Over Cassini's 13-year mission, CIRS has surpassed the capabilities of previous spectrometers on Voyager 1 and 2.

CIRS, featuring two interferometers sharing a telescope and a scan mechanism, spans a broad wavelength range in the mid and far infrared. It examines the atmospheres of Jupiter, Saturn, and Titan, along with the rings of Saturn and surfaces of icy moons, studying temperature, composition, structure, and dynamics. As the mission concludes in 2017, CIRS will leave behind a wealth of scientific results, encompassing over 160 million spectra. The instrument has played a pivotal role in understanding seasonal changes on Saturn and Titan, the characteristics of Saturn's rings, and the compositions and temperatures of icy moon surfaces.

The scientific objectives of CIRS on the Cassini mission encompassed studying the composition, thermal structure, and dynamics of Saturn and Titan's atmospheres, the thermal characteristics and composition of Saturn's rings, and the composition and temperatures of icy moon surfaces. Observations were global, covering various parameters, with an anticipation of discovering time variability as Saturn and Titan progressed through their seasons. The Cassini mission, a collaborative effort involving NASA, ESA, and the Italian Space Agency, has significantly contributed to our understanding of Saturn and its celestial companions. Comprising two interferometers sharing a scan mechanism, CIRS utilizes a common telescope.

The telescope's focal plane divides the field image between the far-infrared and mid-infrared interferometers. In the far infrared, a Martin–Puplett type polarization interferometer covers the spectral range from 10 to 600 cm^{-1}. Meanwhile, in the mid-infrared, a conventional Michelson interferometer covers two spectral segments: 580–1100 cm^{-1} and 1050–1500 cm^{-1}. These interferometers operate on distinct principles.

In the polarization interferometer, the beam splitter separates the beam into two orthogonal polarization components. The phase between these components varies during the scan, modulating the polarization state of the recombined beam. In contrast, the conventional Michelson interferometer splits the amplitude of the electric field into two approximately equal parts (see Fig. 5).

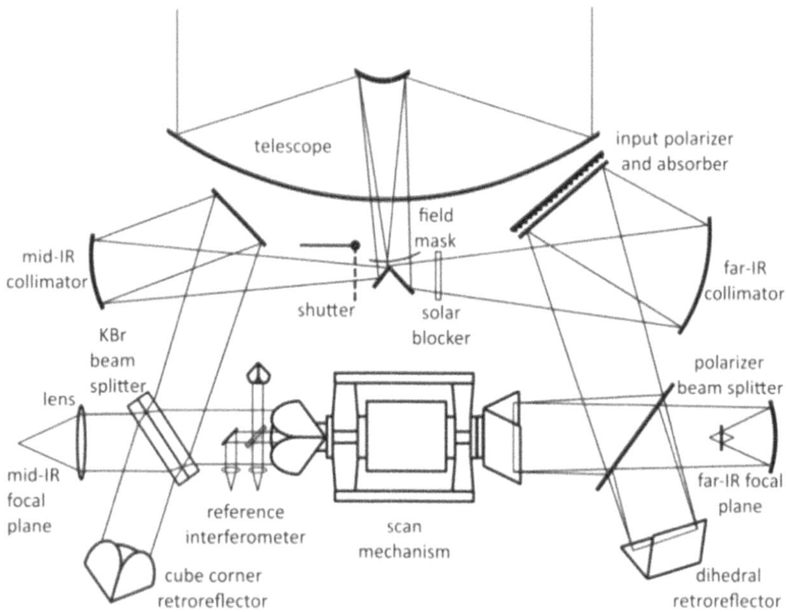

Fig. 5. The conceptual diagram of CIRS depicts two interferometers, one for the mid-infrared and the other for the far-infrared, sharing a telescope and a scanning mechanism. Within the mid-infrared beam, a reference interferometer is introduced in front of the moving retroreflector. Calibration procedures involve referencing views of deep space and observations with the shutter closed.

3.2 SPICAV on Venus Express

Spectroscopy for the Investigation of the Characteristics of the Atmosphere of Venus (SPICAV) [6], is a suite of three spectrometers operating in the UV and IR ranges, with a combined mass of 13.9 kg, installed on the Venus Express (VEX) orbiter. It is designed to probe the atmosphere of Venus from surface level up to the outermost hydrogen corona at altitudes exceeding 40,000 km. Derived from the successful SPICAM instrument aboard the Mars Express (MEX), SPICAV features enhancements, including a new IR high-resolution spectrometer called Solar Occultation IR (SOIR), optimized for solar occultation studies.

The instrument comprises three spectrometers and a straightforward data processing unit facilitating communication between these channels and the spacecraft. One of the spectrometers is a UV instrument (118–320 nm, resolution 1.5 nm), identical to its counterpart on MEX. It is utilized for nadir, limb, and vertical profiling through stellar and solar occultation observations. In nadir mode, SPICAV UV analyzes albedo spectra to retrieve sulfur dioxide (SO_2) concentrations and map the distribution of UV-blue absorbers, aiding in cloud structure and atmospheric dynamics investigations. On the nightside, it studies NO bands and emissions from electron precipitation. During stellar occultations, the UV sensor measures vertical profiles of CO_2, temperature, SO_2, sulfur monoxide (SO), clouds, and aerosols, providing crucial data for atmospheric modeling from 60 km to 160 km altitude, crucial for future aerocapture and aerobraking missions. Additionally, it examines the upper atmosphere, studying the ionosphere and hydrogen corona, contributing to understanding atmospheric escape mechanisms and long-term evolution.

The SPICAV VIS-IR sensor (0.7–1.7 μm, resolution 0.5–1.2 nm) features an innovative acousto-optical tunable filter (AOTF). On the nightside, it observes thermal emission penetrating through the clouds, complementing observations by VIRTIS and the Planetary Fourier Spectrometer (PFS) on VEX. In solar occultation mode, this channel investigates the vertical structure of water vapor (H_2O), carbon dioxide (CO_2), and aerosols. The optical scheme of the UV and IR channels of SPICAV is shown in Fig. 6.

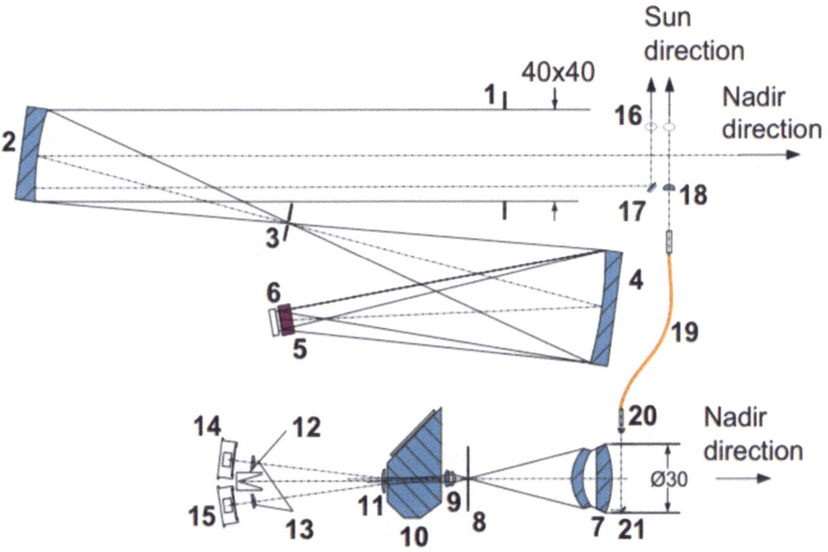

Fig. 6. Optical scheme of the UV and IR channels of SPICAV. 1—mechanical stop of the UV channel aperture; 2—off-axis parabolic mirror;3—slit (can be removed by a mechanical actuator, see text); 4—concave UV grating; 5—intensifier; 6—CCD; 7—IR channel objective; 8—IR FOV diaphragm; 9, 11—collimating lenses; 10—AOTF crystal; 12—light trap for undiffracted light; 13—detector proximity lenses; 14—"extraordinary" beam detector; 15—"ordinary" beam detector; 16—solar opening (closed by shutter when not looking toward the Sun); 17, 21—flat mirror; 18—IR solar entry; 19—optical fibre; 20—fibre collimator.

The Solar Occultation Infrared (SOIR) spectrometer is a groundbreaking addition to the Venus Express (VEX) mission, operating within the wavelength range of 2.2–4.3 μm with an exceptional spectral resolution of $\lambda/\Delta\lambda > 15000$, the highest among instruments aboard VEX. As shown in Fig. 7, this innovative design combines an echelle grating with an acousto-optical tunable filter (AOTF) crystal, enabling the selection of one spectral order at a time. The primary objective of SOIR is to conduct precise measurements of deuterated water (HDO) and water vapor (H_2O) during solar occultations, aiming to understand the escape of deuterium (D) atoms from Venus' upper atmosphere and provide insights into the planet's water evolution. Additionally, SOIR investigates isotopes of carbon dioxide (CO_2) and minor atmospheric constituents, facilitating the sensitive detection of potential new species in Venus' upper atmosphere.

SOIR also endeavors to capture nightside emissions, offering a sensitive assessment of HDO in the lower atmosphere, which can be compared with upper atmospheric ratios. This capability may lead to the discovery of new minor atmospheric constituents.

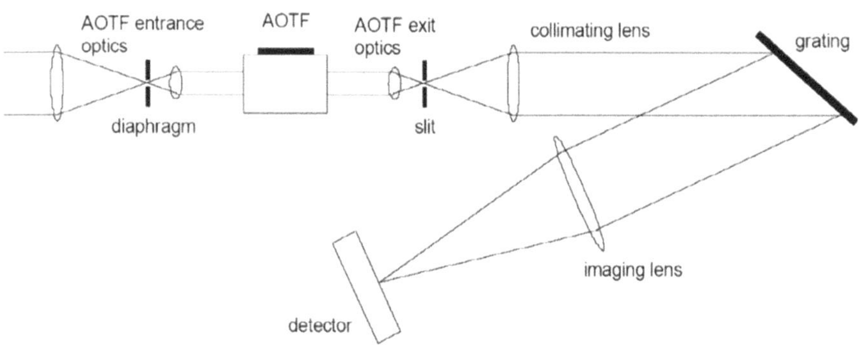

Fig. 7. Simplified optical scheme of SOIR.

3.3 Texas Echelon Cross Echelle Spectrograph (TEXES)

TEXES, the Texas Echelon Cross Echelle Spectrograph [7], is a cutting-edge mid-infrared spectrograph designed for high-resolution observations of interstellar molecules. Operating within the wavelength range of 7–25 μm, TEXES offers three spectrographic modes: a high-resolution cross-dispersion mode (R ~ 100,000), a mid-resolution long-slit mode (R ~ 15,000), and a low-resolution long-slit mode (R ~ 2000).

In the high-resolution mode, TEXES employs a 36-inch long R10 grating echelon with a 7 mm groove spacing, cross-dispersed with a 7-inch long R2 echelle. The mid-resolution mode bypasses the echelon with an Offner relay, utilizing the echelle independently. In the low-resolution mode, a first-order grating is inserted over the echelle.

In designing the TEXES spectrograph, critical decisions were made concerning the echelon grating's parameters, aiming for very high resolution. The diffraction-limited resolution of a grating is determined by the path-length difference between its ends, with a 1-m-long grating potentially achieving a resolution of R = 100,000. However, practical limitations, including unacceptable light loss at the spectrograph entrance slit, led to the consideration of a multi-pass configuration with a smaller grating. Yet, this approach presented efficiency challenges due to light loss on each pass and issues with scattered and diffracted light.

Supporting the echelon grating structure involves attaching the 100 echelon facets to a $2 \times 2 \times 40$ in aluminum bar using thin aluminum support tabs that are clamped to the bar. The overall stability is ensured by three rigid supports and 20 lever-arm supports. These supports not only bear the weight of the structure but also compensate for any changes in orientation or flexure of the base plate. This design aims to maintain the alignment and stability of the echelon grating, allowing it to function effectively within the TEXES spectrograph.

Imaging spectroscopy in the infrared range, particularly utilizing the TEXES imaging spectrometer at the Infrared Telescope Facility (IRTF), has proven instrumental in monitoring minor species' behavior in planetary atmospheres. Ground-based instruments like TEXES offer superior resolving power compared to space-borne counterparts, enabling detailed examination of molecular lines in the Martian atmosphere and stratospheric emissions in giant planets.

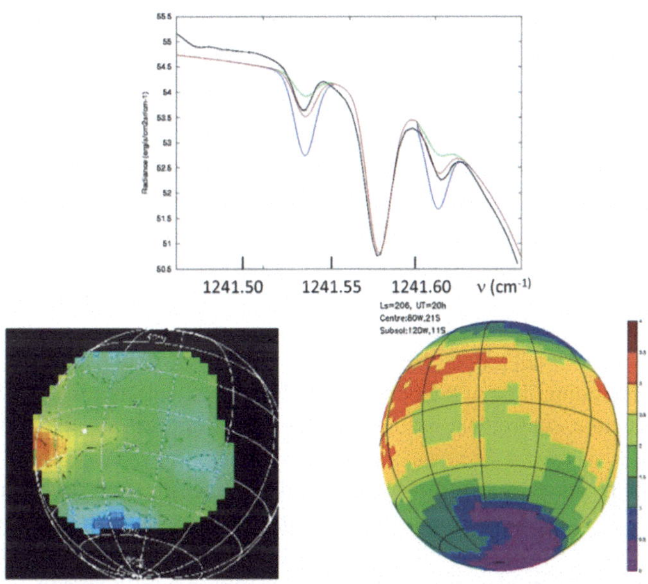

Fig. 8. (Color online.) Top: the TEXES spectrum of Mars (black curve) around 1241.6 cm^{-1} showing a doublet of H_2O_2 transitions (at 1241.53 and 1241.61 cm^{-1}) bracketing a CO_2 transition (at 1241.58 cm^{-1}). Line positions correspond to rest frequencies. Models: $[H_2O_2] = 20$ ppb (green), 40 ppb (red, best fit), and 80 ppb (blue). The spectrum is integrated in a region of the Martian disk where the H_2O_2 abundance is maximum. Bottom: left: map of the line depth ratio of H_2O_2/CO_2 retrieved from the TEXES data recorded in June, 2003 (Ls = 206°). Right: GCM synthetic map of the H_2O_2 mixing ratio under the same observing conditions.

TEXES has played a crucial role in detecting and monitoring various compounds on Mars [8], including hydrogen peroxide (H_2O_2), a key molecule linked to the absence of organics on the Martian surface. Figure 8 shows the results of TEXES detection of hydrogen peroxide in the Martian atmosphere. Observations extended to Venus and Jupiter, revealing spatio-temporal variations of sulfur dioxide (SO_2) above Venus clouds, challenging existing models. Ground-based imaging spectroscopy, exemplified by TEXES, allows for high-resolution mapping of minor species across planetary disks, facilitating the study of daily variations and transient phenomena.

3.4 Tuneable Heterodyne Infrared Spectrometer (THIS)

THIS [9] operates in the mid-infrared range, spanning from 7 to 13 mm, with an extension to 17 mm in development. It boasts a remarkable spectral resolution exceeding 10^7 and a bandwidth of two times 3 GHz (equivalent to 30 km/s at 10 mm). Primarily utilized at the McMath-Pierce Solar telescope on Kitt Peak, Arizona, and tested on smaller European telescopes, THIS demonstrates its versatility.

To assess its suitability for high-resolution observations, particularly for weak signals requiring extended integration times, the frequency stability of the system is crucial. Simulating observing conditions, molecular absorption lines were recorded, revealing a peak-to-peak frequency variation of 2 MHz, a mean value of 1065.9 MHz, and a standard deviation of 0.3 MHz over 4500 s. These variations prove sufficiently small for observations lasting up to 1 h. The instrument's achievable spectral resolution is contingent on factors such as the frequency stability of the LO (local oscillator), line width of the LO, frequency resolution and stability of the back-end spectrometer, and IF (intermediate frequency) processing.

For effective operation, the frequency stability of the AOS and IF processing relies on ambient temperature control. Routine measurements with a high-stability comb-generator spectrum, featuring a known base frequency at the observation's outset, ensure absolute frequency control.

In Fig. 9, the system comprises two main sections: an aluminum cube housing optical components of the receiver (80 kg, dimensions: $80 \times 60 \times 45$ cm^3), containing optics for adapting to various telescopes, beam combining optics, detector, and LO. Additional components, including the (AOS) back-end and necessary electronics, are in two 19-inch racks. Remote control via a computer facilitates convenient operation for data acquisition.

IR heterodyne spectroscopy proves invaluable for investigating the composition of planetary atmospheres, particularly for long-term monitoring and spatially resolved observations that are often challenging for space missions. Recent studies include examining the ozone cycle on Mars and studying the distribution of Ethane in Titan's atmosphere, both before and during the Huygens probe insertion. Historically, these observations relied on a CO_2 laser as a local oscillator (LO). THIS, by overcoming this limitation, widens the range of detectable molecules at heterodyne resolution, encompassing CH_4, C_2H_2, and H_2.

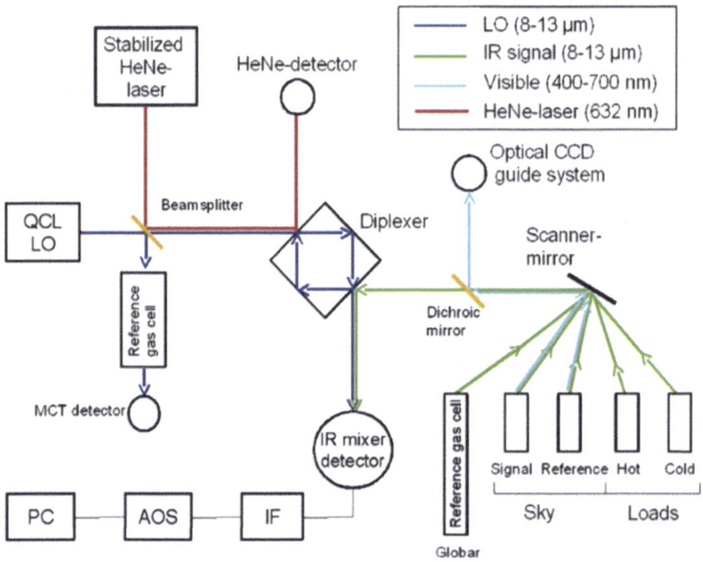

Fig. 9. Schematic view of the setup.

4 Optical Design Considerations

Various optical design factors must be considered in developing high-resolution spectrometers optimized for planetary science. These include spectral coverage, resolution, slit geometry, efficiency, dispersion, calibration capabilities, and stray light control.

Broad spectral coverage is highly desirable to measure multiple chemical species, but often requires cross-dispersion or multiple separate channels. Extending into the UV or IR requires specialized optics. For Fourier transform spectrometers, the maximum optical path difference determines resolution. For grating instruments, the optical configuration, ruling density, groove profile, and grating size affect resolution and dispersion.

The slit geometry impacts the balance between spectral, spatial, and flux throughput resolution. Tall slits enhance spatial information from compact sources, while wide slits maximize collected photons. Image slicers can enhance throughput without sacrificing spectral resolution. Anti-reflection coatings, optical substrates, smooth surfaces, and efficient gratings maximize throughput.

For calibration, onboard wavelength standards like atomic emission lamps allow tracking of instrument drifts. Stable thermal control minimizes drifts, while calibration stars monitor transmission. Baffles, optical stops, and coatings control stray light. No single design can optimize every parameter, so design priorities must be strategically balanced.

5 Technological Advancement of Compact High-Resolution Spectrometers

Fine spectral resolution demands long exposure times to achieve adequate signal-to-noise ratios, which, in turn, limits their temporal resolution and makes the observation of time-varying phenomena challenging. Additionally, the narrow slit widths required for these instruments can complicate absolute flux calibration. The stringent stability requirements impose the need for extremely careful thermal and mechanical design, increasing the complexity and cost of these instruments. To overcome these challenges, researchers are actively exploring several innovative solutions. Emerging technologies such as transition edge sensors (TESs) present notable advancements in compact high-resolution spectrometers, facilitating enhanced detection efficiency and accelerated spectral acquisition. These innovations mark significant progress in recent years within the field of spectroscopy.

5.1 Micro-electro-Mechanical Systems (MEMS)

Micro-electro-mechanical systems (MEMS) technologies [10] allow integration of multi-component optical systems onto microchips using lithographic manufacturing techniques. This enables miniaturized gratings, tunable Fabry-Perot interferometers, and micro-optical benches. MEMS systems provide good stability and low mass, but can suffer from higher stray light compared to traditional instruments.

Mini-spectrometers often face measurement errors arising from dark current, thermal sensitivity, stray light, nonlinearity in photoelectric conversion, and electronic circuitry. Despite their mobility, CCD spectrometers have proven valuable in laboratory settings and are gradually supplanting stationary counterparts. Nevertheless, the evolution of miniaturization has transcended traditional mini-spectrometers, propelled by the emergence of optical sensors integrated with micro-electro-mechanical systems (MEMS) technology.

5.2 Transition Edge Sensors (TESs)

Transition Edge Sensors (TESs) [11] are superconducting devices utilized for detecting electromagnetic radiation, commonly employed in high-resolution spectrometers and other astrophysical applications. Their principle of operation relies on the relationship between the resistance of superconductors and temperature. TESs are typically composed of superconducting materials such as niobium or titanium. They operate near their critical temperature, slightly above it, where the resistance of the superconducting material is relatively low but non-zero. The sensitivity of TESs arises from their highly responsive resistance-temperature characteristics. When exposed to radiation, TESs absorb energy, causing a temperature rise and consequently inducing a drastic change in resistance. To maintain stability around the operating point, TESs are often incorporated into a feedback loop. The feedback circuit adjusts the current passing through the TESs to keep them at the desired operating point. Detection of radiation occurs as the absorbed energy alters the resistance, which can be quantified by monitoring changes in current or voltage in the detection circuit. Overall, TESs offer exceptional sensitivity to faint radiation, making them invaluable sensors in fields like high-resolution spectroscopy for studying celestial objects and phenomena. One significant application of TESs is in the analysis of long-term trends and spatially resolved observations of planetary atmospheres. Unlike space missions, TESs allow for extended monitoring and detailed examinations of various solar system objects. Recent studies have delved into understanding the ozone cycle on Mars and mapping the distribution of ethane in Titan's atmosphere.

TESs play a pivotal role in addressing challenges posed by the Earth's atmosphere. Their high spectral resolution in the mid-IR range enables the study of trace gases with sophisticated retrieval algorithms. This capability allows scientists to deduce vertical concentration profiles up to 100 km altitude from a single, fully resolved molecular line.

Beyond planetary science, TESs find application in exploring the interstellar medium, stellar atmospheres, and even our Sun. The high-resolution spectra obtained with TES can provide valuable understanding of processes occurring in different astrophysical environments, enriching our overall understanding of the universe.

In conclusion, the versatility and high-performance capabilities of Transition Edge Sensors make them indispensable tools for unraveling the complexities of planetary atmospheres and beyond, fostering a deeper comprehension of celestial bodies and their atmospheres.

5.3 Acousto-Optic Tunable Filters (AOTFs)

Acousto-Optic Tunable Filters (AOTFs) have proven to be instrumental in probing the complexities of planetary atmospheres, with notable applications, particularly aboard the Venus Express mission. Venus Express, equipped with cutting-edge AOTF technology [6], has significantly contributed to our understanding of Venusian atmospheric dynamics. AOTFs operate on the principle of acousto-optic interaction, allowing for tunable wavelength selection. On Venus Express, they play a pivotal role in spectroscopic measurements, enabling the analysis of specific spectral bands associated with key atmospheric constituents. The optical design of the AOTF instrument is shown in Fig. 10.

One of the notable advantages of AOTFs is their ability to provide high spectral resolution in a compact and lightweight design—crucial considerations for space missions. On Venus Express, this technology facilitates the precise identification and analysis of molecular features, aiding in the characterization of Venus's atmosphere.

The AOTFs on Venus Express have been employed to study various aspects of Venusian atmospheric composition, including the distribution of trace gases and the dynamics of cloud layers. These observations contribute essential data for unraveling the mysteries of Venus's atmospheric processes.

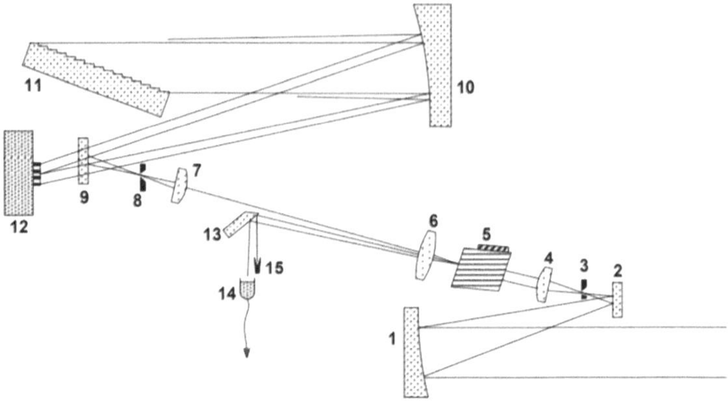

Fig. 10. The optical scheme of the spectrometer: 1- entrance telescope (off-axis parabolic mirror), 2, 9, 13- flat mirrors, 3- field diaphragm, 4- collimator short-focus lens, 5- acousto-optic tunable filter, 6- additional long-focus lens, 7- focusing short-focus lens, 8- slit of the high-resolution spectrometer channel, 10- collimator for echelle grating (off-axis parabolic mirror), 11- echelle grating, 12- linear InGaAs array, 14- InGaAs photodiode: low-resolution channel detector, 15- light trap for undiffracted AOTF light beams.

6 Conclusion

Future priorities for high-resolution planetary spectrometers include extending into the UV and far-IR, enhancing spatial mapping capabilities, and developing robust calibration procedures. Larger telescopes like the Extremely Large Telescope will provide greater light-gathering power and spatial resolution. Space missions will employ high-resolution instruments optimized for exoplanet atmospheres and astrobiology. The James Webb Space Telescope Near-Infrared Spectrograph [12] pioneered extremely stable high-resolution infrared spectroscopy from 0.6 to 5 μm in space.

In conclusion, high-resolution spectroscopy has become an indispensable tool for planetary science and atmospheric research. Steady improvements in spectrometer technology, sensor materials, data analysis, and optical systems will empower new discoveries for years to come. The future looks bright as high-resolution spectrometers continue revealing the secrets of planetary atmospheres.

References

1. Shakun, A., Korablev, O., Moshkin, B., et al.: Fourier transform spectrometers for remote sensing of planetary atmospheres and surfaces. CEAS Space J. **9**, 399–409 (2017)
2. Chen, J., Li, X., Chu, Q., et al.: An ultrahigh-resolution spectrometer using parallel double gratings. Results Phys. **45**, 106258 (2023)
3. Meng, Q., He, F., Zhao, W., et al.: The overview of the planetary atmospheric spectral telescope (past) in the scientific experimental system in near-space (sense). Int. Arch. Photogramm. Remote. Sens. Spat. Inf. Sci. **42**, 1419–1423 (2019)
4. Zhang, X., Fang, X., Li, T., et al.: Multi-channel hyperspectral imaging spectrometer design for ultraviolet detection in the atmosphere of Venus. Remote Sens. **16**(6), 1099 (2024)

5. Jennings, D.E., Flasar, F.M., Kunde, V.G., et al.: Composite infrared spectrometer (CIRS) on Cassini. Appl. Opt. **56**(18), 5274–5294 (2017)
6. Bertaux, J.L., Nevejans, D., Korablev, O., et al.: SPICAV on Venus express: three spectrometers to study the global structure and composition of the Venus atmosphere. Planet. Space Sci. **55**(12), 1673–1700 (2007)
7. Lacy, J.H., Richter, M.J., Yu, W., et al.: Texas echelon cross echelle spectrograph. In: Infrared Astronomical Instrumentation. SPIE, vol. 3354, pp. 436–447 (1998)
8. Encrenaz, T.: High-resolution imaging spectroscopy of planetary atmospheres. C.R. Geosci. **347**(3), 145–152 (2015)
9. Sonnabend, G., Sornig, M., Krötz, P., et al.: Ultra high spectral resolution observations of planetary atmospheres using the Cologne tuneable heterodyne infrared spectrometer. J. Quant. Spectrosc. Radiat. Transfer **109**(6), 1016–1029 (2008)
10. Gilewski, M.: Micro-electro-mechanical systems in light stabilization. Sensors **23**(6), 2916 (2023)
11. Smith, S.J., Adams, J.S., Bandler, S.R., et al.: Performance of a broad-band, high-resolution, transition-edge sensor spectrometer for X-ray astrophysics. IEEE Trans. Appl. Supercond. **31**(5), 1–6 (2021)
12. Bagnasco, G., Kolm, M., Ferruit, P., et al.: Overview of the near-infrared spectrograph (NIRSpec) instrument on-board the James Webb Space Telescope (JWST). In: Cryogenic Optical Systems and Instruments XII. SPIE, vol. 6692, pp. 174–187 (2007)

A System Detection Performance Evaluation Method Based on Temporal Clutter Model

Boyuan Bian[1](✉), Libing Jin[1], Xiaoman Li[1], Fujia Kong[1], and Zhitao Zhou[2]

[1] Beijing Institute of Space Mechanics and Electricity, Beijing 100094, China
boyuan_b@163.com

[2] Shanghai Institute of Satellite Engineering, Shanghai 201109, China

Abstract. In order to solve the problem that the influence factors of the temporal clutter model of staring infrared camera in geostationary orbit are complex and the accuracy needs to be improved, a temporal clutter model is established by considering the camera noise, the clutter caused by line-of-sight (LOS) dither, and the clutter caused by LOS drift. Based on the angular displacement sensor data of satellite vibration experiment, the LOS dither spectrum of camera is analyzed. Based on the interframe offset of the continuous observation image sequence under typical background, the LOS drift rate of the camera is calculated. Combined with the image sequence of staring infrared camera in geostationary orbit, the temporal clutter intensities are calculated and analyzed. The results show that the relative deviation between model calculation and statistical results is less than 15%. Based on the camera index parameters and background characteristic data, the camera noise and background clutter are discussed respectively. The detection performance of the system is quantitatively evaluated and the impact of various factors on detection performance is intuitively reflected. Thus, the detectability of the target can be quickly judged while the detection background, camera parameters, camera LOS movement characteristics and background suppression algorithm remain unchanged.

Keywords: Geostationary orbit · Staring infrared camera · Temporal clutter · System detection performance

1 Introduction

The core of infrared point target detection is to effectively improve image signal-to-clutter ratio (SCR) through image preprocessing, background suppression and threshold detection. Under the staring detection system in geostationary orbit, the target shape and detailed features disappear, leaving little information available for detection. Complex ground background and camera noise will drown the target, easily leading to false alarms. These characteristics make the infrared point target detection task difficult and increase the requirements for SCR. In the temporal infrared point target detection, due to the slowly changing characteristics of the background, clutter caused by camera noise and line-of-sight (LOS) movement are important factors affecting temporal clutter. Clutter

H. P. Urbach et al. (Eds.): ISSOIA 2023, SPP 191, pp. 382–392, 2024.
https://doi.org/10.1007/978-981-97-6718-2_33

caused by LOS movement is the comprehensive result of background characteristics, camera parameters, camera LOS movement characteristics and background suppression algorithms.

Many scholars have conducted temporal clutter research on staring infrared cameras. Pohlman [1, 2] discussed the main sources of noise and clutter in staring cameras, including photon noise caused by fluctuations in incident photons, noise from detectors and electronic components, clutter caused by relative movement between the field of view and the background, and clutter caused by background temporal changes. Myers [3] simulated the strong clutter background for the first-order and second-order temporal difference algorithms under the short-wave and medium-wave infrared spectral bands, and analyzed the clutter equivalent target and camera spatial resolution, LOS movement characteristics. Fraedrich [4] evaluated the clutter suppression capability of the temporal difference algorithm and analyzed its relationship with parameters such as camera LOS movement characteristics, detection distance, instantaneous field of view, and frame period. Rapier [5, 6] proposed a rapid estimation method of clutter equivalent targets under different ground resolutions, LOS drift rates, dither statistics and integration times under the conditions of N-order temporal difference algorithm and different spatial power spectral density models.

Analyzing the influencing factors of temporal clutter and quantitatively modeling them can effectively predict the clutter intensity during on-orbit detection, quantitatively evaluate the impact of clutter on system detection performance, and help optimize the design of staring infrared camera.

2 Influencing Factors of Temporal Clutter

The image SCR can be expressed as:

$$SCR = \frac{I_s}{SET} \tag{1}$$

where, I_s is the radiation intensity projected by the target onto a single detector pixel. With the temporal background suppression algorithm, SET is the temporal clutter intensity, which describes the temporal fluctuation degree of the background.

During the imaging process of staring infrared cameras in geostationary orbit, there are many factors that cause fluctuations in detector pixel output, ultimately affecting temporal clutter. These factors mainly include camera noise, camera LOS movement, and background suppression algorithms.

2.1 Camera Noise

The entire process of pixel signal generation, readout and quantification is accompanied by camera noise. Camera noise that affects temporal clutter mainly includes photon noise n_{ph}, readout noise n_{read}, dark current noise n_{dark} and circuit noise n_{brd}. It is generally expressed by the electron number method as:

$$n_s = \sqrt{n_{\mathrm{ph}}^2 + n_{\mathrm{read}}^2 + n_{\mathrm{dark}}^2 + n_{brd}^2} \tag{2}$$

Based on the target, background radiation characteristics and camera parameters, the camera noise equivalent target intensity can be expressed as:

$$NET = \frac{n_s \cdot R^2}{A \cdot \eta \cdot \tau \cdot T_{\text{int}}} \cdot \frac{hc}{\lambda} \tag{3}$$

where, λ is the central wavelength, η is the quantum efficiency, h is Planck's constant, c is the speed of light, τ is the system transmittance, A is the entrance pupil area, T_{int} is the integration time, R is the detection distance.

2.2 Camera LOS Movement

Camera LOS movement can be caused by factors such as satellite orbit, attitude changes and platform vibration [7]. For a geostationary orbit staring infrared camera, the camera integration time is T_{int} and the LOS movement period is T. Based on the relationship between the LOS movement period and the integration time, the camera LOS movement can be divided into high-frequency movement ($T_{\text{int}} > T/2$) and low-frequency movement ($T_{\text{int}} < T/2$) [8]. High-frequency LOS movement can be called dither. The intensity of the dither-clutter is related to factors such as background radiation intensity gradient, integration time, and LOS dither spectrum. It can be approximated by the angular displacement sensor data of the satellite dither experiment. Low-frequency LOS movement can be called drift, which is approximately represented by image shift. The intensity of the drift-clutter is related to factors such as background radiation intensity gradient and LOS drift rate. It can be approximated by geometric registration and offset measurements of the image. Therefore, the clutter caused by the LOS movement is a combination of the clutter component caused by high-frequency dither and low-frequency drift.

2.3 Background Suppression Algorithm

Clutter is feature-based and relative. Changes in the algorithm used in the target detection process will lead to changes in the feature space or feature values, which will also change the clutter. For sequence images from geostationary orbit staring infrared cameras, based on the slowly changing characteristics of the background and the inter-frame correlation at higher imaging frame rates, the most basic and commonly used temporal background suppression method is the temporal difference method. The previous frame image is used as the background of the current frame, and then the difference operation is performed. Its advantages are fast and suitable for situations with high real-time performance.

3 Temporal Clutter Model Establishment

3.1 Temporal Clutter Model

Through the analysis of factors affecting temporal clutter, it can be seen that for geostationary orbit staring infrared cameras, temporal clutter mainly consists of three parts: camera noise, clutter caused by LOS dither, and clutter caused by LOS drift. Therefore, the temporal clutter intensity can be expressed as:

$$SET = \sqrt{NET^2 + CET_J^2 + CET_D^2} \tag{4}$$

where, NET is the camera noise equivalent target intensity, which is an important indicator for measuring the performance of point target detection cameras and can be measured experimentally. CET_J is the clutter intensity caused by LOS dither, and CET_D is the clutter intensity caused by LOS drift.

Due to the inhomogeneity of the background radiation spatial distribution, the clutter intensity caused by dither in the entire focal plane [9] is:

$$CET_J = \sigma_J \cdot \sqrt{< \nabla I^2 >} \cdot R \tag{5}$$

where, σ_J is the effective dither amplitude, R is the detection distance, ∇I is the background radiation intensity gradient statistic, and < > represents average value. For the GSD of 3 × 3 pixels, the displacement of the instantaneous field of view (IFOV) is shown in Fig. 1.

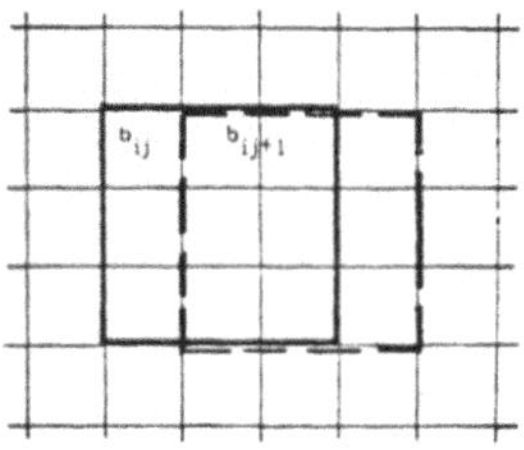

Fig. 1. Schematic diagram of the IFOV (3 × 3) before and after LOS moving by one pixel

$$\nabla I^2 = \left[\frac{1}{GSD}\sum_{p=i}^{i+k-1}\left(b_{p,j+k} - b_{p,j}\right)\right]^2 + \left[\frac{1}{GSD}\sum_{p=j}^{j+k-1}\left(b_{i+k,p} - b_{i,p}\right)\right]^2 \tag{6}$$

$$\sigma_J^2 = \int_0^{\infty} \omega(f)S(f, s, T_{\text{int}})df \tag{7}$$

$$S(f, s, T_{\text{int}}) = \frac{1}{(\pi f T_{\text{int}})^2}\left[1 - \cos(2\pi f T_{\text{int}})\right]\left[1 - \cos(2\pi f s)\right] \tag{8}$$

where, b is the pixel radiation intensity, f is the frequency, s is the frame period, $\omega(f)$ is the power spectral density of dither, and $S(f, s, T_{\text{int}})$ is the differential algorithm transfer function, According to the clutter model caused by dither, the clutter model caused by drift can be established equivalently. The angular displacement caused by the LOS drift is equivalent to the effective dither amplitude σ_D.

$$\sigma_D = vT_{\text{int}} \tag{9}$$

where, v is the LOS drift rate.

Then the clutter intensity caused by LOS drift of the entire focal plane is:

$$\begin{aligned} CET_D &= \sigma_D \cdot \sqrt{< \nabla I^2 >} \cdot R \\ &= v \cdot T_{\text{int}} \cdot \sqrt{< \nabla I^2 >} \cdot R \end{aligned} \tag{10}$$

4 Temporal Clutter Model Verification

In order to verify the established temporal clutter model, a satellite dither experiment was carried out. Based on the angular displacement sensor data, the camera LOS dither spectrum was analyzed and the effective dither amplitude was calculated. Then the inter-frame offset of the continuous observation image sequence was measured, and the camera LOS drift rate was calculated. Finally, combined with the on-orbit observation image sequence, the model calculation and actual measurement of temporal clutter were comparatively analyzed. The camera parameters are shown in Table 1.

Table 1. Camera parameters

Item	Value
Average wavelength /μm	2.85
Quantum efficiency	0.6
Planck constant /J·s	6.626 × 10–34
Speed of light /(m·s-1)	3 × 108
System transmittance	0.6
GSD /m	600
Detector size /μm	15 × 15
Optical aperture /m	0.35
Distance from sensor to Earth /km	35786
Integration time /s	0.4
Frame period /s	1

4.1 LOS Dither Spectrum Analysis and Effective Dither Amplitude Calculation

Among the movable parts on the satellite, the cryocooler and the momentum wheel are the main sources for the LOS dither of the geostationary orbit infrared camera. Based on the angular displacement sensor data of the camera's main mirror position, the dither spectrum of camera LOS is obtained caused by cryocooler and momentum wheel. In the

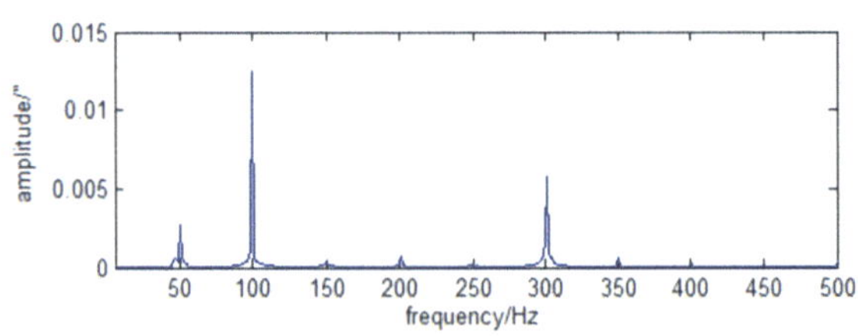

Fig. 2. Dither spectrum of the camera LOS in the x direction caused by cryocooler

x and y directions perpendicular to the camera's LOS, the camera LOS dither spectrum caused by the cryocooler is shown in Figs. 2 and 3, and the camera LOS dither spectrum caused by the momentum wheel is shown in Figs. 4 and 5.

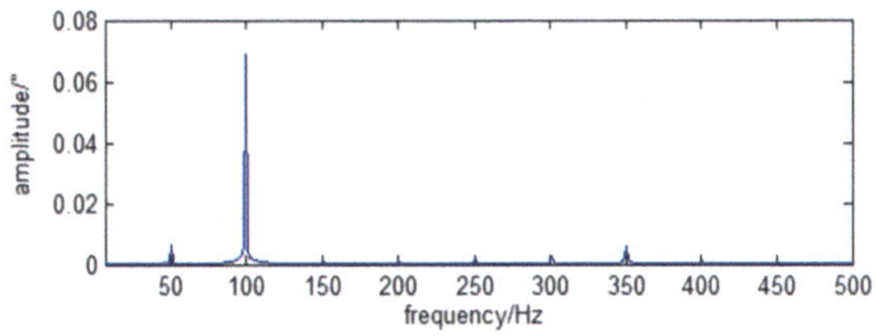

Fig. 3. Dither spectrum of the camera LOS in the y direction caused by cryocooler

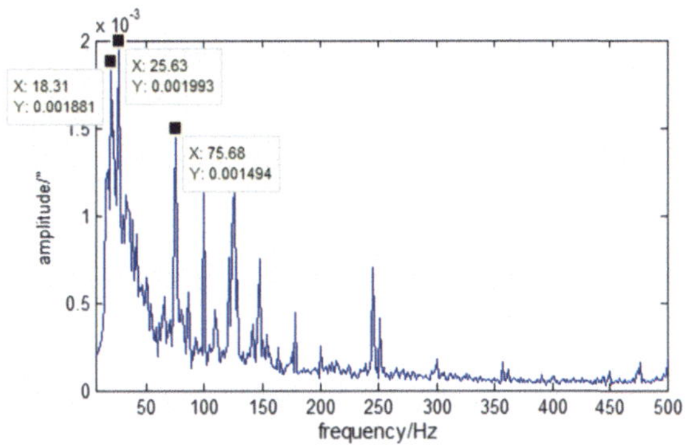

Fig. 4. Dither spectrum of the camera LOS in the x direction caused by momentum wheels

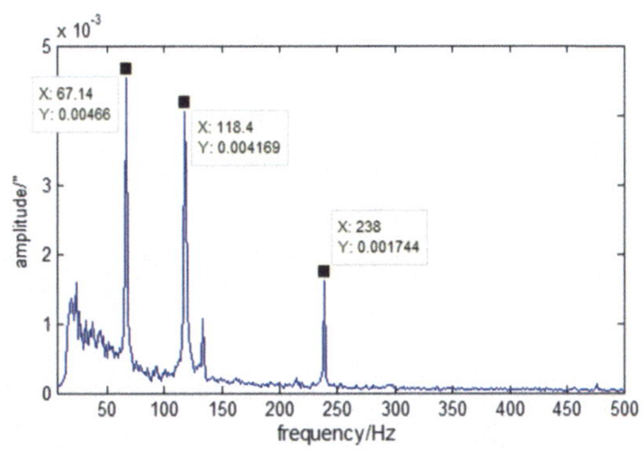

Fig. 5. Dither spectrum of the camera LOS in the y direction caused by momentum wheels

It can be seen from Fig. 2 and 3 that the camera LOS dither caused by the cryocooler exhibits typical harmonic characteristics, and the frequency is the operating frequency of the cryocooler and a multiple of the operating frequency. It can be seen from Figs. 4 and

5 that the camera LOS dither caused by the momentum wheel exhibits typical harmonic and broadband noise characteristics.

According to the camera LOS dither spectrum and combined with the camera parameters, the effective dither amplitude caused by different disturbance sources can be calculated, as shown in Table 2.

Table 2. Effective dither amplitude caused by different disturbance sources

	RX Direction	RY Direction	Total effective dither amplitude
Effective dither amplitude caused by Cryocooler /μrad	0.00027	0.0016	0.00163
Effective dither amplitude caused by Momentum Wheel /μrad	0.0294	0.0337	0.0447

It can be seen from Table 2 that the LOS effective dither amplitude caused by the cryocooler is one order of magnitude lower than that caused by the momentum wheel. Therefore, according to the effective dither amplitude and dither-clutter model established in Sect. 3, it can be seen that compared with the cryocooler, the disturbance of the momentum wheel is the main factor affecting the intensity of the dither- clutter.

4.2 Calculation of LOS Drift Rate

In order to calculate the LOS drift rate during camera imaging, based on the image sequence, the camera LOS offset for 10 consecutive moments is obtained through image geometric registration, as shown in Table 3. Due to the limitation of registration accuracy, we measured the total offset of consecutive multiple frames of images, and then divide it by the number of frames to obtain the average offset at each moment.

Table 3. Camera LOS offset

Direction	1	2	3	4	5
X	0.004	0.017	0.011	0.016	0.015
Y	−0.087	−0.081	−0.082	−0.071	−0.080
Direction	6	7	8	9	10
X	0.018	0.012	0.008	0.030	0.016
Y	−0.110	−0.078	−0.083	−0.119	−0.067

As can be seen from Table 3, the average offset in the X direction of the camera LOS is 0.015 pixels, the average offset in the Y direction is 0.086 pixels, and the root of the square sum of the average offsets in the X and Y directions is 0.087 pixels. Combined with the camera parameters, the camera LOS drift rate is 1.45 μrad/s.

4.3 Comparison of Model Calculations and Measured Statistical Results

Under the same observation conditions and system conditions, based on the image sequence obtained by the geostationary orbit staring infrared camera, the temporal clutter intensity of three typical background areas (200 × 200 pixels) was calculated and counted. Figure 6 shows different background areas, Fig. 7 shows the radiation intensity gradient statistical distribution (unit: W/sr/m) of different backgrounds, and Fig. 8 shows the measured statistical temporal clutter intensity distribution (unit: W/sr) under different backgrounds. Tables 4 and 5 show the model calculation and measured statistical results of clutter intensity under different backgrounds. The noise equivalent target intensity of the camera (NET) is 22.0 W/sr.

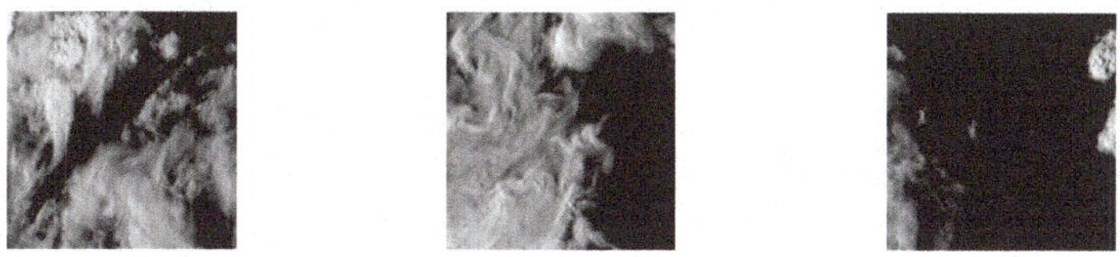

(a) Background 1 (b) Background 2 (c) Background 3

Fig. 6. Background area

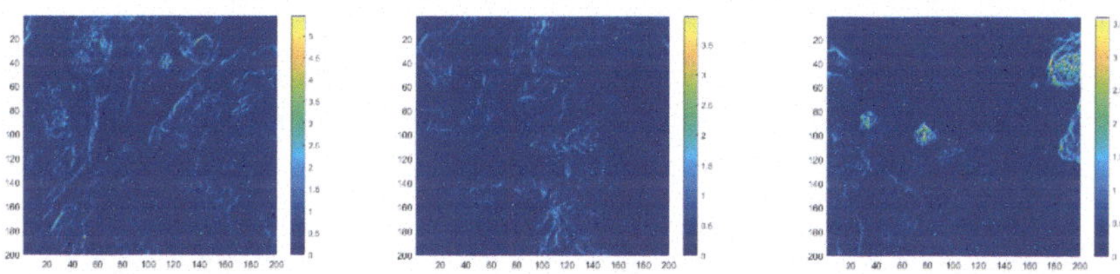

(a) Background 1 (b) Background 2 (c) Background 3

Fig. 7. Background radiation intensity gradient statistics distribution

It can be seen from Tables 4 and 5 that:

(1) The model calculation of temporal clutter intensity under different backgrounds is similar to the measured statistical results, and the relative deviation is less than 15%. Under the same system conditions, the results are consistent with the characteristic that the greater the background radiation intensity gradient, the greater the temporal clutter intensity.

(2) Under the camera parameters, LOS dither spectrum and drift rate studied in this article, the radiation intensity gradient of the strong clutter background (such as background 1) is large, and the clutter caused by drift dominates the temporal clutter, followed by clutter caused by camera noise and dither respectively. The radiation

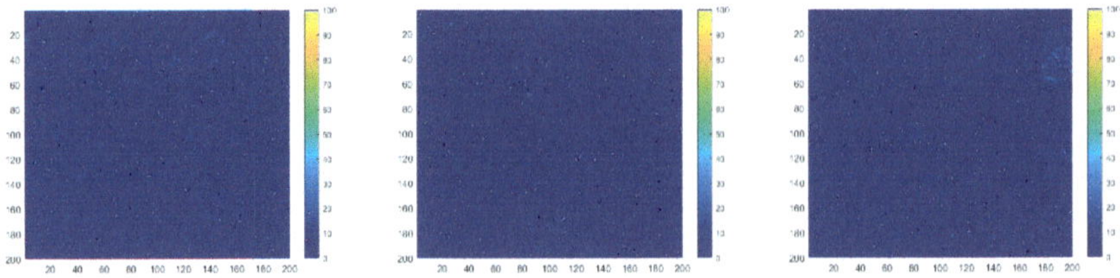

(a) Background 1 (b) Background 2 (c) Background 3

Fig. 8. Actual measured temporal clutter intensity distribution

Table 4. Clutter intensity for different backgrounds

Background	RMS Background Radiation Intensity Gradient Statistics /(W·sr-1·m-1)	*NET* sensor noise /(W·sr-1)	CET_J Calculated from the Model /(W·sr-1)	CET_D Calculated from the Model/(W·sr-1)
Background 1	1.3	22.0	2.1	27.4
Background 2	1.0	22.0	1.6	19.6
Background 3	0.7	22.0	1.1	15.2

Table 5. Temporal clutter intensity of different backgrounds

Background	RMS Background Radiation Intensity Gradient Statistics /(W·sr-1·m-1)	*SET* Calculated from the Model /(W·sr-1)	*SET* Measured /(W·sr-1)	Relative Deviation
Background 1	1.3	35.2	31.7	11.0%
Background 2	1.0	29.5	31.0	4.8%
Background 3	0.7	26.8	29.2	8.2%

intensity gradient of the weak clutter background (such as background 3) is small, and camera noise dominates the temporal clutter, followed by clutter caused by drift and clutter caused by dither. For medium clutter background (such as background 2), the noise equivalent target intensity is close to the intensity of clutter caused by drift, and the intensity of clutter caused by dither is the smallest.

5 System Detection Performance Evaluation

Effective evaluation of system detection performance can intuitively reflect the impact of various factors in the system on detection performance, thereby effectively guiding camera design. The camera design process needs to start from the detection system performance and SCR requirements, converting them into equivalent target intensity requirements for camera noise and allowable background clutter, and further decomposing them into requirements for the camera parameters. The above process can be specifically described in the following aspects.

(1) Since the system response mainly comes from the detection background and camera noise, according to formula (4), the system temporal clutter intensity (SET) is used to characterize the detection system performance. The smaller the SET value, the higher the detection system performance.
(2) Minimum target radiation intensity detectable by the system is expressed by:

$$I_s = TCR \cdot SET \tag{11}$$

where, TCR is the signal-to-noise ratio threshold. The TCR is related to the target detection algorithm. In turn, based on the TCR requirements and target intensity, the requirements for SET in target detection performance can be calculated.

(3) Noise equivalent target intensity (NET) is used as an indicator to evaluate camera performance, and its value is only related to camera parameters. For fixed camera parameters and a certain integration time, NET remains unchanged.
(4) For geostationary orbit staring infrared cameras, when the temporal point target detection method is used, the camera LOS movement converts the spatial variation of the background radiation into temporal clutter. The clutter equivalent target intensity (CET) caused by the LOS movement is used as an index to evaluate the background characteristics and the LOS stability. It reflects the coupling influence of the background and the LOS movement. CET is expressed as:

$$CET = \sqrt{CET_J^2 + CET_D^2} \tag{12}$$

$$CET = \sqrt{CET_J^2 + CET_D^2} \quad (12)$$

Using the image captured by the geostationary orbit staring infrared camera as the background data, a 200 × 200 pixel area of the image captured under the same observation conditions and system conditions was selected as the background. The images of three typical scenes are shown in Fig. 6, and the calculation results of NET, CET and SET of the three typical scenes are shown in Table 6.

Table 6. NET, CET and SET in three typical scenes

Background	NET (W/sr)	CET (W/sr)	SET (W/sr)
Background 1	22.0	27.5	35.2
Background 2	22.0	19.7	29.5
Background 3	22.0	15.2	26.8

For the target detection process, when the SCR is greater than 6, the point target is generally considered detectable. It can be seen from formula (11) that: in the above three scenes, when the radiation intensity of the target in the image after background suppression is 211.2W/sr, 177W/sr and 160.8W/sr, the target can be detected.

6 Conclusions

In this paper, the temporal clutter model for the design of staring infrared camera in geostationary orbit is analyzed and verified. The results show that the relative deviation between model calculation and statistical results is less than 15%. Based on the camera performance indicators and background characteristics, camera noise and background clutter are discussed separately. Using NET, CET and SET as parameters, the detection performance of the system can be quantitatively evaluated and the impact of various factors on detection performance can be intuitively reflected. The detectability of the target can be quickly judged while the background, camera parameters, camera LOS movement characteristics and background suppression algorithm remain unchanged.

References

1. Pohlman, R.T.: Staring mosaic sensor noise sources. In: SPIE Proceedings, vol. 95, pp. 138–147 (1976)
2. Pohlman, R.T.: Noise effects in a staring mosaic sensor mounted on a moving platform. In: SPIE Proceedings, vol. 124, pp. 10–16 (1977)
3. Myers, K.N.: Performance of a staring infrared mosaic sensor against a high reflectance background. In: SPIE Proceedings, pp. 209–217 (1984)
4. Fraedrich, D.S.: Analytic evaluation of frame difference processing on terrain clutter at MWIR wavelengths. In: SPIE Proceedings, vol. 781, pp. 26–32 (1987)
5. Rapier, J.L.: Background clutter leakage estimation in a mosaic sensor whose field of view moves relative to the background. In: SPIE Proceedings, vol. 124, pp. 23–32 (1977)
6. Rapier, J.L.: Scaling methods for rapid estimation of background clutter leakage. In: SPIE Proceedings, vol. 819, pp. 14–21 (1987)
7. Hu, K., Huang, X., Zhang, Y., et al.: Satellite platform jitter detection and image geometric quality compensation based on high-frequency angular displacement data. J. Electron. Inf. Technol. **40**(7), 1525–1531 (2018)
8. Yu, D., Lian, M., Zhou, F., et al.: Effect of micro vibration on image quality of high earth orbit (GEO) remote sensing satellites. Sci. China **49**(1), 74–86 (2019)
9. Bian, B., Zhou, F., Li, X., et al.: Study on dither-caused clutter for staring infrared camera in geostationary orbit. Infrared Technol. **45**(7), 775–783 (2023)

Design and Implementation of a Light and Small Space-Borne Laser

Ming Qi(✉), Pingping Luo, Huan Peng, Yongtao Deng, Peibei Meng, Wenzong shi, and shuo Jiang

Beijing Institute of Space Mechanics and Electricity, Beijing 100094, China
9698918@163.com

Abstract. Aiming at the special requirement of space-borne laser application, a design scheme and realization method of light and small size Nd: YAG solid-state laser are presented. In this method, the flexible support scheme is adopted to isolate the external factors such as forced displacement, assembly stress, thermo-elastic deformation and the influence of laser opto-mechanical. Considering the factors of structural mechanical stability, thermal stability and lightweight, the design scheme is analyzed and calculated by using Hyperworks software, the simulation results show that the effect of 92% forced displacement and 89% thermo-elastic deformation on the relative position of the optical element is removed. In addition, the laser adopts the fully-sealed design scheme to realize the pollution control of the sensitive device, and the leakage rate of the whole machine is less than 1×10 e-8 pa·m^3/s. Finally, the design of the space-borne laser has been finished. The energy of the single pulse is more than 30 mJ, the repetition rate is 10 Hz, and the weight of the space-borne laser is less than 400 g.

Keywords: Space-borne Laser · Lidar · Flexible Support · Simulation Analysis · Sealing · Feat Transfer

1 Introduction

With rapidly development of laser technology, space-borne lidar systems are becoming more and more important for space information detection such as three dimensional imaging, high resolution mapping, space remote sensing, environmental monitoring and so on [1]. Space-borne laser is the core of the laser system, not only requires high power, but also requires higher beam quality and longer lifetime [2–4]. In the design, the vibration and shock of satellite launching, Space Radiation Environment, temperature change and pollution inside the laser cavity should be taken into account. The requirement of space-borne lasers for size and weight is becoming more and stricter [5]. In the traditional laser design, the support base is used to install the precision optical mechanical system in the laser resonant cavity, and the base is also used to seal the laser cavity. The external forced displacement and thermo-elastic deformation directly affect the relative position relation of the optical elements in the laser resonator, the support base must have enough stiffness to resist the influence of external deformation on the laser resonator

H. P. Urbach et al. (Eds.): ISSOIA 2023, SPP 191, pp. 393–401, 2024.
https://doi.org/10.1007/978-981-97-6718-2_34

cavity. In this paper, the results of a small space-borne laser are presented. A flexible unmounted cantilever support is used to isolate the effect of external deformation on the precision optical mechanical system in the laser cavity, the weight and volume of the laser are greatly reduced. The design scheme of this paper has been applied to the development of space-borne laser, and has passed the space environment test.

2 Overall Design Scheme of Space-Borne Laser

In order to meet the design requirements of narrow pulse width and high energy, the overall scheme of Nd:YAG Q-Switching is adopted, as shown in Fig. 1 The laser consists of diode stack array (LD), optical coupler, Nd:YAG, Q-switch, polarizer, plate, output coupler. LDS emits a wide area of pump light, which is compressed by the optical coupler and converted to a size matching the aperture of the laser crystal (Nd:YAG) to improve the utilization of pump light. A laser resonator is composed of a 1064 nm high reflection film and an output coupler.

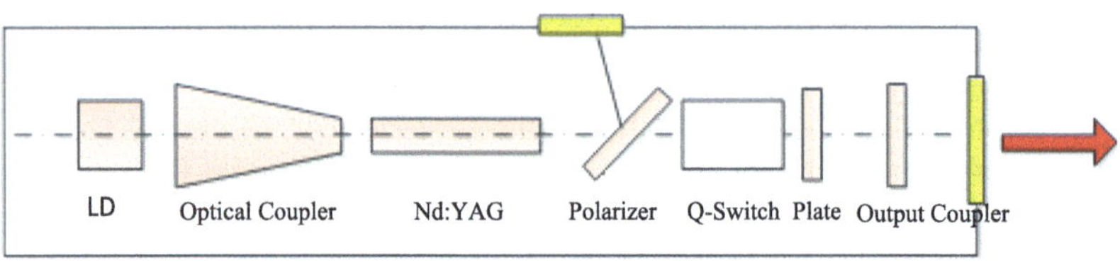

Fig. 1. Overall scheme of the laser

The laser design parameters are shown in Table 1.

Table 1. Technical indicators of Laser.

Parameter	Value
Center wavelength	1064.4 nm ± 0.2 nm
Line width (half height width)	≤ 200 pm
Single pulse energy	≥ 30 mJ
repetition frequency	10 Hz
pulse width (FWMH)	5 ns ~ 7 ns
weight	≤ 400 g

3 Structure Design of Flexible Support

The whole structure of the laser is shown in Fig. 2, and the inner section is shown in Fig. 3. The whole structure is mainly composed of the laser shell and the support base; the inner components including LD, optical coupler, Nd:YAG, Q-switch, polarizer, plate, output coupler are all installed on the structure base.

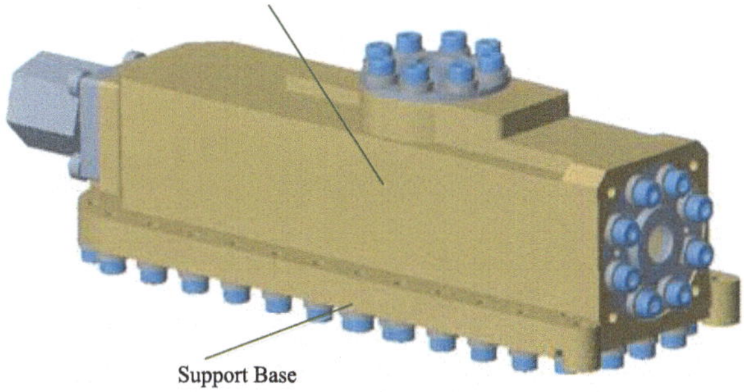

Fig. 2. The whole mechanism diagram of the Laser

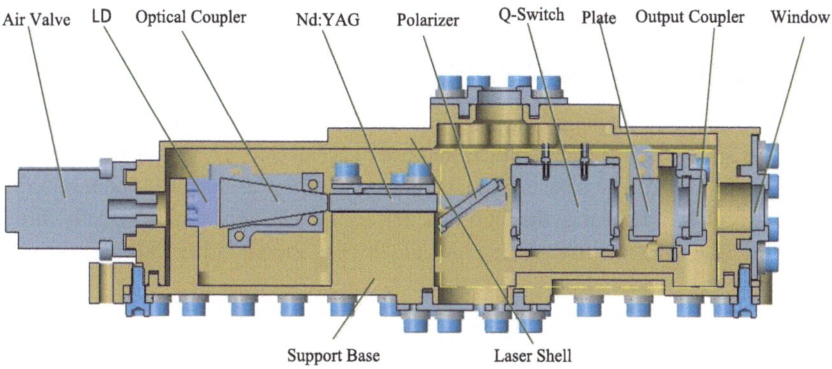

Fig. 3. Profile of the inner side of the Laser

According to the space environment and the characteristics of the space-borne laser, the following problems and solutions are considered in the structure design.

3.1 High Stability Optical Machine Structure Design

A laser resonator is composed of a laser output coupler and a full-reverse mirror. The relative position between the output coupler and the full-reverse mirror is very sensitive; it directly affects the characteristics of the laser resonator, and ultimately affects the performance of the laser. In order to reduce the influence of thermo-elastic deformation caused by external forced displacement and temperature gradient on the laser resonator, the laser adopts the design scheme of sensitive optical isolation [6, 7], as shown in Fig. 4. The laser crystal (Nd:YAG) coated with high reflective film and the output coupler are installed on the support base cantilever, only a cantilever support of the support base cantilever is connected with a support base plate, and the support base plate is connected with the outside. The thermal-elastic deformation caused by external forced displacement and temperature change will lead to the deformation of the support base plate, and the deformation will not be transferred to the base cantilever, thus effectively isolating the influence of external deformation on the laser resonator.

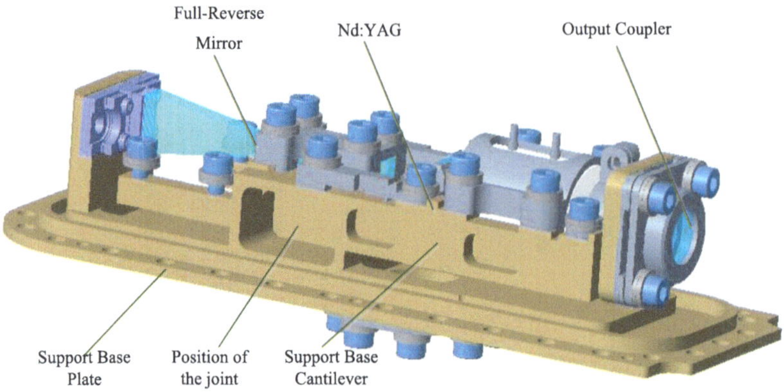

Fig. 4. Schematic diagram of optic-mechanical structure design

3.2 Thermal Design of Laser

The main heating elements of the laser include LD and laser crystal (Nd:YAG). The contact thermal resistance between LD and laser crystal (Nd:YAG) and base is reduced by filling indium foil. The base is made of aluminum alloy, and the LD and laser crystal (Nd:YAG) transfer the heat to the thermal interface of the support base directly through the support base. The heat transfer path of the laser is shown in Fig. 5 [8].

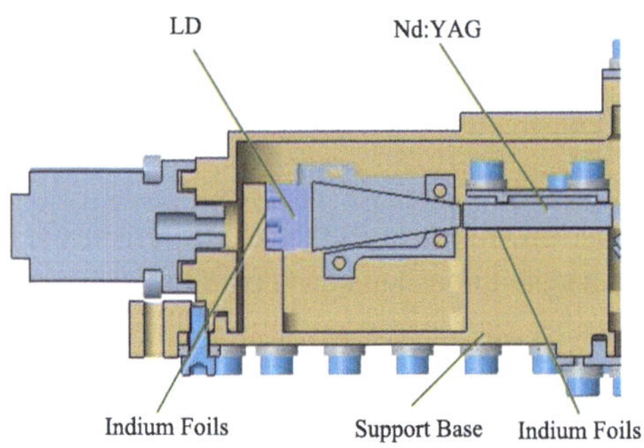

Fig. 5. Heat transfer path diagram of Laser

3.3 Laser Pollution Protection Design

The inner side of the laser has a high energy density and is very sensitive to pollutants. The main source of pollution comes from non-metallic material volatiles. The condensable volatiles of non-metallic materials will cause pollutants to appear on the surface of the optical elements inside the laser cavity. The pollutants will interact with the laser, increase the loss of the resonator, and cause the performance of the laser to decline obviously, as time goes on, more and more pollutants will lead to film damage, laser performance continues to decline, eventually leading to the failure of the entire laser.

Measures to protect against pollution include:

- The cavity of the laser is sealed to prevent external environment from polluting the inner side of the laser.
- In order to reduce the use of non-metallic materials, the inner side of the laser adopts non-adhesive design
- the non-metallic materials with low volatility are selected
- the non-metallic materials are degassed by high temperature vacuum to reduce the volatilization of non-metallic materials
- The cleanliness of the environment is strictly controlled during the whole process of laser production and assembly.

3.4 Laser Seal Design

The inner side of the laser for pollution protection, the entire laser uses a sealed design, the inner side filled with clean air. The sealing design of space-borne laser needs to consider the problem of low pressure discharge. The Q-switch inside the laser will produce high voltage, which is about 3000 V ~ 4000 V, according to the Paschen law of electric field discharge, the maximum leakage rate of the whole machine is $1.76 \times 10^{-8} \text{Pa} \cdot \text{m}^3/\text{s}$ [9, 10].

In order to reduce the amount of nonmetal, all-metal seal is adopted in the space-borne laser. The main sealing parts include: The laser shell and the support base seal, the air valve and the shell seal, the electrical connector seal, the output window seal. The electric connector seal and the output window seal are welded, and the other positions are sealed by metal indium. As shown in Fig. 6, the designed indium filling volume is larger than the volume of the filling area. The indium is squeezed and filled into the gap. The sealing leakage rate of the weld was less than $1 \times 10^{-9} \text{Pa·m}^3/\text{s}$ and that of the metal indium was less than $3 \times 10^{-9} \text{Pa·m}^3/\text{s}$.

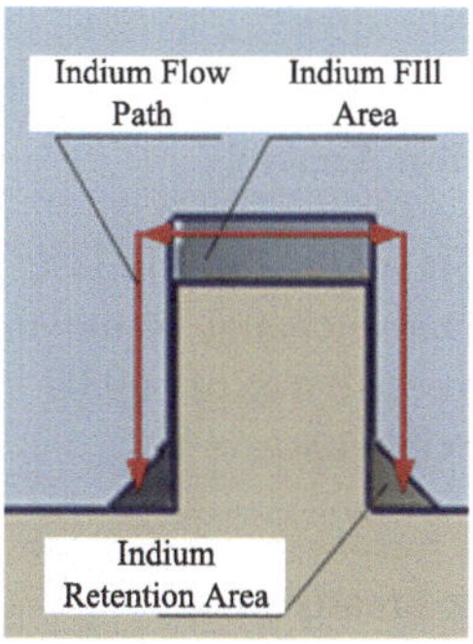

Fig. 6. Schematic diagram of indium seal

4 Simulation Analysis of Space-Borne Laser

HyperWorks is used to simulate the space-borne laser with finite element method. Because the relative position between output coupler and the full-reverse mirror is very sensitive, it directly affects the characteristics of the laser resonator, simulation analysis focuses on the evaluation of the changes in the relationship between the two locations.

4.1 Analysis of Forced Displacement

As the mounting surface, the support base plate is forced to shift by 0.05 mm, and the satellite-borne laser forced displacement deformation cloud is shown in Fig. 7. The laser is designed for optical sensor isolation. When the support base plate is deformed by 0.05 mm, the relative angle between the output coupler and the front surface of the full-reverse mirror changes 13.3″, and 89% of the forced displacement deformation is isolated by the flexible base.

Fig. 7. Laser-induced displacement distortion cloud

4.2 Thermo-Elastic Deformation Analysis

A temperature gradient of 5 °C is applied to the laser according to the requirements of the space environment. The relative change of the angle between the output coupler and the front surface of the full-reverse mirror is 9.07″, and 89% of the thermo-elastic deformation is isolated by the flexible base.

4.3 Thermal Analysis of Laser Steady State

Laser heating devices include: LD heat consumption of 3W, laser crystal(Nd:YAG) heat consumption of 2W, both through the support base heat transfer to the outside of the laser [11, 12], the support base through thermal control temperature, the maximum temperature of 25 °C. The thermal steady-state temperature cloud of the laser is shown. The overall temperature difference of the laser is less than 5 °C, and the highest temperature is located on the LD surface.

Fig. 8. Laser thermo-elastic deformation cloud

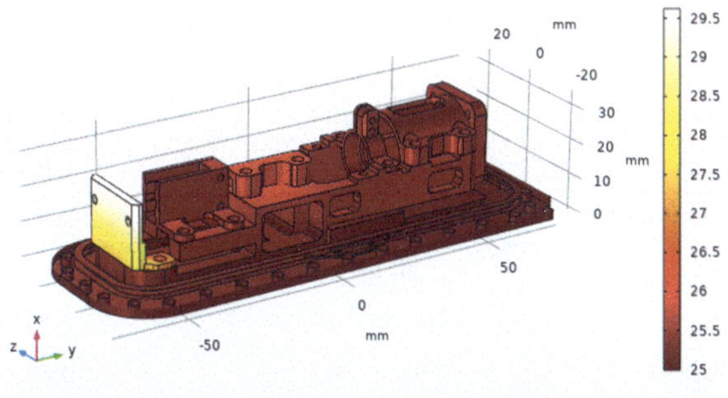

Fig. 9. Thermal steady-state temperature cloud of laser

5 Test and Verification of Space-Borne Laser

The laser is shown in Fig. 10.

Fig. 10. The real picture of the laser

The pump pulse width is 200 μs, the operating temperature is 20 °C, the laser pulse energy is 35 mJ, the pulse width is 6.4 ns, and the repetition frequency is 10 Hz. The far-field spot of the laser is shown in Fig. 11. TEC was used to control the base temperature at 15 °C, 25 °C, 30 °C, 35 °C and 40 °C respectively. The output energy of the laser was 32.6 mJ, 32.8 mJ, 32.6 mJ, 31.4 mJ and 30.6 mJ respectively, meet the target requirements. The laser index conformance is shown in Table 2.

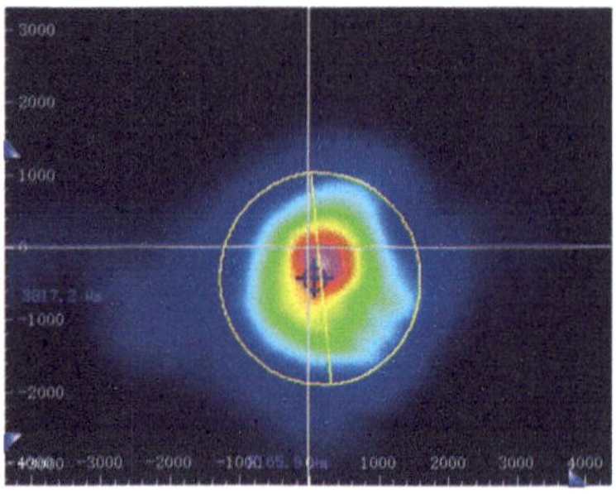

Fig. 11. Far-field facula of laser

Table 2. Laser verification results.

Parameter	Specification	Actual Value
Center wavelength	1064.4 nm ± 0.2 nm	1064.4 nm
Line width (half height width)	≤ 200 pm	200 pm
Single pulse energy	≥ 30 mJ	35 mJ
repetition frequency	10 Hz	10Hz
pulse width (FWMH)	5 ns ~ 7 ns	6 ns
weight	≤ 400 g	38 g

6 Conclusion

In this paper, a structure design scheme of space-borne laser is presented. Through the flexible design of stress decoupling, the influence of external forced displacement and thermo-elastic deformation on laser resonator is isolated. The design scheme is that the space-borne laser cavity has high opto-mechanical stability and can greatly reduce the structure weight. The effect of 92% forced displacement and 89% thermo-elastic deformation on the relative position of the optical element is removed. At the same time, a design method of indium seal space-borne is introduced, which greatly improves the seal reliability of laser resonator. Finally, the design target of single pulse energy 35 mJ, repetition rate 10 Hz and weight less than 400 g is achieved.

References

1. Zhu, M., Cheng, Y., Tan, C., Liu, X., Lu, C., Tan, W., Yang, X.: Development of foreign spaceborne laser. Infrared Laser Eng. **41**(12) (2012)
2. Bi, J.-z., Ma, X., Chen, W.: Experimental study of space-borne all-solid-state laser. Infrared Laser Eng. **41**(12) (2012)
3. Gheng, S., Wang, L.: The semiconductor laser and its applications in military. LASER & INFRARED. **33**(4) (2003)
4. Bi, J., Xia, W., Liu, D., Yang, Z., Chen, W.: Research and progress of spaceborne solid-state lasers. Infrared Laser Eng. **37**(5) (2008)

5. Cao, H., Zhang, X., Zhao, C., Xu, C., Mo, F., Dai, J.: System design and key technolongies of the GF-7 satellite. Chinese Space Sci. Technol. **40**(5), 1–9 (2020)
6. Paul, R., Yoder, J.R.: Opto-Mechanical Systems Design. China Machine Press, pp. 195–199 (2013)
7. Wang, N., Hui, S.: The flexible support structure design of the glued primary-secondary mirror of the aviation dual-band camera. Chinese J. Sci. Instrum. (z1), 77–81 (2014)
8. Zhao, Z., Yu, B., Su, Y.: Thermal optical analysis on space camera based upon thermal balance test. Spacecraft Recovery Remote Sens. **35**(4), 51–58 (2014)
9. Zhang, H., Zong, Y., Xin, T., Li, Q.: General test conditions for low pressure discharge of spacecraft units. Spacecraft Environ. Eng. **33**(6) (2016)
10. Zhang, Q., Meng, F.Z., Lü, J.L.: Research on low pressure discharge and multi-paction of solid state power amplifiers in satellite-borne payload. Semiconductor Technol. **38**(5), 388–391 (2015)
11. Fu, D., Yin, C., Wu, C.: A study of thermal/ structural/ optical analysis of a space remote sensor. J. Astronaut. **22**(3), 105–110 (2001)
12. Yan, Y., Chen, S., Zhang, W.: Review of Thermal-structural-optical Integrated Analysis of Space Remote Sensor. Optical Technique **31**(6), 913–920 (2005)

A New Four-Axis Angular Displacement Measurement System Based on Laser Gyro

Qiang Liu[1](✉), Yang Liu[1], Lijia Yu[1], Hualiang Sun[1], Guochen Wang[2], Ning Yang[1], and Jiande Zhang[1]

[1] Shandong Institute of Aerospace Electronic Technology, Yantai 264000, China
jim2954@163.com

[2] National University of Defense Technology, Changsha 410000, China

Abstract. With the development of remote sensing satellites, the influence of satellite platform micro-vibration on the imaging quality and application of high-resolution satellites is becoming more and more obvious. In this paper, a new four-axis angular displacement measurement system on orbit is proposed, which can be used to measure the micro-vibration of satellite platform. By using high-frequency and high-precision four-mode differential laser gyro, the problem of satellite platform high-frequency micro-vibration measurement is solved. Through a new system architecture, a four-axis measurement system is realized, and any axis can be backed up. By optimizing the application strategy, the high-performance measurement is ensured and the reliability of the system is improved. By designing the magnetic shield structure, the ability of the system to resist the interference of alternating magnetic field is improved, and the lightweight design is realized at the same time. For four-axis measurement systems, the calibration method of mounting matrix is realized. Through performance test, the system measurement accuracy is better than $0.003''$ (3σ), and the anti-interference ability of alternating magnetic field is improved, which is more conducive to long-term use on orbit in complex environment and has obvious effect on image compensation and geometric correction.

Keywords: Platform micro-vibration · Four-axis angular displacement measurement system · Four-mode differential laser gyro

1 Introduction

In recent years, high-resolution earth observation satellites and related technologies have been rapidly developed, and become the focus of high-tech fields. However, in the process of imaging the earth, due to the operation of the solar panel, the high-speed rotation of the momentum wheel, the work of the propeller, etc., the satellite will cause flutter, affecting the pointing accuracy of the satellite and the stability of the attitude control, resulting in the decline of the satellite imaging quality.

On-orbit micro-vibration of satellite is characterized by steady-state sinusoidal response and angular jitter of attenuation oscillation, and its frequency is concentrated

H. P. Urbach et al. (Eds.): ISSOIA 2023, SPP 191, pp. 402–411, 2024.
https://doi.org/10.1007/978-981-97-6718-2_35

in the range of 0 Hz–800 Hz. The influence of low-frequency micro-angular vibration on image is geometric deformation, and the influence of high-frequency micro-angular vibration on image is image blur [1]. Accurate and real-time measurement, analysis and compensation of satellite micro-angular vibration is a direct means to improve the quality of satellite imaging.

At present, there are many methods for measuring satellite micro-angle vibration, such as laser gyro based measurement and acceleration sensor combined measurement A, among which the measurement method based on laser gyro has become the main means of in-orbit micro-angular vibration measurement due to its high accuracy, good dynamic characteristics, wide measurement frequency and good temperature characteristics, especially in the moving imaging mode [2]. According to the in-orbit evaluation, the measurement method based on laser gyro can improve the geometric mass affected by the satellite, and the length deformation accuracy can be improved by about 0.5 pixels [3].

Based on high-frequency and high-precision four-mode differential laser gyroscope, a new type of four-axis angular displacement measurement system is proposed in this paper, the reliability of the system is improved. In addition, the system can improve the ability of anti-alternating magnetic field interference and realize miniaturization design by designing dot-matrix magnetic shield structure.

2 Research Status

In the 1980s, the United States conducted a series of studies on the micro-vibration problem of the Hubble Telescope. In the 1990s, NASA conducted the whole satellite micro-vibration test of LANDSAT-4 and other satellites, focusing on the micro-vibration caused by the operation of spaceborne moving parts such as flywheel assembly and solar wing drive mechanism [4]. Goes-N, a new generation of geostationary orbit environmental satellite in the United States, uses a high-precision three-dimensional angular velocity sensor to measure micro-angular flutter. The pointing and positioning system of Japan's Advanced Land Observing satellite ALOS integrates data collected by star tracker, gyro, angular displacement sensor, etc., and achieves good pointing measurement accuracy [5].

In our country, a lot of researches have been carried out in the field of satellite micro-vibration, and many kinds of micro-vibration measurement systems have been developed, including line displacement and angular displacement measurement. A mechanical environment measurement system was used in the space station and spacecraft to measure the micro-vibration environment of the space station and spacecraft, and a combination of accelerometer and angular displacement sensors was used on remote sensing satellites, the on-orbit micro-vibration environment is obtained, which improves the positioning accuracy and improves the image quality.

3 System Design

3.1 System Composition

The four-axis angular displacement measurement system is composed of sensor module, signal processing module, control and data processing module, power supply module. The sensor module consists of four high frequency and high precision spatial four-frequency differential laser gyros, which are used to detect the angular motion in this axis. The power supply module converts the primary power supply to the secondary power supply for the sensor and circuit module. The signal processing module is used to filter and amplify the analog signal of the laser gyro output to improve the signal to noise ratio. The control and data processing module is used to control the normal operation of the sensor, sample the angular displacement signal, calculate the signal and digital filter (Fig. 1).

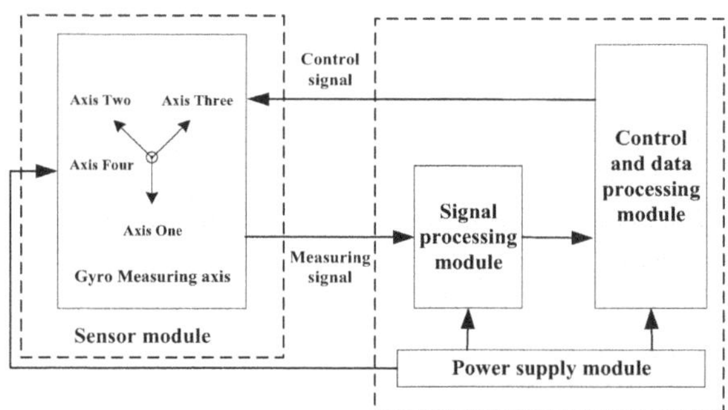

Fig. 1. Four-axis angular displacement measurement system.

3.2 Principle of Laser Gyro and Error Analysis

Based on optical sagnac effect, four-mode differential laser gyro is a kind of optical angular displacement sensitive element with excellent performance. It has the advantages of wide dynamic range, good linearity of scale factor, fast start-up, shock resistance, etc., and can obtain high Angle measurement accuracy. Laser gyro is a kind of ring laser, which uses two beams of traveling waves moving opposite each other in a ring cavity to measure the diagonal velocity or angle signal (Fig. 2).

The frequency difference between the forward and backward traveling waves in the ring resonator is:

$$\Delta \nu = \nu_{ccw} - \nu_{cw} = S\Omega$$

S is the scale factor of the ring laser gyroscope. By integrating the above formula with time t, the number of periods accumulated by the frequency difference of forward and backward traveling waves in time t can be obtained.

$$N = \int_0^t \Delta \nu \, dt = S\theta$$

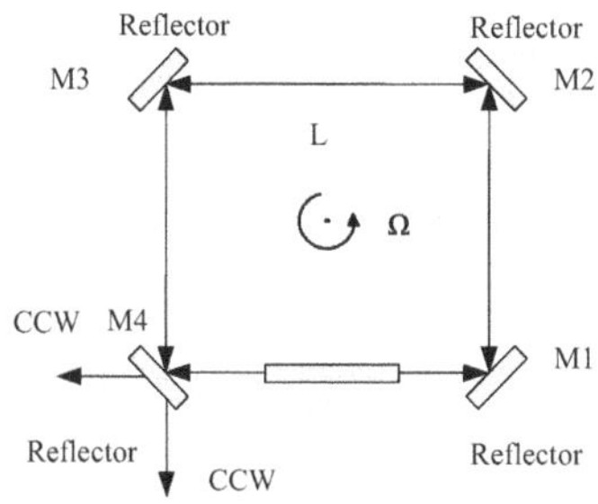

Fig. 2. Ring laser.

$\theta = \int_0^t \Omega dt$ is the total rotation angle of the ring cavity relative to the inertial space in t-time. The frequency difference and the number of pulses N can be detected by the method of optical beat frequency.

$$\theta = \frac{\lambda L}{4A} N$$

As an inertial device, the angle measurement error of laser gyro is composed of random walk, bias stability and quantization error:

$$\Delta\theta = \sqrt{\Delta\alpha^2 + \Delta\beta^2 + \Delta\chi^2}$$

$\Delta\alpha$ is the error due to random walk, the formula is as follows:

$$\Delta\alpha = ARWC\sqrt{t}$$

$\Delta\beta$ is the Sampling error, According to the actual test data, it is generally $0.001''$.
$\Delta\chi$ is the error formula caused by zero-bias stability is as follows:

$$\Delta\chi = (A/2\pi) \times \sqrt{100} \times \sqrt{t}$$

The measurement system's random walk is better than $0.002°/\sqrt{h}$, the zero-bias stability is better than 0.02°/h, the sampling interval is 0.1 ms (10 kHz sample rate), and its systematic error is better than $0.006''$ (3σ).

3.3 Gyro Mounting and Magnetic Shielding Design

In the system, the four laser gyros are installed in pyramid mode. Although the gyros are not orthogonal, the position relationship of any three laser gyros can be projected onto the prism of the system through the installation matrix calibration. Through the design, any combination of three laser gyros can be orthogonal under the prism coordinate system, that is, any laser gyro can be used as the backup of other three laser gyros, which greatly improves the reliability of the system and realizes the long life design.

Laser gyro is sensitive to the external alternating magnetic field, and the equipment structure needs to be designed with magnetic shielding material. Due to the high density of magnetic shielding material, it will increase the weight of the product. To solve this problem, the equipment structure adopts the design technology of lattice hollow, combined with 3D printing technology, and adopts the honeycomb structure inside the housing, which not only ensures the stiffness of the structure, but also improves the effect of the magnetic shield structure against the alternating magnetic field, and minimizes the volume of the equipment (Fig. 3).

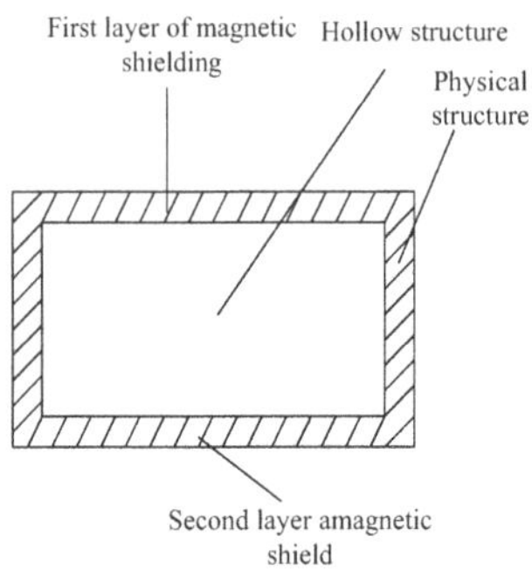

Fig. 3. Magnetic shield structure.

3.4 System Workflow

In order to ensure the best efficiency of the system, the four-axis angular displacement measurement system work flow is: After the system is powered on, the four laser gyros are self-checked to determine whether the gyros fail, and the failed gyros are eliminated.

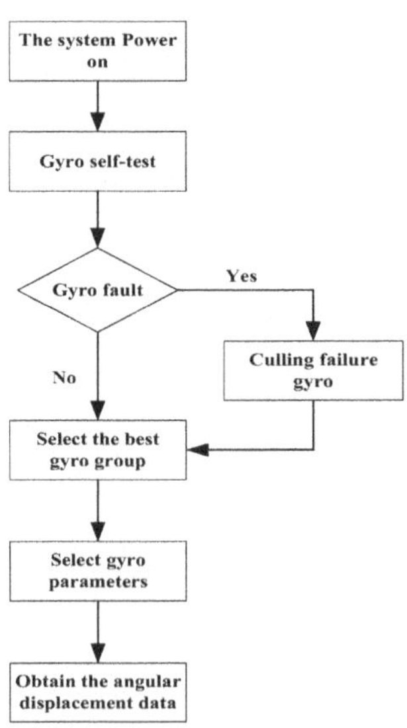

Fig. 4. Four-axis angular displacement measurement system workflow.

The three gyros with the best performance are selected for use, and the corresponding installation matrix and other parameters are selected. The angular displacement measurement data is obtained by combining the output data of the gyros (Fig. 4).

3.5 Calibration Technology of Installation Matrix

The four-axis angular displacement measurement system has an external prism to characterize the measurement axis. The four laser gyros in the system are combined in a group of three, there are four combinations, and four installation matrices can be obtained through installation matrix calibration.

The calibration of installation matrix is realized by three-axis turntable, and the four-axis angular displacement measuring system is installed on the three-axis turntable, the relationship between prism coordinate system and turntable coordinate system and between gyro coordinate system and turntable coordinate system can be obtained by rotating turntable in specific direction.

The relationship between prism coordinates and turntable coordinates is as follows:

$$C_l^t = \begin{bmatrix} \cos\alpha_{11} & \cos(90-\alpha_{22}) & x_l \\ \cos(90+\alpha_{11}) & \cos\alpha_{22} & \times \\ \cos(90-\beta_1) & \cos(90-\beta_3) & y_l \end{bmatrix}$$

α_{11} and α_{22} is the azimuth of a prism.

$$x_l = [\cos\alpha_{11}\ \cos(90+\alpha_{11})\ \cos(90-\beta_1)]$$

$$y_l = [\cos(90-\alpha_{22})\ \cos\alpha_{22}\ \cos(90-\beta_3)]$$

The relationship between gyro coordinates and turntable coordinates is as follows:

$$C_g^t = \begin{bmatrix} 360 & 0 & 0 \\ 0 & 360 & 0 \\ 0 & 0 & 360 \end{bmatrix} \begin{bmatrix} N_x^1 & N_y^1 & N_z^1 \\ N_x^2 & N_x^2 & N_x^2 \\ N_x^3 & N_x^3 & N_x^3 \end{bmatrix}^{-1}$$

$N_x^1, N_y^1, N_z^1 \ldots N_z^3$ is the cumulative value of the gyro output pulse obtained from the rate test.

The accumulated value of the gyro output pulse for the rate test. According to the above formula, the relationship between gyro coordinate system and prism coordinate system is obtained.

$$C_g^l = (C_l^t)' * C_g^t$$

4 Experiment and Analysis

4.1 The Installation Matrix Test

By calibrating the installation matrix of the four-axis angular displacement measuring system, four sets of installation matrix data can be obtained (Fig. 5).

Fig. 5. Installation matrix calibration.

Table 1. Results of validation of the installation matrix.

	Axis	Angular (°)	Angular velocity (°/s)	Angle measurement (°)	Angular velocity measurement (°/s)	Angular deviation (°)	Angular velocity deviation (°/s)
Rotate along the X-axis	X	360	5	359.831814	4.997958896	0.168186	0.002041
	Y	0	0	−0.33260398	−0.0043420839	0.332604	0.0043421
	Z	0	0	0.07170911	0.002217594	−0.07171	−0.00222
Rotate along the Y-axis	X	0	0	−0.3297249	−0.0046089913	0.3297249	0.004609
	Y	−360	−5	−360.12192	−5.001709077	0.12192	0.001709
	Z	0	0	−0.3907443	−0.002147177	0.390744	0.002147
Rotate along the Z-axis	X	0	0	0.152786	0.0058	−0.15279	−0.0058
	Y	0	0	0.371922	0.017597728	−0.37192	−0.0176
	Z	−360	−20	−360.0899	−20.0035098	0.0899	0.00351

In order to verify the correctness of the calibration of the installation matrix, rotation was carried out along the three coordinate axes of the prism coordinate system respectively, and the test results were analyzed, as shown in Table 1.

The results show that the angular error is less than 0.17° and the angular velocity error is less than 0.0021°/s when rotating along the x-axis. When rotating along the y-axis, the angular error is less than 0.13° and the angular velocity error is less than 0.0018°/s. When rotating along the z-axis, the angular error is less than 0.09° and the angular velocity error is less than 0.0036°/s.

The angular error and angular velocity error are very small, which verifies the correctness of the installation matrix calibration.

4.2 Performance Test

The static test and dynamic performance test of the four-axis angular displacement measurement system were carried out respectively. The static test was carried out on the static platform, and the dynamic test applied the angular vibration of 158 Hz, 0.015″, the test results are as follows (Figs. 6, 7, and 8):

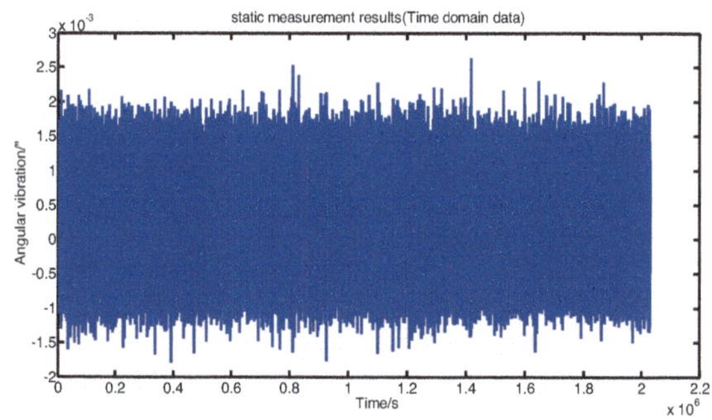

Fig. 6. System static measurement results (Time domain data).

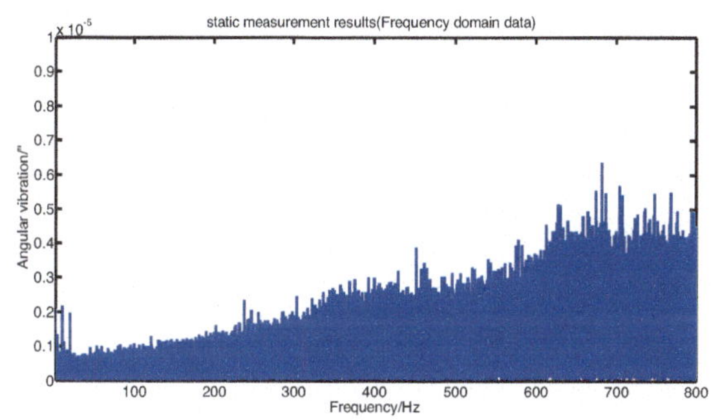

Fig. 7. System static measurement results (Frequency domain data).

It can be seen from the static test results that the time domain error of the system is <0.003″ (3σ), the frequency domain error is <1.3 × 10−5″, and the frequency domain error of each frequency point is evenly distributed without obvious mutation, indicating that the noise level of the four-axis measurement system is well controlled.

It can be seen from the dynamic test results that the frequency error is 0.1 Hz and the amplitude error is 0.00163″, which verifies the good dynamic measurement performance of the four-axis measurement system.

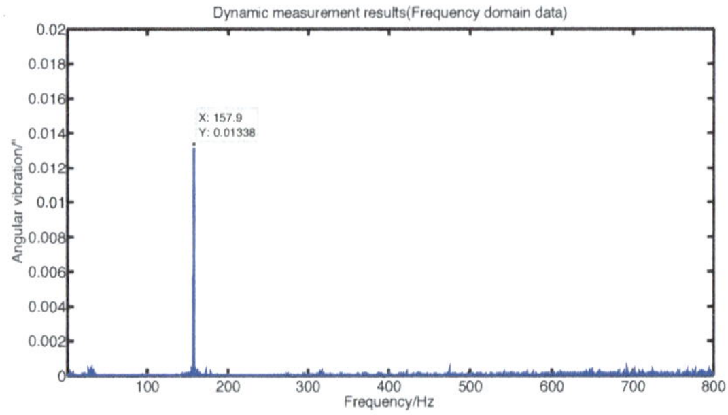

Fig. 8. System dynamic measurement results (Frequency domain data).

4.3 Magnetic Field Shielding Performance Test

Without magnetic shielding, the effect of 10 gauss magnetic field on zero bias of the gyroscope is about 0.001″ (10 kHz sampling rate).

Applying a 10 gauss magnetic field to the system, the test results are as follows (Table 2).

Table 2. Magnetic field shielding performance test.

Gyro sequence number	Zero position before adding magnetic field	Zero position after adding magnetic field	Zero change
No. 1 gyro	−0.00033″	−0.00033″	<0.00001″
No. 2 gyro	−0.00123″	−0.00123″	<0.00001″
No. 3 gyro	−0.00164″	−0.00164″	<0.00001″
No. 4 gyro	0.00130″	0.00130″	<0.00001″

It can be seen that after applying 10 gauss magnetic field, the zero bias of the gyroscope does not change, the magnetic field immunity is improved by more than 100 times, indicating that the magnetic shield design of this system has achieved good results.

5 Conclusion

In this paper, a new satellite-borne four-axis angular displacement measurement system is proposed, and a new architecture is designed to realize the four-axis mutual backup, improve the reliability of the system. By optimize the magnetic shield design of the system, the magnetic field anti-interference characteristics is improved, and the lightweight design is realized. The installation matrix calibration of four-axis system is realized. Static measurement noise of the system $<0.003''$ (3σ), and the dynamic measurement error $<0.00163''$. The four-axis angular displacement measurement system is ideal for satellite flutter measurement.

References

1. Wang, M., Zhu, Y., Fan, C.: Development of platform jitter geometric analysis and processing for high-resolution optical satellite imagery. Geomat. Inf. Sci. Wuhan Univ. **43**(12), 1900–1908 (2018)
2. Wang, Y.L.: Research on the key technology of attitude measurement data processing for optical remote sensing satellite. Geomat. Inf. Sci. Wuhan Univ. **47**(3), 482–482 (2022)
3. Hu, K., Huang, X., Zhang, Y.J., You, H.J.: Satellite platform jitter detection and image geometric quality compensation based on high-frequency angular displacement data. J. Electron. Inf. Technol. **40**(7), 1525–1531 (2018)
4. Masterson, R.A., Miller, D.W., Grogan, R.L.: Development and validation of reaction wheel disturbance models: empirical model. J. Sound Vibr. **249**(3), 575–598 (2002)
5. Iwata, T., et al.: High-bandwidth pointing determination for the advanced land observing satellite (ALOS). In: ISTS 2004-d-10, pp. 385–394 (2004)

A Comprehensive Optimization Error Analysis Method for Error Analysis of On-Orbit Complex Geometry Measurement System

Ruixue Zhang[1(✉)], Qiang Cong[1], Ding Yuan[1], Gan Geng[2], Fanxin Sun[1], and Jian Ding[1]

[1] Beijing Institute of Spacecraft System Engineering, Beijing, China
zrx17@tsinghua.org.cn

[2] China Academy of Space Technology, Beijing, China

Abstract. With the development of space science and technology, the need for high precision and large-scale geometric measurement of spacecraft in orbit is becoming more urgent, which can be denoted as on-orbit measurement. Generally, an on-orbit measurement system is complex, which has many parameters and factors that affect the measurement precision. The error distribution or allocation process impacts the product quality, the production cost and scrap rate. Tight error tolerances allow to assure product performance; loose error tolerances reduce production costs. The tolerance allocation of a complex system is performed under uncertainty. Traditional error analysis and allocation methods are difficult to balance between scientific rationality and realizability. In this paper, we propose a comprehensive optimization error analysis and distribution method. The theoretical error transfer model between each parameter and the geometric quantity to be measured is established by exploiting the error transfer coefficient. The analysis and evaluation of the measurement system are finally realized by combining the distribution law of the errors of each parameter obtained by the experiments.

Keywords: Or-orbit Large-scale Geometric Measurement · Error Analysis · Error Allocation

1 Introduction

Recently, spacecraft are taking on more and more tasks, facilitating the development of large, lightweight and complex space structures [1, 2]. With the development of space science and technology, there is an increasingly urgent need for high-precision large-scale geometric measurement of in-orbit spacecraft in frontier fields such as satellite communication, earth observation, global navigation and space science research [3–6]. On the one hand, challenges associated with aerospace projects require increasingly complex aerospace systems, which in turn demand automation and error tolerance. The automation equipment required by aerospace intelligence requires a high-precision perception of the environment and external parameters as input. In the recent national plan, China's aerospace will be deployed in accordance with the country's decision on the

H. P. Urbach et al. (Eds.): ISSOIA 2023, SPP 191, pp. 412–423, 2024.
https://doi.org/10.1007/978-981-97-6718-2_36

construction of space power, and put forward high requirements for the automation and intelligence of high-precision measurement equipment in the extraterrestrial environment. On the other hand, in order to carry out more accurate remote sensing detection to the earth, whether it is visible or infrared remote sensing, or active remote sensing based on SAR (synthetic aperture radar) imaging, it is necessary to accurately determine the position and attitude relationship between sensors or formation satellites in orbit. For example, formation flying satellites, used for synthetic aperture radar satellites, rendezvous and docking, etc., need to determine their relative positions and orientation with each other precisely.

The on-orbit geometry measurement system is generally complicated, as its measurement accuracy and precision are affected by many parameters or factors. Moreover, each direct measurement parameter in the indirect measurement system has different constraints, such as machining accuracy, assembling process and environment change. All aspects of manufacturing such as machine investment cost, manufacturing cost, the performance of the product, quality of manufacturing, and the reliability of the machining process directly influence tolerance allocation. Tolerance cost and machining time play crucial roles while performing tolerance and error allocation in complex measurement systems [7, 8]. Hence, the main obstacle in the realization of the on-orbit measurement is to analyze and assign the error of the measurement system. As for the complex measurement systems, most of the researchers are still focusing on this research topic to balance between the performance of the systems, such as the precision and accuracy, and the realizability, such as the cost, process capability.

Traditionally, several methods exist for error analysis and distribution of measurement systems [9]. The equal-tolerance method evaluates each error component based on the same tolerance. This method is only suitable for simple systems or conditions, as it is difficult to reflect the difference of each error source. The equal-influence method considers the error transfer coefficient of each error source to be measured, which is more scientific and reasonable. Nevertheless, it needs to pay attention to the engineering realizability and process difficulty. The experiment-based method explores the influence of different parameters through experiments. However, the repeated tests are time-consuming and expensive, which is not applicable to complex measurement systems. Therefore, there are urgent demands for a scientific and practical error analysis and distribution method.

In this paper, we propose a comprehensive optimization error analysis and distribution method that combines the advantages of the equal-influence method and the experiment-based method. The theoretical error transfer model between each parameter and the geometric quantity to be measured is established by exploiting the error transfer coefficient. The analysis and evaluation of the measurement system are finally realized by combining the distribution law of the errors of each parameter obtained by the experiment-based method. In addition, the error results are simulated using commercial modeling software, and the simulation results are consistent with the theoretical analysis, proving our method's effectiveness.

2 Theory and Principles

2.1 Principles and Composition of Measurement System

Our measurement system is based on the interferometric baseline measurement of the two-antenna InSAR (Interferometric Synthetic Aperture Radar) satellite, where the interferometric baseline is the line between the phase centers of the SAR antennas at both ends. In this system, the indirect measurement method is adopted to achieve the high precision measurement of the interferometric baseline length (baseline length B) and the angle between the baseline and the inertial plane (baseline angle α) in orbit. Initially, the measurement system model is established. The composition of the long baseline measurement system is shown in Fig. 1. The key processes to be measured and solved are identified as the relative pose relationship between coordinate systems, including: star tracker coordinate system (ST CS), +Y measurement coordinate system (Meas CS), −Y measurement coordinate system, +Y target coordinate system (Target CS), −Y target coordinate system, +Y antenna mechanical coordinate system (Mecha CS), −Y antenna mechanical coordinate system, and +Y/ −Y antenna phase center (Ant PC). The reference coordinate system is based on the inertial coordinate system (Inertial CS). The +Z direction is the direction pointing to the Earth's center. The +X direction is the direction of the satellite motion, while the +Y direction is determined according to the right-hand rule. Among them, the phase centers of the antenna can be considered as two points, which only have position information. There is a six-degree-of-freedom relationship between position and attitude in other coordinate systems, including the relative position (represented by t_x, t_y, t_z) and attitude (represented by θ_x, θ_y, θ_z) information.

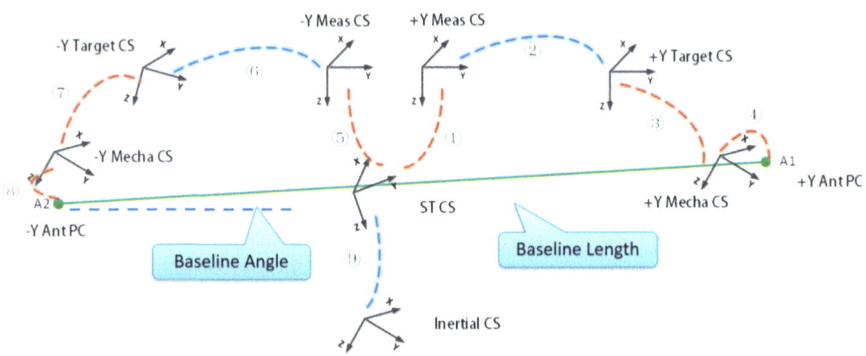

Fig. 1. The composition of the baseline measurement system of an InSAR satellite. Interferometric baseline is the line between the phase centers of the SAR antennas at both ends.

Since the antenna is extended tens of meters or even hundreds of meters by the support arm, the fundamental frequency of the support arm is relatively low, resulting a flexible baseline during the in-orbit operation of the spacecraft. Therefore, a high-precision six-degree-of-freedom measuring device is required to conduct real-time measurement of the coordinate system relationship between the two sides of the support arm. The position and pose relationship between coordinate systems obtained from real-time measurement in orbit is represented by the blue line in Fig. 1, while the position and pose relationship obtained from ground calibration that cannot be measured in real-time in orbit is represented by the orange line.

Then, the parameters in the measurement system can be sorted out based on the established coordinate systems. As can be seen from the above, in the long baseline measurement system, 45 parameters need to be obtained in order to achieve baseline information, in which 24 position parameters and 21 attitude parameters are independent of each other. Error analysis of the test baseline is to analyze and allocate the systematic errors and random errors of these 45 parameters. In order to obtain the baseline error, we need to consider the error propagation model between the baseline information to be measured and the parameters.

To establish the error transfer model between the parameters to be measured, it is necessary to obtain the functional relationship between the parameters to be measured and the parameters. Baseline length and baseline angle are measured indirectly by the calibration results and real-time measured results between the coordinate systems. In the process of coordinate system transformation, a uniform rotation matrix **R** is usually used to represent the attitude conversion between coordinate systems. And the relationship between **R** and the Euler angles (θ_x, θ_y, θ_z) of the two coordinate systems can be expressed by the Eq. (1) and Eq. (2), where $\mathbf{R_x}$, $\mathbf{R_y}$ and $\mathbf{R_z}$ represent the rotation matrix of the new coordinate system relative to the original coordinate system in the three axis directions, respectively. Equation (1) indicates that the rotation of the axis from the original coordinate system to the new coordinate system revolves around the *X* axis, *Y* axis and *Z* axis in turn, and the positive direction of the rotation angle is taken counterclockwise.

$$\mathbf{R} = \mathbf{R}_z\mathbf{R}_y\mathbf{R}_x \tag{1}$$

$$\begin{aligned}
\mathbf{R}_x &= \begin{bmatrix} 1 & 0 & 0 & 0 \\ 0 & \cos\theta_x & -\sin\theta_x & 0 \\ 0 & \sin\theta_x & \cos\theta_x & 0 \\ 0 & 0 & 0 & 1 \end{bmatrix}, \mathbf{R}_y = \begin{bmatrix} \cos\theta_y & 0 & \sin\theta_y & 0 \\ 0 & 1 & 0 & 0 \\ -\sin\theta_y & 0 & \cos\theta_y & 0 \\ 0 & 0 & 0 & 1 \end{bmatrix} \\
\mathbf{R}_z &= \begin{bmatrix} \cos\theta_z & -\sin\theta_z & 0 & 0 \\ \sin\theta_z & \cos\theta_z & 0 & 0 \\ 0 & 0 & 1 & 0 \\ 0 & 0 & 0 & 1 \end{bmatrix}, \mathbf{T} = \begin{bmatrix} 1 & 0 & 0 & t_x \\ 0 & 1 & 0 & t_y \\ 0 & 0 & 1 & t_z \\ 0 & 0 & 0 & 1 \end{bmatrix}
\end{aligned} \tag{2}$$

According to the satellite layout, the values of each coordinate system can be obtained, as shown in Table 1. Since the phase center can be abstracted as a point, its coordinate in the mechanical coordinate system of the +Y antenna is $\mathbf{T_4A_1}$, where the homogeneous coordinate of the point is set as $\mathbf{A_1} = (0,0,0,1)^{\mathrm{T}}$. The coordinates of the phase center of the +Y antenna in the ST coordinate system can be obtained by sequential conversion, as shown in Eq. (3). Similarly, the homogeneous coordinate of the phase center of the −Y antenna is set as $\mathbf{A_2} = (0,0,0,1)^{\mathrm{T}}$, and its coordinate in the ST coordinate system is shown in Eq. (4). Then, baseline vector $\mathbf{B_L}$ to be measured in the inertial coordinate system is expressed by Eq. (5), and the expression of the baseline length, baseline angle is Eq. (6), where $\mathbf{N_1} = (0,0,1)^T$ is the unit normal vector of the XOY plane of the inertial frame, and $\mathbf{B_L}$' is the $\mathbf{B_L}$ non-homogeneous expression. By substituting the values of the parameters in Table 1 determined by the satellite layout into the equations, the baseline length and baseline angle are calculated to be 117.0828 m and 1.9975°, respectively.

$$\mathbf{A}_1' = \mathbf{T}_1 \cdot \mathbf{R}_1 \cdot \mathbf{T}_2 \cdot \mathbf{R}_2 \cdot \mathbf{T}_3 \cdot \mathbf{R}_3 \cdot \mathbf{T}_4 \cdot \mathbf{A}_1 \tag{3}$$

$$\mathbf{A}_2' = \mathbf{T}_5 \cdot \mathbf{R}_5 \cdot \mathbf{T}_6 \cdot \mathbf{R}_6 \cdot \mathbf{T}_7 \cdot \mathbf{R}_7 \cdot \mathbf{T}_8 \cdot \mathbf{A}_2 \tag{4}$$

$$\mathbf{B}_\mathbf{L} = \mathbf{R}_9 \cdot (\mathbf{A}_2' - \mathbf{A}_1') \tag{5}$$

$$B = \|\mathbf{B}_\mathbf{L}\|_2,\ \alpha = \arcsin\left(\frac{\mathbf{N}_1 \cdot \mathbf{B}_\mathbf{L}'}{|\mathbf{N}_1| \cdot |\mathbf{B}_\mathbf{L}'|}\right) \tag{6}$$

Table 1. Parameter values in the current layout.

Parameters	value	Parameters	Value
t_{x1}/mm	500	θ_{x5}/°	−1
t_{y1}/mm	500	θ_{y5}/°	1
t_{z1}/mm	500	θ_{z5}/°	−1
θ_{x1}/°	1	t_{x6}/mm	100
θ_{y1}/°	1	t_{y6}/mm	−58000
θ_{z1}/°	1	t_{z6}/mm	−100
t_{x2}/mm	100	θ_{x6}/°	0.5
t_{y2}/mm	58000	θ_{y6}/°	−0.5
t_{z2}/mm	−100	θ_{z6}/°	0.5
θ_{x2}/°	0.5	t_{x7}/mm	−500
θ_{y2}/°	0.5	t_{y7}/mm	−500
θ_{z2}/°	0.5	t_{z7}/mm	500
t_{x3}/mm	−500	θ_{x7}/°	33
t_{y3}/mm	−500	θ_{y7}/°	0.1
t_{z3}/mm	500	θ_{z7}/°	−0.1
θ_{x3}/°	33	t_{x8}/mm	50
θ_{y3}/°	0.1	t_{y8}/mm	−50

(*continued*)

Table 1. (*continued*)

Parameters	value	Parameters	Value
$\theta_{z3}/°$	−0.1	t_{z8}/mm	50
t_{x4}/mm	50	$\theta_{x9}/°$	2
t_{y4}/mm	50	$\theta_{y9}/°$	−2
t_{z4}/mm	−50	$\theta_{z9}/°$	2
t_{x5}/mm	−500		
t_{y5}/mm	−500		
t_{z5}/mm	500		

2.2 Error Propagation Theory

The previous section defines parameters related to measurement in the measurement system and their relationships. In this section, error propagation rules between parameters and measurement are studied. In most of the complex on-orbit measurement systems, such as the long baseline on-orbit measurement system in this paper, the target quantity to be measured cannot be measured directly but needs to be measured indirectly. The quantity measured indirectly is a function of the measured values obtained directly, and its error is also a function of the error of the measured values. In indirect measurement, if the indirect measurement value y and the direct measurement value $x_1, x_2, \ldots x_n$ have the expression as $y = f(x_1, x_2, \ldots x_n)$, the systematic error of y can be denoted as Eq. (7), where $\partial f/\partial x_i$ is the error propagation coefficient.

$$\Delta y = \frac{\partial f}{\partial x_1}\Delta x_1 + \frac{\partial f}{\partial x_2}\Delta x_2 + \ldots + \frac{\partial f}{\partial x_n}\Delta x_n \tag{7}$$

Suppose the parameters are independent, and the random error of y is expressed as Eq. (8).

$$\sigma_y = \sqrt{\left(\frac{\partial f}{\partial x_1}\right)^2 \sigma_{x1}^2 + \left(\frac{\partial f}{\partial x_2}\right)^2 \sigma_{x2}^2 + \ldots + \left(\frac{\partial f}{\partial x_n}\right)^2 \sigma_{xn}^2} \tag{8}$$

According to the expression of baseline vector (Eq. (5) and Eq. (6)), the partial derivatives of baseline length and baseline angle with respect to each parameter can be theoretically solved. For example, to obtain the error propagation coefficient of t_{y1} to baseline length B, it is necessary to obtain the solution of Eq. (9). Since each parameter is independent of each other, when solving the partial derivative of each parameter according to the definition of partial derivative, the values of other parameters in Table 1 can be substituted as known constants. Taking the partial derivative of t_{y1} as an example, Eq. (10) can be obtained.

$$\begin{aligned} &\frac{\partial\left(\left\|\mathbf{R}_9 \cdot \overrightarrow{\mathbf{A}_2'\mathbf{A}_1'}\right\|_2\right)}{\partial t_{y1}} \\ &= \frac{\partial(\|\mathbf{R}_9 \cdot (\mathbf{T}_5 \cdot \mathbf{R}_5 \cdot \mathbf{T}_6 \cdot \mathbf{R}_6 \cdot \mathbf{T}_7 \cdot \mathbf{R}_7 \cdot \mathbf{T}_8 \cdot \mathbf{A}_2 - \mathbf{T}_1 \cdot \mathbf{R}_1 \cdot \mathbf{T}_2 \cdot \mathbf{R}_2 \cdot \mathbf{T}_3 \cdot \mathbf{R}_3 \cdot \mathbf{T}_4 \cdot \mathbf{A}_1)\|_2)}{\partial t_{y1}} \end{aligned} \tag{9}$$

$$\frac{\partial(\|\mathbf{B_L}\|_2)}{\partial t_{y1}} = \frac{\partial\left(S(t_{y1})\right)}{\partial t_{y1}}, \ S(t_{y1}) = \sqrt{1.0000t_{y1}^2 + 233156.5412t_{y1} + 13591544641.8846} \tag{10}$$

The relationship between $S(t_{y1})$ and t_{y1} is shown in Fig. 2. The deflection value of $S(t_{y1})$ to t_{y1} at the initial value of 500 mm is the curve tangent slope of the curve in Fig. 2 at the specific t_{y1}. After substituting $t_{y1} = 500$ mm, Eq. (10) is obtained to calculate the deflection result. That is, the error propagation coefficient $k_B(t_{y1})$ of t_{y1} to baseline length B is 1.0000.

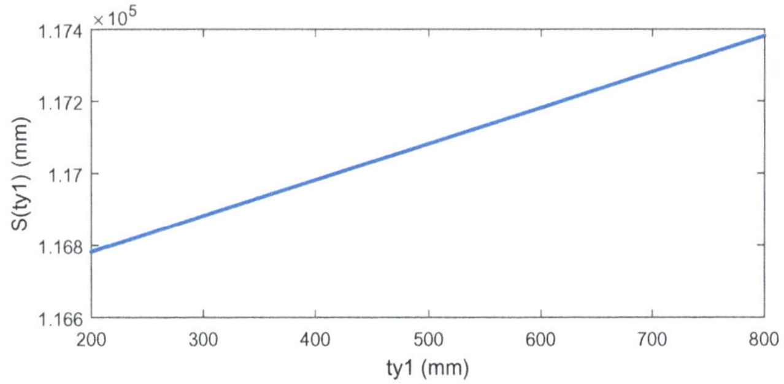

Fig. 2. The relationship between $S(t_{y1})$ and t_{y1}.

Using the same method, the error propagation coefficient $k_\alpha(t_{y1})$ of t_{y1} to baseline angle α can be obtained as 1.14e-8 asec/μm. Similarly, the error propagation coefficients of other parameters on baseline length and baseline angle can be obtained, as shown in Table 2, where the unit of k_B is μm/μm for the position parameters and μm/asec for the pose parameters, and the unit of k_α is asec/μm for the position parameters and asec/asec for the pose parameters.

The error of each parameter can be divided into systematic error and random error. According to the error propagation coefficient of each parameter to the baseline length and baseline angle, as well as the error propagation rule expressed in Eq. (7) and Eq. (8), the relationship between the systematic error of baseline length and baseline angle (ΔB, $\Delta\alpha$) and the systematic error of each parameter is as follows:

$$\begin{aligned} \Delta B =& \sum_i k_B(t_{xi})\Delta t_{xi} + k_B(t_{yi})\Delta t_{yi} + k_B(t_{zi})\Delta t_{zi} \\ &+ k_B(\theta_{xi})\Delta\theta_{xi} + k_B(\theta_{yi})\Delta\theta_{yi} + k_B(\theta_{zi})\Delta\theta_{zi} \\ \Delta\alpha =& \sum_i k_\alpha(t_{xi})\Delta t_{xi} + k_\alpha(t_{yi})\Delta t_{yi} + k_\alpha(t_{zi})\Delta t_{zi} \\ &+ k_\alpha(\theta_{xi})\Delta\theta_{xi} + k_\alpha(\theta_{yi})\Delta\theta_{yi} + k_\alpha(\theta_{zi})\Delta\theta_{zi} \end{aligned} \tag{11}$$

The relationship between the random errors of baseline length and baseline angle (σ_B, σ_α) and the random errors of each parameter is as follows:

Table 2. Calculated Error Propagation Coefficient between B, α and Each Parameter.

Parameters	k_B	k_α	Parameters	k_B	k_α
t_{x1}	0.0088	0	θ_{x5}	6.7865	0.5001
t_{y1}	1.0000	0	θ_{y5}	0.0610	0.0032
t_{z1}	0.0003	0.0018	θ_{z5}	4.0433	0.0173
θ_{x1}	6.6940	0.4910	t_{x6}	0.0087	0
θ_{y1}	0.1768	0.0031	t_{y6}	0.9998	0
θ_{z1}	8.8495	0.0170	t_{z6}	0.0170	0.0018
t_{x2}	0.0262	0	θ_{x6}	2.5176	0.0048
t_{y2}	0.9995	0	θ_{y6}	0.0375	0.0040
t_{z2}	0.0173	0.0018	θ_{z6}	2.2032	0.0001
θ_{x2}	2.2964	0.0038	t_{x7}	0.0002	0
θ_{y2}	0.0448	0.0039	t_{y7}	1.0000	0
θ_{z2}	2.0854	0	t_{z7}	0.0082	0.0018
t_{x3}	0.0351	0	θ_{x7}	0.0740	0.0006
t_{y3}	0.9991	0	θ_{y7}	0.0021	0.0004
t_{z3}	0.0257	0.0018	θ_{z7}	0.2420	0
θ_{x3}	0.0626	0.0006	t_{x8}	0.0016	0
θ_{y3}	0.0038	0.0004	t_{y8}	0.8431	0.0009
θ_{z3}	0.2309	0	t_{z8}	0.5377	0.0015
t_{x4}	0.0334	0	θ_{x9}	0	0.9994
t_{y4}	0.8239	0.0010	θ_{y9}	0	0.0075
t_{z4}	0.5657	0.0015	θ_{z9}	0	0
t_{x5}	0.0088	0			
t_{y5}	1.0000	0			
t_{z5}	0.0003	0.0018			

$$\begin{aligned}\sigma_B &= \sqrt{\sum_i k_B^2(t_{xi})\sigma_{txi}^2 + k_B^2(t_{yi})\sigma_{tyi}^2 + k_B^2(t_{zi})\sigma_{tzi}^2 + k_B^2(\theta_{xi})\sigma_{\theta xi}^2 + k_B^2(\theta_{yi})\sigma_{\theta yi}^2 + k_B^2(\theta_{zi})\sigma_{\theta zi}^2} \\ \sigma_\alpha &= \sqrt{\sum_i k_\alpha^2(t_{xi})\sigma_{txi}^2 + k_\alpha^2(t_{yi})\sigma_{tyi}^2 + k_\alpha^2(t_{zi})\sigma_{tzi}^2 + k_\alpha^2(\theta_{xi})\sigma_{\theta xi}^2 + k_\alpha^2(\theta_{yi})\sigma_{\theta yi}^2 + k_\alpha^2(\theta_{zi})\sigma_{\theta zi}^2}\end{aligned} \quad (12)$$

In summary, we get the error propagation coefficient between each parameter and the error to be measured. Moreover, the error propagation model of the systematic error and random error of the baseline length and the baseline angle is obtained.

3 Simulation and Experiment

This section verifies the correctness of the long-baseline on-orbit measurement system model proposed above. Figure 3 shows the simulation measurement results of baseline length and baseline angle obtained after the baseline measurement system is established by replacing the parameter values in Table 1 with commercial software Spatial Analyzer. It can be seen that the simulation results are consistent with the theoretical calculation results of baseline length and baseline angle in Table 2, which verifies the accuracy of the long baseline measurement model proposed in this paper.

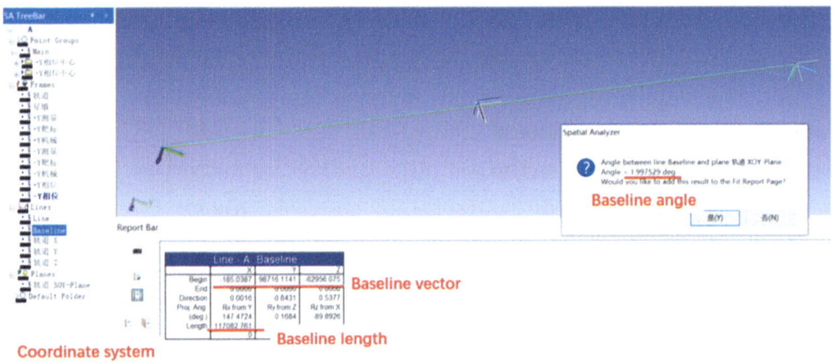

Fig. 3. Long baseline measurement model simulation results. The simulation results is in accordance with the theoretical baseline results, which validates the correctness of our modeling.

In order to further verify the accuracy of the proposed error propagation model in the previous sections, the distance measurement function of PROE software is used to verify the change of the baseline length B when different parameters are changed. Herein, two parameters t_{y1} and θ_{x1} are taken as examples, as shown in Fig. 4. In the case of initial values, the baseline measurement distance obtained by PROE modeling software is consistent with the theoretical value and the results of Spatial Analyzer, verifying the accuracy of the baseline measurement model in this paper once again. For the verification of error propagation coefficient, by changing the value of t_{y1} by 1000 mm, it can be obtained that the length of the baseline to be measured also changes by about 1000 mm. That is, the error propagation coefficient k_B is 1. In addition, by changing the value of θ_{x1} by 0.2°, the baseline length B changes by about 5 mm, which means the error propagation coefficient k_B is 6.9 μm/asec. PROE software was also used to verify the error transfer coefficients of other parameters subsequently, and the results consistent with the error propagation coefficients obtained by theoretical numerical calculation. This simulation verifies the accuracy of the model used in this paper to calculate the error propagation coefficients of parameters of the baseline measurement system.

Finally, the measurement error is evaluated by submitting the measurement precision of the commercial equipment and our developed six-degree-of-freedom high-precision measuring instrument. The measurement error of the ground calibration equipment can be obtained by sending the ground measuring equipment for inspection and passing the

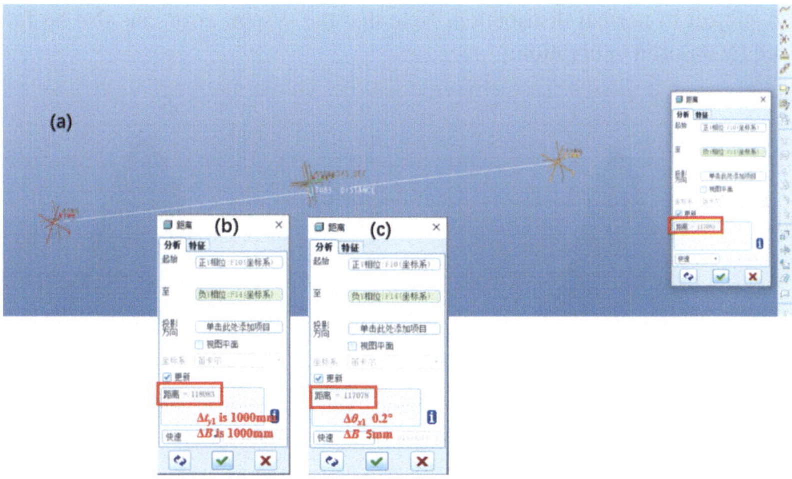

Fig. 4. The simulated error propagation coefficients k_B of parameters t_{y1} and θ_{x1}.

verification process and certificate results of National Institute of Metrology. Since the systematic error of the baseline to be measured depends on the systematic error of the parameters of each coordinate system, the systematic error of the length and angle of the baseline to be measured can be obtained by substituting the systematic error of each parameter into the Eq. (11). Similarly, as the random error of B and α depends on the random error of each coordinate parameter, the random error of the length and angle of the baseline to be measured can be obtained by substituting the random error of each parameter into the Eq. (12). The systematic errors of baseline length and angle to be measured are 483 μm (ΔB) and 2.83 asec ($\Delta\alpha$), and the random errors of baseline length and angle to be measured are 28.3 μm (σ_B) and 0.51 asec (σ_α), respectively.

The Monte Carlo method is used to substitute the above initial values, and the error values obtained by tests and verification are used to conduct 10,000 simulations. As a result, the random error distribution of baseline length and baseline angle as shown in Fig. 5 and Fig. 6 could be obtained. The simulation results were consistent with the theoretical results, and the measurement results of baseline length and baseline angle

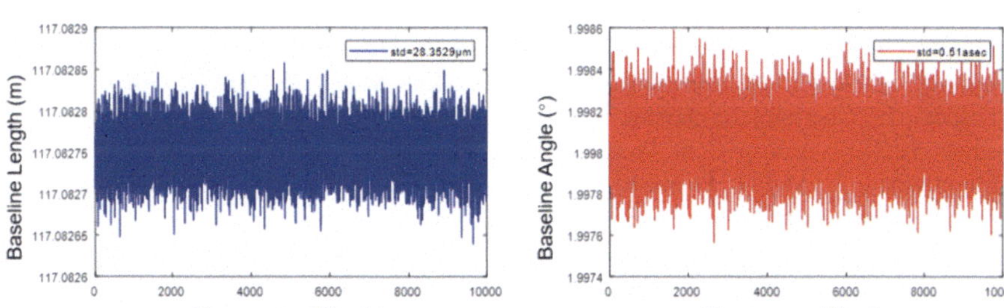

Fig. 5. A 10,000 simulation results of baseline length B and baseline angle α by Monte Carlo.

are also subject to normal distribution. Note that the system error can also be further corrected by on-orbit calibration.

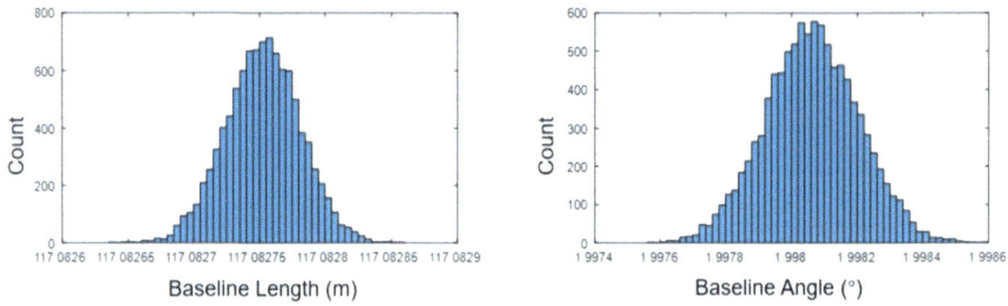

Fig. 6. Error distribution of Monte Carlo simulation results for baseline length and baseline angle. Their random errors are subject to normal distribution

4 Conclusion

To sum up, aiming at the problem that traditional error analysis and allocation methods are difficult to balance the scientific rationality and engineering realizability, we propose an innovative comprehensive optimization method combining equal influence method and experiment-based method for error analysis and allocation. In this paper, an InSAR satellite long baseline measurement system model and error propagation model are established, and data analysis are carried out to explore the error allocation values of each parameter. Compared with the conventional methods, the proposed error analysis and allocation method based on error propagation rule and engineering capability is more scientific and reasonable, which greatly reduces the experiment time input and cost for complex measurement system.

Funding. National Natural Science Foundation of China (62305021).

References

1. Kanimba, E., Yang, T., Huxtable, S.T., Li, L., Tian, Z.: Thermomechanical analysis of a bio-inspired lightweight multifunctional structure. Adv. Eng. Mater. **22**, 2000371 (2020)
2. Fosness, E., Guerrero, J., Qassim, K., Denoyer, S.J.: Recent advances in multi-functional structures. IEEE Aerosp. Conf. Proc. **4**, 23–28 (2000)
3. Flores-Abad, A., Ma, O., Pham, K., Ulrich, S.: A review of space robotics technologies for on-orbit servicing. Prog. Aerosp. Sci. **68**(8), 1–26 (2014)
4. Duren, R., Litty, E., Wong, E., Shaffer, S.: Metrology, Attitude, and Orbit Determination for Spaceborne Interferometric Synthetic Aperture Radar (IFSARK) (1998)
5. Li, W.J., Cheng, D.Y., Liu, X.G., Wang, Y.B., Shi, W.H., Tang, Z.X.: On-orbit service (oos) of spacecraft: a review of engineering developments. Prog. Aerosp. Sci. **108**, 32–120 (2019)
6. Song, Y., Li, C., Zhao, H., Xia, S., Li, X., Fan, X.: Review on on-orbit assembly of large space telescopes. In: SPIE Conference on Applied Optics and Photonics China (2019)

7. Dantan, J.Y., Eifler, T.: Tolerance allocation under behavioural simulation uncertainty of a multiphysical system. CIRP Annal. Manuf. Technol. **70**(1), 127–130 (2021)
8. Hussein, H.M.A.: A novel methodology for simultaneous minimization of manufacturing objectives in tolerance allocation of complex assembly. Appl. Sci. **11**(19), 9164 (2021)
9. Fuller, W.A.: Measurement Error Models. Wiley (2009)

Design Method for High Temperature Stability Thermal Control of Remote Sensor

Hui Zhao(✉), Pan Lu, Xinyang Song, Feng Yu, Jiong Zhang, Juan Gao, Lifeng Jin, and Tingting Wu

Beijing Institute of Space Mechanics and Electricity, Beijing 100094, China
zhaohuicheliang@163.com

Abstract. The high precision thermal control of remote sensor is an important condition for ensuring the imaging quality, achieving long life and reliability of the remote sensor. To ensure the imaging quality and pointing accuracy in orbit, the main components of the remote sensor need to maintain high temperature stability throughout its entire life cycle. With the improvement of the resolution ratio of remote sensor, the optical tolerance is further tightened, and the requirements for thermal stability in optical and mechanical system are further improved. The article proposed a thermal control design method to improve temperature stability under the constraints of limited allocation weight and power resources. The idea of integrated external suppression with inside thermal precision control is adopted. The methods include changing the time of the descending point, optimizing working modes, using camera's door, increase the temperature control threshold of the main structure. Through thermal simulation calculation and in orbit flight verification, the main components of the remote sensor can maintain temperature stability of the $\pm$ 0.2°C throughout its entire life cycle. It is indicating that the method can improve the temperature stability, beneficial for ensuring high-quality imaging of the remote sensor and can be applied to high-precision and high stability temperature control design of future remote sensors.

Keyword: Temperature stability; Precision thermal control

1 Introduction

Over the past few decades, China's space optical remote sensing satellite images have been widely used in many fields and have played an immeasurable role [1]. Thermal deformation can cause deformation of the load inside structure and the external support structure of the load, which has a significant impact on positioning accuracy and elevation calculation accuracy. Therefore, the design of dimensional thermal stability is crucial in the entire task [2]. However, with the further deep of applications, users' requirements for image quality are becoming higher and higher. Remote sensors are developing towards high resolution ratio, large aperture, high energy consumption, high heat flux density, variable orbital maneuvers, and high temperature control accuracy [3]. In order to achieve higher image quality, high-resolution ratio remote sensors need to meet higher geometric accuracy [4], thus higher demands on the constant temperature control of remote sensors are required.

H. P. Urbach et al. (Eds.): ISSOIA 2023, SPP 191, pp. 424–433, 2024.
https://doi.org/10.1007/978-981-97-6718-2_37

The traditional thermal design of remote sensors often adopts passive thermal control or active temperature control methods. Especially for key components with high temperature range requirements, active temperature control heating circuits are usually directly arranged on the controlled object to achieve temperature control [5, 6]. As a large-scale system with complex structure, the thermal coupling between various components of the remote sensor is very complex. A single structural component exchanges heat with the surrounding structure through heat conduction, radiation, and during orbital. The remote sensor needs to undergo long-term alternating heating and cooling of the sun, planets, and space [7]. Its temperature change is greatly affected by the overall temperature of the surrounding environment. When facing with complex thermal coupling structures, variable external heat flow environments, limited weight and power distribution, and high temperature control accuracy requirements, the traditional temperature control methods are difficult to meet the temperature control requirements of the developing high-resolution ratio remote sensors.

This article proposes a thermal control design method to improve temperature stability, using the idea of combining external suppression with inside precision control. The thermal control design method for high temperature stability is elaborated in detail. The difference in temperature control accuracy between the design method and traditional thermal control technology is given through simulation analysis and verified through in orbit flight. The main components of the remote sensor can maintain temperature stability of ± 0.2 °C throughout its entire life cycle.

2 Overview of Thermal Control Design Methods for High Temperature Stability

With the improvement of remote sensor resolution ratio, optical tolerances are further tightened, and the requirements for thermal stability in optical and mechanical systems are further improved [8]. The article proposes a thermal control design method to improve temperature stability, using external suppression combined with inside precision control. The method is adding camera's door mechanisms, optimizing working modes, refining insulation and temperature equalization under the constraints of limited allocation weight and power resources. Thus, the temperature stability of the optical-mechanical system of the remote sensor has been improved.

$$P + Q + Q_e = \mathrm{mc}\Delta T + Q_t \tag{1}$$

$$P + \alpha_s A_p(q_1 + q_2 + q_3) = mc\Delta T + A\varepsilon\sigma(T_h^4 - T_c^4) \tag{2}$$

$$dq_1 = S\cos\beta_s dA = S\phi_1 dA \tag{3}$$

$$dq_2 = \rho S dA \iint_{A_{E'}} \frac{\cos\eta\cos\alpha_1\cos\alpha_2}{\pi l^2} dA_E = \rho S dA\phi_2 \tag{4}$$

$$dq_3 = \frac{1-\rho}{4} S dA \iint_{A_E} \frac{\cos\alpha_1\cos\alpha_2}{\pi l^2} dA_E = \frac{1-\rho}{4} S dA\phi_3 \tag{5}$$

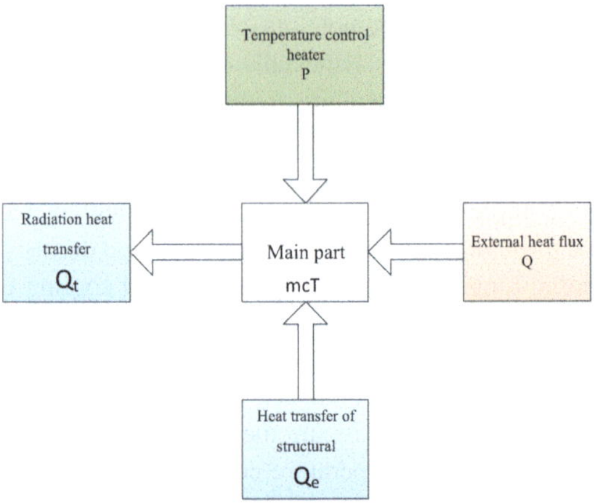

In the formula
q_1 is solar radiation external heat flux.
q_2 Earth's albedo external heat flux.
q_3 Earth's infrared external heat flux.
dA_E differential surface area of the Earth.
$A_{E'}$ the area of the Earth under direct sunlight.
l distance between dA and dA_E;
ϕ_1 solar radiation coefficient.
ϕ_2 Earth's albedo coefficient.
ϕ_3 Earth's infrared coefficient.

The focus of thermal stability design is to reduce short-term temperature fluctuations caused by variation in thermal environment and intermittent operation of internal heat sources. On the one hand, it needs to suppress the influence of external heat flux. A door mechanism has been added in the inlet of the camera, which can be closed during non-imaging periods to suppress temperature fluctuations caused by changes in external heat flux. During imaging periods, the camera's door is turned on, the temperature fluctuates less due to the short time affected by Q (external heat flux). Meanwhile it is beneficial for saving temperature control power consumption also. Optimizing the working model like imaging sway angle and duration can reduce the changes ϕ_2 and ϕ_3 of q_2 and q_3 during imaging, thereby reducing temperature fluctuations; The local time of descending node has been changed to dawn and dusk. The camera will receive less q_2 (heat flux of earth's albedo), and the temperature fluctuations are suppressed. The heat shield structure is added in structural, with multiple layers of insulation between the heat shield and the structure to reduce the impact of Q(external heat flow).

On the other hand, it needs precision control of internal temperature. Designing reasonable heat dissipation path can minimize temperature changes made by short term operation of the internal heat source as much as possible. And large temperature difference insulation structure is used to reduce the impact of Q_e from the internal heat source. Kevlar composite graphite material can improve overall temperature consistency, reduce temperature difference, eliminate high-temperature hotspots, improve temperature uniformity, and improve the efficiency of compensating power. Increasing the temperature control threshold of indirect temperature control can improve temperature stability, which use the duty cycle of the temperature control heat P to control temperature.

The thermal design concept of combining external suppression with internal precision control is adopted. High temperature control accuracy requirements can be achieved through the coupling of multiple methods mentioned above, under the constraints of complex thermal coupling structures, variable external heat flow environments, limited weight and power distribution, the basis of existing technological level, process level, and requirement of reliability of aerospace products. The high stability thermal control design method can fully leverage its advantages and lay the foundation for obtaining high-quality images, under the strict optical and mechanical requirements.

3 Application of Thermal Control Design Method for High Temperature Stability

The implementation path and temperature control measures of the high temperature stability temperature control design method are detailed as follows. Through thermal simulation calculation and analysis, it is predicted that the temperature distribution under this thermal control measure. It is verified the reasonable application of the high temperature stability temperature control design method can improved the temperature stability of the remote sensor.

3.1 Changing the Time of the Descending Point

From the Fig. 1, it can be seen that the heat flow outside the remote sensor's inlet will vary with the time of the descending point. The peak external heat flux can be reduced by $100 w/m^2$ when the local time of the orbital descending point is changed from 10:30 to 9:30. And the peak external heat flux can be reduced by $150 w/m^2$ when the local time of the orbital descending point is changed from 10:30 to 7:30. The reduction of the external heat flux fluctuation is beneficial for reducing the temperature fluctuation of the remote sensor structure.

By comparing the Table 1 of the temperatures of the main parts of the remote sensor at different descending points, it can be concluded that when the local time of the descending point changes from 10:30 to 9:30, the temperature fluctuations of the main structure of the remote sensor decrease. Among them, the temperature fluctuations of the three mirrors decrease the most, at 1.3 °C. The absorption rate of the main mirror is small, and the stability change is minimal, less than 0.1 °C. The temperature difference of the main optical structure decreases. When advance the time of the descending point, the overall temperature level of the remote sensor decreases, and the long-term compensation power consumption will increase.

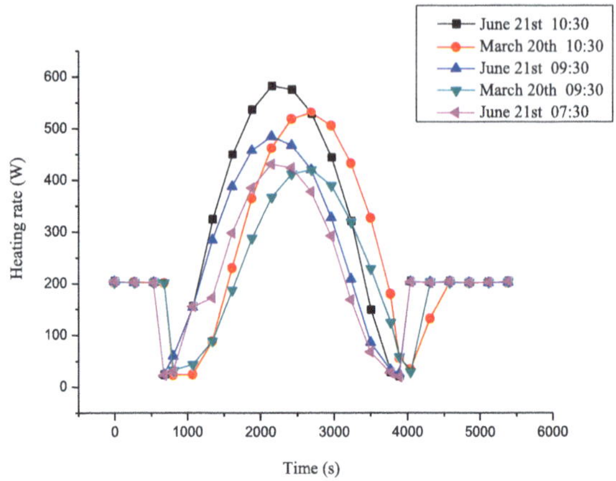

Fig. 1. Curve of heat flow outside the remote sensor's inlet with the time of the descending point.

Table 1. Temperature at different time of the descending point.

Part	Time of the Descending node 10:30	Time of the Descending node 9:30	Temperature stability variation ΔT
Primary mirror	23.1 ~ 23.8	22.8 ~ 23.5	−0.06
Secondary mirror	17.6 ~ 20.4	16.8 ~ 19.1	−0.5
Secondary mirror Three Poles	17.8 ~ 21.8	17.2 ~ 21	−0.2
Three Mirrors	18 ~ 22	17.5 ~ 20.2	−1.3
Four mirrors	18.4 ~ 21	18.1 ~ 20	−0.7
Folding mirror	16.4 ~ 24.5	16.3 ~ 22	−2.4
Lens barrel	18.2 ~ 22	18.2 ~ 21.6	−0.4
Principle bearing Structural	17 ~ 24	17 ~ 23.5	−0.5

3.2 Optimize Imaging Working Mode

The change in the imaging sway angle of the remote sensor will change the heat flow outside the inlet. As shown in the Fig. 2, when the sway angle changes from 0° to 10°, the temperature stability can be reduced by 0.1 °C, and when the sway angle changes from 10° to 20°, the temperature stability change is less than 0.1 °C. The slope of temperature stability gradually decreases with the change of sway angle. Change the working time of imaging reduce the temperature stability by 0.1 °C shorten the imaging time from 5 min to 3 min. The slope of temperature stability gradually decreases with the change of imaging time. Optimizing the imaging working mode, increasing the swing angle and shortening the imaging working time can improve the temperature stability of the remote sensor.

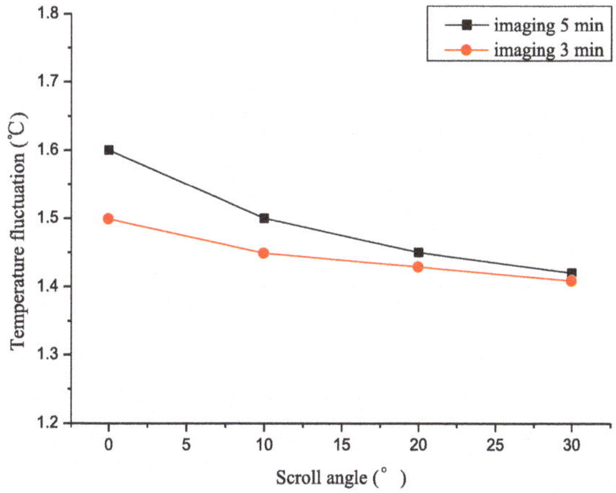

Fig. 2. Curve of temperature stability with sway angle and imaging time.

3.3 Adding Camera's Door

A camera's door mechanism designed at the inlet can reduce the impact of adverse external heat flow fluctuations in the orbit when closed. Camera's door reduce the impact of severe fluctuations and significant periodic changes of external heat flux and complex structures. The camera's door is to turn on for imaging in a short period of time, which can reduce the input of the total heat flux of the remote sensor, reduce temperature fluctuations, and ensure the full cycle temperature stability of the remote sensor. The Fig. 3 shows the heat flux curve outside the inlet in different seasons. Under different seasons, the difference in heat flux of the inlet is 100 W/m^2. The heat flux of the year changes significantly causing significant temperature fluctuations at various positions of the remote sensor within the year. The main mirror fluctuation caused by seasonal changes in heat flux within the year reaches 0.4 °C, which also affects the imaging of optical systems. In the case of camera's door shut down, the impact of external heat flow fluctuations on the remote sensor is significantly reduced, and the temperature fluctuations of various components are reduced. The temperature fluctuation of the main mirror is only 0.1 °C. In the camera's door solution, the long-term temperature stability of the remote sensor is higher. At the same time, camera's door can reduce the radiation heat leakage from the inlet to the cold space, reduce the demand for compensation power consumption, and save long-term power consumption.

At the same time, due to the camera's door shut down, it is adverse to heat dissipation. If the temperature of each component in the shaded area does not return to the equilibrium temperature, heat will accumulate, and the temperature of each component will gradually increase. According to the temperature stability index requirements of the remote sensor, the working mode can be selected for maximum number of consecutive tracks of imaging.

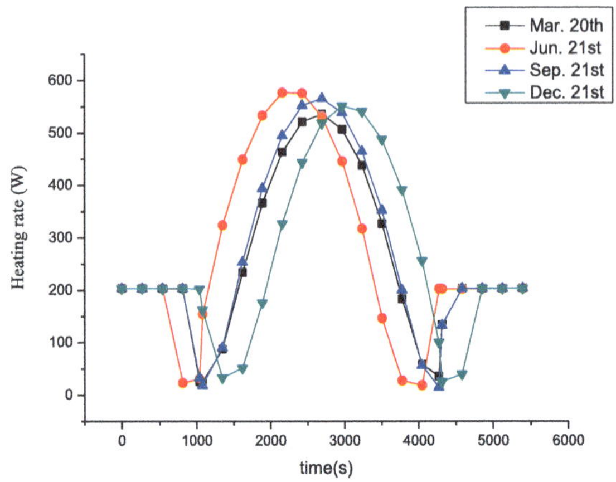

Fig. 3. Curve of heat flow outside the remote sensor's inlet with different dates.

If the temperature change of the continuous imaging exceeds the temperature stability requirements, camera's door can be turned on in the shaded area to dissipate heat (Fig. 4).

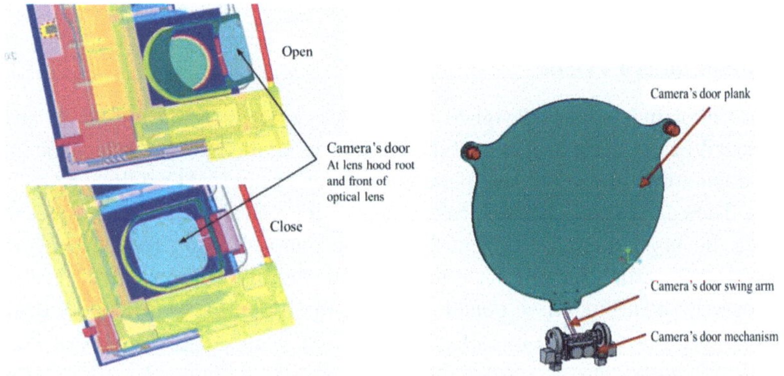

Fig. 4. Structure diagram of camera's door.

3.4 Increase the Temperature Control Threshold of the Main Structure

Some optical key components control the temperature with auxiliary structural components around the controlled object. The thermal control auxiliary structure provides the environment for the controlled object, while weakening the influence of the temperature field around, achieving high-precision constant temperature control. As shown in the Fig. 5, increasing the temperature control threshold of optical key components is beneficial for improving temperature stability. When the temperature threshold increases from 19 °C to 25 °C, the temperature fluctuation can be reduced by 1.92 °C. As the temperature control threshold increases, the slope of temperature stability gradually decreases with the change of temperature threshold. However, the temperature of the thermal control auxiliary structure cannot increase infinitely, the design of temperature threshold should also consider the overall temperature distribution. Auxiliary structural temperature and the surrounding structure temperature interact with each other by radiation and conduction heat exchange, which will be limited by other structural temperature index.

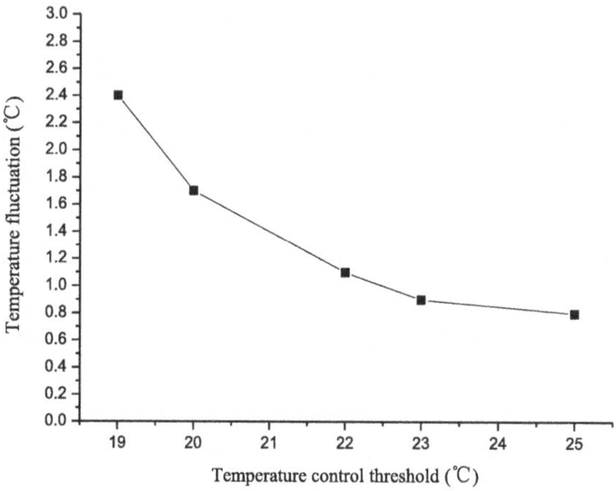

Fig. 5. Curve of temperature stability with temperature control threshold.

4 In-Orbit Flight Verification

Applying the high temperature stability thermal control design method to the thermal control design of actual remote sensor can obtained the temperature curve of in orbit flight of Fig. 6. The temperature stability of the main optical and mechanical structures of the remote sensor is better than ± 0.2 °C throughout its entire life cycle.

The specific thermal control measures are as follows: using camera's door is good for suppressing the external heat, reducing the impact of external heat flow on the remote sensor, improving the temperature stability and saving thermal control compensation power consumption. An internal and external dual radiation heat shield is installed for the secondary mirror and mirror barrel which directly facing the cold space and the external environment changes sharply. This shield effectively reducing the uneven temperature and rapid heating caused by exposure to sunlight. Minimizing the thermal coupling between the spacecraft and the remote sensor as much as possible by installing insulation pad is internal precision temperature control measure. Adding high thermal conductivity graphite to the support structure can improves overall temperature consistency. The focal plane circuit box with black paint can enhance heat exchange radiates and dissipates heat towards the cabin panel. Insulation should be taken at appropriate positions to strengthen the insulation between key components. The temperature control heating power on the components and the auxiliary temperature control plat should reasonably allocated. Key positions are controlled by using high temperature thresholds to improve temperature stability.

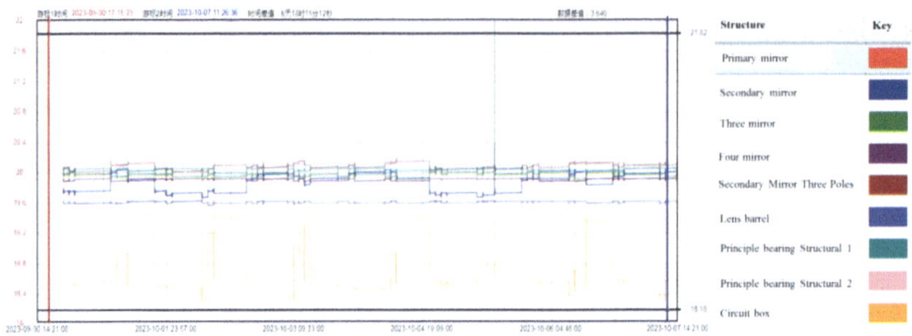

Fig. 6. Temperature curve of main optical and mechanical structures of the remote sensor on orbit

5 Conclusion

With the improvement of remote sensor resolution ratio, optical tolerances are further tightened, and the requirements for thermal stability in optical mechanical systems are further improved. The difficulty of temperature control for remote sensor temperature stability is constantly increasing when faced with complex thermal coupling structures, variable external heat flow environments, limited weight and power distribution, and high temperature control accuracy requirements. The article proposes a thermal control design method to effectively improve temperature stability which adopting the idea of combining external suppression with internal precision control under the constraint of limited weight and power consumption resource allocation. Multiple methods is coupled used, such as adding camera's door mechanisms, refining insulation and temperature equalization, precising temperature control, and optimizing in orbit working mode. The article provides a detailed explanation of the implementation approach and temperature control method of high temperature stability thermal control in thermal design. After simulation analysis and in orbit flight verification, the main components of the remote sensor can maintain temperature stability of $\pm 0.2\,^{\circ}\mathrm{C}$ throughout its entire life cycle. Tt is predicted that using this design method can improve the temperature stability of remote sensors. High temperature stability thermal control method is beneficial for ensuring high-quality imaging of the remote sensor and has good application prospects in future high-precision and high stability temperature control design of remote sensors.

References

1. Jiali, W.U., Tao, Y.U., Xingfa, G.U., et al.: Status and application trend of chinese earth resources satellites. J. Remote Sens. Inf. **6**, 96–101 (2008)
2. Yin Y.Z., Fu, W.C., Lu, Q.R., .et al.: Thermal design of XX-7 satellite based on dimensional stabilityacquirement. J. Spacecraft Eng. **29**(3) (2020)
3. Song, X.Y., Gao, J., Zhao, Z.M, et al.: Application of indirect thermal control technology for constant temperature control of HR optical remote sensors. J. Spacecraft Recovery Remote Sens. **2** (2015)
4. Cao, H., Zhang, X., Zhao, C., Xu, C., Mo, F., Dai, J.: System design and key technolongies of the GF satellite. J. Chin. Space Sci. Technol. **40**(5), 1–9 (2020)

5. Wengang, Y.A.N.G., Lei, Y.U., Rongli, C.H.E.N., et al.: Precise thermal control design and validation for high resolution spacecamera. J. Acta Photonica Sinica **38**(9), 363–2367 (2009)
6. Megahed, A., El-Dib, A.: Thermal design and analysis for battery module for a remote sensing satellite. J.Spacecraft Rockets **44**(4), 920–926 (2007)
7. Zhenming, Z., Bing, W.A., Juan, G.: Preliminary research on the thermal design methods of the geosynchronous orbit stating camera. J. Spacecraft Recovery & Remote Sens. **31**(3) (2010)
8. Guo, N., Bo, Y., Xia, C., et al.: Temperature control with high stability for the assembly base of space optical cameras. J. Spacecraft Recovery Remote Sens. **41**(4) (2020)

Intelligent Enhancement for Low-Light Earth Remote Sensing Based on Physical Process Modeling

Wei Jiuzhe[1,2], Wang Zhi[1], Wang Yun[1], and Wang Xiaoyong[1](✉)

[1] Beijing Institute of Space Mechanics and Electricity, Beijing, China
w8320@126.com

[2] Nanjing University of Aeronautics and Astronautics, Nanjing, China

Abstract. Earth remote sensing is advancing towards achieving all-day availability and enhanced intelligence. However, the presence of low light conditions during nighttime poses a significant challenge as visible remote sensing imaging information becomes heavily intertwined with complex noise components, often leading to its submergence in noise. This issue severely hampers the effective utilization of data. Traditional denoising methods typically rely on assumptions of a universal noise model, which fails to address the intricate noise problem unique to remote sensing applications. Deep neural networks (DNNs) based on learning offer promising solutions; however, obtaining real training sample pairs for space-based low-illumination remote sensing images remains challenging. To overcome these obstacles, this paper proposes a method for enhancing low-light remote sensing based on high-level modeling of physical processes. By meticulously modeling, measuring and calibrating imaging physical processes and detector noise characteristics, we establish an extensive collection of high-fidelity training sample pairs consisting of both noisy and noise-free inputs. These samples are then utilized to effectively train the network model framework, enabling stable and high-performance capabilities specifically tailored for low-light detection using specific hardware configurations.

Index Terms: — Low-light remote sensing · intelligent enhancement · imaging process modeling

1 Introduction

The application of low-light remote sensing imaging technology has the potential to significantly improve the temporal information acquisition capability of existing space-based RS imaging systems [1]. Noise is an inherent factor in all imaging systems. In the case of low-light remote sensing, which is used for detecting ground objects under low illumination conditions such as night or morning/evening, it often results in a high proportion of noise(signal-to-noise ratio (SNR)<20dB). This coupling of signal and noise can obscure scene features and compromise image quality, thereby affecting the utilization of image remote sensing data.

H. P. Urbach et al. (Eds.): ISSOIA 2023, SPP 191, pp. 434–442, 2024.
https://doi.org/10.1007/978-981-97-6718-2_38

SNR Enhancing has been a long-standing problem since the 1950s [2–5]. The space-based imaging system operates within a complex electromagnetic, mechanical, and thermal environment where defects and inconsistencies in COMS detectors, and other detector components as well as circuit data flow processes contribute to noise in the final output image. The common methods for improving SNR can be categorized into filter-based, model-based, and learning-based approaches. In recent years, deep learning methods have emerged as a powerful tool for fitting complex mathematical models to image relationships and learning the potential mapping from noisy images to clean ones. This approach has shown significant improvements in both quantitative evaluation metrics and subjective perception. Establishing high-fidelity training sample pairs with and without noise using deep learning is crucial for addressing the challenge of enhancing SNR in real systems. To tackle this issue, researchers have modeled detector-generated noise based on physical mechanisms and calculated noise parameters individually [6, 7]. Alternatively, they have directly collected noise data and synthesized it into ideal images [8, 9] to enhance the accuracy of sample set construction, which has yielded promising training results for digital cameras. The aforementioned findings have examined noise modeling; however, the practical impact of common time-varying factors in remote sensing systems, such as variations in detector working temperature, integration time, and real-time gain adjustment, has not been taken into account. For space-based remote sensors that cannot be directly measured and operate under complex conditions, it is imperative to establish a more refined noise model. This will directly influence image reconstruction quality, local details within the image, and restoration of texture features.

The present study proposes a physics-based noise generation model specifically tailored for low-light remote sensing cameras to address the aforementioned issues. The model leverages the actual physical characteristics of the camera under low-light conditions to construct an accurate noise generation process. Moreover, each stage of the camera's workflow is meticulously modeled using experimental testing and mathematical statistics in order to simulate the generated noise. By employing a specially designed algorithm, optimal estimation of unknown parameters associated with each noise component is achieved. This low-light noise model effectively approximates the true level of noise present in the camera's output. Finally, The proposed approach involves the development of a hybrid U-Net and transformer model for denoising, which is trained on a carefully constructed sample set. Comparative analysis with other conventional methods or networks is conducted to validate the effectiveness of our method through experimental evaluations.

2 Noise Modeling of Imaging Process

The physical link of low-light remote sensing imaging primarily encompasses the transmission of low-light scene information and atmospheric attenuation, degradation in imaging optics, discrete sampling by the detector, and photoelectric conversion until a digital image is generated. The physical process at the back end of imaging optics for low-light remote sensing imaging can be described as follows [10, 10]: 1) Photons stimulate photoelectrons upon reaching the photosensitive region of a pixel, with their arrival following a Poisson distribution; 2) A quantum efficiency relationship exists between the number of photons reaching the photosensitive surface and the number of excited photoelectrons per unit time, which varies among pixels due to their specificity; 3) Dark electrons contributed by thermal-generated electrons or defects within the pixel are unrelated to signal electrons. These dark noise electrons depend on both working temperature and integration time of the detector and may interfere with subsequent processes; 4) Column noise caused by amplifier inconsistency and nonlinear response during signal charge-to-voltage conversion is introduced; 5) Linear errors and quantization errors occur due to digital rounding when converting voltage signals into digital signals. The aforementioned process can be illustrated by Fig. 1. Typical processes are described below.

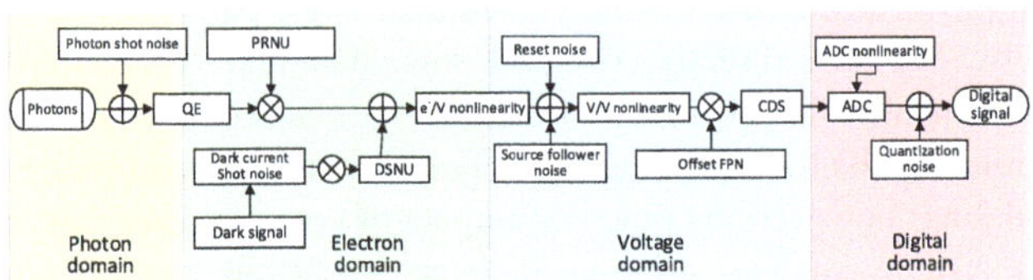

Fig. 1. Imaging noise transfer process

2.1 Photon-Electron Conversion Process

2.1.1 Photon Shot Noise

The process of photon capture introduces uncertainty due to random fluctuations in particle properties during photodiode detection. This inherent uncertainty gives rise to photon shot noise, which can be described as a Poisson process. In the case of high light levels (exceeding 1000 photon arrival events [12], the Poisson distribution can be approximated by a Gaussian distribution. However, such an approximation is not suitable for low-light conditions. The topic of approximating the Poisson distribution with Gaussians is extensively discussed in [13]. For the purpose of low-light imaging, the aforementioned process is modeled as follows:

$$I_{ph.shot} = \rho(I_{ph}) \tag{1}$$

The average number of photons at a specific sensor position after simulating the optical imaging link is denoted as I_{ph}, without providing details on its calculation process. ρ represents the Poisson distribution function, while $I_{\text{ph.shot}}$ refers to the discretized and randomly distributed photon signal.

2.1.2 Photo Response Non-Uniformity(PRNU)

PRNU refers to the spatial variation of pixel output under uniform illumination conditions. It primarily occurs due to changes in photodiode surface area and substrate material during detector manufacturing [14], resulting in different photoreceptors producing varying photocharge responses to the same input photon. PRNU is signal-dependent, and remains fixed over time.

$$I_e = QE \cdot I_{ph} \cdot shot \cdot (1 + N_{PRNU}) \quad (2)$$

The quantum efficiency (e^-/photon) for a given wavelength is represented by QE, and the PRNU factor matrix is denoted as N_{PRNU}.

2.1.3 Dark Current Shot Noise

The dark current shot noise, arising from the dark signal, exhibits a linear relationship with integration time. Even in the absence of light, a dark current is generated upon exposure initiation. With increasing exposure time (t_I), the amount of dark signal per pixel also increases. The dark current shot noise, characterized by random electron arrivals, follows a Poisson distribution at low levels of dark current(temperature dependent). The standard deviation of this Poisson distribution can be expressed as follows.

$$\sigma_{d.shot} = \sqrt{t_I \cdot D_R} = \sqrt{S_{dark.e^-}} \quad (3)$$

The D_R represents the average dark current, $S_{dark.e^-}$ denotes the dark signal at a specific integration time and temperature.

2.1.4 Dark Current Non-Uniformity (DSNU)

Similar to the origin of PRNU, DSNU exhibits a spatial random and fixed pattern noise structure when exposed to the same dark current electron. Calibration is also necessary for its acquisition. Based on the shot noise of dark current, it can be characterized as follows.

$$\sigma_{d \cdot FPN} = t_I D_R \cdot D_N \quad (4)$$

where t_I is the integration time, D_R Is the average dark current described in Eq. 1, and D_N is the dark current FPN factor. Additionally, there exist outliers or dark spikes [15], where certain pixels exhibit significantly higher responses to the dark signal compared to the average. This phenomenon of " spikes" or "outliers" can be attributed to the pool-frenkel effect [16], which refers to an increase in emissivity caused by a defect in the presence of an electric field.

2.2 Charge-Voltage Conversion Process

In the actual process of photosensitive imaging, the charge-to-voltage conversion is not perfect and can introduce uncertainty to the measured signal due to sensor node reset noise, source following noise, offset fixed mode noise, and gain nonlinearity. Correlated double sampling(CDS) is commonly used in high-level CMOS detectors to eliminate reset noise and source follower noise.

2.2.1 Offset FPN

Because pixels in the same column of a detector share a column amplifier, inconsistencies in bias voltage and integrating capacitor size lead to column bias offset FPN. Which exhibits a distribution law of column stripes. This component can be calibrated according to the average response value of each column pixel in a flat field.

2.2.2 Nonlinear Response

The nonlinear response can be divided into two categories: V/e – nonlinearity and V/V nonlinearity. This part of the error is difficult to be stripped out and considered in the actual detector calibration process, but can be considered into the overall system response non-uniformity (including the response nonlinearity of the ADC link) for calibration.

2.3 Voltage-To-Digital Conversion Process

An Analog to Digital converter (ADC) converts a voltage signal into a discrete digital signal. The nonlinearity and noise involved in this process are: integral linear error(ILE), differential linear error(DLE), quantization error and ADC offset. The quantization error is the main factor.

2.3.1 Quantization Error

Quantization error, also known as rounding error, occurs due to limited quantization accuracy in ADCs. Rounding during conversion leads to quantization errors that follow a uniform distribution pattern. Consequently, low-light remote sensing scenarios can significantly amplify the proportion of quantization errors.

3 Noise Calibration

During the actual calibration process, the noise model can only be inverted through digital signals, and its estimation relies on mathematical models adjusted by input signals and the temporal-spatial correlation between multiple frames and single frames of noise. The portion of noise that is relatively small or difficult to decouple from other noises in the temporal-spatial domain is considered as an overall distribution solution. In this paper, the sequence for calibrating model parameters is as follows: 1) offset FPN (column noise); 2) PRNU; 3) DSNU; 4) dark current shot noise calibration; 5) other time-independent readout noise calibration; 6) photon shot noise estimation. This study utilizes a low-light integrating sphere in the laboratory with light illuminance set at 0 ~ 0.2lux, camera focal length set to infinity, and hundreds of original images captured under different integration times ranging from 0.01 to 2000 ms at a focal plane temperature equilibrium point of 25 °C (Figs. 2 and 3).

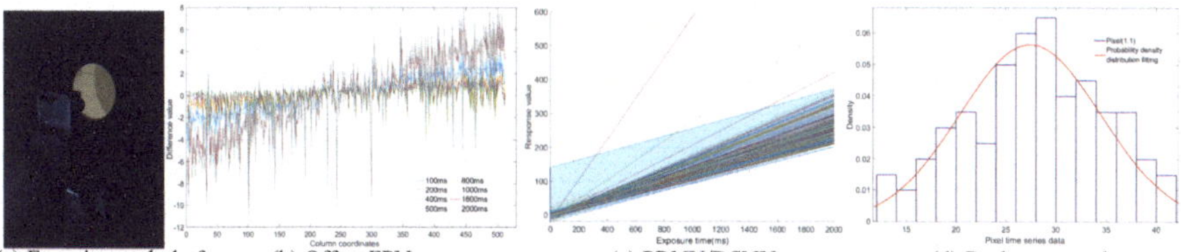

(a) Experimental platform (b) Offset FPN (c) PRNU/DSNU (d) Dark current shot

Fig. 2. (a) Experimental platform comprises an sCMOS detectors, an image acquisition device, and an integrating sphere. (b) The offset FPN after calibration, it obtained by averaging uniform field images and adjusting it based on the column statistical mean. The integration times show a column-wise upward trend due to dark noise accumulation during pixel storage under the rolling shutter operation mode. (c) The PRNU/DSNU statistics each pixel's average is calculated to eliminate time-dependent random noise influence. Then, the relationship between gray values of different pixels and integration time is calculated and fitted. Most pixels fall within a close confidence interval, with only a few outliers showing heightened sensitivity or unresponsiveness to light response resulting in "spikes" in the output image.(d) The calibration process for dark current shot noise involves establishing relationships between dark current shot noise power and integration time as well as working temperature using single-pixel behavior variations over time series data.

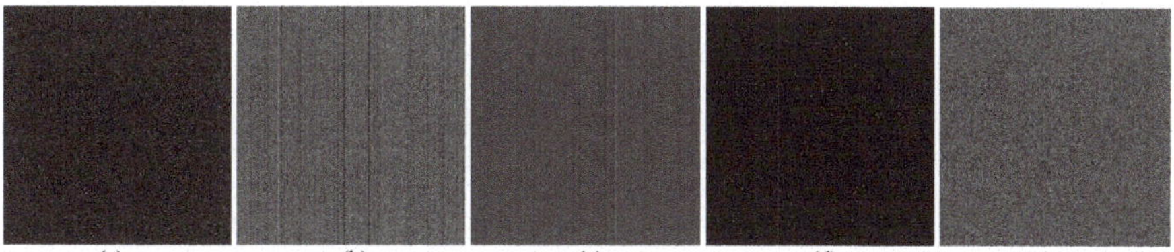

(a) (b) (c) (d) (e)

Fig. 3. (a) the original single imaging image (b) the average result of the image sequence (c) the column noise separated by the statistical method (d) the PRNU image separated by the statistical method (e) the noise image reorganized, which was calibrated by our method and then simulated and reorganized.

4 Enhancement Algorithm

4.1 Network Architecture

Due to the specific application scenarios of space-based low-light imaging, this paper integrates the U-Net network [21] with the Transformer network. The schematic diagram of the proposed network architecture is illustrated in Fig. 4. Following preprocessing, the fidelity noise maps in the sample set are cropped to a size of 512×512 pixels.

Subsequently, an input mapping layer converts these images into patch blocks for further processing. The Transformer block then extracts informative features from these noisy images, which are subsequently downsampled and passed on to subsequent modules through additional Transformer blocks. By leveraging skip connections, the decoder effectively decodes feature maps and establishes connectivity with the encoder module. Ultimately, a denoised image of size 512×512 pixels is generated as output by the final layer of our proposed network.

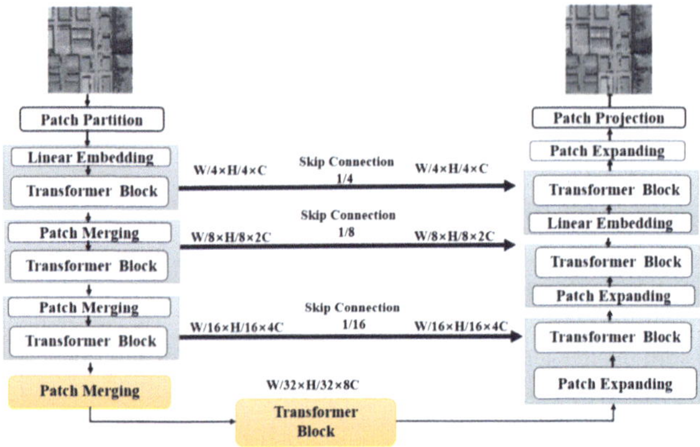

Fig. 4. Denoising network architecture

4.2 Experimental Analysis

The experiments in this paper were conducted using the PyTorch framework deployed on a server equipped with four NVIDIA RTX3090Ti 16GB graphics cards. The epoch was set to 200, the learning rate was 1×10^{-6}, and the optimizer used was Adam. The model was saved every 20 epochs during training.

4.3 Experimental Results Analysis

The dataset generated based on the noise parameters calibrated by the method proposed in this paper was tested against the denoising algorithms presented in this paper and other denoising algorithms. The PSNR and SSIM values obtained are shown in Table 1.

Table 1. PSNR(dB)/SSIM results of various algorithms

Image class	grassland	Railway,	Building
BM3D	30.86/ 0.6645	31.20/ 0.8402	31.24/ 0.8154
DnCNN	34.28/ 0.9038	34.86/ 0.8698	33.58/ 0.8025
FFDNet	34.66/ 0.8843	35.23/ 0.8840	36.05/ 0.9046
CBDNet	38.77/ 0.9423	38.98/ 0.8865	39.21/ 0.9542
Our	39.59/ 0.9768	39.24/ 0.9265	39.82/ 0.9698

The visual results of the experimental data are shown in Fig. 5.

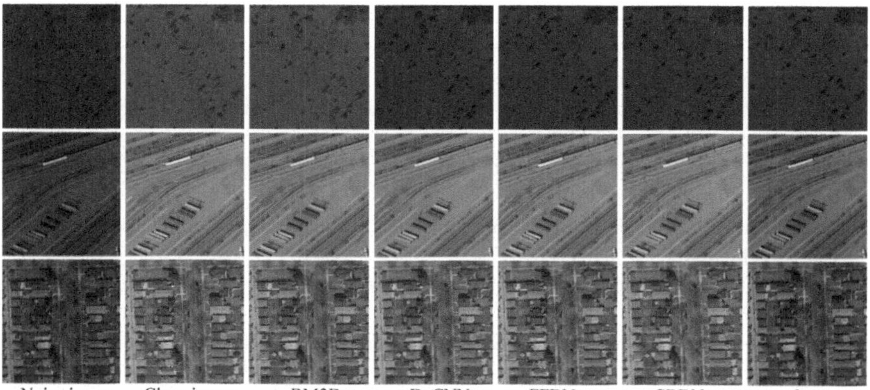

Fig. 5. Results comparing

5 Conclusion

A more complete method for noise modeling and calibration of earth RS physical process has been described in the previous section. Based on the quantitative data and visual analysis above, it is evident that by fine-tuning and optimizing the network and high-fidelity datasets construction, we achieve better results than other denoising networks, which is a specific model that is strongly correlated with the corresponding sensor. After removing the noise, our algorithm can better restore the true physical features of the image, demonstrating the effectiveness of the method in the field of space-based earth RS image processing.

References

1. Wang, X.: Development and prospect of space optical technology. Spacecraft Recovery Remote Sens. **2018**(04), 79–86 (2018)
2. Schowengerdt, R.A.: Remote Sensing: Models and Methods for Image Processing(Second Edition). Academic Press, San Diego USA (1997)
3. Niu, Z., Zhu, C.G., Wang, C.Y.: Analysis of atmospheric cross radiation in a slant viewing angle. J. Remote Sens. Beijing China **1**(2), 88–93 (1997)
4. Rohith, G., Kumar, L.S.: Paradigm shifts in super-resolution techniques for remote sensing applications. Vis. Comput. N. Y. USA **37**(7), 1965–2008 (2021)
5. Watson, K.: Processing remote sensing images using the 2-D FFT; noise reduction and other applications. Geophys. USA **58**(6), 835–852 (1993)
6. Wei, K, Fu, Y, Yang, J, et al.: A physics-based noise formation model for extreme low-light raw denoising. In: Proceedings of the IEEE/CVF Conference on Computer Vision and Pattern Recognition, pp. 2758–2767. Springer, USA (2020)
7. Feng, H., Wang, L., Wang, Y., et al.: Learnability enhancement for low-light raw denoising: where paired real data meets noise modeling. In: Proceedings of the 30th ACM International Conference on Multimedia, pp. 1436–1444. Springer, Lisbon POR (2022)
8. Zhang, Y., Qin, H., Wang, X., et al.: Rethinking noise synthesis and modeling in raw denoising. In: Proceedings of the IEEE/CVF International Conference on Computer Vision, pp. 4593–4601. Springer, London UK (2021)

9. Kim, Y., Soh, J.W., Park, G.Y., et al.: Transfer learning from synthetic to real-noise denoising with adaptive instance normalization. In: Proceedings of the IEEE/CVF conference on computer vision and pattern recognition, pp. 3482–3492. Springer, USA (2020)
10. Kolehmainen, T.T., et al.: Simulation of imaging system's performance. Opt. Model. Perform. Predictions **5178**, 204–212 (2004)
11. Wach, H., Dowski, E.R., Jr.: Noise modeling for design and simulation of computational imaging systems. Vis. Inf. Process. XIII **5438**, 159–170 (2004)
12. Mendenhall, W., Sincich, T.: Statistics for Engineering and the Sciences. Prentice-Hall, Inc., Upper SaddleRiver, NJ, USA (2006)
13. Hubbard, W.M.: The approximation of a poisson distribution by a gaussian distribution. Proc. IEEE **58**(9), 1374–1375 (1970)
14. Hornsey, R.I.: Part III: Noise in Image Sensors. University of Waterloo (1999)
15. Harrell, W.R., Frey, J.: Observation of Poole-Frenkel effect saturation in SiO_2 and other insulating films. Thin Solid Films **352**(1–2), 195–204 (1999)
16. Simmons, J.G.: Poole-Frenkel effect and Schottky effect in metal-insulator-metal systems. Phys. Rev. **155**(3), 657–660 (1967)
17. Chen, J., Chen, J., Chao, H., et al.: Image blind denoising with generative adversarial network based noise modeling. In: Proceedings of the IEEE conference on computer vision and pattern recognition, pp. 3155–3164 (2018)
18. Dosovitskiy, A., Beyer, L., Kolesnikov, A., et al.: An image is worth 16 $\times$16 words: transformers for image recognition at scale. In: International Conference on Learning Representation (ICLR) (2021)
19. Yu, S., Wu, W., Dou, R.: Design and implementation of remote sensing image processing system with intelligent vision processor. Spacecraft Recovery Remote Sens. **44**(05), 54–64 (2023)
20. Hashimoto, F., Ohba, H., Ote, K., et al.: Denoising of dynamic sinogram by image guided filtering for positron emission tomography. IEEE Trans. Radiat. Plasma Med. Sci. **2**(6), 541–548 (2018)
21. Jiang, X., Zhao, R., Liu, Y., et al.: Image denoising algorithm based on improved u-net neural network. Sci. Technol. Eng **22**(09), 3629–3635 (2022)

Author Index

H. P. Urbach et al. (Eds.): ISSOIA 2023, SPP 191, pp. 443–445, 2024.
https://doi.org/10.1007/978-981-97-6718-2

SPRINGER NATURE

GPSR Compliance

The European Union's (EU) General Product Safety Regulation (GPSR) is a set of rules that requires consumer products to be safe and our obligations to ensure this.

If you have any concerns about our products, you can contact us on ProductSafety@springernature.com

In case Publisher is established outside the EU, the EU authorized representative is:

Springer Nature Customer Service Center GmbH
Europaplatz 3
69115 Heidelberg, Germany

The manufacturer's authorised representative in the EU is Springer Nature Customer Service Centre GmbH, Europaplatz 3, 69115 Heidelberg, Germany. If you have any concerns regarding our products, please contact ProductSafety@springernature.com

Printed and bound by CPI Group (UK) Ltd, Croydon, CR0 4YY
01/10/2026
02225857-0014